True artist and true friend

True artist and true friend

A BIOGRAPHY OF HANS RICHTER

CHRISTOPHER FIFIELD

CLARENDON PRESS · OXFORD
1993

Oxford University Press, Walton Street, Oxford OX2 6DP
Oxford New York Toronto
Delhi Bombay Calcutta Madras Karachi
Kuala Lumpur Singapore Hong Kong Tokyo
Nairobi Dar es Salaam Cape Town
Melbourne Auckland Madrid
and associated companies in
Berlin Ibadan

Oxford is a trade mark of Oxford University Press

Published in the United States
by Oxford University Press, New York

© Christopher Fifield 1993

All rights reserved. No part of this publication may be reproduced, stored in a retrieval system, or transmitted, in any form or by any means, electronic, mechanical, photocopying, recording, or otherwise, without the prior permission of Oxford University Press

British Library Cataloguing in Publication Data
Data available
ISBN 0-19-816157-3

Library of Congress Cataloging in Publication Data
Fifield, Christopher.
True artist and true friend: a biography of Hans Richter /
Christopher Fifield.
Includes bibliographical references and index.
1. Richter, Hans, 1843–1916. 2. Conductors (Music)—Germany—
Biography. I. Title.
ML422.R514F5 1993 784.2'092—dc20 {B}
ISBN 0-19-816157-3 (cloth)

Set by Best-set Typesetter Ltd.,
Hong Kong
Printed in Great Britain by
Biddles Ltd, Guildford and King's Lynn

*To Eleonore and Sylvia
in memory of their
grandfather Hans Richter*

FOREWORD
by Sir Georg Solti

A FEW years ago in Chicago, I acquired a treasure, the notebooks of the legendary conductor Hans Richter, who did so much to promote the music of Wagner, Brahms, Bruckner, and Elgar. I have always felt an affinity with Richter for there are similarities in our lives. He was a fellow Hungarian, whose early conducting engagements were at the Budapest opera-house and in Munich. For twenty-five years he conducted in Vienna, but for over thirty he regularly worked in London, and ended his working days in Manchester as Hallé's successor. His appearance in London must have been like a sudden burst of light on the English music scene. According to George Bernard Shaw, who was not always generous with praise, Richter produced noble results. He was also a frequent conductor at my old opera house, Covent Garden, where he gave the first performances of Wagner's *Ring* in English.

When Christopher Fifield asked me to write the foreword to his biography of Richter, I turned once again to my treasure, the conducting notebooks. To protect them they are as I received them, wrapped in tissue paper in a transparent perspex box lined with silver paper. There are six books covering his entire career, three bound in leather and three backed with linen on boards now covered with beige wrapping paper for protection. The earliest book has a blue-edged label with the printed address of a Budapest stationer stuck on the front, on which is stencilled the dates 1865–1884 with the words 'Dirigier Buch' written in Indian ink in the middle. The ink, however, has faded little and in a firm hand Richter has written details of his first concert in his home town of Raab. For the next forty-seven years he kept a methodical record in the six notebooks of every performance he gave. He obviously loved numbers and marked every tenth performance, starting again at every thousandth. The final entry was made in August 1912, a few months before I was born.

As his career progressed the performances became so frequent that the record is astonishing to read. On 2 December 1892 he conducted Mascagni's opera *L'amico Fritz* at the Vienna Court Opera. Two days later he spent Sunday morning conducting Mozart's Mass in F major at the Court Chapel followed by the midday concert with the Vienna Philharmonic orchestra consisting of Mendelssohn's Overture *Ruy Blas*, the first Viennese performance of Bruch's new Third Violin Concerto, and Schubert's 'Great' Symphony in C major. That week he conducted *Don Giovanni*, *Lohengrin*, and *Fidelio*, and on the 14th travelled to Budapest for a concert in aid of the widows' and orphans' fund. On the following day he was back in Vienna for *Tristan und Isolde*, and three days later, on another Sunday (18th), he directed a Mass by Michael Haydn at the Chapel in the morning followed by another lunch-time Philharmonic concert, this time the first performance of Bruckner's massive Eighth Symphony in the Musikverein.

When, I wonder, did he have the time to study his scores? His schedule in England was equally arduous. Between the beginning of May and the end of July 1904 he conducted thirty performances of opera (twenty-six at Covent Garden, four at Bayreuth). They were *Tristan und Isolde*, *Don Giovanni*, *Die Meistersinger*, *Lohengrin*, *Le nozze di Figaro*, and *Tannhäuser*. Amongst them was the inaugural concert of the newly formed London Symphony Orchestra with a vast programme, the *Meistersinger* Overture, Bach's Suite in D major, the Overture to *Die Zauberflöte*, Elgar's Enigma Variations, Liszt's First Hungarian Rhapsody, and Beethoven's Fifth Symphony. The last week of the period was spent in Bayreuth, where he conducted Wagner's *Ring* on four consecutive nights.

My view of Richter is essentially one-dimensional because my only direct point of reference is the notebooks, but how I admire the vitality of this man. He must have had phenomenal energy, both physical and mental, for he travelled a great deal and gave these enormously demanding programmes which he had to prepare and study. Apart from the programmes he was also an educator and promoter of new music. He moulded public taste in England, taking the Hallé on tour from Manchester to all the major provincial music centres, Leeds, Birmingham, and Newcastle, as well as to smaller places such as Rotherham, Middlesbrough, Blackpool, Hanley, and Burnley, where he gave a performance of Bruckner's Fourth Symphony. His programmes included the standard works of Bach, Handel, Mozart, Beethoven, Schubert, and Schumann as well as those of his contemporaries Brahms, Bruckner, Wagner, and Liszt. He also performed the new music of Bruch, César Franck, Saint-Saëns, Strauss, Dvořák, Sibelius, Humperdinck, Grieg, and the British composers Stanford, Parry, and Coleridge-Taylor, and he did more than any other to support and promote the emerging Elgar. They became close friends and Richter conducted the first performances of the Enigma Variations, *The Dream of Gerontius*, and the First Symphony, which the composer dedicated to him. In that hectic spring of 1904 he also organized an Elgar Festival in London, at which all the composer's major works were heard—yet another similarity in our lives, a mutual love of the music of Elgar.

Christopher Fifield has devoted years to researching Richter's remarkable life. Not only does his book, which I welcome, give a detailed account of the man and the musician, it also provides an insight into the musical life of two vital centres, London and Vienna, from the years leading up to the turn of the century until the First World War, the period which laid the foundations for music-making in Europe as we know it today.

Georg Solti

PREFACE

DURING my studies as a music student at Manchester University and the (then) Royal Manchester College of Music in the 1960s, I often encountered the name Hans Richter. The first occasion was during the preparations for a performance of Bach's Third 'Brandenburg' Concerto which I conducted. The Breitkopf and Härtel score and parts came from the vaults of the College orchestral library and each carried Hans Richter's signature in the top right-hand corner. I also went regularly to rehearsals and concerts given by Sir John Barbirolli and the Hallé Orchestra, which Richter had taken over five years after the founder-conductor's death. Above all there were still many teachers and players in Manchester thirty years ago who had tales to tell of the formidable Richter.

Hans Richter (1843–1916) was one of the first career conductors to gain international fame. Prior to this conductors were invariably composers, such as Berlioz and Wagner, or performing composers such as Hans von Bülow and Liszt. Richter was not a composer; on 13 January 1871 he completed an arrangement of Wagner's *Siegfried Idyll* for piano duet and solo violin, but he also wrote three works of his own, a short two-part piece for four horns (*Im Walde* and *Nachtruhe*), a Romance for horn and piano (*Am See*), and a Concert Overture in F minor, written in 1878. He was able to play all instruments except the harp, but he started his professional career as a horn-player. He could sing to a standard that kept the curtain up at an early performance of Wagner's *Die Meistersinger* in Munich when he stepped in for the indisposed Kothner.

Once under way Richter's career focused on three major musical centres of the nineteenth century, Vienna, Bayreuth, and London. His conducting work began in Budapest and ended in Manchester over forty years later, his influence in both cities lasting long after he had departed. His name can be found in biographies and textbooks on music relating to the last quarter of the nineteenth century and the first decade of the twentieth. He features in the lives of Wagner, Brahms, Bruckner, Dvořák, Elgar, Stanford, Parry, Tchaikovsky, Sibelius, Bartók, and Glazunov. Not only was he personally acquainted with all of them, but he also gave first performances of some of their works. He also appears in biographies of singers and instrumentalists, often brought by him before the public at an early stage in their careers. It is time, however, to look at Hans Richter in his own right, to explore his personality and to detail his life and work. A primary source for this information comes from the man himself. He meticulously listed, in six books entitled *Dirigierbücher* (conducting books), all the 4,351 public performances he gave in his professional life between 1865 and 1912. Each entry details the venue, programme, soloists, and (in the case of operas) the number of total performances he had reached with each particular performance. That he had the time to list them in such detail and in such a neat and legible manner reveals a basic facet of Richter's nature; he was equally meticulous and systematic at his rehearsals. I am much obliged to Sir Georg Solti (a compatriot of Richter, whose repertoire he also shares to a large degree) for access to these books.

Whereas the six *Dirigierbücher* have proved an invaluable source and a quick reference to establish Richter's whereabouts at any given time, the absence of diaries from his effects has proved a handicap, all the more so because it is known that he kept a detailed journal throughout his life. He had six children, and one of these, Edgar, had custody of the diaries until at least 1960 (they are mentioned in a letter to his sister Mathilde in London from America, where he had emigrated after the Second World War) after which no trace remains. They certainly survived the war (unlike much other material) and, like the *Dirigierbücher*, may well have been sold. Some extracts have, however, survived. Before the war Edgar copied all information from the diaries relevant to a period of Richter's activities in England for Mathilde, who stayed behind in 1911 when her father retired to Germany. These were the years 1877–1900. Unfortunately Edgar did not continue into the Manchester/Hallé years (which would also have included the founding of the London Symphony Orchestra, the first *Ring* in English in 1908, and Richter's annual appearances from 1903 at Covent Garden), presumably because Mathilde was in her father's company for that period. Similarly he copied (in 1938) all information relevant to Wagner (the years 1866–83) for Otto Strobel's use. Strobel, curator and archivist at the Wagner Museum at Wahnfried in Bayreuth, wished to write a biography of Richter, and Edgar (who had been forced to abandon his career as an operatic tenor) was on the staff there. Mathilde and Edgar were particularly keen to see a biography of their father in print, and whereas Edgar approached Strobel, Mathilde made various approaches in England to writers such as Herbert Thompson, H. C. Colles, and Neville Cardus, but for many reasons the idea was not taken up.

Preface

A more important handicap is the absence of any aural or visual record of Richter. Although he retired in 1912 he never made a recording for posterity. Those of Richter's scores which have survived in the family are mainly untouched (all the Elgar full scores carry the composer's deeply felt dedications), for he usually did no more than seek out printing errors or highlight metronome markings in blue pencil. There are additional clarinet parts written in his score of Beethoven's Second Symphony to reinforce the woodwind line at moments where the composer rested that instrument, but little else. According to biographies of the cellist Pablo Casals, whom Richter befriended and did much for from 1904, he bought scores and orchestral parts from the conductor's widow at the end of the First World War when he was setting up his own orchestra in Barcelona.

I am indebted to Manfred Eger, curator of the Wagner Archive in Bayreuth, for allowing me access to material held there. Similar thanks go to the Nationalbibliothek, Stadt und Landesbibliothek, Hof- und Staatsarchiv, and Theatermuseum in Vienna, as well as the Wiener Hofkapelle, Gesellschaft der Musikfreunde (Dr Clemens Hellsberg), and the Vienna Philharmonic Orchestra (Dr Otto Biba). In England I am similarly indebted to the British Library Manuscripts Division, the Royal College of Music, the National Sound Archive, the Henry Watson Music Library in Manchester and the Brotherton Library in Leeds (Philip Morrish), the Hallé Orchestra, and the Royal Northern College of Music. In America I owe thanks to Columbia University Library, Pierpont Morgan Library, New York, Harvard College Library, and the Boston Symphony Orchestra, in Germany to the Bayerische Staatsbibliothek and the Stadtsbibliothek in Munich, to Dr Jan Kralik who guided me to the Prague Music Archive and their holdings of the Richter–Dvořák correspondence, and to Maria Eckhardt, Director of the Ferenc Liszt Memorial Museum in Budapest. To all these, and particularly to Stewart Spencer for his endless patience in putting up with my queries, I express my sincere thanks for invaluable help. I gratefully acknowledge the British Academy, the Worshipful Company of Musicians, and University College London for their financial help.

Among many individuals who have assisted me with either their knowledge or from their private holdings are Robert Elliot, Marjorie Cox, Grant Longman of the Bushey Museum Trust, Dr Francis Jackson, Jerrold Northrop Moore, Dr Gareth Lewis, Christopher Dyment, David Robinson, Chris de Souza, Reg Cane, Jeremy Dibble, Kate and Laura Russell, Wilfred Stiff, Andrea Vogel, Philip Wults, the late Paul Richard, Feri Gyenes, Paul Cummings, Michael Kennedy, and the late Winifred Christie, who at the age of 95 shared her memories of a bygone age and allowed me access to letters written to her grandmother Marie Joshua from, among others, Eugen d'Albert.

I am extremely grateful to the descendants of Hans Richter. His grandson, David Loeb, generously loaned me many photographs in his possession, including several taken by his father Sydney. There are also two ladies who have now become close friends; both are Richter's granddaughters and both have made me part of the Richter family. The operational methods of the researcher can be simple. It was a matter of consulting the L–R London telephone directory that set me on my way. The recipient of that phone call, Caroline Loeb, immediately referred me to her aunt Sylvia, the first of the Richter granddaughters I was to meet. Sylvia Loeb entered into the spirit of the project with boundless energy and total co-operation by providing me with letters written by her grandfather and subsequently returned by the recipients to her family after his death, as well as newspaper cuttings and photographs. Through her I contacted her cousin, Eleonore Schacht-Richter of Würzburg in Germany, without whom this book would also have been impossible to write. Her tireless enthusiasm, her depth of knowledge, and profound love for her grandfather have inspired me and to her and other relatives (Wolfram Dehmel of Hassfurt and Peter Dehmel of Regensburg) I express my sincere gratitude. The family provided me with many of Richter's letters, all of which are written in a beautifully legible hand. I have translated all those used in this biography, whereas Richter's letters written in English have been left in his own inimitable style without correction. My final thanks go to the editorial team at Oxford University Press, to David Elliot for help in proof-reading, and to my wife for her patient reading of the manuscript and invaluable suggestions.

C. F.

CONTENTS

Foreword by Sir Georg Solti	vi
Preface	ix
List of Plates	xv
List of Figures	xvii
Chronology	xix

1.	1843–1865: Childhood and Years of Study	1
2.	1866–1867: Tribschen	10
3.	1868–1869: Munich	25
4.	1870–1871: Brussels; Tribschen	39
5.	1871–1874: Budapest	52
6.	1874–1875: Budapest and Bayreuth	64
7.	1875: Vienna	84
8.	1876: Bayreuth	105
9.	1877: London	119
10.	1878–1879: Vienna	131
11.	1879–1880: Friends and Enemies	141
12.	1880–1881: London and Vienna	156
13.	1881–1882: Richter and d'Albert	167
14.	1882: Richter and d'Albert	177
15.	1882–1883: The Master's Death	187
16.	1884: More Opera in London	204
17.	1885–1886: Vienna, London, and Birmingham	215
18.	1887–1888: Return to Bayreuth	235
19.	1889–1900: Vienna	249
20.	1897–1900: Richter and Mahler	268
21.	1889–1890: England	280

22.	1891–1895: England	287
23.	1895–1900: England	296
24.	1890–1899: Bayreuth	311
25.	1894–1899: Richter's Diary	323
26.	1899–1900: Hallé Orchestra	332
27.	1900–1902: England	347
28.	1903–1904: England	359
29.	1904–1906: England	377
30.	1906–1908: England	390
31.	1908–1909: England	404
32.	1909–1911: England	420
33.	1911–1914: Retirement	433
34.	1914–1916: The Last Years	447
35.	Finale	458
	Notes	475
	Appendix I. Works Conducted by Hans Richter	499
	Appendix II. Cities and Towns where Richter Conducted	505
	General Index	507

LIST OF PLATES

1 (*a*) Hans Richter's mother Josefine
  (*b*) The boy Hans with his father Anton Richter
  (*c*) Hans as an Imperial Chapel chorister
  (*d*) Hans Richter, second from right, as a teenager, with his uncle (Anton's brother) and cousins
2 (*a*) Hans Richter in 1868 after his appointment as Music Director in Munich following Hans von Bülow's departure
  (*b*) Hans Richter with his fiancée Marie von Szitányi, Budapest 1874
3 (*a*) Richter (centre) with members of the 'Nibelungen Chancellery' at Bayreuth in 1872 (left to right, Hermann Zumpe, Demetrius Lalas, architectural assistant Karl Runckwitz, and Anton Seidl)
  (*b*) Hans Richter's children (left to right, Hans, Mathilde, Ludovika, Marie, Richardis, and Edgar) about 1884
4 (*a*) Hans Richter with the score of Wagner's *Die Meistersinger*, London 16 June 1898. The photograph is dedicated to Pedro Tillett, nephew of the conductor's agent Narciso Vertigliano
  (*b*) Richter outside Birmingham Town Hall during the Triennial Music Festival in October 1909, photographed by his future son-in-law Sydney Loeb. The musical quotation is the bassoon part taken from the introduction to the Prisoner's Chorus from Beethoven's *Fidelio*
5 (*a*) Richter at The Firs, Bowdon, in 1908
  (*b*) Richter and his daughter Mathilde on board ship crossing the Channel
6 (*a*) Richter with George Bernard Shaw and his wife Charlotte at Bayreuth in 1908
  (*b*) Richter with (left to right) Eva, Isolde, and Siegfried Wagner, Daniela and Blandine von Bülow, Bayreuth 1890
7 (*a*) Richter holidaying at Baracs, Hungary
  (*b*) With Marie and his granddaughter Eleonore in the garden of Zur Tabulatur in Bayreuth, 1914
8 (*a*) The familiar sight of Richter with shopping bag during his retirement in Bayreuth. The musical quotation from *Die Meistersinger* was the family whistle
  (*b*) Richter in his Bayreuth study on his seventieth birthday, 4 April 1913

LIST OF FIGURES

1. Dedication to Camillo Sitte from Hans (Johann) Richter quoting the opening of Beethoven's 'Eroica' Symphony, Vienna 1864 — 21
2. Richter, under the watchful eye of Wagner and the applauding Liszt, conducting in Budapest in 1871. *Borsszem Janko* (Johnny Peppercorn) was a satirical magazine of the day and was clearly commenting on the young man's concert programmes — 57
3. *Der Floh*, 9 May 1875 — 89
4. *Illustrated Sporting and Dramatic News*, 26 August 1876 — 111
5. Hans Richter returning to Vienna with the financial rewards of his first visit to Covent Garden. Cartoon by Hans Schliessmann — 207
6. Brahms, Johann Strauss, and Richter playing cards. Silhouette by Otto Böhler — 224
7. *Wiener Journal*, 1 February 1890 — 251
8. Richter conducts a Bruckner symphony, possibly the première of No. 8, 18 December 1892. Silhouette by Otto Böhler — 261
9. Hans Richter conducts Wagner's *Götterdämmerung* at the Court Opera in Vienna — 276
10. Richter at Bayreuth in 1889. The notice in his hat reads: 'Please do not ask me for dress rehearsal tickets as I do not have any.' — 316
11. Hans Richter's autograph at the 1896 Bayreuth Festival, the year in which the *Ring* was staged for the first time since its première in 1876. Above his signature are the opening bars of *Das Rheingold*, below the closing bars of *Götterdämmerung* — 321
12. Hans Richter conducting at the Vienna Court Opera — 365
13. Richter conducting the *Dream of Gerontius* at the Elgar Festival, Covent Garden 1904 — 371
14. A postcard from Richter to the Hallé Orchestra's principal horn Franz Paersch dated 27 February 1904 inviting him

	and the principal bassoon Otto Schieder to a cold supper at Bowdon. The musical quote is Siegfried's horn call	385
15.	Richter conducting the first *Ring* in English, *Daily Graphic*, 28 January 1908	402
16.	Frank L. Emanuel's drawing in the *Manchester Guardian* of Richter at rehearsal with the Hallé Orchestra in 1904	419
17.	*Punch*, 25 March 1903	467

CHRONOLOGY

1843 Born Raab, Hungary, on 4 April 1843.
1854 Entered the Piaristen Gymnasium in Vienna and became a choirboy in the Imperial Chapel.
1859 Entered the Vienna Conservatoire.
1862 Horn-player at the Kärtnerthor Theatre, Vienna.
1866 In response to a request from Wagner for a copyist, Richter is sent by Heinrich Esser to Tribschen, Lucerne.
1868 Chorus Master for the première of *Die Meistersinger* in Munich. Appointed successor to von Bülow as Court Music Director by King Ludwig.
1869 Withdraws from the première of *Das Rheingold* and is sacked by the king.
1870 Conducts the Brussels première of *Lohengrin*.
1871 Appointed Music Director in Budapest.
1872 Foundation-stone laying ceremony at Bayreuth.
1875 Marries Marie von Szitányi. Appointed as conductor at the Court Opera in Vienna and to the Philharmonic Orchestra.
1876 Conducts the première of the complete *Ring* at the first Bayreuth Festival.
1877 First London appearance at the Wagner Festival. Conducts first performance of Brahms's Second Symphony. Vize-Hofkapellmeister at the Court of Emperor Franz Josef.
1879 Returns to London to establish the annual Richter Concerts.
1880 Conducts first performance of Brahms's Tragic Overture. Dvořák dedicates his Sixth Symphony to Richter. Honorary member of the International Mozart Foundation, Salzburg.
1881 Conducts first performance of Bruckner's Fourth Symphony and Tchaikovsky's Violin Concerto. Richter's 1,000th public appearance (9 January).
1882 Conducts first performance in England of *Die Meistersinger* and *Tristan und Isolde* at the Theatre Royal, Drury Lane, London.
1883 Death of Wagner. Richter conducts first performance of Brahms's Third Symphony.
1884 Appointed conductor of the Gesellschaft der Musikfreunde in Vienna. Conducts first performance of Stanford's opera *Savonarola* at Covent Garden.
1885 Appointed conductor of the Birmingham Triennial Music Festival. Conducts first performance of Bruckner's Te Deum. Honorary Mus. Doc. from Oxford University.
1886 Conducts first Viennese performance of Bruckner's Seventh Symphony.

1887	First appearance at the Lower Rhine Music Festival. In June he conducts first performances of three new British symphonies, Parry's Second, Cowen's Fifth, and Stanford's Third.
1888	Returns to Bayreuth and conducts *Meistersinger*. Richter's 2,000th public appearance (23 December).
1891	Honorary member of the Gesellschaft der Musikfreunde.
1892	Conducts first performance of Bruckner's Eighth Symphony.
1893	Accepts then declines the Music Directorship of the Boston Symphony Orchestra. Promoted to Imperial Hofkapellmeister.
1895	Death of Hallé; Richter offered Hallé's post in Manchester, but declines until he is able to retire from Vienna on full pension. Richter's 3,000th public appearance (17 November).
1896	Conducts the first production of the *Ring* at Bayreuth since its première in 1876. Death of Bruckner.
1897	Death of Brahms. Mahler appointed Director of the Vienna Opera.
1898	Resigns as conductor of the Vienna Philharmonic concerts.
1899	Conducts first performance of Elgar's Enigma Variations.
1900	Leaves Vienna and takes up the Hallé offer. Conducts first performance of Elgar's *The Dream of Gerontius* at the Birmingham Triennial Festival.
1902	Honorary Mus. Doc. from Manchester University.
1903	First appearance at Covent Garden with the German Opera season.
1904	Conducts Elgar Festival in London. London Symphony Orchestra formed, Richter conducts inaugural concert. Honorary member (4th class) of the Royal Victorian Order.
1906	Richter's 4,000th public appearance (14 May).
1907	Honorary Commander of the Royal Victorian Order.
1908	At Covent Garden Richter conducts first *Ring* cycle in English. Conducts first performance of Elgar's First Symphony, which the composer dedicates to him.
1909	Resigns conductorship of the Birmingham Triennial Festival.
1911	Resigns from the Hallé Orchestra and retires from the concert platform. Moves to Bayreuth.
1912	Final (4,351st) public appearance conducting *Meistersinger* at Bayreuth Festival on 19 August.
1916	Dies in Bayreuth on 5 December.

1

1843–1865

Childhood and Years of Study

SEVENTY-FIVE miles south-east of Vienna, and about half-way to Budapest, lies the town of Györ. In Roman days the town was named Arabona; more recently, when part of the Austro-Hungarian Empire and as a town with a modest manufacturing industry centered chiefly on tobacco and cutlery, it was called Raab. It lies where the rivers Rába and Rábcza flow into the Little Danube. Here, by the Bishop's Palace, with its fifteenth-century Dóczi Chapel, stands the Cathedral, founded in the twelfth century and rebuilt between 1639 and 1645. Its Héderváry Chapel contains stained-glass windows and a fifteenth-century silver bust of St Ladislaus. It was in Raab, on 4 April 1843, that Hans Richter was born. His family came from former Austria-Silesia, in the area of Freudenthal, the earliest recorded ancestors being Georg and Anna Richter, parents of Melzer (Melchior) Richter (1656–1720). Melzer's son Melchior (1692–1742) was, like his father, a farmer, but his grandson Josephus Richter (1726–1787) was first a labourer then a tailor in Breitenau and Markersdorf. Josephus's son Anton Franciscus Richter (1762–1819, grandfather of Hans) became a schoolmaster in Probstdorf, in the district of Vienna, and from his first marriage with Theresia Knöbel he had a son Anton born in 1802.

Anton spent the last twenty-two years of his life in Raab, but for ten years from 1822 he sang bass in the service of Count Nikolaus Esterházy, whose family in Eisenstadt was famous for its patronage of Haydn. Anton Richter was a gifted organist, singer, string-player, and composer and on 22 June 1832, from among eleven short-listed candidates, he was appointed Succentor or Subcantor (effectively the choirmaster) at Raab Cathedral, but he was soon elevated to the top musical post of Kapellmeister. He threw himself whole-heartedly into his duties, composing prolifically, and there are 118 extant compositions (64 sacred, 54 secular) in the archive of the Gesellschaft der Musikfreunde (Society of the Friends of Music) in Vienna and 30 in Győr Cathedral. He was also very conscientious in recording, over a twenty-year period from the start of his appointment in 1832, details of the musical life of the Cathedral and its choir, all of which give an invaluable insight into the workings of such institutions. There are lists of personnel, of new compositions, of works performed, of instruments donated or purchased, maintenance records of the Cathedral organ, as well as his own comments on the well-being and the problems of his musicians and musical standards for which he was held responsible. He even produced a handbook for his musicians after he sensed a feeling of insecurity and ill discipline among those performing liturgical music. He was supported by such acquaintances as Franz Liszt and Otto Nicolai, who came on occasional visits, and was also fortunate in having for ten years (1838–48) a music-loving bishop in Johann Sztankovits, later godfather to Hans. In 1846, with his bishop's moral and material support, Richter founded a male-voice choir in Raab and in the following year created the town's first music school. The school celebrated its eightieth anniversary in 1927 by erecting a plaque on its wall in double memory of its founder and his famous son, and to this day a triennial Anton Richter competition is held for students of woodwind instruments. In August 1846 Bishop Sztankovits was host to Nicolai. 'Throughout my stay Cathedral Kapellmeister Richter of Raab behaved in a very friendly and respectful manner towards me. Everything went as well as I could wish for; they were three very pleasant days!'[1]

Anton Richter also taught singing, and on 2 February 1842, at the age of 40, he married one of his students. She was Josefine Czasensky, born in Tabor, Bohemia, in 1822, 20 years old when she married and twenty years younger than her husband. Anton and Josefine had four children. Besides Hans (the eldest, born 1843) there were three other children, two of whom (Joseph, born 1845, and Antonia, born 1847) died in infancy and another (Marie, born 1844) when she was 14. Josefine's grandfather Ignaz Steyer had been a musician at Raab Cathedral and her father Albert a horn-player and conductor with a military band.

The birth of Hans Richter was greeted with joy by his parents. Anton wrote to his brother-in-law Johann Schöpfleuthner on 5 April 1843 to report the good news of the previous day.

At last dear God has presented us with the long-awaited, dear, good boy; yesterday in the evening of the 4th at five minutes to ten he gave his first cry, but was immediately quiet and has remained *bis dato* really peaceful and sleeps most of the time. Permit me to describe the hitherto unknown joy of fatherhood, you already know it after all. The birth itself was happy, though for me shattering! Mother and child are well. Today at three this afternoon is the baptism and, according to our and the bishop's wishes, he will be named Johann Baptist Isidor. . . . The little chap is really sweet, and his mother is pleased that he has a little dimple on his chin, just like me.[2]

Hansi, as the young baby was called by his doting father, had a 'genuine Richter nature' and when his equable mood was disturbed the result was 'a short cadenza from his alto voice'.[3] As a child he soon showed musical promise and from the age of 4 received piano lessons from his mother. He had perfect pitch and proved a useful assistant to the local organ tuner, who often took him with him to act as a human tuning fork. The child naturally tried to take up the organ, but his feet did not reach the pedals until the age of 10, and even then these had to be especially built up for him. He also sang unofficially (either soprano or alto) in the cathedral choir, and in 1850, at the age of 7, played the all-important timpani part in Haydn's C major *Paukenmesse* (so-called because of the prominence given to that instrument in the Agnus Dei). With an innate sense of rhythm and his assured and confident style he made a great success of the occasion, and years later would observe that 'the melody is the flesh but the rhythm is the bones'.[4] He repeated his role as orchestral timpanist on 15 August 1851 in one of his father's Masses, but his final childhood appearance was as pianist in Hummel's Piano Quintet in E flat in one of his mother's concerts on 14 June 1853; he was 10 years of age.

Josefine Richter began her career as an opera singer and from Anton's records it appears that in April 1852 both he and his wife were offered contracts at the opera house in Hamburg, where the Raab conductor Josef Wurda was then employed. Although Anton travelled there to discuss terms, he could not agree a salary with the opera house and the idea was abandoned. Two years later, on 2 January 1854 at the age of 52, he suffered a stroke and died; apparently he had been greatly affected by the deliberate out-of-tune singing of one of his mischievous choristers. In the few years they had together Anton had encouraged his son's obvious musical gifts. 'All my family, including my grandfather, were musicians,' recalled Richter in 1899, 'except one, who was an organ builder. [My father] had a good library, which included Berlioz's treatise on instrumentation. [He] composed a Requiem Mass without violins in the accompaniment. A Nonet by him, for strings and four wind instruments, was performed before the Queen of England at one of her private concerts.'[5]

The settled tranquillity of family life was shattered by Anton's death. Josefine was now compelled to pursue her career as an opera singer and did so with success, with guest appearances in Budapest, Braunschweig, and Hanover. She then took a two-year contract as a member of the ensemble in Leipzig, where she also took singing lessons with Friedrich Schmitt, who strengthened her voice to a more dramatic sound. She was soon able to present herself not only as a coloratura soprano, but also for such roles as Fidelio, Donna Anna, Elisabeth, Valentine, and Agathe. She extended her guest appearances to Munich, Berlin, Karlsruhe, Augsburg, Basle, Zürich, Amsterdam, and even Moscow, where, in March 1864, she scored a particular triumph. Her greatest success, however, took place seven years earlier, on 28 August 1857, when she sang Venus in the first Vienna performance of Wagner's *Tannhäuser* at the Josefstadt Theatre. In view of the nature of the role and the fact that she, the widow of a Cathedral Kapellmeister, was singing in the city where her son was, at that time, a member of the Court Chapel choir, Josefine thought it prudent to protect him from scandal by using another name, Lieven. Having decided to give up her operatic career in 1865, she moved to Vienna and devoted herself entirely to teaching singing.

On 4 March 1867 she married Anton von Innffeld and, apart from a short time spent in Munich from December 1868 as teacher of singing at the new Royal Music School (founded with the strong encouragement of Wagner), she settled in the Austrian capital for the rest of her life. Her stay in Munich was not without curious incident. In a letter to Richard Pohl, Hans von Bülow wrote, 'Yesterday a new production of *Fidelio* with Frau Richter (I conducted from memory).'[6] This letter appears in the collected edition of von Bülow's letters edited by his second wife Marie, who wrote as a footnote, 'Mother of the conductor Hans Richter.' It would seem that Josefine was attempting a come-back to the stage at the age of 46, for she did indeed appear as Leonore under the name of Frau Innffeld-Richter on 20 December 1868 at the Imperial Theatre with Heinrich Vogl as Florestan and Max Schlosser as Jacquino. She was evidently not a success; a review in the Munich *Unterhaltungs-Blatt* had nothing kind to say, describing her inadequate vocal powers, ham acting, and lukewarm interpretation.[7] In a letter to Wagner of 11 August 1869 (see Chapter 3), as the *Rheingold* affair was brewing, Richter complained of Intendant Baron Perfall's treatment of his mother. Whether this had anything to do with any consequences of this unsuccessful attempt at resuming her singing career (for she made no further appearances as Leonore), or whether Perfall dismissed her from her teaching post in Munich out of spite at her son's behaviour during the

Rheingold affair, can only be the subject of speculation. When Hans conducted *Fidelio* on 4 February 1869 (to positive critical acclaim from the same paper which had so recently damned Josefine[8]), it was unfortunately not an occasion when a mother could be heard singing under the baton of her own son. In 1887 Josefine produced, and had published, a method of teaching singing based on her own experience.[9] She died five years later in 1892. Unlike Hans's father, his mother had lived long enough to witness her son's rise to fame and to enjoy many hours of his music-making with both orchestra and opera in Vienna, the city she eventually made her home.

For Josefine Richter to pursue her stage career as an opera singer it was necessary to make arrangements for the two surviving children. In the case of Hans she was able to enter him in the Piaristengymnasium, a boarding school in Josefstadt, Vienna, where he stayed for four years and studied religion, Latin, Greek, German, history, geography, mathematics, and natural history. At the same time she also entered the boy for the choir of the Imperial Chapel (she and her husband had done so the previous summer but at that time he had been rejected for singing out of tune and for being physically too weak). The entrance examination took place at the Löwenburgische Konvikt (or theological seminary) on 9 August 1854 and thirty-two children competed for only two places. One of the alto competitors was rumoured to be guaranteed a place by the Emperor Franz Josef's mother Archduchess Sophie, who regularly visited the coffee house in Ischl owned by the child's father. When young Johann's turn came he was undaunted by such favouritism, and, when asked by Hofkapellmeister Ignaz Assmayer to sing the arpeggio of a certain key, having first been given the keynote, he responded indignantly, 'Well if you do that, it's easy!' He also soon realized that the piano on which he was being tested was identical in pitch to the family piano back home in Raab, and with this perfect pitch and his other musical gifts, including the sight-reading of part of a Mass, he was accepted and enrolled at the seminary, or in today's terms became a member of the Vienna Boys' Choir. Here, in addition to the subjects taught at school, he studied French, Italian, handwriting, drawing, and dancing.

Johann possessed a good alto voice and was thoroughly schooled in classical church music as well as receiving a good all-round education. He was soon asked to take the alto solos in addition to singing in the choir. The only other child accepted from the group examined that summer's day in 1854 was Josef Sucher, who, as a treble soloist, joined Hans and remained a friend until his death in 1908. Sucher later became a fine conductor of opera in Berlin and married the soprano Rosa Sucher. When

Johann's voice broke in the summer of 1858 his career with the boys' choir was ended, but he was awarded an annual sum of 150 gulden for the next three years to enable him to fulfil his most ardent wish, for the advanced study of music at the Vienna Conservatoire. By leaving the Court choir or Hofkapelle he was not, however, bidding it a final farewell for he was to return as its conductor in 1877, twenty-three years after he had first joined as a child.

He could have remained at the Gymnasium, even if his career with the boys' choir was at an end, but 'I said to myself, I do not wish to become a doctor, nor a lawyer, nor a philosopher, so what is the point of continuing studies at school?'[10] Instead he enrolled at the Conservatoire in Vienna (known then as the Konservatorium der Gesellschaft der Musikfreunde). The records show his name as enrolled from 1859 to 1865, though he entered the school in the autumn of 1858 and left in 1862, continuing with horn lessons as an external member of the school. His primary study was horn under Wilhelm Kleinecke (a member of the Court orchestra), and his subsidiary subjects were violin under Carl Heissler (a member of the Court Chapel orchestra), piano with Professor Ramesch, orchestral studies with Josef Hellmesberger (the Director of the Conservatoire), and theory and composition with the eminent Simon Sechter (teacher of Schubert and Bruckner). Richter distinguished himself each year from 1860 to 1865 as Kleinecke's best pupil. On one occasion during his studies with Sechter (by now an old man of over 70) Richter took with him a score of Wagner's *Tannhäuser*, borrowed from his mother, for the theorist's opinion. After a few days his teacher returned it with the comment, 'Yes, well I'm afraid that's how many compose these days.'[11]

The timetable was a full one. Horn lessons were two hours on Monday, Wednesday, and Friday afternoons; on Mondays and Wednesdays these were followed by an hour-long violin lesson. He saw Sechter and Ramesch (for three hours) on Tuesdays, Thursdays, and Saturdays, and Hellmesberger (for two hours) on Tuesdays and Thursdays. He also studied viola, trumpet, and timpani, using every available chance to play chamber music with his fellow students. He found the time to play the piano for his mother at a concert given by public demand in the Richters' home town of Raab on 16 September 1859. The programme was shared with the band of Archduke Max's 8th regiment of lancers (uhlans). Together mother and son performed operatic arias from *Fidelio* and *Robert le Diable*, a song by Heinrich Proch 'in Austrian dialect' entitled 'Morgenfensterl', and one in Hungarian written for the soprano Anna de la Grange by its composer Béni Egressi. Though already enrolled as a student at the Conservatoire in Vienna, Johann was still described in the programme as a pupil of the Imperial Chapel. He also gained experience and supplemented his meagre income as an extra player in the various opera orchestras in the capital, such as those of the Kärnerthor or Burg theatres. He played timpani for three months under Franz von Suppé at the Theater an der Wien and viola for a year under Jacques Offenbach at the Quay Theatre. He immersed himself in the performing life of the city and participated fully in musical activities at the Conservatoire. He gained both repertoire and playing experience, and diligently prepared himself for his chosen professional career as a conductor, for this was (and had been for some years by now) the goal upon which he had set his sights. Fellow student Franz Fridberg, in an article in the *Berliner Tagblatt*, recorded a memory of Richter in his Conservatoire days.

If there was no trombonist, Richter would lay down his horn and seize the trombone; next time it would be the oboe, the bassoon or the trumpet, and then he would pop up among the violins. I once saw him manipulating the contra-bass, and on the kettledrums he was unsurpassed. When we—the Conservatory orchestra—under Hellmesberger's leadership, once performed a mass in the church of the Invalides, Richter sang. How he did sing! At times he helped out the bassi in difficult passages, at others the tenors, and I believe he even sang with the soprani. I got to know him on that day, moreover, as an excellent organist. It aroused uncommon merriment among us fellow performers when he stood there, and with a self-important look, emitted, over the whole orchestra and chorus, his *Crucifixus* into the body of the church.[12]

Having graduated with flying colours from the Conservatoire, Richter auditioned on 4 August 1862 as a horn-player for Director Matteo Salvi of the Kärnerthor Theatre (later the Hofoperntheater or Court Theatre) and was accepted; his contract began on 1 September that year. A few weeks prior to this first permanent professional post he had passed an unofficial examination as a Kapellmeister given him by Heinrich Esser (first Kapellmeister at the Kärnerthor Theatre who watched and guided the early stages of Richter's career) and Franz Lachner. Richter was new to Lachner, who was based in Munich as Music Director at the Opera, but the visiting conductor gave the student a glowing recommendation, describing him 'as an able pianist, a skilful score-reader and above all a basically sound and educated musician. With his significant talents he will soon master the routine necessary for a conductor.'[13] Esser fully endorsed his colleague's opinion 'with complete conviction'.[14]

Richter remained a member of the Kärnerthor Theatre until 31 March 1866, during which period he observed Wagner on more than one occasion.

As he himself described it, 'I had seen him conducting, and had worshipped him at a distance; but I had never spoken to him, though I had always longed to do so.'[15] Heinrich Esser himself was a thorough workmanlike professional and, 'while never a true blue Wagnerian',[16] conducted *Lohengrin* in May 1861 especially for the composer (due to his exile from Germany in 1848 Wagner had not yet heard the work). *Der fliegende Holländer* followed and Wagner was so encouraged that he tried to cast *Tristan und Isolde* and have it staged in the capital, but to no avail. It was considered unplayable in 1863 after two years' effort and seventy-seven rehearsals, and it was left to the tireless von Bülow in Munich to première it two years later in 1865. Wagner was also in Vienna for Christmas and the New Year of 1863, when he conducted three concerts of orchestral and vocal extracts from his operas at the Theater an der Wien. Later, in December 1863, he was back in the city and participating in a concert given by Tausig. Wagner's contribution was again some of his own pieces, but he also gave a startling (and in his view authentic) interpretation of Weber's Overture to *Der Freischütz*, though not without a considerable struggle to get the orchestra out of its bad habits.

There is little information about Richter's four-year career as a horn player in Vienna. Hardly any letters survive from those years; the one or two that do were birthday greetings to his mother.

Much loved Mother,
I fulfil my sacred duty by wishing for your honoured birthday every conceivable good, which you deserve in the greatest amount. May God keep you healthy and happy both for my well-being and for the happiness of your other relatives; you can be assured that I shall do my utmost to ensure your happiness. Meyerbeer's *Dinorah* was performed by us on the 11th. It was very well received, but that was largely thanks to Murska, Beck, and Eppich from Graz who is a splendid tenor, an excellent actor even if with little voice. After March the Italian opera company are coming; I'm really fearful of this foreign opera trash.

You need not worry at all about my health, thank God I am quite well, for beer is the best medicine. Also I already have 50, yes *fifty*, gulden deposited in the bank and set aside for clothes; if only my blessed grandmother had known that her spendthrift Hans would one day learn to save! When are you coming to Vienna? I don't have any more to write about, hence I remain your grateful son,

Johann Richter[17]

There are the signs of later prejudices and attitudes in this letter: a preference for the German rather than the Italian operatic style, an awareness of the need to harness his finances, and a hedonistic disposition which at this age favoured beer, though later included a good wine. The young man was becoming impatient to obtain a first conducting post for himself, especially with his glowing testimonials from the influential Lachner and Esser. He had already made his professional début as a conductor when, on 19 September 1865, he returned to Raab. This initial entry in the first of his six conducting books records the event and a second in Pressburg during the following winter.

After several attempts at conducting at the Vienna Conservatorium (and also in some churches), I conducted my first public concert in my home town Raab. It was presented by the Raab Music Association. Concert on 19 September 1865. Auber, Overture to *La Muette de Portici*, Mozart, 'Jupiter' Symphony (C major).

During the winter '65/66 I also conducted in a charity concert in the Pressburg theatre, the Overture to *La Muette de Portici* and, for Frau Millerschek and Herr Calori, a *pas de deux* from *Monte Christo*.[18]

Many years later, in April 1898, a postcard was sent to him signed 'Batka' from Pressburg. This was probably a friend from Richter's youth of whom he had enquired for further details of this second Pressburg concert. The postcard is stuck into the inside cover of the first volume of the conducting books, and reads:

Your first concert was here at the Association with Miss Tellheim, Messrs Walter, Zamara, Zellner, Risegari on 9 April 1866. Old Wawra wrote in the *Pressburger Zeitung* on 11. 4. 1866, 'Mr R[ichter] also rendered a performance of the duet from Gounod's *Faust*, in which he played the piano accompaniment from memory and proved his true dedication during the evening. Mr R[ichter] is almost appointed a Kapellmeister in Augsburg, where he must go in September.' But Man proposes, God disposes![19]

Batka's final comment was made with hindsight for he knew what actually happened to Richter in the year 1866, the most significant year of his life. The outbreak of the Austro-Prussian War on 14 June 1866 put paid to any further ideas of going to Augsburg. It was, however, not the war but a letter written to Kapellmeister Heinrich Esser in Vienna on 16 August 1866 that changed and determined the course of Hans Richter's life. The source of that letter was Tribschen on Lake Lucerne in Switzerland and its author was the man whom the student Richter had admired from afar and so longed to meet, Richard Wagner.

2
1866–1867

Tribschen

DURING the summer of 1866, having completed the first act of *Die Meistersinger*, Richard Wagner wrote to his publisher Schott to complain about the dirty condition in which the manuscript of the Prelude had been returned to him by the engravers. On completion the full score of the opera was destined to be donated to his patron King Ludwig of Bavaria, and so Wagner decided he needed a copyist to make a fair copy of what he had written and to keep pace with him as he composed and scored the remaining two acts. Schott could not help, but Wagner sent a letter on 16 August 1866 to Hofkapellmeister Heinrich Esser in Vienna which produced a different result. He specifically asked for:

a very intelligent copyist with a perfectly complete understanding of music. If you are able to recommend an individual from among the many younger or older needy musicians of Vienna, you would do me a favour by letting me know; I would then take this adjutant, initially for half a year, with full board and expenses paid, so that under my supervision . . . he could make an exact copy. Accordingly I would be glad to know his terms: this engagement would begin in October.[1]

Esser recommended Hans Richter, but, as he told Schott, he 'hoped in so doing for no retribution from heaven and prayed for forgiveness from God for sending such a young and unblemished soul to Wagner'.[2] Towards the end of October Wagner wrote a letter to King Ludwig in which he informed him of the impending arrival from Vienna of:

an able musician I have engaged as my secretary, because I need a very gifted, thorough person with enough understanding of a score to enable him to copy it correctly and at once. I have seen at first hand how disgracefully my original manuscript was treated by engravers and copyists, and as I now view my own handwriting with quite different eyes from before, because it all now belongs to my gracious King, I take the minutest precautions that it shall reach him in the purest state.[3]

Richter's journey from Vienna to Tribschen is well documented in a letter he wrote to his mother:

The happiest of men writes to you. I have arrived safely in Lucerne. *Vienna*: Although I left Vienna in relatively good spirits, I later fell into a more depressed mood. All at once, I thought to myself, I had to leave my dear ones, and bitter regret overcame me when I thought how often I had hurt you; I was very sad. *Salzburg*: In the evening I dashed to the theatre to visit a colleague from Vienna, and I found him. On another day he [the violinist Julius Blau] took me to Mozart's birthplace (house and memorial), to Haydn's little room, etc. *Munich*: . . . I visited Lachner. He was extremely friendly and was very pleased when I told him about you; he sends greetings. Schmitt was very friendly, but he gets as angry as I do. In my room he paced up and down, and took great pleasure (and in this I heartily joined him) in raging against present day singers (except for Fistelhuber) and the blockheads who resist his methods. He has high regard for you. 'Yes, your Mama', he says, 'she slaved night and day, until she had the matter in hand, but the others are lazy.' I was also with Mallinger; she was overjoyed to find an acquaintance from Vienna.

On my journey to *Augsburg* I met with a Mr Schöner, a salesman, who was a very nice man. We spoke about the theatre, and I told him I am your son, but only after he had praised you heartily. 'Yes, since [Josefine] Richter we've had no such comparable singer.' He was quite astonished, but very pleased, and told me he had danced with you at the *Die Harmlosen* association. The Hofmanns expected me. I didn't see the old lady as she was ill, but father and daughter fetched me. The daughter took me to the grave of my dear Marie [Hans's sister who had died in Augsburg in 1858]. This stop was the saddest for me; first because of the memory of Marie and also because I was really feverish with excitement and tension. The next day I set off for *Lindau* in the company of a very charming young travelling salesman, who was very musical. We parted in Lindau; he remained and I went across the Bodensee to *Romanshorn*. Magnificent! Wonderful! Then off to *Zürich*. There I sought out my friend, the tenor Reinhold. He and the bass Roth were extremely friendly towards me. The orchestra there thinks highly of you. I had no time for our Zürich acquaintances as Reinhold would not let me go. The present Kapellmeister was in Breslau when you were there. On Sunday I conducted a ballet rehearsal of *Faust* for him; in the evening Reinhold sang Faust very well. Tell Aicher or Kleinecke, they will also be pleased. The next day Reinhold and Roth accompanied me to *Lucerne*.

On Tuesday [30 October 1866] at 11 o'clock I introduced myself to Wagner. I had not expected that he would be so exceptionally friendly. He still remembered me half and half [probably from one of Wagner's visits to Vienna in 1862–3]. At six in the evening I moved in. A few minutes later Wagner came to me. He spoke with me for a long time, and in a very friendly manner. He is pleased to have a musician around him, and sees it as a sign of respect for him that I have taken this somewhat adventurous engagement here on Lake Lucerne. He has already promised to help my progress in the future. I live on the second floor; a charming room with a view over the lake and the Rigi mountains. Our house is half an hour from Lucerne and lies twenty to thirty paces from the lake. It is occupied by Wagner, the Bülow family, me, and the servants. I have already made a very nice boating trip. The surroundings are indescribably beautiful, the roses are still in flower, and this pure air! Today I'll make another short trip and tomorrow (Thursday) I shall start work. I already have the score.

I shall not starve for I am allowed as much as I wish to eat and drink. Wagner's housekeeper [Vreneli or Verena Weidmann, later Frau Stocker], who has provided me with everything, seems to be a very good woman. I have everything I need here in the house. You do not need to send me money, you can keep it till I return to Vienna. Do not forget to send me your new address, but write it clearly. Mine is c/o Richard Wagner in Lucerne, Landhaus Triebschen [sic]. Just write Hans Richter, not Kapellmeister or anything. I will have much to do, so I shan't be able to write often, therefore do not worry; I am healthy, the cough has almost gone. Greetings and kisses to everyone. I remain your grateful son Hans.[4]

Richter's diary gave a slightly fuller account of his arrival at Wagner's home that Tuesday morning.

At first they did not want to let me in as he never receives visitors, but as I was explaining the purpose of my visit he came out of his room. ... The occupants of Tribschen are Wagner, Baroness Bülow with the children Lulu [Daniela], Boni [Blandine], and Loldi [Isolde]; the housekeeper Vreneli, her niece Marie who is the children's governess, and Agnes who takes care of them; Marie the cook, Steffen the servant, Jost the houseboy, and 'ego'. Furthermore there are two peacocks, two cats, one horse, the dogs Russ [a Newfoundland] and Koss [a fox terrier], and a number of mice. These then are the staff and the livestock.[5]

The children were very young. Lulu, aged 6, and Boni, aged 3, were the children of Cosima's marriage with Hans von Bülow. Loldi (Wagner's child though never acknowledged as such by her mother) was just eighteen months old. The relationship between Hans Richter and these three children, and later with Eva and Siegfried (born in 1867 and 1869 respectively), was always extremely close throughout his life. They became the siblings he had lost in his own childhood. The relationship with Cosima, however, was not so easy. She was naturally protective towards Wagner and probably considered Richter as a threat. She was still commuting between her husband in Munich and Wagner in Tribschen and would do so until the final break with von Bülow in July 1868. At the time of Richter's arrival she knew she was vulnerable, living as she was with a public, controversial figure many years her senior (in 1866 Wagner was 53 years old and Cosima 29). Like Wagner, Cosima wrote to King Ludwig of Bavaria to report Richter's arrival and it is clear that she did not relish the prospect of an addition to the household:

I find it almost disagreeable that Tribschen awaits a guest, in fact a young musician who has the task of copying our friend's score, so that the manuscript, dedicated to your noble self, will not be ruined by the printers. It is absolutely necessary because the *Meistersinger* Prelude has already been in the greatest danger; I hope also that the new arrival will be unassuming and quiet.[6]

A few days after Richter was installed at Tribschen Cosima seemed happier and more reassured: 'The apprentice Hans Richter, or Jean Paul as we call him, behaves quite well. He turned down the offer of a Music Directorship in order to come here and "learn something". He is modest and diligent.'[7] Richter's rejection of such a high post (the one he nearly took in Augsburg earlier in the year) is untrue. The newspaper quoted on Batka's 1898 postcard referred only to 'a Kapellmeister's post', and his inexperience would have precluded him from being offered anything else. By the beginning of December Hans was playing *Lohengrin* to Cosima each day, but it was not until Christmas Eve, some two months after his arrival, that he was permitted to join the couple socially. Thereafter he was taken fully into the family circle. From October until Christmas he worked hard at the score and was occasionally asked to play to the couple. 'Last Wednesday [14 November 1866]', he wrote to his mother, 'we had our first tea and music evening. I had to play the tenor aria [Prize aria] from *Meistersinger* on the horn; this gave much pleasure.'[8] Wagner's own daily regimen of hard work was a source of inspiration to his young apprentice and it seems that master and pupil became genuinely close. The older man was determined that his amanuensis should never return to his seat in the orchestra as a horn player, and often promised to help him further his career as a conductor when he left his service. On Christmas Day he told his mother:

For Christmas I received from Wagner an exquisite silk shawl (or whatever it is you call the thing that you tie around your neck), a fur cap, and, from Baroness Bülow, a beautiful leather briefcase. I am so happy to be in the proximity of such a man! I have successfully finished the first act of the opera. Wagner was very pleased.[9]

Hans enlisted his mother's help in obtaining for Cosima a collection of butterflies as a gift for Wagner. A great conspiracy of secrecy was necessary to prevent the composer from spoiling the surprise which he received on 7 February 1867, his name-day. 'The butterflies arrived healthy and happy, if dead,' Hans thanked his mother. 'The baroness and Wagner were very pleased and wish, by May or June, to have received another such collection to give it a symmetry. . . . The baroness is so kind and teaches me French.'[10] The French lessons persisted throughout 1867 and Hans was able to practise the language by conversing with the servants, some of whom only spoke French.

Richter's life at Tribschen is also detailed in a small collection of letters he wrote to his friend Camillo Sitte. The two men, born a fortnight apart, had known one another since the 1850s when they were both pupils at the Piaristen School in Vienna, and they were to remain lifelong friends. During their further years of study in the Austrian capital (where Sitte became an architect) the two young men became the leading lights of a group of artistic intellectuals who were distinctly Bohemian in both outlook and manner. The group bore the name *Zwack*, which colloquially means to tease or pester, behaviour which came easily to them. Their colleagues were the scholar of German Ludwig Blume, whose sister married Camillo Sitte, the philologists Adolf Lichtenfeld and Karl Lindemayr, the musicians Johann Faistenberger (later a famous singing teacher in Vienna and timpanist in the Philharmonic Orchestra), Leopold Landskron (subsequently an eminent piano teacher at the Conservatoire and joint editor of a musical journal), and the opera conductor Josef Sucher. On account of Sitte's chosen profession, Richter informed him of a visit on New Year's Day 1867 to Zürich in Wagner's company to see Gottfried Semper, a long-time friend of the composer and then professor of architecture at the Zürich Polytechnic, who had a model of a new Festival theatre ready for inspection. Because King Ludwig thought this building was destined for his capital, Richter called it 'the Munich' theatre, and his complicated layman's description would nowadays be summarized as a false proscenium arch.

The main building is an amphitheatre, the auditorium naturally the same only without boxes and galleries. The first row of seats is level with the stage, the orchestra is also semicircular and invisible to the public because it lies deeper and the audience seats rise higher and higher such that their view goes over the orchestra. This would not be possible if the stage were not constructed in the way that it is. This is not immediately adjacent to the two side walls which border the auditorium and form the front of the stage, but instead there is a second wall situated some one and a half paces behind these other two walls and decorated exactly like them except that it is somewhat smaller in scale, so that the stage is very much like a picture in a frame. This also prevents that unattractive habit that many singers have of stepping outside the set. On both sides of the main building is provision for a concert hall, rehearsal hall, and dance hall. My description is really quite confusing because I don't understand the technical terms, but as I cannot do better you'll have to put up with it![11]

Among the party visiting Semper in Zürich was Hans von Bülow, who had arrived from Basle. Richter's diary refers to a performance at Tribschen on 30 December 1866 by von Bülow of the Piano Sonata in F minor by Schumann. He told his mother that he no longer wished to hear any other pianist play (he was yet to hear Liszt!), and he told Sitte, 'Friend, that is the greatest. That feeling, TECHNIQUE, memory, interpretation!'[12] Wagner also made his young secretary study writings on music and orchestration, though Richter felt that his knowledge of Berlioz's treatise on the subject, which his father owned, was sufficient. He was also encouraged to use Wagner's library to the full and by the New Year of 1867 he was deep into Goethe's Italian journeys. On 7 February he took part in a concert of chamber music in Lucerne, organized by Music Director Arnold, in which Hans played horn in Beethoven's Horn Sonata, Op. 17 and viola in Mozart's Trio in E flat for piano, viola, and clarinet. When he gave Wagner a private performance of the Beethoven work two days later Hans reported to his mother that the composer was 'very pleased. I am very happy. On this occasion he promised me that I will be given a post for the production of the new opera [*Meistersinger*] in which I can display my abilities. Bülow will conduct it. I know of no other worthy artist for it. I have so much to learn! Thank God I am at the source of musical perfection!'[13] When the Horn Sonata was repeated in Basle on 26 March things did not go so well, as he told Camillo:

They still have the higher pitch [a quarter tone higher]. At the rehearsal with Bülow's own piano we were perfectly in tune, so everything went splendidly, but his concert piano, because of the Basle orchestra, is still tuned in the old way and no one thought of that. The Horn Sonata was therefore a mess. Fortunately I had a makeshift crook with me and in the Brahms Trio for piano, violin, and horn [Op. 40] I made amends for what might have been worse. This quite infamous stroke of fate strengthens my resolve even more to give up the horn completely, even though I'm thought of very much as a first horn. You'll know I'm right not to give in; on the other hand I can't take it amiss when they seek to persuade me. On the contrary I must take it as proof of their confidence in me as an artist.[14]

In April Cosima was away from Tribschen for a week and as a result Hans appears to have become closer to Wagner; an entry in his diary for Sunday 21 April speaks of a jolly evening meal together in which the composer opened up and reminisced about his life. Wagner spoke of the year 1834 in which, at the age of 21, he conducted his first opera, Mozart's *Don Giovanni*, how he had prepared it and the difficulties he had encountered with conducting Donna Anna's recitative [presumably 'Crudele? Ah, no, mio bene' preceding 'Non mir dir' in Act II]. He went on to speak of how he had met his first wife, Minna Planer, in the summer of that year in Lauchstädt, and of her singing the role of the amorous fairy in *Lumpaci Vagabundus*. The spring of 1867 was a difficult time for Wagner. Cosima was in the throes of leaving von Bülow, with whom Wagner still had strong musical connections, and in Richter, a young man barely older than he was at the period about which he was reminiscing, he found an easy and receptive listener. Wagner, on the other hand, was not so easy to talk to. His new secretary was often obliged to accompany him on his afternoon walk during which there were usually long periods of silence. One day Richter was unable to prevent himself from asking the composer which opera he preferred, *Tannhäuser* or *Tristan*. In kindly fashion, Wagner told him not to ask such a stupid question. From then on Richter spoke only when he was spoken to on these walks, but he also took Wagner's trust in him to heart and his life-long reverence begins to show. 'I'm not here surreptitiously to get myself a good job', he told his mother, 'but to learn, and thank God I am able to. Nowhere else could I have learnt but here under the eyes of this genius. What I have learnt here will prepare my way throughout the world.'[15] This last sentence was prophetic and recurs during various press interviews he gave throughout his life. In another letter he wrote: 'I'm fine. Really! I'm happy in the knowledge that I am loved by the greatest man of all time. My grateful respect has no bounds.'[16]

Wagner invited Hans to travel with him in May to Schloss Berg on Lake Starnberg, where the composer was to be a guest of King Ludwig for two months. The plan fell through, however, as Ludwig's engagement to Archduchess Sophie Charlotte foundered at this time (it was broken off in October 1867) and the king felt in no mood for entertainment. On the other hand he was eager for *Meistersinger*; Wagner completed the orchestral sketches on 5 March and two weeks later, on 22 March, began orchestrating the second act. The opera was scheduled for performance as part of the king's engagement celebrations in the autumn, but Ludwig's domestic upheavals forced its postponement until the following year, no doubt to Richter's relief. The two men did, however, undertake another trip together instead, but it was to Munich, where Wagner supervised the final rehearsals of *Lohengrin* in June with von Bülow as stage director and conductor.

Letters to both his mother and Camillo Sitte described the musical problems and opera house politicking that surrounded the staging, not only of *Lohengrin*, but of any Wagner opera. Wherever he went Wagner aroused controversy. Whether his opinions were expressed in pamphlets or verbally, he invariably created factions which either earnestly supported him or vilified him; extreme, polarized positions were taken on his behalf or against him, and moderation never prevailed. The *Lohengrin* demanded by Ludwig had to be a 'model' performance. It had been two years since the première of *Tristan* in Munich, and the monarch was becoming very anxious to have more of Wagner's music performed for him. It soon became apparent, however, that the composer and his patron were on a collision course, particularly with regard to the casting of the opera. Wagner's choice of tenors was limited. Albert Niemann refused to sing the role uncut and Ludwig Schnorr von Carolsfeld (the first Tristan) was dead. Wagner had to persuade the king to accept the 60-year-old Joseph Tichatschek from Dresden. The final rehearsal was postponed once because the Telramund (Franz Betz) was hoarse and then a second time because, due to the death of an Austrian princess, the Court was in mourning for a week. Wagner had rapidly lost interest by now; he was embarrassed by Tichatschek's tremulous voice (remarkable though it was for his age) and full of foreboding that Ludwig, peering through his opera glasses, would see little that would pass for youthfulness in the Knight of the Grail (Wagner had desperately tried to soften the blow by warning the king that Tichatschek's singing was 'like a painting by Dürer, but that his outward appearance was more like a picture by Holbein'[17]). Sure enough the King, who had already overruled Wagner by casting Mathilde Mallinger as Elsa, objected to Tichatschek and to Frau Bertram-Mayer as Ortrud. The performance took place on 16 June with a new Lohengrin and Ortrud (Heinrich Vogl and his future wife Therese Thoma), but Wagner had returned to Lucerne a day earlier, leaving the luckless von Bülow to placate the discarded singers and quickly coach the replacements.

Richter told his mother that 'the rehearsals for *Lohengrin* were for me of infinitely great value'.[18] Judging by a report in the newspaper *Die Signale* it is clear why the apprentice conductor would benefit so much by being allowed to observe Hans von Bülow at work. Von Bülow had been virtually responsible for laying the foundations for a schedule of operatic rehearsals that has endured to this day. Coaching of individual singers and ensembles around the piano, preparing the orchestra with sectional rehearsals, piano-accompanied production rehearsals, a *Sitzprobe* with singers and orchestra (where no staging takes place but the singers are able to concentrate solely on the music), stage rehearsals with orchestra, leading to a final rehearsal and the performances. Von Bülow perfected his system with *Tristan* and *Die Signale* described the 'usual priming-needle rehearsals' for the Munich *Lohengrin*.

Last week there were a million little rehearsals: (a) for first and second violins, violas, etc. alone, (b) for winds and horns alone, (c) for trumpets, trombones, and percussion alone; (d) for the off-stage band. Then there was a four-hour ensemble rehearsal for the strings, and now for each act there is a three-hour full orchestral rehearsal in the presence (though not with the participation) of the singers. All the stage rehearsals are for the present with piano, next week everything will be together. This new system, which is actually only tiring for the conductor, should prove very effective; Herr von Bülow spends ten hours a day in the theatre and probably also spends the night there so that, by early morning, he is always first on the field of battle![19]

Hans told a friend, the opera singer Alexander Reinhold, that 'the rehearsals for *Lohengrin* are fully underway, and I am present all the time.... Bülow is the Master of all conductors. Now that's what I call studying!'[20] He told Camillo Sitte that he had

got to know the opera really well. Bülow prepares it wonderfully. Although the orchestra is nothing like as good as the Vienna orchestra, it does achieve excellent results. Once again it was proved to me that a good conductor is able to accomplish splendid things with even ordinary musicians. There was one rehearsal which I found especially interesting, the piano rehearsal which Wagner took with Fräulein Mallinger (Elsa), Frau Bertram-Mayer (Ortrud), and Tichatschek (Lohengrin). You should have seen the Master! He takes all the production rehearsals. The scenery was magnificent and, together with the costumes, true to the period in which the action is set. After great preparation the final dress rehearsal took place, for which more than a hundred tickets had been issued. The king was present from beginning to end, and it all went splendidly. Unfortunately the old master Tichatschek did not sing at the performance. He possesses a fault which he will never correct in this life (he has his 60th year under his belt). His acting was too wooden. This might have accounted for the lack of consideration shown him if only his substitute had been a phenomenal actor, but he too (Herr Vogel) is quite an insignificant singer with no poetic interpretation of the role, and it was the greatest disappointment as far as the opera was concerned that Herr Tichatschek was not permitted to sing. Vocally he outshines all our young tenors; he has a noble sound and, what was remarkable, he improved towards the end of the opera, which stood him in good stead for the Narration of the Grail by Lohengrin in Act III. Then there was his wonderful declamation, his diction, everything rhythmically short, so that one did not have to haul the words from his mouth with a tow rope as with most singers. . . . Frau Bertram-Mayer did not sing either, but I feel less sorry for her. She acted very well in fact, but her voice has an ordinary sound in the middle range. . . . Betz from Berlin is an eminent singer and actor. The orchestra was very good, but the chorus was best. Yes, you may be surprised. Compared with Vienna the chorus numbers are weaker, but the members went for their task with an enthusiasm unique for such people. One can, with hindsight, call this a model performance.[21]

The reasons for Wagner's dissatisfaction with Mathilde Mallinger were clarified by Hans to his mother. 'She sings in tune and in time but without warmth, and she does not grasp the musical expressiveness of this opera at all. Yes, there's a difference between Italian ear-tootling and the music of German opera. With the former an effective final cadenza is the main thing, but the latter has an integrated development without the slightest neglect of any secondary figuration.'[22] Once again his bias towards German music and his lukewarm opinion of Italian music became apparent. Earlier in the year he had advised his mother on the repertoire her pupils should be studying. 'Just Mozart, Weber, Beethoven, Schubert Lieder, Mendelssohn, Wagner, and in God's name also Meyerbeer and Gounod's *Faust*. Only no Verdi; one cannot learn anything from him. By studying just a little representative amount of music, I can see what these modern Italians are scribbling down as music. My hate and disdain for them are mounting to a frenzy!'[23] In later life these opinions of French and Italian music were transposed.

Richter's mother married her second husband, Anton von Innffeld, on 4 March 1867. Hans was absolutely delighted for Josefine, whose happiness and well-being concerned him constantly, but he was also genuinely fond of his stepfather. 'I'm not going to rack my brains for specially chosen words,' he told the newly weds in his letter of congratulation. 'To you, my dear mother, and to you, my father-friend, I cry a hearty "Good luck!" You know best of all what you, dear Anton, have found in my mother, and I thank heaven that Fate has placed her in the hands of the honourable gentleman that you are, a man who commands my fullest respect and love.'[24] He bombarded them with requests to send him this and that— winter coats, summer jackets, even the crook for the F horn that rescued him in the Brahms trio in Zürich, but never money. He lived on savings to supplement his board and lodging with Wagner, and his needs were few. Wagner gave him 100 francs for his name-day (24 June). 'That was noble,'

Richter told his mother. 'He said it was recompense for the journey [to Munich], but he had already paid me that as well as giving me twenty florins spending money in Munich.'[25]

In this same letter Richter reveals that Wagner had promised him the job of training the chorus for *Meistersinger*. It was the first step along his professional path. In spite of his resolve never to resume his career as a horn player, he did not give up playing altogether. One of the visits he paid during his stay in Munich (on 24 May 1867) was to Franz Strauss, first horn in the opera orchestra under von Bülow and father of the composer Richard, then just 3 years old. The paths of Richter and Strauss senior would cross again before long in connection with the première of *Meistersinger*.

After working all day at copying, Richter would often row out to a small island on the lake. It would not be unusual for him to take his horn with him and, in the solitude of his surroundings, play phrases from *Meistersinger* amongst his repertoire. Many years later in 1885, after the degree ceremony at Oxford at which he received an honorary doctorate in music, he was approached by an academic. During the course of the conversation it emerged that this professor was holidaying in Lucerne at the time of Richter's apprenticeship with Wagner. He asked Richter if he knew the identity of a mysterious horn player whose music used to waft across the lake at dusk. Richter confessed that it was he, adding that the professor could consider himself the first member of the public to hear music from what was then the unperformed opera *Meistersinger* by Richard Wagner.

On one occasion the composer himself used his secretary's expertise as a horn player during the course of the composition of the opera's second act Finale. Asked by Wagner if the passage in the brawl scene (21½ bars before the beginning of scene vii where the first horn takes up the music of Beckmesser's Serenade) was playable, Richter replied in the affirmative but added that the result would be strange and somewhat nasal. His answer could not have delighted Wagner more, for this was just the effect sought by the composer, though in the finished score he strengthens the first horn by also giving the frenetic semiquaver passage to principal oboe, clarinet, and bassoon, all of which, as reed instruments, provide an added edge to reinforce this nasal quality. Asked by Wagner to demonstrate on the horn, Richter had to play the passage again and again at an increasingly faster tempo until the composer was satisfied.

Quite what Richter made of Wagner's private life, which he witnessed more or less at first hand, is difficult to tell. On 17 February 1867 Wagner's daughter Eva was born (his second child by Cosima), but Richter

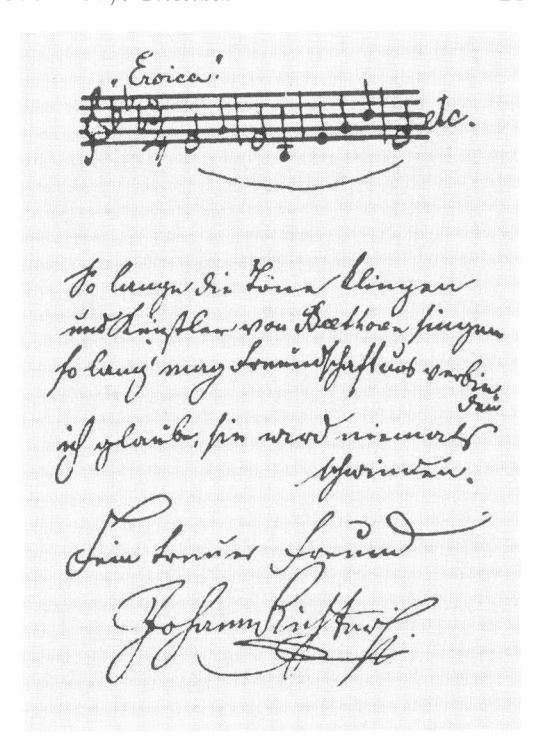

FIG. 1 Dedication to Camillo Sitte from Hans (Johann) Richter quoting the opening of Beethoven's 'Eroica' Symphony, Vienna 1864.

calls her von Bülow's fourth daughter in his diary. He notes too that the child was born at nine in the morning and that von Bülow arrived at two in the afternoon. Two days later, on the 19th, the child's baptism took place at three in the afternoon. Hans was a witness and also deputized for the absent godfather Emil Merian. Five days later von Bülow left, followed on 2 March by his acknowledged children Daniela and Blandine, with their governess Agnes. There was much journeying between Tribschen and Munich throughout the year, much of it done by Cosima to maintain the charade of her marriage. Someone who finally had to be brought into the affair was Cosima's father Franz Liszt. Wagner dreaded the inevitable meeting with his former close friend for he knew that he would have to give a complete account of himself. Liszt arrived in Munich on 21 September 1867 and on 1 October Wagner sent Richter with an introduction to meet him. This came about in a way typical of Wagner's impetuous nature. On 29 September, as Hans recounted to Camillo Sitte,

We were seated at the piano, Wagner and I, playing a duet version of Bach's preludes and fugues from the *Well-Tempered Clavier*. My friend! That wasn't the old pedant, the father of fugue and counterpoint! No, that was the prototype of

Beethoven's C minor Symphony, the work of the greatest composer, the founder of German music. It sounded quite different from what I was used to hearing. Oh, this Wagner! It is impossible to describe what demonic power lies in these pieces when they are interpreted by my noble master. When we got to the C sharp minor Fantasy I could restrain myself no longer, the tears poured from my eyes. Wagner too was quite moved by the power of Bach's sounds. Time and again he called out, 'He is the greatest master.' Then he said I should hear them played by Liszt. Hardly had he said this when he explained to me that on Tuesday I should make preparations to travel to Munich. And that's how it happened.[26]

Hans heard Liszt play at von Bülow's house on the following day, Wednesday 2 October. 'It is indescribable', he told his mother, 'how this master has these sounds within his power.... On Wednesday 9th Liszt came from Basle on a visit. How uplifting it was for me to be in the company of the two greatest living composers. Liszt played a lot from Wagner's new opera. This continued late into the night. On Thursday morning he travelled back to Munich.'[27] It is worth returning to the letter to Sitte for more detail about these few extraordinary days.

Wagner had given me a written introduction to Liszt. He doesn't look at all clerical; he wore quite an ordinary black suit, and his manner was kindness itself. In the evening at Bülow's I was lucky to take part in hearing the master of all masters. My friend! After I'd heard Bülow and Tausig I couldn't imagine anything more perfect than their playing. But when one hears Liszt play, one realizes that he is the creator of pianists and everyone else is his pupil. He played me four preludes and fugues by Bach, among them the aforementioned C sharp minor Fantasy, then to end his own Prelude and Fugue on the name Bach. That evening was a musical landmark on the path of my artistic career. I was just in time, for the next day he left for Stuttgart. I awaited him the next morning at the station where we chatted for a long while. I received several cigars from him (he's a heavy smoker, and as a matter of fact uses an excellent brand), one of which he wanted me to keep and smoke in your presence as a 'Cigar of Harmony'. I'm afraid, however, that it's gone up in smoke as has my hope to be with you soon. I remained in Munich for a week. On Wednesday 9 October Liszt came to us at Lucerne. What a reunion between both these intimate friends and masters! In the evening, after we had all eaten together, Liszt played the first act of *Meistersinger* from the proof copy, and the second and third acts from Wagner's sketches of the work. His ear hears everything with incredible accuracy, but to follow with one's eye how he finds his way around the smeared and unclearly written sketches is an impossibility, even for me who until now fancied my ability as a sight-reader. My friends, if I didn't know that Mother Nature had given me a drop of talent, and that I take my art as seriously as both these masters, my courage would disappear when I am with them. But for me the association with such men is only uplifting and stimulating in the awareness of my serious endeavours. On Thursday morning Liszt travelled back to Munich. My friends, I can tell you that I am proud that Liszt kissed me more than a dozen times. It's his habit to salute people to whom he is well disposed like that. I'll see him again next summer for he promised that he would come to Germany each summer for a longer period to visit his friends.[28]

The impressionable young man was so elated at being in the presence of the two men during those heady days that he appeared to be totally unaware of Wagner's discomfort during the first six hours of Liszt's visit on 9 October when the two were closeted together to discuss the Wagner–von Bülow–Cosima triangle. The evening's music-making that followed dispelled any difficulties, however, and Wagner was able to record in his annals that the meeting was 'dreaded but agreeable'.[29] Liszt was no less affected by the strain of the occasion, although he considered the visit to be the best course of action. Nevertheless he also felt as if he had visited Napoleon on St Helena and in a letter to the Princess Wittgenstein described Wagner's appearance as 'very changed. He has become gaunt and his countenance lined. But his genius has not become weaker. *Die Meistersinger* left me astonished by its essence, boldness, strength, fervour, and its inexhaustible riches. No man but he could have written such a masterpiece.'[30]

Two weeks after Liszt's departure from Tribschen Wagner completed the instrumentation of *Meistersinger*. It happened on 24 October, and Richter's diary entry is simple: 'On the stroke of eight the end of *Meistersinger*.' The actual hour of completion is at variance with Carl Gianicelli's fuller account in his obituary notice of Richter.

As work on the instrumentation of the third act neared its conclusion Wagner reckoned [on 21 October] that he would be finished within three days, and indicated six o'clock as the precise hour on the third day at which the completion of the work would take place. The day came and to Richter's astonishment the Master called him in the early afternoon, at which time he would normally be working, for a walk. Richter assumed that the Master had forgotten that he wanted to complete *Meistersinger* on that day, but refrained from reminding him. After they had returned home shortly before six, Wagner summoned Richter to his workroom—the first time this had ever happened. There lay the final sheets of the *Meistersinger* score, complete but for the last C in the double basses. The Master, in his pleasure at completing the great work, had worked hurriedly and as the church clocks of Lucerne struck six he took his quill and wrote in the last note of the score. He then rang for the servant, who brought a half-bottle of champagne, and Richard Wagner and Hans Richter toasted 'to *Die Meistersinger*'. The accuracy of this story is easily verified; in the handwriting of the *Meistersinger* score it is clear that the last C has been added later.[31]

Sir Adrian Boult, in a letter to Richter's son-in-law Sydney Loeb, wrote: 'I wonder if I have told you before of an interesting point Richter told his friend [Ernst] Schiever, the violinist in Liverpool, who told my mother, that he was usually an hour longer each day over the job than Wagner over the actual scoring.'[32] Richter's work as copyist of the full orchestral score, subsequently sent to Schott for printing, was completed on 2 November 1867. At the end he signed it 'H. Richter scrp.', an abbreviation of *scripsit* or 'he wrote it'. Two days later Wagner left for a well-earned rest in Paris and, on his return, presented his scribe with a beautiful travelling case. Hans now found himself with little to do, but he had not long to wait before his next task. On 1 December he left Tribschen after thirteen months and travelled to Munich. Two months earlier, during his visit to the city to meet Liszt, steps had been taken to secure this next move. As he told his mother:

> My future has also been half decided. I am to be taken on as chorus master in December [at the Court Opera in Munich]. Wagner is against this, he wants me as Music Director. For the present I shall start in December to train the chorus for *Die Meistersinger*, but also be free from working on all other operas except this one; for this I shall receive a special gratuity which pleases me very much for I do not wish to remain a chorus master.[33]

To Wagner's annoyance Richter was passed over for the vacant Music Directorship, but by writing both to Court Secretary Düfflipp and to the Intendant of the Court Theatre, Baron Carl von Perfall, he secured him the post at the Court Opera of répétiteur as well as chorus master for *Meistersinger*, as in his original plan, and other operas such as Nicolai's *The Merry Wives of Windsor*, Gluck's *Armide*, Spohr's *Jessonda*, and Rossini's *William Tell*. To supplement his income Richter also secured some teaching duties for himself at the new Royal Music School, which was headed by the indefatigable von Bülow. Richter's departure from Tribschen was a matter of sincere regret to Wagner, who told Mathilde Maier that 'the cultured person in my company is a Viennese musician who can play the horn, violin, and piano, who is totally organized, and is also a child-like, good, and handsome man. I shall now have to give him up to a post in Munich.'[34] Richter would return to live with Wagner at Tribschen between June 1870 and April 1871, but first came two turbulent years in Munich.

3

1868–1869

Munich

RICHTER'S first encounter with any anti-Wagner sentiment in Munich occurred soon after his arrival in December 1867. In spite of his appointment at the Court Opera he had to pass an audition as repetiteur set for him by Franz Lachner, General Music Director of the city and an opponent of Wagner. Lachner had met Richter on a visit to Vienna in February 1866 when he was pleased to give him a glowing reference as a potential Kapellmeister. Now Lachner was to test him again, but with Richter's colours firmly fixed to the Wagnerian mast, the young man no longer found a friend in the General Music Director. A message was sent to Richter to appear at ten o'clock the next morning at the opera-house and accompany a production rehearsal of Nicolai's *The Merry Wives of Windsor*. He was ordered not to bring a vocal score but to play from the full score provided. Fortunately he knew the work from his horn-playing days at the Vienna Opera and he duly came through the test with flying colours. He played the work with astonishing facility and earned the admiration and respect of both singers and observers, of whom there were many because word had got about of a possible public humiliation. Lachner's plan had misfired but he grudgingly conceded that in Hans Richter 'a Wagnerian had been found who also knew other music'.[1]

Richter was lodging in Munich with a répétiteur named Ludwig Eberle who told him of a small town in Bavaria with a charming eighteenth-century opera-house in which he had worked with a visiting company from Bamberg. The house had a particularly large and fine stage and Eberle recommended it to Richter. The town's name was Bayreuth. When the idea of building a festival opera-house in Munich specifically to stage Wagner's *Der Ring des Nibelungen* was thwarted by the politics surrounding the scandal of the Wagner–Cosima–von Bülow affair and the impossible position in which it placed his patron King Ludwig, Richter suggested Bayreuth to Wagner as a possible alternative venue because it was still on land belonging to the king. Wagner remembered the town from his youth and warmed to the idea immediately, and though the opera-house which was already in the town proved unsuitable for his needs, Bayreuth was chosen as the site for his own purpose-built house. Richter always claimed the credit for having suggested the town of Bayreuth to Wagner.

Work on the preparation of *Meistersinger* for the première in June now began to gather pace. Von Bülow's system of piecing together the various components of the cast was once again put into practice. Hans embarked on a monumental series of sixty-six rehearsals with the chorus, which had to be augmented by amateurs from the city to make up the numbers of male choristers. He began, not with learning the music, but by training them to act convincingly and enunciate clearly in anticipation of Wagner's declamatory style of setting his texts. He also coached the two tenors, Max Schlosser who took the part of David and Franz Nachbaur who sang Walther von Stolzing. The latter was not permitted to leave Darmstadt, where he was under contract, until 1 June, so Hans had to travel there (on 20 April) and work with him. 'I was sentenced to six weeks in Darmstadt,' he wrote.[2] Von Bülow put it another way. 'We have sent our excellent young repetiteur to Darmstadt where, for a fortnight, he has been giving the Moor a good soaping even if he cannot scrub him white.'[3] Wagner also told Richter to 'go for it', and added that 'the finest and only reward for me is to find a man who gives me true joy. Once again I have found him in you, you good man, and believe me that is worth more than any stroke of good fortune. Now stay worthy of me; I don't think you will feel too sorry that your destiny brought you to me.'[4]

Other members of the cast included Franz Betz from Berlin as Hans Sachs, Mathilde Mallinger as Eva (Elsa of the previous year's *Lohengrin*), and Sophie Diez, Munich's 'house' mezzo-soprano, as Magdalene. Rehearsals for *Meistersinger* brought Richter once again into contact with Franz Strauss. This eminent horn player was no friend to Wagner or his music and took every available opportunity to disparage it. During a stage rehearsal of the opera with the ninety-piece orchestra Strauss stood up and told von Bülow that the passage in the Finale of Act II, about which the composer had sought his copyist's advice, was unplayable, a risky thing to say with Hans Richter about. Sure enough the young chorus master appeared from the wings, leaned down into the pit, and asked to borrow Strauss's horn, whereupon he proceeded to demonstrate to the full company (and to von

Bülow's delight) the error of the player's ways by performing the passage perfectly. Franz Strauss never forgave him. Many years later Richter told the journalist F. G. Edwards in answer to an enquiry, 'I am very sorry I cannot tell you any particular interesting story of Richard Strauss as a boy. I was in Munich from 1867–69. Richard Strauss' father was an excellent horn player at that time, but with regard to his behaviour towards the work of Wagner, it is better not to mention it. Strauss' son may be happy that he has not his father in his orchestra. All this is, of course, quite privately and confidentially, and you may forget it.'[5]

On 20 June (the day between the final dress rehearsal and the first night) Wagner presented Richter at dinner with a set of gold buttons in the form of doves of the Holy Grail, which he himself had received from King Ludwig three years earlier. The first performance of *Meistersinger* took place on the day on which the action of the opera is set, 21 June or Johannistag, and it was a triumph for all concerned. After the première a heated discussion took place at an inn among supporters and opponents of Wagner's music. Richter was there and to take the heat out of the moment he reached for his horn and blew the Nightwatchman's call from the end of the second act, and argument was transformed to laughter. Apart from the implacable Eduard Hanslick from Vienna who despised Wagner's music, the composer enjoyed a good press; not that he stayed in Munich to read it. Before the second performance he was back at Tribschen and hard at work on *Siegfried*, which he had put in abeyance twelve years earlier. Richter, meanwhile, remained in Munich and monitored his chorus throughout the five other performances of *Meistersinger*. The run did not end without further drama, however, when Wilhelm Fischer, who took the part of Kothner (one of the Mastersingers), was taken ill on the day of the last performance, 16 July. With the prospect looming of cancelling the performance (there was no understudy system in those days) Richter stepped into the breach. Although he had never been on stage before, he said he knew the part and would sing it. It was his one and only performance as a singer apart from his boyhood days as an alto soloist with the Vienna Court Chapel choir. His versatility as a musician knew no bounds, and Wagner wrote that 'astonishing a deed as it was, it did not surprise me in the least for I know that in your place I would have done exactly the same'.[6] The king made Hans an *ex gratia* payment of 200 florins for his contribution to the preparation of *Meistersinger* on 3 July 1868.

Wagner intimated in the same letter that he had asked Düfflipp to appoint Richter as Royal Music Director at a salary of 1,200 florins, and that failing this he would try and secure a position for the young man elsewhere. He told Richter that he would have nothing more to do with the Munich theatre and that if the young man found himself without work, he was free to return to Tribschen. Wagner's intervention paid off. Baron Perfall, on behalf of the king, announced his appointment on 9 September as provisional Court Music Director for three years with duties as both conductor and repetiteur. On 25 August Hans conducted Rossini's *William Tell*. 'Send my washing very soon', he told his mother, 'because I want to wear the beautiful shirt when I conduct; 25 August is the king's birthday. . . . Also send the red box containing my buttons, I forgot them.'[7] He backdated his appointment from this date, for he wrote in his conducting book, 'My activities as an "appointed" conductor begin at the Court and National Theatre as Royal Bavarian Court Music Director.'[8] Von Bülow, away in Wiesbaden at the time, asked Peter Cornelius to 'write and tell me about Richter's début as my vice-baton'.[9] Hans did well and remained in the post for one year and three days, conducting on average three times a month. His repertoire for that period consisted of *William Tell*, *La Dame blanche* (Boïeldieu), *Das Rothkäppchen* (Dittersdorf), *Le Premier Jour de bonheur* (Auber) (in von Bülow's opinion 'the best product of the French school in many decades'[10]), *Le Postillon de Lonjumeau* (Adam), *Der Rothmantel* (Krempelsetzer), *Le Prophète* and *Les Huguenots* (Meyerbeer), *Fidelio* (Beethoven), *Das graue Männchen* (Ballet), *Ruy Blas* (Max Zenger), *Le nozze di Figaro* (Mozart), and on 27 July 1869, just a month before he resigned the post, his first performance of *Meistersinger*, the opera he would make his own and conduct 141 times throughout his life.

His musical patrons were satisfied with their protégé. Wagner's congratulations on the appointment were included in a letter dated 7 September. Hans von Bülow wrote to Joachim Raff: 'On Sunday it's *Les Huguenots*, otherwise rehearsals for *Die Meistersinger* with Sigl (Beckmesser). I'm leaving the coaching of Hans Sachs (Kindermann) to Richter, whose appointment as Court Music Director was announced yesterday.'[11] Von Bülow developed a high regard for Richter and reported to Wagner in December that the young man was working very capably and earning the respect of his colleagues. 'Don't feel insulted that, as a Christmas present, I have given Hans Richter the seal in the shape of your bust that you gave me. On it I have had engraved the words "Ehrt eure deutschen Meister!" [Honour your German master—the final chorus of *Meistersinger*] I think I have led the way as a good example.'[12] A little later von Bülow was calling Richter 'Hans II' when he told Joachim Raff that were he himself to be advanced, then Richter would be appointed First Theatre Kapellmeister.[13]

Wagner's letters to Richter (of which 100 were published in 1924) are full of a variety of topics, from news of the occupants (human and animal) of Tribschen to advice on the second series of performances in Munich of *Meistersinger*:

Sigl [as Beckmesser] was said to be correct but somewhat dull; apparently the serenade under the window—namely its passage work—dragged even more than with Hölzl. Urge him to sing these things in a brisker tempo; such moments must only be comical, not boring. Beckmesser is a virtuoso who enjoys his singing skills too well.... You seem to be well liked; that's very necessary, particularly in Munich.... Very soon I'll send you the first act of *Siegfried* to copy. I'd be grateful if you would get on with it straight away.... I'm beginning to recover, am busily writing out a clean copy of the *Siegfried* score and hope to finish the composition of this, my second *opéra comique*, on my next birthday, when, on my reckoning, I'll be 106 years old.[14]

Hans had already started work on copying *Siegfried*; his diary entry shows 23 November 1868 as the day he began. *Der Ring des Nibelungen*, of which *Siegfried* is the second opera, occupied Wagner from 1848, when he wrote the first prose draft of *Der Nibelungen-Mythus als Entwurf zu einem Drama*, to 1874, when he completed the full score of the third act of *Götterdämmerung*. The tetralogy begins with the prologue *Das Rheingold*, and it was this work that precipitated Hans Richter's departure from Munich in the summer of 1869. King Ludwig, always eager to have Wagner's operas performed for him, told Hans von Bülow on 25 February 1869 that he wished to see *Tristan* in the spring and *Rheingold*, which Wagner had completed in May 1854 but which had not yet been staged, on his birthday, 25 August. Its lack of performance was a consequence of Wagner's stipulation that his *Ring* (which at that time was incomplete) should only be given in its entirety, that it should on no account be in the repertoire of any German theatre whose operational methods he loathed, that it should only be performed before a selected and invited audience, and only then in a theatre especially constructed in the countryside for such a purpose. Needless to say the king, equally stubborn when it came to his own wishes, ignored Wagner's protestations and called the performance for his birthday. In spite of an assurance given by the composer to von Bülow that he would attend the occasion (the conductor reported this to Hans von Bronsart in a letter dated 20 April 1869), in the end Wagner decided not to do any more than issue instructions and advice from Tribschen. As von Bülow told Court Secretary Düfflipp, 'Herr Music Director Richter has returned from Lucerne, with exact musical instructions for the performance of *Das Rheingold* provided by the composer. Would you be so good and take steps to ensure that [Intendant] Herr von Perfall does not set himself up in opposition to these instructions, as he did with *Die Meistersinger*, and once again make my task so difficult as a consequence?'[15]

As events turned out von Bülow resigned his post on 8 June. His health was very poor; a tumour in his throat had been diagnosed a year earlier at the time of the *Meistersinger* première and in April 1869 he had caught a chill in Regensburg, where he was giving a piano recital. His duties at the Munich Music School were onerous and he now faced the task of reviving *Tristan* for the king with a cast in whom he had no faith ('Mr and Mrs Vogl—good devourers of notes but otherwise? . . . The result will be at best a *fiasco d'estime*').[16] Cosima and the children had left him for good (on 16 November 1868) and in June 1869 she bore Wagner his only son, Siegfried. This was a further humiliation for the cuckolded von Bülow, who was also exhausted and in need of total rest. 'I am saving myself, my health, and my future human and artistic existence,' he told Hans von Bronsart. '*Periculum in mora*. . . . Here I am literally half dead from work and three-quarters dead from worry—*pour le roi de Bavière*. . . . My only concern is to be able to get away from here quickly and without a scandal.'[17] Matters came to a head with his orchestra during a stormy rehearsal of *Tristan* on 15 June (though some players, to whom he was grateful, sprang to his defence and evidently realized the worth of the man in their midst whom they were about to lose forever). 'The performance', he told Richard Pohl, 'was surprisingly correct and even beautiful. . . . I have already given Richter *Die Meistersinger* (because of new building work in the theatre it will be the last performance of the season and takes place on the 27th with Betz and Mallinger) and have also given him the preparation of *Das Rheingold* for 25 August. He can get it done better than I because he is unbroken, fresh, healthy, and ambitious.'[18] Richter telegraphed a worried Cosima, 'Bülow came to life during performance, afterwards cheerful, calls of "Bülow stay!", success tremendous.'[19] Although the king had given his Music Director leave until 1 October, during which time he was to reconsider his decision to resign, von Bülow did not need it. He plainly saw what was coming as far as *Rheingold* was concerned and there was no question of second thoughts; his mind was made up and he left, spending most of the next three years in Italy.

The new building work mentioned by von Bülow involved rebuilding the stage to accommodate the special machinery for *Rheingold*, and lowering the pit to modify the acoustics. Meanwhile Richter had already been travelling back and forth between Munich and Tribschen to receive instructions from Wagner about either the preparation of *Rheingold* or the copying of *Siegfried* (Cosima noted in her diary on 5 April, 'I spent the day on the sofa listening to R[ichard] going through *Das Rheingold* with Richter, and embroidering').[20] There was one last happy occasion, the calm before the storm, when Richter was invited by Cosima to Tribschen for Wagner's birthday on 22 May. Early in the morning he blew birthday greetings in the form of Siegfried's horn call, and a string quartet from Paris played three Beethoven quartets. Others involved in the production of *Rheingold* also undertook the journey, Reinhold Hallwachs, the stage director, the machinist Karl Brandt (who made a very favourable impression upon the composer), the singers Franz Betz, Otto Schelper, and Karl (Max) Schlosser (the first two eventually pulled out as Wotan and Alberich respectively and Schlosser changed roles from Loge to Mime at the last minute), and the scene painter Jank. As problems with the staging of the opera began to emerge and as other crises arose, so the date of the performance began to be repeatedly put back and the king had to make do with Spohr's *Jessonda* on his birthday. The sequence of events which led to Richter's departure can be seen by quoting Cosima's diary (R. denotes Richard Wagner).

4 July . . . As R. is playing me what he has written, Richter comes in, to our astonishment. Dismal news from Munich, the theatre manager a coward, all the rest so crude that it defies description. Dear good Richter weeps as he recalls his happy days in Tribschen.

8 July . . . In the evening said farewell to R[ichter], he will resign his position if Hans [von Bülow] really goes.

25 July . . . In the morning a letter from Richter, who has requested his dismissal. He is being asked to stay on, R. answers that he should insist on his dismissal.

26 July, Costumes for *Das Rheingold* arrived, very silly and unimaginative.

5 August . . . Richter reports that they could not get the right singer for Alberich, and he asks whether they might accept the wrong one. 13 August . . . Richter describes the conditions in Munich as ever more horrible.

28 August . . . Then came telegram after telegram and letter after letter all reporting that the dress rehearsal of *Das Rheingold* had been appalling, ridiculous to the highest degree, and that stupidity had joined hands with malice to ruin everything.

29 August . . . Richter announces his decision not to conduct. . . . R. telegraphs a bravo to Richter. . . . In the evening news from Richter, he really has been suspended.

2 September . . . *Rheingold* impossible, the return of Richter to the conductor's desk would be the signal for the resumption of the old witch hunt against us and

the king; and besides this the staging of the work is so abominable that the mechinist [Brandt] is demanding three months to put it right.

10 September . . . Late in the afternoon Richter arrives. *Das Rheingold* is to be done in a fortnight, with all the roles changed; the singer R. rehearsed for Loge is to sing Mime, the orchestra has been reduced etc. etc.

14 September . . . Richter has left for Paris.[21]

There are two telegrams and one letter extant from Richter to Wagner written during the *Rheingold* affair. These and another source, Judith Gautier, provide a more detailed insight into the sequence of events. Richter's letter (dated 11 August) is as follows:

Your last telegram gave me firm proof that you do not wish to worry about anything that is going on in Munich. I understand that quite perfectly and in the interest of the magnificent creations you still wish to give the world, I must indeed express the wish that you, my honoured Master, are not prevented from working by the crowd of unworthy people who are here. But now I do not know if I should come with Wotan, Alberich, and Loge or not?

Things look awful here! Perfall always wants to understand things better, gives his instructions regarding the [stage] machinery, and above all he wants to get himself noticed as an 'intelligent boss' in every possible way. So meetings, conferences, and discussions are held; he takes his pleasure by debating and 'encapsulating' matters (his favourite expression). Then when everyone comes out of these meetings, they all know even less than when they went in, and so further post-conference meetings have to be held. What is going to happen regarding the costume designs I do not know, i.e. how far they will be altered.

There are only two men here who really know what they are doing and how you want the stage machinery to be; they are the Brandts (especially the elder from Darmstadt), but their hands are tied. During the holidays the others built some machines which have since proved impractical as they are not at all easy to handle in the time allowed in the score. Brandt was the only one who had thought about that aspect and who spoke with me in private. Now they are all falling over one another in their hurry to get something finished which is usable within the fixed time limits.

In the rehearsal Hallwachs could hardly get a word in edgeways about the machinery, for it's the Intendant's [Perfall] way that something always occurs to him only after he's heard someone else's opinion. It's like Bedlam here. A few days ago the Intendant and I had a bad meeting. He called me and accused me strongly of ignoring his position as being in charge and of allowing myself to be commanded only 'from Lucerne'. He described this as an 'unheard of situation' and as 'tactlessness', that the most important issues, as for example costumes and stage machinery, should be handled by a Music Director, and told me to write to you that the future welfare of my post is dependent on my 'good standing' with my

Intendant. In other words he told me to stop acting the role of an intermediary and that he would walk out of the conference if I were to read out the letter about changes in costume design. Naturally I guaranteed him nothing and in fact regarding my post I told him bluntly that to have it suited me only as long as I enjoy the honour of carrying out tasks for my Master. In any case his indignation subsided very soon because his cowardice always gains the upper hand in the end. At present I cannot describe at all how abominably he also treated my mother. If perhaps you do not think it desirable that I come to Lucerne now, I will, with your permission, take the liberty of doing so at the beginning of September.

For the moment I have undertaken to offer my resignation to the king, either personally or in writing, within the next few days, and to tell him the whole truth about the mess in the theatre, and to ask him to have the goodness to allow me my salary until the winter when my mother has obtained a secure position once again, and I have completed the piano score of *Siegfried*. Please, most honoured Master, give me your kind permission.

I would have liked to have reported better news to you on these pages, but I am given very little opportunity here to do so. On the other hand I have experienced very many good things, namely from members of the orchestra.... To their honourable credit I must report that they have tackled the study of your masterpiece with the greatest enthusiasm, the singers as much as the musicians. I have told them frankly that in view of the muddle taking place on stage it is now a matter of the highest honour that we demonstrate the great work to the public as perfectly as possible. I have already had over five hours of long rehearsals with the orchestra. Nobody complained, not even those who always grumble. No one can escape the magic of *Das Rheingold*.[22]

A witness to the latter stages of the *Rheingold* affair was Judith Gautier, the French author and music journalist, daughter of the writer Théophile Gautier and married to Catulle Mendès. Both ardently followed the Wagnerian cause. Judith visited Tribschen with the poet Villiers de l'Isle Adam in August 1869 *en route* to an exhibition in Munich. It was whilst there that she caught her first glimpse of Hans Richter in a restaurant. She described him as having 'a golden beard and gold-rimmed glasses [which] glistened in the sun. He had an expressive face which fairly radiated happiness and enthusiasm.'[23] Richter noticed Judith's enthusiastic applause for the restaurant musicians' rendering of the *Meistersinger* Overture and immediately came over and introduced himself. 'The Master', he said, 'wrote to me to put myself at your service and to act as your guide in Munich.'[24] From the restaurant Hans took his new guest to Countess Schleinitz's home, where Liszt was in attendance. Judith quickly became aware of Richter's anxiety over the staging of *Rheingold*. Although he professed himself satisfied with his orchestra and singers, he was deeply suspicious of the management in general and Intendant Perfall in particular, in spite of Wagner's initial support for his appointment. On 25 August Richter told Judith that 'Perfall will not allow anything to be seen of his stage arrangements, and he has the expression of a traitor'.[25] She described Hans 'like St Christopher with the child Jesus, who would bear the whole weight of the undertaking upon his robust shoulders'.[26]

There are several technical problems in *Rheingold*. The first is to re-create the opening scene of the Rhine in which the three Rhinemaidens are to be found swimming. Then there is the change of scene as Wotan and Loge descend, and later ascend, from Valhalla to Nibelheim and back. This is the change (mentioned in Richter's letter to Wagner) where the timing is finite and governed by the music. Then there is the portrayal of the giants Fafner and Fasolt, the special effects of transforming Alberich into a dragon or a toad when he demonstrates the power of the Tarnhelm, and, at the conclusion of the work, the entry of the gods into Valhalla over the rainbow bridge built for them by Donner. The scene changes were handled awkwardly and Valhalla appeared tiny upon a miniature mountain. The journey down to Nibelheim was accompanied by loud hisses as steam and smoke were poured on stage, drowning the orchestra. At the end Richter, 'red with anger, throws down his baton; usually amiable, [he] looks positively fierce'.[27] Just a week before the final rehearsal a confident Wagner had urged Otto Wesendonck to attend it because it would be 'very respectable, nothing having been spared to fulfil all . . . technical requirements',[28] yet there are those, including Ernest Newman, who have since accused Wagner of anticipating an embarrassing disaster and of manipulating events accordingly in order to organize a *putsch* against the theatre. At the pre-dress rehearsal on 26 August (at which Liszt embraced Richter for his sterling musical achievement with the opera) there were neither costumes nor scenery and the Rhinemaidens shuffled about in straw hats and very little else before the closed curtains. Gautier's account of the final rehearsal was vivid.

A frightful oil lamp suspended from the highest moulding was supposed to represent the Rhine gold. It only recalled the lantern which is placed by night atop a street obstruction. Each Rhinemaiden was depicted as a mannikin with dangling arms and hair hanging before its face. It was precipitated head first from above, and half way down remained suspended, balancing from the end of a string. . . . Soon after the mannikins were withdrawn and the true singers, standing upon supports, half concealed by the jutting out of the paper rocks, appeared and agitated their arms to represent swimming. Then they went away, and the puppet Rhinemaidens returned and capered desperately about the smoking lamp. What absurdity! They would not dare to present anything so bad at the Punch and Judy show of the Champs-Elysée.[29]

This dress rehearsal on 27 August is the key to the whole business. It was held before the king and a select audience which included many notable artists and musicians from all over Europe such as Liszt, Saint-Saëns, Joachim, Serov, Pasdeloup, Levi, Klindworth, Hanslick, Pohl, Henry Chorley, Manuel Garcia, Turgenev, and Pauline Viardot-Garcia. Perfall came before the curtain at the start of the rehearsal to ask for the audience's indulgence regarding any inadequacies they might notice during the event, not an act to inspire anyone's confidence in what they were about to witness. Richter registered his opinion publicly in the manner of his former mentor von Bülow by ostentatiously rapping his baton on the conductor's music desk, but he did not enjoy sufficient status to act in such a manner and it did not go unnoticed. It was not the most tactful behaviour of someone in his relatively junior position. Someone who struck a happier note was the English critic Charles Ainslie Barry. He wrote a review for the *Guardian*, the first time Hans Richter's name appeared in the English press. Barry, or C.A.B. as he signed himself, became a life-long friend of the conductor and was responsible for many programme notes for Richter's concerts in England. He wrote that 'Herr Hans Richter conducted. Though but quite a young man, he is said to have a practical knowledge of every instrument employed in the orchestra; as a conductor he has certainly especial talent, and, as was fully proved by their playing, had drilled the band to a remarkable state of efficiency.'[30]

Richter sent a telegram on the day after the dress rehearsal which rang alarm bells at Tribschen. Scenery and props were little more than symbolic, he wrote, with cardboard mountains and a lantern for the Rhinegold. The journey to Nibelheim was accompanied by too much smoke and the grinding gears of stage machinery. Even the final shaft of light to represent the rainbow bridge illuminated Wotan's nose rather than fulfilling its intended purpose. In spite of the excellence of the musical state of affairs the conductor urged Wagner to do everything in his power to prevent the performance, even suggesting that the composer's many friends present in Munich should not be disappointed by the cancellation of the opera. He proposed a quick substitution for *Rheingold* of '*Meistersinger, Lohengrin, Tannhäuser* and possibly *Tristan*'.[31] Ernest Newman, in his *Life of Wagner*, cynically considers this an attempt by Richter to increase his own repertoire and conducting experience, but a fairer interpretation would be to regard it as naïve.

Two telegrams were sent by Richter on 29 August 1869. The first reported that the opera had been postponed and that he had refused to conduct it, but that nevertheless his musical honour was saved. The second followed five hours later and said that, because he had rejected the staging, he was now suspended from his post by the theatre authorities and that another conductor was being sought to take over. Richter ended his telegram, 'Liszt shares my opinion.'[32] Approaches were made to Herbeck (Vienna), Lassen (Weimar), Levi (Karlsruhe), and even Saint-Saëns (Paris), but they all refused to take Richter's place in the pit. All attempts which Wagner then undertook to save the situation were thwarted by Perfall and Dürflipp, who were both adamant that Richter had gone too far and that he had given the composer an exaggerated account of the state of affairs on stage. Wagner travelled to Munich on 1 September, hoping for a technical rehearsal with sets, costumes, and lighting followed by a full orchestral rehearsal conducted by Richter. Both ideas were rejected, though a technical rehearsal was held that evening without either composer or conductor present. The king (warned of Wagner's arrival in Munich) fled to his country retreat at the Hochkopf to avoid any contact with him and even threatened to withdraw his retainer and ban his works forever from the Munich stage. After two days of fruitless negotiations with the theatre officials, Wagner returned to Tribschen (on 2 September) and Richter followed five days later.

On 2 September Franz Betz (Wotan), who had described the production to Wagner as laughable and urged him to come and see for himself, walked out and was eventually replaced by August Kindermann. The king's response to Richter's action had been one of fury and indignation. He told Düfflipp to try and get von Bülow to change his mind, make an exception, and return to conduct.

If he knows he is doing this for me, he'll certainly do it; remind him of his indebtedness to me, which he has acknowledged so often in the past. I expect the performance to proceed on Wednesday and all the changes indicated by me to have been undertaken. Among other things, the ageing of the gods did not come off and can be fully corrected by the simplest means, e.g. the burning of spirits mixed with salt. Schlosser's performance as Loge was quite wrong. . . . If Richter or indeed anyone else in the theatre set out to defy my explicit orders, these weeds must be mercilessly uprooted. I order you to take steps against such infamy. . . . This is my command. Amen.[33]

The following day the king wrote to Düfflipp again, expressing his fullest confidence in him and condemning 'the behaviour of the theatrical riff-raff as criminal and shameless. It is an open revolt against my orders and this I cannot tolerate.' He then turned his guns on Richter, who 'may on no account conduct any longer and is dismissed forthwith. That is agreed. Members of the theatre must obey my orders and not Wagner's whims. . . . Richter must jump to it and Betz and the others be brought to heel. I have never encountered such insolence.' He ended by reiterating his confidence in Düfflipp (and by implication Perfall) and signed off, 'Vivat Düfflipp! Pereat Theaterpack!'[34] ('Long live Düfflipp! Perish the theatricals!')

Wagner's reply (13 August 1869) to Richter's long letter of warning had urged the young man to resign, stressing that the king would understand his wish to present Wagner's work only according to the composer's intentions and not at the dictates of the theatre administration! He also warned Richter that if the king did not eventually accede to their artistic wishes, he must understand that things would go pretty badly for him in Munich. Richter's letter has been given in detail in order to present his side of events because Newman takes a very critical and uncharitable view of Richter's part in the proceedings. He denies a conductor (particularly a young one) the right to take a firm stand when the total presentation of an opera is in jeopardy, but Richter maintained (just as any maestro would today) that what one sees on stage is as important as what one hears from the pit. As to his ambition and that it was all a machiavellian plot to jump into poor von Bülow's shoes the moment he had left Munich, that, and Newman's implication that a huge bluff was under way to make the king back down and impose the Richter/Wagner duo on the theatre, is patent nonsense. Richter was far and away the most gifted conductor of his generation and there were simply no others on the spot who possessed one iota of his talent. He may have been an unknown young man of 26 at the time, but even without Wagner's help (and there is no denying that he shot to the top because of it) Hans Richter would have carved out a career for himself as a figure of international fame.

Wagner had had enough of the routine of German opera-houses, and was already bent on creating and running his own theatre to present his works according to his own concept. A month after the *Rheingold* affair his pamphlet *Ueber das Dirigieren* (On Conducting) appeared, and in it he bitterly attacked the current system and the slavish manner in which it was served by German Kapellmeisters. More importantly, it is also evident that the first mention of resignation came from Richter (on 11 August) and not Wagner, who nevertheless supported him in his reply two days later. These were hardly the well-laid schemes of an ambitious young conductor.

Newman finds no fault with Perfall, but the man was a bureaucrat with limited musical ability, who operated through committees. From Richter's letter to Wagner it appears that Perfall even dragged Richter's mother into the affair by sacking her from her post as a teacher at the Munich Music School, which was nothing more than vindictive revenge on the young man. The truth is that both sides got themselves so deeply entrenched on matters of principle that the situation became hopeless. Richter believed unequivocally in Wagner's musical ideas and the rest of his professional life bears witness to his talents.

The man who did conduct *Rheingold*, which eventually took place on 22 September 1869, was Franz Wüllner, a choral conductor and pedagogue with no operatic experience. He bore the brunt of Wagner's unleashed fury. 'Get your hands off my score! That's my advice, sir, or may the devil take you! Go and beat time amongst your glee clubs and choral societies, or if you must have an opera score in your hand, go and choose one written by your friend Perfall. . . . You two gentlemen will have to spend longer at school before understanding how to deal with a man such as I!'[35] Wüllner's work was virtually done for him by the meticulous preparation carried out by Richter, but even so he took three hours to perform the opera where Wagner's chosen conductor had taken two and a half. By the time the première took place the staging problems were all but solved, but, because few of the musical celebrities who had been present at the dress rehearsal could make a second journey within a month to Munich, the audience was more representative of the general public and consequently less comprehending of and sympathetic to Wagner's work. There were no arias or set pieces and there was too much recitative for their taste and understanding. The press (who had been having a field day throughout the whole affair) seemed no more enlightened. Nevertheless, Brandt did his work well, and Wagner did not forget him in 1876 when it came to staging the complete *Ring*. The king, however, wanted to forget Hans Richter, whose last appearance in Munich had been on the day after the dress rehearsal. To celebrate Goethe's brithday on 28 August he conducted Beethoven's *Egmont* Overture and Gluck's *Iphigenia in Aulis* Overture with Wagner's ending. It was to be thirty-nine years before he conducted in Munich again, long after the main characters were all dead (including the king, who committed suicide in 1886). When he returned it was to conduct *Meistersinger* in 1908, the opera's fortieth anniversary and his 125th performance of the work.

4

1870–1871

Brussels; Tribschen

HANS RICHTER invited Judith Gautier to visit his mother with him before he left Munich for Lucerne. She lived 'in a little village somewhere in the neighbourhood of Munich'.

Frau Richter was a professor of singing, and it was the lesson hour when we entered the little house where she lived. Scales and trills of remarkable shrillness struck our ears while we waited. . . . Frau Richter was still a young woman of attractive presence and manner. She spoke very regretfully of the events which had led to the dismissal of her son and she seemed to fear that he would never again find so good a position. They brought us beer and pretzels. The talk languished a little at first, but when Richter told us his mother had invented a method of singing which increased the power of the voice five-fold, she at once became interested and animated. In fact the pupils we had heard just before had seemed to us to have a very unusual volume of tone. Frau Richter's method consisted in throwing the sound when singing against the roof of the palate which then forms a sort of drum, increasing the resonance and the force of the tone to an astonishing degree. Richter sat down at the piano and sang according to this method. His voice came out in tremendous volume, making the little house tremble to its foundation. . . . Our amiable hostess explained her discovery in detail, illustrating meanwhile in a voice that sounded like a bell. 'The curious thing about it', said Richter, 'is that this system which my mother has found, does away with all fatigue. One is able to use the voice indefinitely in this way.' And Richter, to prove the truth of his assertion, sang us the entire third scene from the *Rheingold*.[1]

Hans left Munich on 7 September 1869 together with his friend Emanuel Glaser. They spent three days in Zürich where Richter met the young Irish-French composer Augusta Holmès, who had been present at the *Rheingold* dress rehearsal, and who subsequently wrote an article on the event for the Paris newspaper *Siècle*. Richter travelled on to Tribschen, where he arrived in the late afternoon of the 10th. He spent the rest of the evening briefing Wagner and Cosima on developments in Munich, and Wagner probably sent his letter to Wüllner as a consequence of hearing what Richter had to say. On the 12th Richter and Cosima 'drafted a factual account of the situation'[2] as a series of reports and counter-reports were appearing daily in the press. Next day Augusta Holmès visited Tribschen together with her friends Catulle and Judith Mendès. The significance of the presence at Tribschen of the three French nationals is interesting for on the day after their visit (14 September) Richter left Wagner's home with his friend Glaser as a companion to investigate the musical life of Paris. They stayed for just over a month.

Berlioz had died earlier that year (although his brand of Romanticism had already fallen into neglect) and the music of Meyerbeer and Offenbach was beginning to lose its hold over the Paris Opéra. The Théâtre Lyrique was being more adventurous and had staged Wagner's *Rienzi* in April that year, and the conductor Pasdeloup had inaugurated his *concerts populaires* which were continually bringing new music to the public's attention. Composers currently active at the time of Richter's visit would have included Massenet, Franck, Lalo, Gounod, Saint-Saëns, Bizet, and Fauré, and further changes (in particular the foundation of the Société Nationale de Musique) would be rung within a year or two of his visit as a result of the Franco-Prussian War of 1870–1, which was brewing in 1869. During his visit Richter made only one diary entry and that was on 20 September 1869 when he drank china tea with the Mendès couple and the Belgian musician Franz Servais. Servais, together with the pianist Louis Brassin (Director of the Brussels Conservatoire), had conceived the idea of staging *Lohengrin* in Brussels with Richter as conductor. Wagner's response, according to Judith Gautier, was positive if, and only if, Richter could make any money out of the affair, and in that way repay himself for what he had lost as a result of his loyalty to the composer. On 26 October Hans left for Brussels with the task of preparing and conducting the first performance in French of *Lohengrin*, which took place at the Théâtre de la Monnaie on 22 March 1870 'after much torment [but] a great success (after the first act the queen called me into her box)'.[3] Wagner had at first been very sceptical of the venture. A week before Hans left Paris for Brussels, Wagner told him that over a period of eighteen years various proposals to perform his operas had emanated from Brussels, but all had come to nothing.

If they are serious this time about *Lohengrin* then I really am astonished. Belgium belongs to the Barbary States, and I have never felt comfortable with the administration in Brussels. Of course no one will ask me about *Lohengrin*. . . . We can't expect much from an area where they speak French, believe me! In particular there's no money to be had from there! . . . Unfortunately you were mad about Paris. . . . What I hear of Pasdeloup is not very encouraging, things seem to be going badly for him and his enterprise. . . . I don't think we can count on him any more. There are certain things that don't work, and particularly in Paris. Those *concerts populaires* were once a smash hit—but now period! That was Pasdeloup's mission but he's not cut out for anything else. I no longer believe in performances of my operas in Paris. At best the only success would be a state-subsidized international theatre with permanent German operas.[4]

Wagner, who was becoming fonder of Richter and now often called him his *Geselle* or companion, found it difficult to greet the Belgian *Lohengrin* with anything like his usual enthusiasm even though he agreed that it could well do him good and promote his work elsewhere. He felt, however, that it was more important that Richter should get the maximum benefit from it, in particular a 'development to greater independence both as artist and man'.[5] After Wagner's own bad experiences with concert giving he felt himself unqualified to offer advice, although what he had to say on the subject together with his example of Pasdeloup would have given the conductor plenty to think about a decade later when the London Richter Concerts began. 'Don't undertake anything that is uncertain'[6] was the advice received from Wagner which Richter heeded all his life. For the moment Wagner told him to forget the *Rheingold* affair and rather concentrate on performing his earlier works, but he also wrote of greater things to come.

I doubt that Betz will have opened up new opportunities for you, and fully believe that after this recent Munich abomination the general tendency will be against us. There's one thing I do know; something good and confidence-inspiring will come of it in a totally unexpected manner. And do you know what I advise you to do in this belief? Listen! Play the Brussels piece as best you can, then come at once to Tribschen and help me again, as you did with *Meistersinger*, only this time with the *Nibelungen*, but taking a greater share. We shall then perform the *Nibelungen* as it should be—on that you can depend—and you shall receive it all in your hands as your due. You can be assured of this and make your calculations accordingly. Everything else is nothing as far as you are concerned. I shan't say more for I can't say better than this. Believe me![7]

By February 1870 Wagner was writing that Ludwig ('my young man in Munich'[8]) had, in spite of Wagner's protestations to the contrary, set his heart on a performance of *Die Walküre*. This time there was little more than a token resistance by the composer, who had resigned himself to another defeat were he to resist. On the other hand it served to strengthen his resolve to build his own theatre in which his word would be law. He told Richter that his wish that he should return to Tribschen after Brussels was based not on concern for his material well-being but rather on his desire that they be bound together for the future. Meanwhile congratulations were the order of the day as the success of *Lohengrin* reached Wagner's ears. 'So you received a golden laurel wreath and the queen summoned you?' he wrote. 'Next time it will be an emperor who receives you. . . . The Vienna *Meistersinger* is supposed to have been dreadful; you'll probably have already heard about it. In Berlin it should go marginally better. . . . I've great troubles once again in Munich, God knows what they'll do there now.'[9]

A few days after the *Lohengrin* première (there were a further twenty-two performances which Richter did not conduct) Wagner wrote again.

Once more you have held our banner on high! In Munich it was with *Rheingold*, when you refused to conduct an inadequate performance, now again it is because you have steered the little ship, my *Lohengrin*, through all sorts of hurdles and obstacles and brought it safely to harbour. . . . May the triumph that you have achieved in the French language compensate for the sad experience in our own fatherland. I thank you with all my heart.[10]

On 7 April 1870 Richter arrived in Vienna and stayed with his mother and stepfather. On the way there from Brussels he had travelled via Paris (where he visited the Louvre), Tribschen, and Munich. During the couple of days he spent with Wagner (he was greeted with 'rejoicing and happiness'[11]) the composer presented him with a birthday gift, a score of *Lohengrin* in which he had inscribed:

Weil er so gut hat dirigirt ihn,
Kriegt Richter heute auch den Lohengrin
Richard Wagner, Triebschen um die Zeit des Geburtstages desselben 1870

This example of Wagner's doggerel, with its grammatical aberrations that produce a rhyme, can roughly be translated as:

Because Richter did so well in conductin'
Today he gets the score of Lohengrin

Richard Wagner, Tribschen about the time of the birthday of same, 1870

He spent his birthday (4 April) in Munich, where he had a happy reunion with unspecified musicians and also met Liszt, who issued the first of many invitations to him to move to Budapest. The Hungarian paper *Fövárosi Lapoka*, a review of literature, art, and music, had already recommended Richter to the musical authorities in the city on 9 January 1870. Though both Liszt and Wagner were to be influential in finally securing a post there for him in 1871, for the moment both men failed in their attempts. Wagner received a letter from the acting Director of the Opera in Pest, Anton von Zichy, and sent it on to Hans. On its flyleaf Wagner expressed the hope that Richter would conduct *Tannhäuser* there at the end of the year. Zichy's letter regretted that, in spite of Wagner's recommendation of Richter for a post, the incumbent Ferenc Erkel showed no signs of leaving, but that his name would be borne in mind should such a vacancy arise.

From Wagner's letters to Richter in Vienna written in the months of May and June 1870, and from the one extant letter from Richter to Wagner, it would appear that the young man used his time there to teach score-reading and to coach singers, but he also studied the technique of singing and the art of teaching it from his mother, who had now established her own practice in the city. The composer's letters were full of instructions and exhortations, orders and demands, though they were not devoid of Wagner's special brand of humour. Richter wrote to Wagner on the occasion of the composer's birthday: 'Where can I find the words sufficient to express what I feel for you? Nothing is said by mundane congratulations, for really all of us, the whole world in fact, should celebrate the fact that you, great Master, live among us; and what friends and admirers now do, soon the whole world will do—celebrate 22 May 1813 as the first day of a new, truly gigantic, and great era. I can say no more than that my whole life belongs to you, esteemed Master.'[12] Wagner's reply regretted Hans's absence from the birthday celebrations (forty-five soldiers playing the *Huldigungsmarsch* instead of one horn played by Hans). He was then asked to ensure that both opera houses in Vienna and Berlin secured the correct instrument for the Nightwatchman's horn at the end of the second act of *Meistersinger*. Above all he and his friends were on no account to go to Munich to hear Wüllner conduct *Walküre* for King Ludwig (the sequel to the *Rheingold* affair damned by Wagner as the '*Walküre* filth'). To attend would be to show approval of the event; on the contrary an open protest and boycott would be more appropriate. In conclusion Wagner gave Richter his blessing and pointed out that one day the latter would be an archbishop when he himself was pope.[13]

Hans was also used as a messenger, usually because Wagner was too indignant to write personally. If Vienna's new Court Opera (opened on 25 May 1869) wanted to perform *Rienzi*, then Director Dingelstedt should ask him directly and not via his Dresden publishers. Neither would he have anything to do with the centenary celebrations of Beethoven's birth in Vienna later in the year (when it was hoped he would conduct the Ninth Symphony) if the critics Eduard Hanslick and Eduard Schelle were members of the organizing committee. Richter was even obliged to chase up royalty payments due to the composer after performances of his operas (e.g. the Brussels *Lohengrin*). A brief correspondence ensued on the subject of the German language in singing technique and teaching methods. 'No one taught me how to speak German,' he wrote. 'Tichatschek, Mitterwurzer, Mallinger, and many really competent, talented singers speak the German language quite flawlessly when singing without having studied any method beforehand. That is the point; my experience in the methodical treatment of such matters has filled me with such crucial concern—and Schmitt is just the one who provides proof of the rightness of my case.'[14] Why have these teachers not produced successes? he went on to ask, though he was quick to point out that he meant both Richter and (more importantly) his mother as well. He had frequently asked Hans to return to Tribschen, but this letter was more urgent. There was a job to be completed quickly, the copying of the score of the third act of *Siegfried*. Hitherto Richter had not responded to Wagner's pleas for him to return, but this time it was different. Within a week he was at Tribschen; it was 26 June, the day of the king's command performance of *Walküre* in Munich, but in Tribschen all the inhabitants drank punch together that evening in an attempt to forget.

Hans soon fell into the Tribschen routine of work and play. Beethoven's music was to dominate the house for much of 1870 (the centenary of his birth), beginning with a gift of five small volumes of his music from Wagner followed by a play-through of the String Quartet in C sharp minor, Op. 131. His Third, Fifth, and Seventh Symphonies, as well as some by Haydn, were played in piano duet with Wagner, but Richter would also play pieces by Bach as well as Wagner's new and old works. There were also plenty of non-musical activities, as Cosima noted in her diary on 30 June:

Yesterday Richter took the children for a ride in a boat along the Tribschen banks; I did not wish to deprive the children of their enjoyment on account of my nervousness, and, so as not to make things difficult for Richter, I did not go into the boat. Thus I was running around on the bank in the most absurd, indescribable fear, unceasingly working out how best I could leap to the rescue in case of an accident. The ride was calm and pleasant, Rus swam behind the children's boat . . . I suffered on my own behalf and rejoiced on theirs.[15]

As well as copying the third act of *Siegfried*, Richter was also making a copy of Wagner's sketches for *Götterdämmerung* which were sent to King Ludwig for his birthday on 25 August 1870. Since his departure from Vienna in 1866 Hans had acquired an enormous knowledge at first hand of Wagner's works. In four years he had observed von Bülow's preparation of both *Meistersinger* and *Tristan*, and had conducted one performance of the former. He had thoroughly prepared and conducted *Rheingold* and *Lohengrin*, including the final dress rehearsal of the former and the opening night of the latter, and he was now immersed in copying parts of *Siegfried* and *Götterdämmerung*, the surest way to learn and digest an orchestral score before conducting it. The period from October 1866 to April 1871, which includes Richter's two sojourns at Tribschen, was fundamental to his life ahead. It was no wonder that he would be able to conduct so many of Wagner's works from memory and with such authority.

On 10 July 1870 Richter went on a five-day walking holiday with Wagner, Cosima, her two daughters by von Bülow, Blandine and Daniela, and various members of the Tribschen household. They climbed the Pilatus, the mountain overlooking Tribschen and Lake Lucerne, Wagner and Cosima on horseback, the children carried in a sedan chair. This caravan enjoyed excellent weather and stayed in the Hotel am Klimsenhorn at 6,230 feet; the excursion ended with a beautiful moonlit night. The next day they climbed the Klimsenhorn in the morning and Richter, together with Jakob Stocker and a visiting student Lorenz Schobinger-Amrhyn, the Tamlishorn. The following morning Schopenhauer was read to the assembled company followed by another afternoon walk and a move to the Hotel Bellevue am Esel. Unfortunately after ten minutes they were surrounded by a thick mist and the heavens opened. They were soaked to the skin when they arrived at the hotel and spent a rather miserable evening and the following day together. Cosima took to her bed feeling unwell and Hans continued his Schopenhauer studies alone. On the morning of the 14th he returned to Lucerne with Schobinger, who was leaving the party. In the city he discovered that Karl Klindworth, the pianist and arranger for that instrument of much of Wagner's works, had arrived with a message for Wagner. Glasenapp asserts that it came from von Bülow in Berlin and contained his agreement to divorce Cosima and make way for the composer to marry his wife. Gianicelli, on the other hand, says that the message was news of the outbreak of the Franco-Prussian War, but as this occurred only five days later his version can be discounted. At first Hans took Klindworth to Tribschen where, to while away the time until better weather arrived, they played the Norns' trio from Act I of *Götterdämmerung*. In a short while the weather improved considerably, the Pilatus standing suddenly clear and beautiful. Richter and Klindworth decided on impulse to set off for Wagner's hotel immediately. They arrived at the Klimsenhorn in the evening and resumed their journey after a few hours, arriving at the Esel hotel at dawn, 'a splendid morning with the Berner Oberland looking majestically beautiful'.[16] Though it was so early, Richter went into the dining-room under Wagner's bedroom and began to play *Meistersinger* on the piano ('In the morning I heard the *Morgentraumdeutweise*', recalled Cosima).[17] Glasenapp said that Richter played the Prelude to the opera, but Gianicelli confirms Cosima's diary entry that it was the first two lines of Walther's Prize Song. Wagner rushed down in his dressing-gown to greet his friends; he took a special delight in early morning surprises.

Later that day the whole party returned to Tribschen, where the domestic routine and the busy schedule of entertaining were resumed. After supper on 16 July Klindworth played the third act of *Siegfried* and extracts were read from Wagner's autobiography *Mein Leben*. In spite of the declaration of war on 17 July there were several French nationals as Wagner's guests at the time including Judith and Catulle Mendès, Saint-Saëns, and Duparc, a situation Cosima found embarrassing. Music-making, such as on 20 July when Saint-Saëns accompanied Wagner singing the second act of *Walküre*, helped to ease her discomfort.

Another visitor was Nietzsche, who arrived on 28 July, the day that Cosima heard that her divorce had come through. She and Wagner began to make wedding plans straightaway. Much Beethoven was played and Wagner, despite all the momentous events going on around him, completed his essay on the composer. Richter finished the copy of Wagner's sketch of Act I of *Götterdämmerung* for King Ludwig on 18 August. That same evening Hans entertained the household by singing the Pharaohs' duet from Rossini's *Mosè*, an opera in which he had played horn in Vienna. Everyone derived much amusement from this cabaret turn, though the music itself incensed Wagner to such a degree that he maintained that no German should suffer the disgrace of enforced involvement with it. Hans continued to be a companion to the children, though sometimes he was over-zealous in his play. 'I am writing this at Loldi's bedside,' Cosima confided to her diary. 'I am concerned about her; our good Richter is to blame for her illness, because he allowed her to swim and kept her too long in the cold water. I have difficulty in restraining R. from showing Richter his annoyance over his lack of caution.'[18] Next morning (Ludwig's birthday, 25 August) at eight o'clock at the Protestant church Hans Richter was one of two witnesses to the marriage of Richard Wagner with Cosima von Bülow (the other was their friend the writer Baroness Malwida von Meysenburg).

A few days later Hans told his friend Camillo Sitte that Wagner had restrained him from enlisting in the army, preferring instead that he should reserve his energies for another future battle. 'Yes, dear Camillo, I shall conduct the *Ring des Nibelungen* in three years at the most. Where? I cannot nor may not tell you, but not in Munich, Berlin, or Vienna. . . . I shall spend the whole winter here. The peace does me so much good, quite apart from the great intellectual pleasure which I derive from living with the greatest of all Masters. I really don't know why I deserve such indescribable happiness.'[19] Wagner's prediction about the *Ring* may have been a miscalculation of three years, but he kept his promise.

Throughout the autumn Richter worked on two projects, first a copy of Wagner's article 'Beethoven' (which he completed on 28 September) and a fair copy of the full score of the third act of *Siegfried*, which he began on 18 October. Another project came to nothing (much to Richter's relief in the end). In November 1870 Wagner, inspired by the Siege of Paris, decided to write the text of a farce in the style of Aristophanes which he called *Die Kapitulation* and for which he wished Hans to write the music in the style of Offenbach. This parody was intended for performance in small theatres throughout the land. A month later Cosima noted, 'In the evening Richter plays us his music for *Die Kapitulation* and admits to us that he would find it embarrassing to put his name to it; he declares that the reason Betz does not reply to him is undoubtedly that he thinks Richter needs money and has therefore started to compose!'[20] If Hans found no pleasure in this task, a much more agreeable one was soon to take shape. On 20 December he set off for Zürich for the first rehearsal the next morning of the *Siegfried Idyll* or the *Tribschener Idyll* as it was originally known. Many years later the conductor told Theodor Müller-Reuter:

On 4 December 1870 the Master gave me the original score of the completed *Siegfried Idyll*; he wished Frau Wagner to have a beautiful fair copy, but he gave his own original to me. [It is now in the Richard Wagner Museum at Tribschen.] I immediately copied out the orchestral parts and travelled to Zürich, where, with the help of my friend Oscar Kahl, at that time leader of the city orchestra, I engaged the musicians. On Wednesday 21 December the first rehearsal took place at ten o'clock in the morning in the foyer of the old theatre. The Wesendoncks were present. The musicians were superb and the music sounded wonderful.[21]

Cosima, who knew and suspected nothing of the surprise that awaited her, commented only that the children were working in secret and that there was great excitement everywhere around the house. Hans, meanwhile, met the musicians in Lucerne for a two-hour rehearsal of the work on the afternoon of Christmas Eve at the Hotel du Lac. In his diary he listed the gifts he received that evening (mainly clothes, though the previous month Wagner had given him a gold pen), but he also described his fears about the practicalities of the next morning's enterprise. 'Great fear about podia and instruments. At 7.30 p.m. with Friedrich [a Tribschen servant] into town to fetch cello and double bass.'[22] Exactly twelve hours later the music was performed on the staircase to awaken Cosima. Hans played the trumpet (there are thirteen bars) and second viola (seven bars near the end) as well as relaying Wagner's beat from midway down the stairs to those players further down who could not see him. 'Now at last I understood all R.'s working in secret, also dear Richter's trumpet (he blazed out the Siegfried theme splendidly and had learned the trumpet especially to do it) which had won him many admonishments from me.'[23] After lunch the company adjourned to the gallery on the ground floor, where the *Idyll* began and ended a concert which also included the Wedding March from *Lohengrin* and Beethoven's Septet. Nietzsche had joined the household on Christmas Eve and Hans played *Tristan* for him and Cosima on more than one occasion over the Christmas period. The celebrations concluded on New Year's Eve with two Beethoven string quartets; Wagner coached the players in Op. 59 No. 1 but Op. 135 was sight-read by the Zürich players Hegar, Rauchenecker, Kahl, and Ruhoff. After the musicians had left, Richard and Cosima Wagner, Friedrich Nietzsche, and Hans Richter toasted in the New Year of 1871.

The question of a post for Richter in Budapest was becoming more probable. His friend Franz Servais, an instigator of the Brussels *Lohengrin* and devoted accolyte of Liszt, wrote from Budapest in November 1870 that Liszt assured Hans of his 'strongest support in the matter of the Pesth Theatre.... He charges me to tell you that you can count completely on him and he greets you affectionately. He asks if you will accept the post. I told him in confidence that I thought your answer would be "Yes."'[24] Talk of Richter's move to Budapest began in earnest in the New Year of 1871. Cosima records a letter from the Hungarian composer and critic Viktor Langer to Hans offering him the post of conductor there on behalf of the management. 'Richter will have to accept it, but for us it will be difficult to let him go—we look on him after all as our eldest son!'[25] The next day Liszt telegraphed Richter that he should come urgently to Pest, but Wagner persuaded him to await a formal written contract. For the next four months Richter concentrated on completing the copying of *Siegfried*.

Wagner finished his full score on 5 February and Hans completed the copy three days later. There was much playing of Beethoven's string quartets, in particular the late ones, with Hans playing viola instead of Hegar. These events took place mostly on Sundays, one of which Cosima described to Nietzsche.

Richter himself dominates the whole thing with his viola and recently amused us with his cries of 'Don't rush!' The music stand, which he made himself, fell over in the Scherzo of the E flat major Quartet. He felt himself obliged to make good musically what he had failed to do as a carpenter. He picked up both the stand and the music, singing his own and the cellist's part at the same time, and did it all so calmly that the violinists just carried on playing and eventually the cellist came in again. Richter compelled us to silence too. When the Presto was over we all collapsed in laughter, the cellist declaring that he didn't know how he had managed to get back in again, and the violinists for their part didn't know how they had kept going, but Richter's presence of mind and domination had brought it about. With his hands in the pockets of his jacket he bore our merriment with strict aplomb.[26]

By the beginning of April 1871 Richter was anxious to rejoin his mother in Vienna and assist her in her teaching work (though she was still singing within the confines of her own home, as Franz Servais described to Hans in the New Year of 1871, when he mentioned accompanying Josefine in *Walküre* and the first act of *Tristan*).[27] Richter's decision to go was conveyed to Wagner, who promptly borrowed 1,000 florins from Cosima to give to his secretary as a farewell gift. There was a general exodus from Tribschen on 15 April, the Wagners travelling *via* Bayreuth to Leipzig and Richter to Vienna *via* Augsburg and Munich. Having parted company with the Wagners in Augsburg, Hans arrived home on the morning of 19 April. His second and final visit to Tribschen was at an end. The Wagners meanwhile had assessed Bayreuth and concluded that the town, if not the existing opera-house, was the place where Wagner's music-dramas would be staged. Richter still made no move towards Budapest, holding off from a commitment that might preclude him from rejoining Wagner. On the other hand he must have been beginning to doubt the composer's original schedule of 1873 as the year for the *Ring*. In May 1871 Wagner wrote that 'from next Easter [1872] you will be my appointed future Kapellmeister, but there's work to be done beforehand. You'll see. . . . In the late summer there'll be a large conference in Bayreuth; land purchase, building work, etc.'[28] Wagner felt that he had to make amends for the outcome of the *Rheingold* affair. He also realized that he should no longer regard Richter primarily as a copyist but as a burgeoning conductor.

Don't say that we shouldn't worry about you; we have our worries and with good reason. You cannot dispel our feelings that we regard the succession of troubles which have dogged you these past two years with genuine grief. From our observations we can only wish very much that you will take the post in Pest immediately. As both conductor and Kapellmeister you are in your element. I do not doubt that you will then find your equilibrium. Should my enterprise take place in 1873, I must be able to depend on you from Easter 1872. . . . I've sent you the score of *Rheingold* which Schott wants to have typeset. Unfortunately this time I must express the wish that you'll harness your former energies in order to finish work on this score immediately. When you've sorted out the instruments in question once and for all for *Rheingold*, successive scores can then follow the same scheme and be arranged by another competent musician, so that you need have no more to do with it in the future. Another great service you could render me would be to recommend such a musician. I need a capable, musical scribe without fail; he could begin next winter. Send me the rest of *Siegfried* as far as you've got, don't trouble yourself with it any longer—I don't want that! . . . Now let's see how things will go with you. I hope for the best. Strangely enough the opportunity in Pest coincides with Tausig's death. A 100-piece Wagner orchestra is supposed to have been engaged in Berlin which Tausig wanted to train using my ideas. My first thought was that if it comes off you should get it in Tausig's place. But that's the way things go, one moment there's shilly-shallying the next there's a rush, but mostly there's uncertainty.[29]

On 2 August Hans left Vienna by steamer down the Danube for Pest. The next afternoon he was in the flat of Baron Felix von Orczy and within fifteen minutes had completed negotiations for his contract as a staff Kapellmeister at the opera house. It was signed on the morning of 4 August 1871 effective from the end of the month. Meanwhile he paid a short visit to his birthplace Raab before returning home to Vienna for a fortnight. Towards the end of the month he returned and began to prepare his first opera, *Lohengrin*. His first task was to rescue the work, which he found badly disfigured by atrocious cuts made over the years. He had ensemble calls for the singers and sectional rehearsals for the orchestra. These were followed by a rehearsal of Act I, two of Act II, and detailed rehearsal of Act III before a final rehearsal of the complete opera on 5 October. His career as a conductor (which would continue unbroken for the next forty-one years) began when, on the evening of 7 October 1871, he took his place in the pit of the opera house in Pest. His diary records that his reception was warm. There was great applause after the Prelude and the Prayer and after the double chorus in Act II. He was called on stage twice at the end of Act I and again at the end of Act II. He recorded two other facts; the chorus of twenty-two were paid five florins as a bonus that day (with such small numbers they surely earned it), and the performance was attended by the emperor of Brazil, Pedro II, who was also to attend the first Bayreuth Festival five years later.

5
1871–1874

Budapest

NOTHING of importance happened in the musical life of Pest without the knowledge and influence of Liszt, and it was he who played a prominent role in securing Hans Richter his post as Kapellmeister at the city's opera-house. At the end of August he told Viktor Langer, 'Richter's appointment is a vital gain. Baron Orczy has acted well and wisely thereby to secure and promote his musical progress. Richter's task to achieve fullest recognition is made easier for him by being a Hungarian by birth and by his absolutely correct and modest manner, together with his exceptional talent and skill as a conductor.'[1] Liszt's first point needs to be qualified, however, in the context of 'musical progress'. As far as Richter was concerned, this was embodied by only one composer, Wagner. The second point was equally vital. Richter may have been a Hungarian by birth, but his life from the age of 11 had been spent elsewhere, in Austria, Germany, and Switzerland. Above all it was centred on German culture, German music, and the German language, not the sort of background that achieved instant popularity in Hungary. Liszt had already succumbed to Germanization, and in due course Richter and Nikisch would follow the same path. Nationalism, and a certain degree of independence, was the legacy of the 1848–9 War of Independence and of the disastrous defeats of the Habsburg Empire in 1859 by Italy and 1866 by Prussia. Opera at the National Theatre was having to contend with the Vienna Opera, which seduced young Hungarians away from their homeland with more lucrative contracts. The repertoire of Pest's opera-house consisted mainly of German, Italian, and French composers (respectively Beethoven, Weber, Wagner, and Meyerbeer; Rossini, Bellini, Donizetti, and Verdi; Auber, Gounod, and Halévy) though Hungarians were beginning to come forward with their own works, Erkel leading the way. Ferenc Erkel was also the Music Director of the Opera and had maintained high musical standards over many years despite the poor remuneration awarded its orchestral players. He was aided in the pit by his talented sons Gyula and Sándor.

In Hungary Ferenc Erkel (known today as composer of the opera *Bánk Bán* completed in 1861) was the leading conductor of his day, and, as a composer, second only to Liszt. By 1870 much of the success of Hungarian music-making (orchestras and choral societies in particular) as well as teaching methods and institutions owed their existence or survival to Erkel. Consequently he stood accused of nepotism and of grooming a musical dynasty for years to come (his third son, Elek, was a conductor at the People's Theatre and his fourth, László, was a choral conductor). When Hans Richter arrived in Pest he unwittingly became the means by which Erkel could be removed by his opponents. Howls of protest were also raised when a conductor had to be displaced by Richter's appointment, particularly as it was not a junior Erkel who had to go but Károly Huber, who became head of violin teaching at the Conservatoire. Liszt summed the position up when he wrote to Princess Carolyne Sayn-Wittgenstein, 'Erkel represents the old Hungarian regime plus a few compromises, Richter the new state of affairs at its most extreme. His god is Wagner, he knows no other. Consequently he professes absolute Germanism as revealed by his god.'[2] Richter's friend Franz Servais had written early in the New Year of 1871 to warn Hans that an offer of the post in Budapest was imminent, but he also went on to give his friend a brief account of how things stood in the city and what awaited him.

Liszt has arranged everything perfectly for you. . . . Everything is still very secret, no-one knows anything. Moreover Liszt has instructed me to tell you that you would do well to arrive here immediately; that is his advice. . . . Liszt has been appointed Director of the future School of Music here. Apart from the theatre, you will probably also be professor of instrumentation, score-reading, etc. to supplement your salary. I have also spoken to Liszt about it. Now dear friend, don't waste time talking . . . strike while the iron is hot. When you pass through Vienna do not speak of this. Erkel, who conducts at the Pest Theatre, is a fox, therefore you must be extremely prudent.[3]

Richter's fame and reputation had preceded his arrival in Pest, yet not even that could have prepared the musicians and public alike for what lay in store. He took the orchestra and ensemble of the National Theatre apart, and Cosima Wagner commented in her diary, 'Richter . . . makes his singers pay fines when they alter something in the operas!'[4] The choice of *Lohengrin* as his first opera with which to present himself to Pest was shrewd. He knew it from memory, he rehearsed it from memory, and he performed it from memory. It was already in the repertory of the theatre but the performers and orchestral players would never have thought so from the way in which he systematically took it apart and put it back together again to reveal its true beauty. His approach to the Philharmonic concerts, which had been discontinued through lack of finance and interest, was just as thorough. Whereas Erkel had become complacent and lax despite earlier triumphs on the concert platform, Richter brought a standard of discipline and driving energy which transformed the concerts into events to which the public flocked. The number of rehearsals was increased (with extra ones occasionally paid for by Richter). Even stage presentation was thoroughly overhauled with the players now uniformly dressed in white tie and black suits, and the seating arrangements reorganized. The violins were divided either side of the conductor with violas and cellos in the middle; at the back were the double basses to the left, brass and percussion in the middle, and wind-players to the right. Richter insisted that extra players were brought in to supplement the orchestra where necessary for large musical events. In their conductor the players found a thoroughly prepared musician who stood at the head of his orchestra with calm assurance, complete knowledge, and total command. His platform manner was impeccable, combining a detached authority to give clear cues with an emotional response to inspire his men. The result was consistent and flawless playing rooted in a commitment which had long been lacking.

In the forty-four continuous months of Richter's stay in Budapest he conducted eighteen Philharmonic concerts and five under the auspices of the Society of the Friends of Music, which he led from the spring of 1873 when Károly Thern resigned. Richter's concerts show a programming policy which would in time arouse resentment at so much German music. Twelve of the twenty-three events contained orchestral or vocal extracts from the Wagner operas. He performed all but the first symphonies of Beethoven and Schumann, two by Mozart, Schubert, and Volkmann, and one each by Haydn and Raff. Overtures and incidental works including concertos were also mainly by the Germans Bach, Spohr, Volkmann, Goldmark, Henschel, Mendelssohn, Weber, Schuman, Beethoven, and Wagner. The names of Berlioz, Méhul, Cherubini, and Stradella were exceptions, and the only Hungarians represented apart from the Hungarian-German Liszt were László Zimay and Ödön Mihalovich. Richter was conducting orchestral concerts professionally for the first time in his life and much of the music included in these programmes gave him the repertoire experience which he was anxious to learn, but his bias remained unconcealed. Turning to the 207 performances of operas which he conducted in Pest in the years 1871–5, Wagner was represented by *Fliegende Holländer* (15), *Tannhäuser* (13), *Lohengrin* (12), and *Rienzi* (4), Mozart by *Figaro* (8) and, curiously, *Der Schauspieldirektor* (1). Meyerbeer's operas dominated with *Robert le Diable* (15), *Dinorah* (13), *L'Africaine* (13), *Les Huguenots* (9), *Le Prophète* (8), and his incidental music to *Struensee* (3). The other German operas were Beethoven's *Fidelio* (2), Volkmann's *Richard III* (8), and Weber's *Der Freischütz* (18), as well as the incidental music to Mendelssohn's *Midsummer Night's Dream* (4). The Italian operas were Rossini's *William Tell* (20), Bellini's *Norma* (8), Donizetti's *Dom Sébastien* (6), and Verdi's *Il trovatore* (1), and his French repertoire consisted of Gounod's *Roméo et Juliette* (20) and *Faust* (1), and Halévy's *La Juive* (5). Out of his operatic activity in the city 146 (70 per cent) of his performances were of German operas.

Richter was also busy as a chamber music player in Pest, sometimes playing piano, accompanying singers, or playing horn or violin. On 20 December 1871 he played second viola in the Mendelssohn Octet (half the group was Hellmesberger's string quartet from Vienna). He played as duo-pianist in Brahms's *Liebeslieder Walzer* on 6 March 1872 and horn in Brahms's Trio, Op. 40, on 24 November. On 25 March 1873 he played viola in Beethoven's Septet and piano accompaniment for other vocal works in a concert in aid of the Charitable Institution of Authors and on 7 October accompanied the 21-year-old American singer Minnie Hauk in Liszt's *Die Loreley* in a concert given by the National Choral Society. This concert took place on the eve of the celebrations for Liszt's fiftieth anniversary of his artistic career. The centrepiece of this Festival was a performance conducted by Richter on 9 November 1873 of Liszt's oratorio *Christus* with the Buda Choral and Orchestral Academy, the Society of the Friends of Music in Pest, and with the National Theatre orchestra. It was a grand jubilee occasion.

When rumours emanated from Germany that similar celebrations were being planned there, the Hungarian Society of Authors, Artists, and Performers organized themselves quickly to be first to honour Liszt. He arrived in Pest on 8 November to a welcome of epic proportions, with a torch-lit concert by a military band in front of his flat in Fish Square, followed by a reception accorded him by the city. On the morning of 9 November he was presented with a golden laurel wreath, and later that afternoon the performance took place of *Christus* at the Concert Hall. The festivities were concluded when a banquet for 500 people was given on the afternoon of the next day. The final dress rehearsal of *Christus* had been held on the afternoon of 7 November and was besieged by thousands of people who were unable to buy tickets for the performance two days later. Liszt's suggestion that the Committee might recoup some of the heavy financial expense of the celebrations by charging an entrance fee for the rehearsal was rejected, and the event was declared private. The public took matters into their own hands, however, and smashed windows to gain entry into the hall. Nothing could be done to prevent the invasion and it was all Richter could do to steer the rehearsal to a successful conclusion in spite of many an interruption from the over-enthusiastic throng. The first complete performance of *Christus* had been given in Weimar earlier in the year (29 May) under the composer's own direction, but public reaction had been difficult to gauge as applause was forbidden in church. The best sign was that no one left before the end, including Wagner, who was present and whose opinion of the work had hitherto been ambivalent because he considered it was not German enough. The Weimar rendition of the work fell far short of Richter's in Pest, not least because of the amount of rehearsal time set aside for it and the thoroughness with which the conductor approached his task. The success was splendid, the public was enthusiastic, and the composer expressed his gratitude ten days later at a concert conducted by Richter. Liszt presented him with the full score, elaborately bound, in which was inscribed, 'To Hans Richter, in grateful remembrance of his masterful direction of this oratorio at the Festival performance on Sunday 9 November 1873. Most sincerely F. Liszt. 19 Nov. 73 Pest.'

Meanwhile Wagner had not severed any of the bonds which bound Richter to him and his cause. Though he was pleased that Budapest recompensed Hans for the abrupt end to his conducting career in Munich through the *Rheingold* affair, there was still Bayreuth and the *Ring*. 'The question now occurs to me as to who, from Easter of next year on, will take care of the musical preparation, for which I would have engaged you, if you'd been free. I cannot imagine that, right at the start of your work there . . . your theatre management would release you for five three-month periods without terminating your present engagement straightaway. That's the point which gives me much cause for thought at present. I share these worries with you to make my current position quite clear.'[5] Wagner engaged a new copyist from Zürich in November 1871 ('Herr Spiegel . . . no very pleasant addition to our household'[6]) and the poor man, immediately christened the new 'Herr Richter', was compelled by all the children to do what his predecessor had done, join in somersaults! Wagner

1871–1874: Budapest 57

FIG. 2 Richter, under the watchful eye of Wagner and the applauding Liszt, conducting in Budapest in 1871. *Borsszem Janko* (Johnny Peppercorn) was a satirical magazine of the day and was clearly commenting on the young man's concert programmes.

still continued to pressure and blackmail, in an unseemly attempt to dampen the obvious success Hans was having in Budapest. 'Oh! Richter! The Italian Richter is now called Mariani; he conducted a *Lohengrin* in Bologna so well that all Italy is mad about the piece. Yes! You should hear how things are going there, somewhat different from Hungary! In the end must I bring Mariani to Bayreuth??'[7]

By the New Year of 1872 Wagner had made definite plans to move to Bayreuth in May. At first he would use temporary accommodation whilst a new house was built for them by the city, ready for occupation in the autumn of 1873 (this would become Wahnfried). He had also received a tract of land on which to build his new theatre, which Brandt had promised him would be completed by the summer of 1873. He would scour the theatres of Germany for his singers and staff, and he would require them from the autumn of 1872. 'You must therefore spend the whole winter and spring [1872–3] preparing and coaching the individual singers, and also bring my orchestra together. With the performances in the summer you would have to give me nine months. Now see to it how you do it, for I know of no one who could replace you. . . . I see now that it's a good thing if certain friends don't hide themselves away in Tribschen, for they can achieve other things elsewhere. To be a conductor you have to have character and be a clever fellow.'[8] A month later plans were further advanced and Hans was receiving specific orders.

> On 22 May the foundation stone of the theatre in Bayreuth will be laid. I wish to give a model performance of the Ninth Symphony on that occasion in the existing theatre. A chorus of 200 singers selected from the Berlin and Leipzig Choral Societies is already promised; I also want an élite orchestra of 100 men, and am turning to the orchestral leaders of Vienna, Berlin, Dresden, Karlsruhe, etc. to get the best men sent from there. They'll need five days' leave, two for travelling and three for rehearsals and performance. The musicians will not get a fee, just travelling expenses, free board and accommodation in Bayreuth. Now I await the necessary positive responses. If it happens thus—I need you. You must be on the spot to keep order as my General Concert Master. . . . So take leave from about 10 May until after the festival, which takes place on the 22nd. I am counting on you. . . . Fire away![9]

Richter sent practical help beginning with the creation of a Wagner Society in Pest. Musicians from the orchestra in the city would join those sought from other cities, and profits amounting to 1,000 florins from the Philharmonic concert in Pest on 28 February 1872 (extracts from *Meistersinger,* the *Huldigungsmarsch,* and Beethoven's 'Eroica' Symphony) were sent to the Bayreuth building fund. This elicited another stream of orders. 'As far as you are concerned, I am definitely depending on you for the following musicians: two first violins, two seconds, and two violas (good ones), one cello, one double bass, one first clarinet, and one first trumpet (Hellmesberger [in Vienna] is getting me horns). Now you, who always prattle on about a contra-bassoon, must get one for me, Beethoven wishes it! I'll not turn to anyone else for one. So!?!'[10] In the end Pest's contribution was limited to a clarinet, trumpet, double bass, cello, and one second violin, all of whom asked for a meagre fifty florins instead of travel expenses.

On 12 May Hans left Pest for Vienna and met Wagner later that evening at the home of their mutual friend, the doctor Joseph Standhartner. The following day Wagner, Cosima, and Richter set off for Bayreuth, where they arrived at nine the next morning for an immediate meeting with the mayor of the town, Theodor Muncker. Hans had much to do for the next few days; the cellist brothers Grützmacher together with Heinrich de Ahna, a violinist and member of Joachim's quartet, had withdrawn at the last moment and had to be replaced (the ever suspicious Cosima blamed the anti-Wagner Hiller and Joachim respectively for exerting pressure on the instrumentalists). Hans was delegated the responsibility of receiving everyone and ensuring that all preparations for each artist had been made. The first rehearsals took place on 20 and 21 May in the town opera house. Cosima commented that most of the musicians took a while to understand Wagner's interpretative qualities as a conductor, and of the solo singers Marie Lehmann, Albert Niemann, and Franz Betz gave sterling performances, in particular Miss Lehmann. The alto soloist, Wagner's stepniece Johanna Jachmann-Wagner, arrived without knowing her part and Hans was given the task of getting her properly prepared very quickly. He was constantly dealing with crises wherever and whenever they occurred. Because a trumpeter from Berlin had suddenly withdrawn, he played second trumpet in the Symphony, but also had to play triangle in the *Kaisermarsch*, and bass drum in the last movement of the Symphony (presumably from his second trumpet chair!) at the final rehearsal.[11] The next day, 22 May, was Wagner's birthday, and at eleven o'clock in the pouring rain the foundation-stone laying ceremony took place. Hans was unfortunately prevented from attending because he had to oversee preparations at the opera house for the speeches which took place before assembled dignitaries at noon. He was, however, present for the performance of the 'Choral' Symphony, which began at five o'clock that afternoon. 'Splendid! Unique!' were the two words which describe his impression of the event. There is no mention in his diary of whether he was still having to substitute for missing players by playing in the orchestra during the actual performance.

Richter now saw for himself how an élite orchestra of the best players had been assembled from all parts of the country, and how they had been welded by Wagner into a cohesive musical unit within a very short space of time. It was to provide invaluable experience for what lay ahead in staging the *Ring*, though an important difference was the players' familiarity with Beethoven's work compared with their total unfamiliarity with Wagner's epic. These 1872 celebrations were also to prove invaluable to him in forming musical friendships, of which many would endure for years. One such new friend was the leader of the orchestra August Wilhelmj, who was later to lead the first Bayreuth Festival and the London Wagner Festival of 1877. Another was the music dealer from Mannheim Emil Hecker, founder of the first Richard Wagner Association and a tireless worker in the Bayreuth cause. On the day after the performance Richter attended a patrons' meeting in the town at which a formal decision to go ahead with building the Festival theatre was taken, after which he set off for Budapest with his musicians, his old friend Camillo Sitte, and with Anton Seidl.

Seidl was born in Pest in 1850 and at the time of the Bayreuth ceremony was completing his studies at the Conservatoire in the city. He was introduced to Wagner at Richter's instigation and made part of the team set up to prepare for the first Festival. Richter was to Seidl what Esser had been to Richter. Seidl, describing himself as a disciple, called Richter (only seven years his senior) his Jesus Christ and Wagner his law-giving God. With the unfortunate Mr Spiegel having been quickly sent back to Zürich to resume his career as a music teacher, Seidl soon took over work as a copyist in the so-called Nibelungen Chancellery. This was set up in October 1872 at Richter's suggestion, and consisted of a group of aspiring conductors and coaches who were initially assigned the work of copying and correcting orchestral parts but were later to have their duties extended to the musical preparation of the soloists, assistant conducting, and turning their hand to anything demanded of them by Wagner. The resident pianist was the Russian Jew Josef Rubinstein, who had been a fellow student with Hans in Vienna. Richter remained senior to the members of the Nibelungen Chancellery, though over the years men such as Anton Seidl, Franz Fischer, and Felix Mottl (who followed Richter's tracks through Vienna's Imperial Choir and the Conservatoire) began their training in it and later proved their own worth as Wagner conductors. Seidl described their work to Richter, who must have recalled his own first impressions and feelings of awe at meeting Wagner for the first time just six years earlier.

Early on Saturday I went to [the banker] Herr Feustel to change my money; at the same time I asked him when was the best time to visit the Master, and where he lived. He gave me one of his employees as a guide, and having come to the street I began to study my speech. I'd memorized some very fine words when my companion stopped before a lovely two-storey house [today Dammallee 7] and rang the bell. The door opened and we entered. Russ lay across the entrance hall and did not move, just gave a low growl, a warning that he was not to be disturbed. We stepped over him. A woman came; my guide told her we had been sent by Herr Feustel. She went upstairs, and after a while returned to say I should go up. I took off my coat and went up with a beating heart. I entered the salon; a door off stood open to Wagner's workroom. Here I saw him busy putting his books in order. Without looking up he said, 'Come in. Now then, you've come from Herr Feustel. What's so urgent?' I was so nonplussed at this brusque questioning that I forgot my memorized speech and could not utter a word. He came nearer and looked at me; I stammered that I was called Seidl and born in Pest. 'O Jesus, Jesus, Herr Seidl, welcome to my home. I apologize for the disorder but I only moved here yesterday, and I still don't know whether I'm coming or going with this unpleasant task of putting books in order and such like,' and so it went on with much friendly laughter. . . .

There are three of us so far. A bassoonist and trumpeter from Leipzig, who understands all wind instruments because he was once an instrument maker; he is called Eichel. The second is an unbearable fellow, a blockhead by the name of Emmerich Kastner from Vienna. . . . I am the third, another is coming from Leipzig [Hermann Zumpe]. I have given them the string parts to copy. I transpose the tubas. I write the fourth horn and fourth tuba in one part, so that the fifth horn plays the first tenor tuba, the seventh horn the second tenor tuba, the sixth horn the first bass tuba, the eighth horn the second bass tuba, because the fifth and seventh horns are high horn players and the sixth and eighth low horn players. Is that right? . . .

I live quite nicely on the first floor of a house in the Ziegelgasse [now Badstrasse 31]. The flat consists of four rooms. The first is my living room, the second a communal workroom for the Nibelungen Chancellery; here we three work (later we will be four). Wagner sent pictures to decorate this room [including] a large life-size photograph of the King of Bavaria. . . . P.S. The master warms daily to me, wherever he goes he presents me with a smile as his chief of the Chancellery. 'I am the captain of the robbers, you are my lieutenant.' It's all like a dream to me.[12]

Though back in Pest for most of 1873, Richter had won Wagner's trust once again by his presence with the composer during that New Year visit. It would seem that Wagner was only really safe from his own paranoia about his lieutenants when he had shackled them to him. As soon as they returned to leading their own lives, and particularly if they were being successful in areas which had nothing to do with his own cause, Wagner became mistrustful and hurled all sorts of wild accusations at them based on half-truths, reports, rumours, or, more usually, the product of his own fertile imaginings. For the moment, however, Hans was basking in his Master's praise. 'Now I have two arms, my left one is Feustel and you are my right!'[13] Feustel was Friedrich Feustel, mentioned in Seidl's letter to Richter, a banker of influence in Bayreuth whose task it now was to raise sufficient subscriptions for the Bayreuth Festival. Wagner would gladly have brought Hans to Bayreuth then and there had he the money to match the salary he received in Budapest. But he also knew that Richter (thirty years his junior) would be bored in the sleepy town of Bayreuth because he thrived on the daily pressures of theatre life, opera performances, and orchestral concerts, none of which was currently available in the town. From Seidl's reports of his life with Wagner, Richter could see exactly how his place and function within the household had been taken by his compatriot.

Last Saturday we were all together and he was in a good mood. After supper we went into the salon and he asked, 'Now gentlemen, who trusts himself to make music with me?' No one stirred, not even I. 'Herr Seidl, surely you have the courage? How about *Rheingold?*' Thunderstruck and scared stiff, I took the piano score in my hand. He sat down and played the introduction to the second scene (Valhalla motif) and where the voices begin (Fricka), he called me to sit down and play. I did my best whilst he sang. I sat enchanted at the piano, the impression of his singing, his solemn phrasing and expressive feeling made such an impression upon me that I took fire; he had to restrain me at Loge's appearance, which previously had been my Achilles' heel. When the giants' powerful and weighty motif was heard he laughed and said, 'Very good, here come two who want their salary.' I played for about an hour, and he sang at the top of his voice with me. At the end (where Wotan sets off for Nibelheim with Loge) he and his wife [die Meisterin] were very pleased with me. The perspiration was streaming down my forehead.[14]

There are no letters from Wagner to Richter between December 1872 and March 1874. Hans visited Wagner in Bayreuth for three days in the New Year of 1873 at the composer's invitation, which also included a request that he should come and sort out 'a fatal incident concerning the young Seidl',[15] something which Hans played down in his diary as 'unpleasantness with Seidl'.[16] Seidl had written to him because he felt his position was in jeopardy due to intrigues by his colleagues, and also because he had got himself into a fearful muddle when transposing the bass tuba parts. On his visit Hans also met another member of the Chancellery, Hermann Zumpe, and, together with Seidl (whose problems he had evidently solved), the three left Bayreuth. Hans travelled with Zumpe to Leipzig and saw *Fliegende Holländer* with Eugen Gura in the title-role. He then reached Budapest by way of Dresden, Prague, and Vienna. He next saw Wagner for a few days after Christmas 1873 when he paid another social visit.

The year 1873 was Liszt's year as far as Hans Richter was concerned. Relations between Wagner and Liszt were far from close at this period. Liszt was still smarting from the publicity surrounding the von Bülow–Cosima–Wagner triangle, in spite of the hours the two men had spent closeted together in private at Tribschen six years earlier. Although Wagner was present in Weimar when Liszt conducted *Christus* in May 1873, his reaction to the work was rather indifferent. Liszt had not attended the foundation-stone laying ceremony at Bayreuth (much to Wagner's annoyance), and Wagner in turn had stayed away from the Liszt celebrations in Pest. Cosima was caught between her father and her husband, though from her one letter written in 1873 to Richter she might well be thought to have been influenced by her husband's paranoia.

I was in Weimar (the Master too, naturally) for the performance of *Christus*, and whilst there I met with some Hungarian gentlemen, who asked me to use my and Wagner's influence upon you not to allow yourself to be influenced by intriguers against my father. Completely astonished I said curtly that I knew you full well, that certainly much unkindness might be strewn hither and thither, but that I was sure of your opinion of anything concerning us. To that it was asked: (1) Why does Richter not perform anything of Liszt, yet he puts works by Raff and Brahms on the programmes? (2) Why has Richter not conducted Bülow's music for *Caesar* yet he prepared [Berlioz's] *Lear* Overture? (3) Why does Richter belittle Liszt's *Hymnus* and *Szozat*? (Mihalovich says you told him so.) To these three points I gave the reply: I wish Richter were here, for I am convinced that he'd give a simple and truthful response to these accusations that would shame his accusers. Then came the matter of *Christus*. The gentlemen want to perform it in Pest, but know that without your support nothing would happen; they believe you would not co-operate, but I took the view that, in spite of all these misunderstandings, you would; so I now ask you, dearest comrade, as the wife of your Master not to hold back (certainly not on account of scandal-mongers) and, as Wagner's disciple, to work just as energetically for my father in Pest not only in word but in deed.[17]

Cosima was determined that Richter should not be used as a wedge to be driven between Liszt and Wagner by the latter's enemies. The truth of Richter's initial reluctance to support a performance of *Christus* was based entirely on whether or not he would personally have to sustain any financial losses incurred by the event. He had already dipped into his own pocket on behalf of Pest's Philharmonic concerts, but as soon as he realized the extent of the commitment to the jubilee celebrations by the city, he gave the venture his fullest support. The period from April to November 1873 saw the peak of Richter's residency in Pest. It ended with *Christus* but began with a Philharmonic concert on 9 April, which was long remembered for the superb performance and interpretation of Wagner's Prelude and 'Liebestod' from *Tristan* and for his first account of Beethoven's 'Choral' Symphony, just under a year after he heard Wagner conduct it in Bayreuth. Cosima's concern was misplaced, for she could have trusted her instinct that Richter had no divided loyalties between her father and her husband. Before his stay in Pest was over the three men would appear together on the concert platform.

6
1874–1875

Budapest and Bayreuth

THE gap in the Wagner–Richter correspondence of the fifteen months between December 1872 and March 1874 is only partly explained by Richter's new post in Budapest, his preoccupation with his duties there, and the Liszt celebrations throughout 1873. When Wagner wrote in March 1874 his letter, together with the following five to May 1874, was of great significance. Matters concerning the *Ring* were rapidly developing. In March that year Wagner had told him that King Ludwig had finally committed himself financially to aiding the Bayreuth Festival project (the sovereign having abandoned any idea of hosting Wagner's festivals in Munich), and that contracts were now being signed by Karl Brandt as stage director and the Viennese artist Josef Hoffmann as designer. Richter himself would now have to agree to conduct the Festival which Wagner intended to stage in 1876 after various stages of planning, selection, preparation, and rehearsal had been worked out. In view of many rash forecasts made hitherto by Wagner, this letter proved near to the eventual truth.

In this year, 1874, everything must be sufficiently prepared that next summer 1875 we can rehearse with the chosen cast (which must have been musically thoroughly prepared over the winter) on stage in the sets. The orchestra too must be complete and able to be called for rehearsals (read-throughs, part-correction rehearsals, and with the singers [*Sitzproben*]). From these rehearsals it will become clear who is not up to his or her task—player or singer—so that we can make the necessary replacements. Only by being prepared in this way can we be sure that we'll need two months at the most in the year of performance to stage the gigantic work in worthy fashion, for we'll only have two weeks for each part, and, reckoning on delays through illness etc., these rehearsals must bear all the hallmarks of dress rehearsals with necessary rest periods between them.

This is my plan. Who will help me to carry it out however? Can you, from May on at the latest, place yourself at my disposal for three months? That's the question. You would have to do the following: (1) Audition or re-audition the women in particular, of whom I have almost completely lost sight after so long a time, travel from place to place, send me reports, write etc. (2) Select and engage wind-players for the orchestra with me (likewise involving travel). Serious and decisive ordering of the strings with Wilhelmj. (3) Listen to and give an opinion of the singers I have summoned to me individually this summer. How about it? Can you? Don't forget also—amidst your fine troubles with Pest's opera administration—to let me have back my hastily scrawled diagram for setting out the orchestra, or, if you've lost the piece of paper, let me have a drawing based on your memory of it. I no longer remember what I did at the time, and all such worries must be spared me if possible.[1]

An added complication as far as Wagner was concerned was that from the beginning of 1874 Hans was Director of the Pest Opera, the one and only time in his life that he would hold a position of administrative as well as musical responsibility. Ferenc Erkel had written to the Ministry of the Interior[2] asking to be relieved of his post of Director of the National Opera because his health was being affected by the pressure of his responsibility for the day-to-day running of the house. He also felt unable to continue with his career as a composer, which, he reminded the Minister, was equally devoted to producing works for the Opera and for his country. Erkel was granted his wish and given the laureate title of *Generalmusikdirektor* for life. Hans Richter's consequent promotion was now to affect his ability to get leave of absence. Rather than congratulate his former secretary, Wagner now used all sorts of pressures on Richter and virtually implied that the date of the Festival (which he still wanted to take place in August 1875) depended on the conductor's total availability from 1 May 1874 to 31 August 1875. Such an uninterrupted sojourn with Wagner would have cost Hans his position in Pest, and even though the composer would have paid him a salary (stated categorically in his next letter of 12 March 1874), Richter's long-term career prospects would have suffered. At the beginning of April such discussion of the Festival's scheduled data for 1875 became academic when Wagner wrote that Brandt would not after all be ready with his stage machinery that year. It was now to be 1876 and Wagner wanted Hans for the same four-month period (from the beginning of May until the end of August) for each of the three years 1874–6, or one year's work over a three-year period as he put it. Nothing else would be acceptable, 'only I must have you, and that's that!'[3] Plans were made for Richter's arrival in Bayreuth on 12 May 1874, but he was delayed by the Emperor Franz Josef's presence in Pest and Wagner became very petulant.

> The Vienna express arrived here early at nine o'clock: the children were standing all day at the garden fence to await you, there was no end to their questions about your arrival, Fidi [Siegfried] had to be torn by force from the window to go to bed. You would have done well to send us news by telegram! However, you are the Opera Director in Pest! Your intention to come only on the 20th throws my plans into confusion. I had thought to discuss the female personnel with you for a few days, and then send you on your way to have a last inspection bearing our points of view in mind. You must be back here by the beginning of June to work with me with the singers we expect (and with those yet to confirm), of which Betz has already given notice that he will arrive then. Now we must arrange everything around your Pest engagements. I have already thought much about this engagement; it came very much too late! It would have made more sense two years ago. But that's how everything goes in this world. The noble and illustrious can only thrive in the wake of the ship of the ordinary! The Devil take it![4]

Eventually Wagner got his way and Richter stopped work for three months on 12 May 1874 with a performance of *William Tell* (he resumed his duties there by with the same opera on 15 August). On 20 May he arrived in Bayreuth in time for Wagner's sixty-first birthday two days later. He stayed in Bayreuth for three days, long enough to receive instructions, and set off on a tour of several cities on the first of many quests to cast singers and recruit orchestral players for the *Ring*. He was armed with an authorization from Wagner to contract any orchestral player or singer he thought suitable for the composer's purposes. His first destination was Leipzig, where he heard a performance of *Figaro*, and no sooner had it ended than he was off to Dresden to hear *Der Freischütz* (it should have been *Rienzi* but Weber's opera was substituted). Kapellmeister Julius Rietz's handling of the work did not meet with the young conductor's approval ('Shame!' he wrote in his diary[5]), but he heard a promising alto (Miss Kellner) who could act (Richter was trained by Wagner to lay equal emphasis on singing and acting ability when making his judgements), and as well as securing her services, he obtained those of Eugen Gura. From Dresden he travelled back to Leipzig to hear *Il trovatore*, then on to Berlin where he heard Lortzing's *Alessandro Stradella* and *Lohengrin*, the latter an indifferent performance with meaningless cuts, and imprecise chorus, and an out-of-tune orchestral wind section, all under Karl Eckert, who was deemed by Hans to be lazy and unable to hold the opera together. From the Prussian capital Richter travelled to Hanover where orchestral players were recruited, then on to Braunschweig where he heard Lortzing's *Undine* and where the conductor Franz Abt recommended both singers and players for Wagner's project. By no means all of those who agreed to participate did actually do so in the Festival two years later (one who did was Gura, who went on to sing Donner in *Rheingold* and Gunther in *Götterdämmerung*). Hans's trip continued to Frankfurt, where he heard Adam's *Le Postillon de Lonjumeau* ('nothing special'[6]), then on to Mannheim. Here he lunched on 5 June with Kapellmeister Ernst Frank (a fervent admirer of the composer Hermann Goetz) and auditioned the tenor Georg Unger, who sang from *Tannhäuser*. Though no one realized it at the time, Richter had found Wagner his Siegfried when Unger promised to attend at Bayreuth. Hans now journeyed on to Wiesbaden, where he consulted with the future leader of the 1876 Festival orchestra, August Wilhelmj, then back to Wiesbaden to hear three more singers before arriving back in Bayreuth on 7 June.

The events of this tour should also be viewed from another source. The identity of the recipient of a series of letters Richter wrote also provides another possible reason for the fifteen-month gap in the Wagner–Richter correspondence. Hans had fallen in love. The object of his affection was the 19-year-old Baroness Mariska (Marie) von Szitányi, whose mother Mathilde (née von Montbach) had applied to the National Theatre for singing lessons for her daughter. Marie's father, Wilhelm von Szitányi, owned two properties in Hungary, at Pentele and at Baracs. Family legend has it that the request for singing lessons reached Hans and that, having met the girl, he offered his own services as her teacher. History was about to repeat itself, for his own mother Josefine had been a pupil of his father Anton before she married him in Raab in 1842. When Hans was elevated to the post of Director of the Pest Opera early in 1874, his duties became too onerous for the hour-long lessons to continue, and he stopped teaching Marie. He did so reluctantly, describing his dilemma to her mother: 'As I do not have the courage within me to call off in person what has become for me such a lovely hour, I beg you to let these lines speak for me.'[7] He continued to accompany Marie at concerts, and used every possible pretext to see her, such as paying a call in order to help her select her programme of Lieder and arranging rehearsals for a concert on 19 March 1874. The first extant letter from Hans to Marie (there are none from that period from her to him) begged forgiveness for something he allegedly said (the implication is that it was about her mother) which had been reported to Marie and which had obviously caused her to reproach him. Hans asked the date of Marie's birthday, so the relationship was possibly not of long standing despite the ardour of the letter.

My one and only heavenly Marie, you are an angel of goodness. I am racking my brains who could have told you this dreadful tale about me. I hardly consort with people any more and of all those I meet either on the street or in the theatre, none has my trust. If I really used that expression in an unguarded moment, then it occurred only in an excess of joy; you should put my coarse behaviour down to an uncouth student-like nature. Sometimes I even call my own mother 'my old woman', exactly as I call my highly honoured Master 'our old man' in the company of acknowledged, sound Wagnerians. But I will shake off this wild behaviour, an intention I will carry out in all seriousness. Forgive me, my guardian angel!... O Marie, if you turn away from me, I am a lost man. When I have honourably fulfilled my task for my Master, for whom then shall I live and work with enthusiasm if not for you?...

It is extraordinary how fickle Fate can be in dispensing her favours. What others achieve with ease, I must fight for by suffering and hardship. I have known since my earliest youth that I must struggle only for the highest and noblest ideals; I am motivated not by self-interest, but by the perfect love of Art and the truest gratitude towards my Master to propel me on my way to Bayreuth, the Olympia of today; how difficult this journey will be made for me. It is not understood that the experience I can gain there will in turn benefit the artistic conditions of my homeland.[8]

From this letter a picture emerges of Richter's personality. He was seemingly ingenuous and naïve, undisciplined and inexperienced in life's finer points. The effect his father's early death had had upon him now begins to show. Over twenty years after his father's death he could still write to his friend Alexander Reinhold, 'My sincerest wish is that you did not meet with bad news when you arrived home. If, however, your noble heart has suffered a grim loss, so mourn as a *man*; remember that you still have a mother! Believe me, no one feels more for you than I; I too was deprived [of a father].'[9] Hans was mostly in male company in his Vienna days as a music student, and when he left, he immediately cloistered himself in the monastic existence of Tribschen. There he was treated virtually as a servant for two months before gradually being accepted as part of the Wagner family circle. In fact it was with the children that Hans got on best (in 1867 during his first visit the oldest was 7, and in the summer of 1874 the ages of all five ranged from 5 to only 15). His childish nature appealed to them and he in turn derived pleasure from amusing them. This letter to Marie, with the immaturity of its teenage sentiments and youthful ardour, was written by a man over 30 years old driven by an innate talent to perform (and with that gift devoted largely to Wagner's cause) but with little *savoir-faire*. Reading between the lines, Richter's nature probably did not bother Wagner much for he too had no compunction about behaving in a ridiculous manner in public (such as climbing trees), and could also reveal a childish nature. With Cosima (herself a baroness) it was another matter, for she was fundamentally snobbish even though she tolerated her husband's behaviour. Her devotion to him overrode any such disapproval, but with others it was different. For the moment Hans was also intent on keeping his baroness under wraps from the Wagners because she had Jewish blood. Fearing their reaction he decided to tell them in person, whereas in Pest secrecy was of little consequence. A story still circulates among Richter's descendants that during a performance in the National Theatre of *Lohengrin*, precisely at the moment in the third scene of the first act when Lohengrin takes Elsa in his arms and sings, 'Elsa, ich liebe dich!', Hans turned from his seat at the conductor's desk to look pointedly to where Marie sat in a box. This would not have gone unnoticed by the members of his orchestra. A similar incident is described in another letter.

On Saturday 9 [May] I shall conduct for the last time before my holiday, *Lohengrin* or *Roméo* [he in fact conducted the latter, though he also did *William Tell* three days later]. Perhaps I shall be lucky and see you. The reason I have not been turning round so often to look for you lies with you. My musicians have been keeping a watchful eye on me for a long time; when not looking at the score, I have been looking at the stage and among the orchestra. They noted very well that in *Tannhäuser* my eye was searching not for my 'star of eve' but for my 'sun of life'. I no longer look around because the empty box only stares back at me darkly.[10]

A great deal of serious conversation had evidently been taking place before Hans wrote this last letter. His financial affairs as well as his readiness for marital responsibility were considered impediments to any engagement between the couple. No mention is ever made of her father Wilhelm. It appears that the females in the family were the ones Hans had to impress, for, as a provincial musician with a basic education, he was attempting to marry into an aristocratic family. Marie, on the other hand, had received a very strict upbringing (as a child she was permitted to be in her mother's company for only one hour each day) and a very thorough education (she was fluent in both French and English). Hans was so desperate that he even offered to seek his fortune in Germany, and begged Marie to wait a year for him, by which time he promised to be able to provide financial security and proof of his diligence and sincerity. This rather drastic action proved unnecessary, possibly as a result of an audience with Marie's mother on 16 May. Four days later he was in Bayreuth and preparing for his tour of Germany.

By 7 June Richter was back with Wagner and stayed until 29 July at Wahnfried, or Peace from Illusion, as the house had been christened by its owner in May. Much of the first ten days was spent playing through the various acts of the *Ring*. For a few days from 14 June the existing opera-house in the town was occupied by a visiting company from Coburg-Gotha, and Hans attended performances of *William Tell* and *Don Giovanni*. In the Donna Elvira (Helene Stierl) he thought he had found a Valkyrie, but of more importance was the discovery of Friederike Sadler-Grün among the company. She, and not Fräulein Stierl, would go on to sing Fricka and the third Norn in 1876. The Coburg company presented four other operas but Richter was laid up, as he reported to Marie.

I went on a little outing with the children on the 18th to the so-called Hermitage, a wonderful country house with beautiful parkland, waterfalls, and suchlike. The castle was built by Wilhelmine, the last Margravine of Bayreuth and sister of Frederick the Great. It contains a lot of imaginatively furnished rooms and halls; in the park there are splendid grottoes and dense arcades and the fountains are modelled on the famous originals in Versailles. . . . Towards evening our group was enlarged by the appearance of Frau Wagner, together with two guests (Countess Dönhoff and the famous painter Lenbach) who were visiting the Master briefly. When it was time to go home, Frau Wagner invited me to travel with her and her guests, but I preferred to walk back to town with the children. I was not to benefit too well from this; hardly had we set off through the park when I sprained my left foot at a steep point on the path. With great trouble I reached a farmstead lying about 300 paces off. The people very kindly took me in, and gave me fresh water and a bandage. But the poor children! They cried and moaned so much that I, forgetting my own pain, had my hands full just keeping the inconsolable children quiet. After an hour a farm wagon drove up, and brought me and my charges home. Here of course everyone was worried about our long delay and there was great surprise as we were seen driving up. A doctor [Dr Landgraf, Wahnfried's house doctor] was summoned immediately, and he reassured any worries about any nasty after effects. I had not injured anything important, just a stretched tendon which kept me in bed for four days with constant ice-packs. Today I am more or less in good health again, but I must take care of my foot for the next three to four weeks so that a weakness doesn't develop which might later cause problems with very tiring standing on the conductor's podium.

P.S. After I had finished this letter, the Master and his wife came to my room and gave me the enclosed picture with a request to send it to you, my beloved Marie. To my knowledge it is not possible to buy this picture, the Master gave it to only a few. I believe also that it will interest you to see the brilliant creator of *Lohengrin* portrayed as a happy husband.[11]

Hans did not write to Marie again until the beginning of July, and when he did it revealed a love for her that was beginning to conflict with his commitment to Bayreuth, thoughts which, had they been discovered at Wahnfried, would have been considered treacherous and blasphemous. In the meantime singers and musicians were in and out every day. There was Albert Eilers (eventually Fasolt in *Rheingold*), Theodor Kruiss, who auditioned with *Meistersinger* and *Siegfried* ('unclear diction but not without talent'[12]), and Emil Scaria (not a participant in the 1876 *Ring*, but the future Gurnemanz in the 1882 première of *Parsifal*), who sang some of Hagen's role from *Götterdämmerung*. The first two acts of this opera were rehearsed on 28 and 29 June and on this second day an American soprano arrived in the afternoon 'to have her voice tested; I accompanied her in the page's B flat aria from *Figaro* [Cherubino's aria 'Voi' che sapete' though Cosima's diary says she sang an aria by Donizetti!]. In the evening Rubinstein gave a really virtuoso performance of [Beethoven's] Eighth Symphony in Liszt's arrangement.'[13] Scaria was described by Richter as 'really vain and gossipy. He warned the Master about Fräulein [Luise] Jaide [the future Erda and Waltraute] and Fräulein [Marianne] Brandt.'[14] The latter, from Berlin's Court Opera, eventually considered Waltraute too small a role and wanted Fricka instead, but in the end she did not sing in 1876. From Mannheim came two young hopefuls (Fräulein König and Fräulein von Müller) for the roles of the Woodbird and Erda respectively, but they were unsuccessful. Hans reported life at Wahnfried to Marie.

Singers come every day and go through an ordeal by fire by having to show their ability to the Master. Those considered unready to take part in the performances continue on their way; those who prove themselves suitable must remain here for several days to study the role chosen for them so that they can get fully into the spirit of the part, and then one need not fear that in their private study they will acquire bad habits. Until now the following were here: Frau Vitzthum-Pauli and the tenor Kruiss from Hanover, Herr Eilers, a bass from Coburg, Herr Scaria from Vienna. The latter has a lovely and really masculine bass voice. He will sing Hagen, one of the most important roles. Yesterday two novices whom I discovered arrived. Though they have never been on stage before, their voices and gifts are most promising (Fräulein König and Fräulein von Müller) from Mannheim. The poor creatures were terrified of the Master; only after they were convinced of his kindness did they gather enough courage to sing to him calmly. It went quite well. It's also wonderful that Wagner knows how to handle people! Even though he is serious and strict when studying, nevertheless in convivial company he bewitches everyone with his humour and his flashes of wit. Most people have their eyes opened in these study sessions and are completely filled with astonishment and wonder at the greatness of the concept and the workmanship of this masterpiece. I too have had my mind enriched by many beautiful artistic experiences.

> During this month all those who were hitherto absent have been arriving, because the theatres are closed for the holidays. There's a continual coming and going. When I am not rehearsing, I have my hands full with correspondence. Letters and telegrams arrive and are dispatched to all points of the compass. On top of that I have a really difficult and painstaking job to do with proof-reading the vocal and full scores. This major undertaking, and the complete conviction that my presence here is necessary and useful to the Master and his great idea, provides the only consolation for being far away from my beloved angel. In the past, the time I was allowed to spend in the presence of the Master was always too short, every day was like a holiday. Oh how different it all is now! Now I am gripped by a deep longing for my homeland, which is where my adored Marie lives, and this only gets stronger and stronger.[15]

Wagner did not suspect his apprentice of such divided loyalty, or if he did, he sought to mollify him with gifts of scores of Cherubini's opera *The Water Carrier* and Liszt's *Dante* and *Faust* symphonies. Throughout July more musicians and singers came and went, such as the harpist Peter Dubez, summoned from Budapest to check the harp part of the *Ring*, Franz Doppler, ballet conductor in Vienna, and August Knapp from Mannheim to rehearse Donner (his theatre committee eventually prevented his participation). Among those who made first appearances in Bayreuth, and who were all destined to participate two years later, were Franz Betz as Wotan and, in the middle of July, Marie Haupt as Freia, and Georg Unger. Though he was eventually destined for Siegfried, Richter was rehearsing the part of Loge with him at Anton Seidl's house on 14 July, and when he left the following day, the tenor promised to have the part ready by October that year. Later in July (23rd) Alberich made his debut when Carl Hill arrived and made an enormous impression on Richter and the Wagners. 'A genuine artist, full of the purest, dedicated enthusiasm. I know of no one better among the singers.'[16] According to his next letter to Marie, his harpist from Pest brought news of intrigues and rumours circulating in his absence. Richter's devotion to Wagner was now beginning to cause him problems in Pest, where nationalism was gaining the upper hand and his pro-Wagner (i.e. pro-German) musical preferences were contributing to his growing unpopularity. 'He [Dubez] told me that my detractors were using my absence to spread gossip and tales about me, hoping to harm me. I do not know how truthful this news is; I don't really care much either, for I know I have done nothing wrong. My only cause for regret would be if you were to suffer as a consequence of such wickedness.'[17]

On 25 July Hans left Bayreuth. His diary records 'a sad farewell, the children cried',[18] though Cosima's account was even more dramatic. 'The children in mourning rags, tears and wailing, [the dog] Rus as the funeral horse!'[19] He set off home via Munich, where he had business on Wagner's behalf with the instrument dealers Ottensteiner concerning the ordering of Wagner tubas. From there he travelled to Passau, caught the Danube boat to Linz and thence to Vienna and Pest where he arrived on 29 July, Marie's twentieth birthday. He resumed his duties at the theatre with *William Tell* on 15 August, as he described in a letter to Marie's mother. 'The overture was received with thunderous applause; the singers too went at their musical task with enthusiasm and it was noticeable that they had all made good use of their rest period. Yesterday Fräulein [Ilma di] Murska sang Lucia in rehearsal. She surprised us all with her great and secure technical skill; her voice too still sounds fresh [she was then 38 years old]. Time has not changed her personal appearance, she looks the same as she did ten years ago, for the "painting" on cheeks and hair is genuine. Next Thursday [20 August 1874] she will sing Dinorah.' This same letter began with a request by Hans to pay a call on mother and daughter on 21 August, bringing with him three of 'my best and trustworthy musicians' to entertain the ladies with two Beethoven string quartets which they had been diligently practising (Hans himself would play viola). This tactical campaign of wooing Marie with soirées was intensified when he announced in an undated letter of the period that he had arranged the Prelude to Wagner's *Tristan* for piano quartet, and would be arriving with his friends to play it to her. His choice of Wagner's erotic love-tragedy reflected his own increasingly desperate state of mind. He was so *distrait*, he had recently neglected to acknowledge the applause after conducting an overture in his last concert. 'My fate lies in your hands,' he wrote. 'You yourself said that the decision was yours to make. Oh, make it, only say the word! Only through your pure love and by winning your hand, can I serve my art and the world as a worthwhile human being. Be unshaken in your love for me and let nothing shake your trust of me.... The very sound of your name gives me an electric shock to the heart, but then I lose myself again in my dream world.'[20]

Towards the end of August 1874 Hans travelled with his compatriot and former colleague, the elderly Ferenc Erkel, to the Transylvanian city of Klausenburg (today Cluj in Romania), where they were guests of honour at a Festival. They were given a civic welcome and entertained with gypsy music at supper. Richter conducted works by two Hungarian composers, Liszt and Mosonyi. He was billeted with a young German doctor who was in the best of moods, having recently become engaged. His guest felt this was a good omen and he increased the pressure on Marie by reminding her, 'Oh my most beloved Marie! They were blessed hours that I spent with you. Sadness and worry vanished when I looked into your lovely eyes. Your words of comfort "it will not be much longer" allowed me to forget the world around me.'[21]

On his return it was back to work with performances of *Flying Dutchman* and *Les Huguenots*, whilst between rehearsals he fretted outside the shuttered windows of Marie's town house and paid occasional visits to her in the country in the company of his mother. It was on one of these, in September 1874, that Hans asked for Marie's hand and received her mother's blessing. Wagner sent rather brief best wishes to the couple later that month in the first letter written since Hans's departure earlier in the summer. It was largely devoted to the receipt of 1,000 gulden, promised towards the Bayreuth venture as proceeds from a concert. A little later in October Cosima noted in her diary, 'Yesterday a telegram from Richter announcing his engagement; we congratulate him all the more joyfully since he is said to have chosen a good and pretty girl from a good family, with means of her own.'[22] At the same time Hans was addressing Marie as

My dearest beloved bride. Actually I do not want to do more than cover all four sides of this letter with anything other than calling you my bride. I'm so blissfully happy, my adored Marie, that I can call you *my bride*. This morning I could write nothing, just send you countless greetings and kisses, but had to force myself as I had 'business' to attend to. . . . I ordered, in my *best German*, 100 letters and envelopes with and without monogram. In two weeks you will receive these into your lovely little hands. Then I went to the photographer to order pictures, and then to the tailor, where I was measured for clothes of the most elegant fashion. You see that I am obedient to you in everything where you are *right*. . . . At the landing stage I met Archbishop Haynald and General Türr, who both congratulated me—us—in sincere and heartfelt friendship on our engagement. Haynald assured me that, wherever he might be, he would come to my wedding to give us his blessing before God. He did not wait to be asked, but made the offer himself in the kindliest manner. He mischievously also noticed that I appeared to be taking much more care of my beard and hair than in the past. I told him I had received strict orders to look after myself more than hitherto.[23]

In November Marie remained at Pentele where Hans continued to write with news of his work in Pest. He had to attend many committee meetings concerning the founding of the Hungarian Music Academy (Liszt's brainchild). He conducted his first *Fidelio* on the 10th and his and the city's first *Rienzi* on the 24th, the latter meeting with resistance from the anti-Wagner faction. Between them, on the 18th, was his thirteenth Philharmonic concert, which ended with Beethoven's Seventh Symphony. The lovers called it the 'Pentele Symphony', no doubt because the name of Marie's country seat fitted the dotted rhythm of the 6/8 'vivace' of the first movement, though 'the most perfect performance of your favourite movement (the second) shall provide you with proof that I thought of you faithfully and with purest love as I prepared this symphony.... I have found a rare antique for you, the theatre poster of the first performance of *Zauberflöte* at the Theater an der Wien under Mozart's personal direction on 30 September 1791.'[24] He gave a more detailed description of his next Philharmonic concert on 16 December

I must say that the orchestra's performance was perfect. The *Meistersinger* Overture was played with fire and verve; when, at the end, the violins played the motif from Walther's Lovesong with thrilling enthusiasm, ending with the entry of the full orchestra and the strong chords of the main theme of the *Meistersinger* thundering forth, no one could resist its overall powerful impression. Even before the wonderful work ended, the public's applause broke out. I had to come out repeatedly and bow in thanks; once again Wagner was the conqueror. After this uplifting experience came an anticlimax; after the sound of the Wagnerian majestic ocean, the modest little stream of Volkmann's Muse trickled forth. Although the Serenade is nothing at all exceptional, it gave pleasure. First because the public likes to see old Volkmann in his tails and obligatory white kid-gloves take a stiff and awkward bow, and secondly the work was played charmingly by Ruhoff and the rest of the orchestra. The third piece produced a local composer, Herr László Zimay. It's entitled *Honvéd-Csatadal* and is set for baritone solo, chorus, and orchestra. Although its invention is nothing special throughout, this novelty deserved a better reception than it got. The composer was called once. On the whole Beethoven's Second Symphony learnt much from Papa Haydn, but the young lion already has powerful claws. The symphony was masterfully played, noisy applause after each movement, especially after the tender Andante and the witty Scherzo. A good and lovely angel hovered above the whole performance, and that angel was you, my beloved Marie.... As soon as a letter arrives from Wagner, I shall let you know if he can be personally present at our wedding.[25]

Letters were arriving thick and fast from Wagner in November and December 1874, and, whilst acknowledging that his former secretary was more than preoccupied with his own wedding plans, he nevertheless continued to bombard him with a barrage of requests and instructions. *Götterdämmerung* was in the final stages of printing. Orchestral and vocal extracts from it, such as Siegfried's Funeral March and the final Immolation Scene, were being prepared for performance in concerts which Wagner needed to put on to fund his operations in Bayreuth. Two such events were being planned by Wagner in the early months of 1875. They were to take place in Pest and Vienna ('The first in Pest must bring me in a clear 5,000 florins, and I must reckon on 10,000 florins net from a repeat of the same event in Vienna'[26]) and Richter was to be involved with both. Meanwhile bass trumpets, Wagner tubas, tenor, bass, and double bass tubas and trombones had to be ordered and the players found. Then there was the question of the tenor Dr Franz Glatz, a young Hungarian lawyer discovered by Richter and suggested by him as a possible Siegfried for 1876. In spite of his total lack of stage experience, he was gifted with a fine voice trained by Josefine Richter (much to the Wagners' disapproval: 'his voice is powerful, but the consequences of the training he has undergone are dire').[27] Wagner spent a considerable amount of time coaching him but eventually gave up, even though he did sing in one of the two concerts then being planned. He later changed his name to Gassi and remained a member of the Pest National Opera for many years. Another fund-raising concert eventually took place in Vienna on 24 January 1875, just three days before Richter's wedding day. It was promoted by the Academic Wagner Association with the Philharmonic orchestra and the Vienna Male-Voice Choir and two soloists, Dr Emil Krauss (baritone) and Herr Schittenhelm (a tenor from the choral society). The programme consisted of Wagner's *Huldigungsmarsch*, the Prelude and 'Liebestod' from *Tristan*, Wotan's Farewell and the Fire Music from *Walküre*, and ended with Liszt's *Faust* Symphony, in which the organ part was played by Anton Bruckner. Richter scored an unmitigated triumph in Vienna with this concert. In his diary he wrote, 'Dazzling success. Three laurel wreaths. Countless platform calls.'[28] The *Neues Wiener Tageblatt*, presaging his future annual series of concerts in London with the headline 'Richter Concerts', singled him out as the central figure of the occasion. The spelling of his Christian name as 'Hanns' had also been his custom, though by 1874 he was signing himself 'Hans'.

The main interest was reserved for one man who stood up there in the musical pulpit and who led his enlarged forces to victory with his calm confidence, sure hand, and self-assured intelligence, the conductor Hanns Richter. . . . The storm of applause which erupted after every work over conductor, orchestra, and the absent composer appeared only to be a ratification of an earlier verdict. . . . In his autobiography, on which Richard Wagner is occupied at present but which will only appear after his death, the name of Hanns Richter will doubtless hold an outstanding place, for among all who work for him or purely for self-interest, among all those who are kindred spirits or just intrusive hangers-on, there is no one who stands nearer to the composer of the *Nibelungen* than the appointed conductor to the same, Hanns Richter. . . . Richter's manner is both masculine and elegant, serious and yet friendly. He should not desire to conduct a ladies' chorus. The singers would constantly be submerging themselves in his magnificent Barbarossa head and in his moist and fiery eyes instead of in their copies. But what use would that be? The man is already provided for and celebrates his wedding in Buda tomorrow.[29]

Eduard Hanslick in the *Neue freie Presse* began with a dose of his customary vitriol for Wagner by pointing out that any proceeds from the concert would be 'flying off' to the box office at Bayreuth. Meanwhile Kapellmeister Hanns Richter from Pest, 'who is known as being high in Wagner's trust, showed himself to be a very talented conductor, and especially impressed with his calmness, which one cannot praise highly enough. His fiery colleague Sucher threw him a huge laurel wreath from the gallery immediately after the first number; later apprentices from the Wagner Society brought him three or four more.' Hanslick dismissed the *Huldigungsmarsch* as 'an insignificant piece, weak and sentimental. . . . Whereas the "inner melody" in the *Tristan* Prelude only tickles the listener to death like a remorseless sciatica, the *Magic Fire Music* is embroidered with the refined effect of musical colour.' After tearing Liszt's symphonic poem to shreds, Hanslick concluded, 'the concert lasted from half-past twelve until three o'clock. Half dead we managed to get out hardly knowing any longer what music is. Happily there was a barrel organ playing (in tune) the *Blue Danube*.'[30] The unanimously positive reception which Richter received on his début in Vienna marked him out for great things in that city. The timing of his impact on the musical life of the Habsburg capital was impeccable, and it was only a matter of months before he moved there.

Meanwhile he had to return to Pest for his marriage. This was apparently not so straightforward a journey, for his diary records, 'My lucky star!!! Returned home at night. In Marchegg I luckily escaped a collision, for I had already got out of the coach.'[31] Marchegg, with its station restaurant, is a railway junction situated about thirty miles along the line from Vienna to Budapest. It appears that Hans left the train for a meal while the carriages were being shunted, and during this operation a collision occurred. The next entry occurs on 27 January in Pest.

Here begins a new life! My marriage to my most beloved Marie took place at the inner city parish church at eight o'clock in the morning. The orchestra of the National Theatre struck up the Bridal Chorus from *Lohengrin*; it was moving. Abbot Schwendner performed the religious ceremony (in place of Archbishop

Haynald who was prevented by his mother's death from fulfilling his freely given promise to officiate). My witnesses were Baron Augusz and Ministerial Secretary Josef Ribary. At nine o'clock we travelled to Vienna. The Academic Wagner Society sent my Marie a beautiful bouquet; on the platform of Pest station we were surprised by [the florist] Gustav Weber with a little basket of fresh roses. We arrived in Vienna at seven in the evening.[32]

Wagner had been prevented from attending through pressure of work, and Baron Augusz was his replacement. The newly weds travelled to Bayreuth from Vienna on 5 February, arriving for a late lunch at Wahnfried. Hans described their reception as 'heartfelt. We were obliged to stay with him. At table the Master toasted us. The dear children quickly befriended Marie.'[33] Cosima's observation of the same occasion reads, 'At two o'clock Richter with his wife, he still the magnificent creature of old, she to my mind strange, a decidedly Jewish type; the children help us over our awkwardness.'[34] The Richters stayed for two days, with Hans showing Marie the Hermitage and the new Festival theatre still under construction. Cosima still found Marie strange, and the atmosphere became strained when the subject of Franz Glatz's singing was discussed (any criticism by the Wagners construed as criticism of Josefine Richter's teaching methods). Cosima confided to her diary her feeling that Hans would now be pursuing other paths, and that he had grown apart from them. What had happened was that he had gained more independence from them, even if, in the process, he had surrendered it at once to his new bride. In spite of any superficial freeze, there was an underlying warmth of feeling between the two men, manifested by Wagner's wedding gift to the couple. This was a magnificently bound full score of *Walküre*, on the flyleaf of which Wagner inscribed a lengthy dedication prompted by his guilt at the way Richter had suffered on his behalf during the *Rhinegold* affair. His unique rhyming couplets contain both his own peculiar brand of wit and his predilection for word play; thus it is quite impossible to do more than present it in its original German:

> Dem Meister stand der Gesell zur Seite
> als der eine tüchtige Meisterin freite:
> Nun steht der Meister zu seinem Knaben,
> der Richter soll eine Richterin haben.
> Der Meister nimmt die Sache wichtig,
> damit bei Richtern alles steh' richtig;
> und dass es tüchtig beim Trauen fleckt,
> der Meister sich hinter den Augusz steckt:
> den Segen ertheilt für ihn der Baron;

1874–1875: Budapest and Bayreuth 79

    und kommt auch die Walküre, nicht in Person
    zu richten die Richter'sche Eh'stands-Uhr,
    so kommt sie doch wenigstens in Partitur:
    Sie ist nicht von Gluck, doch wünscht sie Euch Glück,
    dass kein böser Spuck Eure Seele berück'!
    Gedenkt dess' noch in fernen Tagen,
    wie Richter und Wagner es einst mochten wagen, eher Werk und Taktstock zu zerschlagen,
    als die Welt mit schlechten Aufführungen zu plagen.
    Dafür nun blüh' Euch Segen und Lohn:
    Das weitere sagt besser Augusz der Baron!

The two concerts under discussion in the letters of December 1874 from Wagner had now in fact been reversed in order. The first was to take place in Vienna on 1 March, but Richter was unable to obtain leave to be present. His absence was sorely regretted, and Cosima showed her disapproval with complaints that he had forgotten to write out the harp parts, and to book tubas for the first rehearsal. Nevertheless the composer left Vienna, having scored a notable triumph, and travelled on to Pest where Richter met him. Still irritated by being without Richter in Vienna and being quick to condemn, Wagner refused Hans's offer of hospitality and instead took rooms in the Hotel Hungaria (in an earlier letter he had also expressed his wish not to intrude on the newly weds). Richter's diary protests that he had indeed sent the instruments off in good time, and that it was the railway that had caused the delay.

Poor Wagner's troubles did not cease. Kept awake for most of the night by an all-night ball taking place in the hotel, he sent his servant Franz Mracek to Richter at seven o'clock the next morning to say that alternative accommodation must be found immediately. Marie rushed to her mother's nearby apartment at once and begged her to move out and allow the Wagners to move in. By ten o'clock that morning the rather bemused Frau von Szitányi had been transported to her sister's flat and an hour later the Wagners' problems were at an end. The florist Gustav Weber, an ardent admirer of the composer, had got wind of what was going on and, during the hour when Frau von Szitányi's flat was unoccupied, managed to transform it and the staircase leading to it into a flower shop to welcome Wagner and his wife. Another visitor to supper at the Richters' that evening (together with the Wagners) was Liszt, a participant in the concert three days later. Wagner was in the best of moods after the meal, even if Cosima complained at the parlous state of her father's work, *Die Glocken von Strassburg* (The Bells of Strasbourg) with its introduction in the Gregorian mode which Wagner quoted in *Parsifal*, rehearsed earlier that afternoon. Richter noted drily that eventually 'the really successful performance gave the lie to that.'[35] The next day they all ate in the Casino and gathered again together at the Richters' apartment in the evening. Liszt played the piano, an *Ave Maria* and the *Souvenir de Mouchanoff* (*Schummerlied in Grabe*), which reduced Cosima to tears. But, as Hans recorded in his diary, 'the men all remained quite cold, which seemed to annoy Liszt for he said angrily, "My playing does not appear agreeable." The Master, who had sat sulkily throughout what was actually a boring performance, replied, "Dear Franz! You were playing a really sentimental piece in order to make the ladies cry." The fun was gone and Frau Wagner, who was little pleased by the Master's frankness, urged him vehemently that it was time to go home.'[36]

The programme for the concert consisted of *Die Glocken von Strassburg* by Liszt, Beethoven's 'Emperor' Piano Concerto, and, to conclude, excerpts from Wagner's *Ring*. Liszt and Wagner conducted their own works, whilst Richter conducted the concerto in which Liszt was the piano soloist. Wagner initially wished Liszt's appearance to be purely as conductor/composer ('ask Liszt to lend me his baton and his bells for the occasion instead of his fingers'[37]), fearing the inclusion of a work by Beethoven would not comply with the purpose of generating funds towards Bayreuth. He was also reluctant to include a choral work because it would reduce revenue by limiting the number of seats available to the general public, but he gave in on that point. The dress rehearsal took place on the morning of 9 March. A comparison of the accounts of both this rehearsal and the performance the following day by Richter and Cosima makes interesting reading. First Cosima's diary.

Dress rehearsal; an ugly hall, bad acoustics, insufficient preliminary rehearsals; my father absolutely overwhelms us with the way he plays the Beethoven concerto—a tremendous impression! Magic without parallel—this is not playing, it is pure sound. Richard says it annihilates everything else.

Concert at seven o'clock. . . . A very full hall, very brilliant, and great enthusiasm as well.[38]

Richter's account is as follows.

Final rehearsal in the morning. Liszt played the E flat Concerto by Beethoven with heavenly beauty. The remaining works suffered under the insecure conducting of both composers. However Liszt conducted really better and more clearly than Wagner.

On 10 March the concert took place: in spite of the high ticket prices the hall was full to overflowing. *Die Glocken von Strassburg* for soloist (Lang), chorus, and orchestra went very well. Even if it was not as perfect as in the dress rehearsal, it was the most successful piece in the concert. After that Wagner was given an enthusiastic reception on his appearance. The Forging of the Sword (Glatz), Siegfried's death and the Funeral March (had to be repeated), and Wotan's Farewell and the Fire Music (Lang). The Master's unclear beat caused some hesitations (I only conducted the orchestral accompaniment to the piano concerto). After the concert we all gathered in the *Hungaria*.[39]

Between the dress rehearsal and concert Hans had his duties to carry out at the National Theatre. On the evening of 9 March he conducted a performance of *Fliegende Holländer* in the presence of the composer and his wife. In his diary Richter reported that 'the Master was satisfied with the chorus and orchestra, and in part with the [principal] singers. Some tempos and cuts annoyed him very much; on the whole he expressed himself favourably to me about my part in the performance.'[40] Cosima's account was quite different. '*Der fliegende Holländer* conducted by Richter; sung in Hungarian and Italian. Great disappointment! Nowhere else has so much been cut in the *Holländer*, and Richter has also introduced cymbals, etc. Astonishment over this Wagnerian *par excellence*!'[41] It is hard to assess quite what did happen at this performance of Wagner's opera. Although the incident was not referred to any further at this time, there was a very difficult period in the Richter–Wagner relationship during the summer of 1875 in which letters flew back and forth between Wagner, Cosima, and Richter. The *Fliegende Holländer* performance was one of several complaints aired at the time, and will be referred to later in more detail. Suffice to say at this point that Richter vigorously denied recomposing Wagner's opera, and justified the cuts in the work as necessary to help a weak singer.

The Wagners left Pest on 11 March for Vienna. From there Cosima wrote a courteous letter of thanks to Marie for the manner in which she and her mother had looked after them during their stay, but even this was not without a sting in the tail. After exhorting her in her wifely duties to support her husband through thick and thin she went on, 'I wish very much that he could use his position to support my father; for example, I was astonished that *Prometheus* had to be played in Pest on the piano, and various impressions make me regret that your dear husband were not more the master of the situation. I count on you, dear Madam, to give him the right support in everything.'[42] Opportunity for Richter to promote Liszt's music in Pest was in any event to become academic. Changes were taking place in Vienna, where, at the Court Opera, Johann Herbeck had resigned. His place was taken by Franz Jauner, currently director of the Imperial

Karlstheater. Jauner had accepted the post on condition that Richter would be appointed his musical right-hand man. This was just the sort of post Hans wanted, musical responsibility without administrative duties, and its offer was undoubtedly precipitated by the outstanding success resulting from his début in the city earlier that year. On 14 April Jauner reported that the Court Director of the Opera, Count Konstantin von Hohenlohe-Schillingsfürst, had agreed to meet Richter's conditions of appointment and that he could start on 1 May 1875, the date on which Jauner's appointment would also commence. The opera with which Richter would make his debut was to be Wagner's *Meistersinger*.

His departure from Pest was regretted by many and misreported by more. It was alleged that he had been driven away by the anti-Wagner, pro-Nationalist faction in the city, and that his successor Sándor Erkel had a hand in his departure. Richter was always quick to counter such versions of events, and stressed that he himself had recommended Erkel to his post. He readily acknowledged that he did not suit the post of Director and vowed, 'I shall never be a Director again, that I know for sure; one never makes any progress as an artist if burdened with the worries of a Directorship.'[43] His final concert with the Pest Philharmonic Society (his eighteenth) took place on 23 April 1875 with works by Wagner, Bach, and Beethoven. He was greeted with a huge ovation from the audience and with a *Tusch* from the orchestra (a nineteenth-century German convention, the *Tusch* was an improvised fanfare, usually played by the brass section of an orchestra as a mark of honour and respect towards the recipient). The conductor's podium was bedecked with flowers (supplied once again by courtesy of Gustav Weber, for whom 1875 was proving a busy year) and upon Richter's entry he was presented with a laurel wreath by the orchestra's leader Dragomir Krancsevics. 'The performance was really fabulous. After the concert members of the orchestra arranged a banquet at the Hungaria Hotel, which my lovely little wife also attended. I received two group photographs of the orchestra as a memento. The first speech was by Herr Spiller, after which my reply, accompanied by tears sincerely shed. Count Apponyi spoke on behalf of the public. During the day I received a silver service from Böhm on behalf of the singers, and the chorus presented me with a laurel wreath adorned with silver.'[44] The root of any controversy about Richter's activities in Pest centred on Wagner. Of the eighteen Philharmonic concerts Richter conducted, twelve contained works by Wagner, whereas Hungary was represented by six works of Liszt (who many Nationalists considered tainted as a German), and one each by Zimay and Mihalovich. Whatever his detractors might say, none could deny the rise in standards of orchestral playing that he had brought to the capital. He was preparing the way for conductors such as Mahler and Nikisch, who followed him and included Budapest as a stepping stone on the path of their careers. The next day (24 April) Richter conducted at the National Theatre for the last time as its Musical Director. The opera he gave as his farewell performance was *Lohengrin*. Once again Weber had been at work, and Richter came into the pit to find it and the conductor's desk decorated with flowers and wreaths. 'I left Pest with a heavy heart.'[45]

7

1875

Vienna

THE opera has always been the central focus of musical life in Vienna. The Court Opera or Hofoper which stands today on the Opernring has its origins in another building, the Kärtnerthortheater, whose site is now occupied by the Hotel Sacher. This original theatre was built in 1763 but by 1857 Emperor Franz Josef had plans laid to demolish the walls, gates, and narrow streets of the inner city and to broaden them into wide boulevards in concentric rings. As part of this plan the Kärtnerthortheater had to go and a new opera-house was built on one of the ring roads. It was run on the Italian *stagione* system; in other words operas were staged in seasons. In the decade of the 1840s the French repertoire dominated the German by two to one (45 French operas to 23 German and 25 Italian), but the amount of Italian works was to increase dramatically and Wagner's operas would spearhead the revival of German operatic fortunes. The Kärtnerthortheater was the theatre in which Hans Richter began his professional musical career as a horn player in September 1862, and from whose orchestra he was plucked by Kapellmeister Esser in March 1866 and sent to Wagner in the autumn of the same year.

By the time he returned to the city eight years later the scene was radically different. Matteo Salvi had directed the last years of the Kärtnerthortheater and had established the works of Gounod and Meyerbeer in the city. His successor was Franz Dingelstedt who staged Gluck's *Iphigenia in Aulis* in Wagner's version, brought Gounod to conduct his *Roméo et Juliette*, and staged Thomas's *Mignon*, both in 1868. Dingelstedt's post fell vacant after the opening of the new opera-house, which took place on 25 May 1869 with *Don Giovanni* (the Kärtnerthortheater closed in April 1870 with Rossini's *William Tell*). As Director he had increased the opera orchestra to 111 men led by three conductors, Heinrich Proch, Heinrich Esser, and Otto Dessoff. Proch and Esser retired in 1870 and, because Dessoff was preoccupied as Director of the Philharmonic concerts, Dingelstedt appointed Johann Herbeck in 1869. It was Herbeck who took over as Director in April 1870.

Herbeck was extremely popular in the city and was both a brilliant conductor and producer (both functions were often combined in one man). He was a devotee of Wagner, and helped found Vienna's Akademische Wagnerverein. Having already conducted Vienna's first *Meistersinger* in February 1870 (a stormy but triumphant event), he then staged excellent productions of *Fliegende Holländer* and the Vienna première of *Rienzi* in the following year. Wagner's continual bone of contention with the city at this time was twofold. The first was to do with his royalties and the second with the drastic cuts made in his operas. Herbeck took three and a half hours with *Meistersinger* compared with five at its uncut première in Munich two years earlier. Wagner bombarded him with telegrams from Lucerne (Dingelstedt had thwarted any attempts to invite the composer to Vienna), in which he protested at the casting of Berta Ehnn as Eva and at the conductor's decision to rescore such stage instruments as the Nightwatchman's *Stierhorn* and Beckmesser's *Stahlharfe* with bass flugelhorn and guitar respectively. Here Herbeck was only doing his best to protect Wagner's work from his enemies, for he regarded these replacement instruments as more reliable in performance.

Herbeck also brought the work of Verdi to Vienna. Having seen *Aida* at La Scala he staged it in 1874, but these were difficult times for the opera-house. On a Black Friday in May 1873 the stock market crashed and the financial consequences took their toll on the operatic life of the city. Herbeck's last première was Karl Goldmark's *Die Königin von Saba* (The Queen of Sheba) in 1875, but during his tenure he had also introduced other German operas such as Weber's *Oberon*, Schumann's *Genoveva*, and Goetz's *Der Widerspenstigen Zähmung* (The Taming of the Shrew), though none did particularly well at the box office. Salvi had increased the size of the orchestra, and both he and Dingelstedt raised the vocal standards of the house by contracting a higher calibre of both resident and guest artists. Among those singers who appeared in Vienna during this era were Berta Ehnn, Louise Dustmann, Gustav Walter, Marie Wilt, Ilma di Murska, Minnie Hauk, Pauline Lucca, Adelina Patti, Amalie Materna, and Emil Scaria. Herbeck inherited and capitalized on these higher standards but, in spite of personal protection from Emperor Franz Josef (who elevated him to the aristocracy), he fell foul of the *Generalintendanz*, a board of management set up as a buffer between the Court and those who ran the day-to-day operations of the theatre. One who intrigued against him was Richard Lewy, first horn in the orchestra but also appointed by Herbeck as *Studienleiter*, a post which today would combine the duties of chief coach with head of music staff. Lewy was a devoted Wagnerian, and it is not without significance that Hans Richter, horn player and Wagnerian *par excellence*, was soon on the staff of the Vienna Court Opera. Herbeck left the Opera and returned to direct the concerts of the Gesellschaft der Musikfreunde, whose operations had been directed in his absence at the opera by Anton Rubinstein and Brahms. Two years later he was dead at only 46, but his influence on Vienna's musical life had been enormous. He believed passionately in Wagner's music but he also responded immediately when shown the manuscript of Schubert's Unfinished Symphony by arranging a performance, and recognized Bruckner's worth by obtaining a teaching post for the composer at the Conservatoire, and with it some financial security.

Richter's concert in Vienna on 24 January 1875 was the event which ensured his move from Budapest to take up an appointment in the city. Vienna's musical scene was ripe for change and Richter knew it, for he confided as much to Marie after he had met the orchestra for the first time at rehearsal.

Afterwards I went to see the theatrical agent Gustav Lewy to talk business about a new guest contract for the singer Adams; I hope this singer can appear with us next Saturday as Raoul in *Les Huguenots*. Then I went to the opera-house. A great revolution is taking place! Everyone is unhappy with Herbeck. He has released the best singer, Frau Materna, without having found a replacement. My close friend, the *Studienleiter* Lewy [Richter used the word *Operinspector*], told me that it could not go on like this much longer! Universal unhappiness with Herbeck. I was asked quite openly if I was inclined to come here. I thanked Herbeck for releasing Adams for the guest engagement. At a stage rehearsal which I watched I met Baron von Hofmann, who was very nice and friendly towards me. . . . I liked him very much, and I could see from his manner that I pleased him too. . . . At three o'clock I had my first rehearsal. You can imagine with what feelings I stood before my former colleagues as a conductor. I made a short but heartfelt speech, in which I emphasized that, as a former pupil of the Vienna Conservatoire and the opera orchestra, I looked back with joy at the period of my studies, and hoped that my former colleagues would recognize a worthy pupil in me. Whereupon unanimous applause! Then we rehearsed industriously. After the rehearsal the musicians praised my assuredness, prudence, and calmness! . . . It would appear that everything is looking good for me.[1]

Herbeck was succeeded by Franz Jauner, a non-musician but experienced theatre administrator, but before he left, he recommended Richter to his successor. Jauner was a shrewd operator. As soon as he arrived at the Court Opera in May 1875, he sent the agent Gustav Lewy on a tour of Europe to scout for new works. From Paris Lewy telegraphed his enthusiasm for the opera *Carmen* by a talented young composer called Bizet, and, almost as an afterthought, added that Verdi was in Paris conducting his newly composed Manzoni Requiem. Jauner's reaction was to book everything Lewy had described. Verdi conducted the Requiem on 11 June 1875 and three other performances together with two of *Aida*. *Carmen* came too, but without its composer, whom Jauner had engaged to conduct. Bizet, only 36 years old, died in June 1875 and when the opera's Viennese première took place on 23 October it was Hans Richter who was conducting in the pit. Jauner also had to be shrewd when it came to dealing with that shrewdest of composers, Richard Wagner. By securing Richter, Jauner had a bargaining chip. Wagner needed Richter for Bayreuth in 1876, and he also needed Amalie Materna for the role of Brünnhilde. The bass Emil Scaria had priced himself out of the market until *Parsifal* in 1882. Jauner laid down conditions for releasing any artists to Wagner's Bayreuth venture, for he wanted a *Ring* for the Vienna Court Opera and got it in 1879.

This was the heady atmosphere which awaited Hans Richter when he arrived in Vienna in April 1875. A pair of telegrams sent to Richard Lewy from Budapest on 13 and 14 April show how very late matters concerning his new post were left. The first said, 'Probably arriving on 1 May. Possible debut with *Lohengrin*. Await promised letter from Jauner. Fraternal greetings.'[2] The next day he telegraphed, 'Everything arranged. Arriving 1 May. Contract signing here or in Vienna?'[3] Jauner replied, 'I ask you to exert everything to make 1 May a red-letter day for the rebirth of our stagnant opera-house, and I am sure that we shall go on together hand in hand artistically and commercially, to achieve our goal of making the Vienna Court Opera viable.'[4] The contract was dated 20 and 21 April in both Vienna and Pest, and engaged him initially for five years as first Kapellmeister. His salary was fixed at 4,500 florins for the first three years, rising to 5,000 for the next two. He remained under contract, with two intermediate renewals, until 31 July 1900, when he received his pension. When he retired (he was released at his own request four months earlier in April 1900) he was still first Kapellmeister of the Court Opera, though from 20 September 1893 he also had the title of Imperial Hofkapellmeister.

Opera schedules in 1875 were either not organized very far ahead, or they could be changed at will, for though his début did take place on 1

May, it was not with *Lohengrin* but with *Meistersinger* and cuts were only made in the last act. His diary entry was simple, 'Welcomed immediately on appearance. Great success.'[5] The critic August Wilhelm Ambros agreed. 'Richter passed the test brilliantly. He conducts with spirit and with a deep understanding. He does not seem to thrash around as if with an unseen opponent, as several other enthusiasts do when beating time. On the other hand he is not one who lets things happen on their own, whose conducting is like grinding a coffee mill. It was very engaging to watch Herr Richter, how he seemed to be everywhere at once with his glance and his hand movements.'[6] His arrival was also greeted in Vienna's general press. The *Illustrirtes Wiener Extrablatt* gave him front-page coverage two days later, a large pencil drawing of him with a full head of hair flowing back from his high forehead, his wire-rimmed glasses, and a full beard. Describing him as Otto Dessoff's replacement and as the youngest Imperial Court conductor, the accompanying text went on to call him 'a musician of outstanding significance; his name strikes a good chord in the world of music, and he has, through his connection with Richard Wagner, become a popular and even a celebrated personality, particularly with the Wagner Association'.[7] The paper went on to describe how Wagner had chosen Richter to conduct his first Bayreuth Festival the following year, but then asserted incorrectly that taking up his post in Vienna now made it impossible for him to accept this honour. It concluded with a rather unkind description of Director Jauner as an insignificant zero before whom Richter was a more credible digit and a man more capable of filling the shoes of the late-lamented Herbeck and Dessoff. Richter was welcome in the city's House of Art. On 9 May the whole of the front cover of the satirical Sunday paper *Der Floh* (The Flea) featured a baton-wielding Hans Richter arriving as Lohengrin, standing fully armoured in a boat drawn by a swan. The boat's figurehead was Wagner, the three Rhinemaidens swam around it, Brünnhilde stood on the cliffs, and Valkyrie rode among the clouds.

On 3 May the Wagners arrived in the city for a concert conducted by the composer three days later at the Musikvereinsaal. The programme consisted of extracts from *Tristan* and the *Ring*. Richter wrote in his diary that Wagner was very friendly towards him, but that Cosima was less so. The concert took place at noon, and that evening the Richters joined the Wagners at a performance of *Around the World in Eighty Days* at the Carlstheater. The opera orchestra's principal viola-player Sigmund Bachrich recalled Wagner's conducting of the final scene from *Götterdämmerung*.

FIG. 3 *Der Floh*, 9 May 1875.

Everything went very well. Then came the difficult place [twenty bars from the end] where the 6/8 passage changes to 3/2. I don't know why it happened, perhaps he overlooked something in the score; in short there was a sudden crisis and things became very difficult for us, the more so as the Master's baton became very unclear, hung in the air, and finally functioned no longer. At such crucial moments the Vienna orchestra is unique. Each and every one knew how to put out their feelers and adjust, how to understand the energetic intervention by the leader and a loud life-saving drum roll; in short we all got back on the rails.[8]

Richter began his duties with *Fidelio*, *William Tell*, *Roméo et Juliette*, *Euryanthe*, *Tannhäuser*, *Lohengrin*, *Don Giovanni*, and *The Merry Wives of Windsor* before taking a holiday in June. He and Marie went to Nasswald in Lower Austria, from where he wrote to Director Jauner on 2 July to ask for official permission to assist Wagner at Bayreuth for the first two weeks of August that year for a series of orchestral rehearsals and from 1 June until the end of August the following year for the inaugural Festival. Wagner had specified these dates in a letter to Richter of 30 June, though he would have preferred him to attend earlier piano rehearsals in July. Richter offered to forgo his salary for those periods and, in conclusion, emphasized the reflected honour upon the Court Opera that such an engagement would bring with it. Jauner, mindful of the debt Wagner would owe him and of his own aim to stage the *Ring* in Vienna, agreed.

At the beginning of July 1875 the singers arrived for their rehearsals but to Wagner's displeasure Richter did not do so until the 17th (Marie, who was now carrying their first child, remained in Nasswald with her mother). Until the end of July Richter participated in the daily piano rehearsals with the singers, which took place for two hours in the morning and from late afternoon until seven o'clock. From the beginning of August there was a full set of orchestral rehearsals for performances planned for the same time the next year. There were two rehearsals daily for twelve days, with the singers joining for each afternoon session. *Rheingold* was spread over two days and thereafter each day was devoted to one act of the remaining operas. All orchestral rehearsals took place in the unique totally covered pit, a new experience for players and conductor alike. On 1 August they started at 5 p.m. by playing the Valhalla motif as the Wagners arrived in the auditorium, and Betz, as Wotan, sang 'Vollendet das ewige Werk'. On the 12th Richter wrote: 'At midday we finished with the rehearsals for the last act of *Götterdämmerung*. I left Bayreuth at five without any farewells. As the orchestra gave me a cheer at the end of the last rehearsal, Wagner came forward, puzzled by the ovation. "Oh Richter", he said. "I nearly forgot all about you!"'[9] There was a definite chill in their relationship at this time for Wagner was easy prey to rumour, gossip, and misunderstandings. Richter was beginning to exert an independence of mind which the composer was not used to from those who served him; the second generation of 'Richters' now in the Nibelungen Chancellery all jumped to it when he snapped his fingers. Gustav Kietz recorded his impressions of Richter on 25 July 1875.

Frau Materna . . . and Kapellmeister Hans Richter had already arrived. It was a particular pleasure for me to get to know Hans Richter better, and the congenial impression, of which I had already heard much, was strengthened by this closer acquaintance. He stood for everything which produced the highest artistic results; one felt the presence of this highly gifted artist everywhere, despite his modest and unassuming manner. He sat at rehearsal with the large score on his knees and I sat beside him and could follow it. As I remarked to him, before Wagner's arrival, that Frau Materna had already completely settled into her role of Brünnhilde, he replied to me with a smile, 'She? Oh, she has no idea yet, but when she has grasped what she has to portray, then you'll be astonished.'

On 3 August he continued.

1875: Vienna

The first orchestral rehearsal is now over. Yesterday everyone assembled after four o'clock in the theatre. Only a few tickets were given out for the auditorium. The orchestra consists of 115 men. 32 violins (Wilhelmj as leader), 12 violas, 12 cellos, 8 double basses, completely new tubas which caused such a stir in Vienna, 6 harps, etc. etc. Hans Richter began with part of the second scene of *Rheingold*, the passage where Wotan catches sight of Valhalla. After it had been rehearsed several times, Betz joined in, singing the part of Wotan with his wonderful voice.

. . . [Wagner] then crossed over the gangplank on to the stage and went to sit at a small desk close to the orchestra pit. On this desk stood a copy of the heavy score, leaning against a crate, on top of which was an oil lamp. Hans Richter was the conductor. Wagner followed the score but was in such an agitated state that he kept waving not only his arms but his legs as well. In spite of the darkness of the auditorium Menzel was able to capture the scene in a characteristic crayon drawing.[10]

If there was someone whom Richter rarely offended it was the archetypal orchestral player, now confined to an airless chasm below stage, out of sight and sound of what was going on above him, and faced with the music of four new operas. The parts were riddled with errors, because some of those in the Chancellery were less than diligent, but Richter had a detailed knowledge of the *Ring* and a phenomenal ear. Writing out a score in full, the best way to get to know it, may be time-consuming, but the result was that later in life Richter conducted most of his Wagner performances from memory: 63 *Rheingold*, 123 *Walküre*, 84 *Siegfried*, and 78 *Götterdämmerung*; 15 *Rienzi*, 55 *Fliegende Holländer*, 85 *Tannhäuser*, 57 *Tristan*, 141 *Meistersinger*, and 198 *Lohengrin*—a staggering total of 899 performances of Wagner operas over forty-four years between 1868 and 1912, an average of twenty each year.

The music of the *Ring* was completely new to all apart from those who may have played in the Munich premières of *Rheingold* and *Walküre*, and the preparation was fraught with difficulties. Although the orchestra consisted of highly qualified musicians, it did not at first understand the significance of the dramatic function of the music. An orchestral violinist from Berlin, Waldemar Meyer, wrote after the 1875 sessions:

The first task was to get the orchestral parts corrected. I have known many famous conductors, but I am sure that none was capable of correcting those parts as efficiently as Hans Richter. He had the score in his head; whenever a wrong note occurred he did not tap his stand and signal the orchestra to stop, but the rehearsal carried on whilst he called out continually, 'second horn F#, third tuba E♭, trumpet C#, first violin D, viola D♭, second violin A♭ etc'. What a difference from new works rehearsed by Taubert, Eckert, and Radecke! As soon as they heard something wrong, everything came to a halt, and at the point where the conductor thought he had heard a mistake, the player had to read out the note in his part so that it could be corrected. One often found the error was in quite a different instrument. By this method we would have needed a whole year to correct the parts of all four works in the Nibelungen operas, whereas it all took just two weeks.[11]

In spite of earning such well-deserved accolades from the critical band of musicians with him in the pit, Richter was not a happy man when he left Bayreuth that summer, and he wrote a long comment in his diary which began:

Even if the news which was reported to me in confidence was maliciously exaggerated, it is certain that the Master and especially his wife did not treat me well. Hill, Wilhelmj, the Eckerts, and several others who were non-partisan have taken up my cause. Even the [Wagner] children took my side. Daniela told me with tears in her eyes that when I was spoken ill of at the dining table Fidi once asked: 'What do you have against Richter, Mama?' Even my poor wife was not spared. The rehearsals were my only pleasure and consolation for so much undeserved wrong, even though they were very tiring and upsetting mainly due to the bad orchestral parts copied with unheard-of carelessness by Seidl and his colleagues. The last rehearsal ended at midday on 12 August and I left Bayreuth without saying goodbye. I replied from Vienna to Wagner's reproachful letter. His reply to my admittedly hasty letter (written with zeal in the first flush of excitement) contained no response to any of my accusations. When I then received Frau Wagner's letter, I telegraphed immediately asking for forgiveness, but only out of my indestructible love for the Master, for I had suffered all sorts of wrongs. But should I become the reason that Wagner would not come to Vienna? I write this half a year after the events, therefore with complete calmness and after our reconciliation.[12]

The trouble had begun with a newspaper cutting from the *Augsburger Abendzeitung* of 15 August, which Wagner enclosed in a letter written the same day to Richter, now back in Vienna. The article (attributed incorrectly by Wagner to a Hamburg paper) reported unrest in Bayreuth and included the statement that even 'reliable sources' in Vienna knew that Richter himself had his own doubts about the project and that his enthusiasm was dimmed. Wagner was furious.

This cannot go on! Read No. 187 of the *Hamburger Zeitung* (an article about 'Dissension in Bayreuth', which is probably known to you in Vienna), and ask yourself if in future I can remain connected with those who are the cause of such dreadfulness, at least according to pub gossip. You have stubbornly stayed away from my house and this incident reveals a somewhat malevolent reason and puts an evil construction upon it. Only later do I discover that rumours arose from your quarter, that I or my wife had accused you of embezzlement. Now that may all be foolishness or madness, but you gave it credibility by staying away from me and my house. Your departure without saying farewell was as uncharitable as it was disgraceful. A wife such as mine deserves only your respect, and I demand that this be paid to her under any circumstances, whether by comedians or musicians.[13]

The letter concluded with Wagner's uncompromising insistence on a penitent declaration of loyalty from Richter if any further association between the two men were to continue. On the newspaper clipping Wagner had written a comment 'That's what happens!', but alongside this Richter had also written with his signature, 'But it's all a dreadful lie; I never said that.'[14] Richter replied immediately (too hastily as he readily acknowledged in his diary). He reminded Wagner of a similar incident in May of that year when he stood accused of the authorship of an article in the *Wiener Tageblatt* critical of the composer. At that time he proved successfully that someone else was responsible and now history was repeating itself. The rest of the letter accepts any criticism from Wagner, in spite of Richter's protestations of continued loyalty to him and his cause, but he draws the line at being criticized by Cosima.

I will submit to anything from you, but not from your wife. One thing has now become certain to me, that your wife has hated me all along, and she wishes my downfall. I don't know how I have deserved this. What I had to put up with after those cuts in *Holländer* [in Pest on 9 March 1875 when he conducted the opera during the Wagners' visit], as if it gave me any pleasure to make them. Everyone was being told how I got every tempo wrong. No day passed when someone didn't say to me, 'Well now, yesterday you were being got at again!' Betz told me that there was a suspicion that I made something out of the rental charge for the hall for the Pest concert. . . . But what hurt me most and made me indignant was the fabrication that I had added a cymbal clash in *Holländer*. My consideration for a weak singer forced me—against my innermost desire—to make cuts; I suffered enough for that. But that I should commit the barbaric act of composing even a dot, that must debase me in everyone's eyes, and the irresponsible spread of this rumour could affect my very existence. Today I no longer stand alone, I now have a dear wife to care for; therefore I cannot be indifferent when I see how attempts are made to humiliate me, suspect me, and in the end to do away with my very existence. Under such circumstances how could I come to your house? With regard to the newspaper and that dreadful article I am completely innocent, and am also prepared to provide any explanation, for in spite of everything I remain truthfully your lifelong faithful H.R.[15]

Wagner's rage was now up for he was as prepared to defend his Cosima as Richter was his Marie. Not surprisingly his relationship with Richter was at its worst at this point. Wagner's bitter response, written in the night according to Cosima's diary for 25 August 1875, was not published by Karpath in 1924, even though Richter noted in his diary in 1915, just a year before his death and in a moment of reflection on his own past, 'This letter of the Master should be published because it so clearly proves how much he loved his wife [die Frau Meisterin].'[16]

I leave my bed, where I cannot sleep, to tell you how your last letter seemed to me. It is despicable, despicable! You wretched man, you find it in your heart to tell me you are certain 'my wife has hated you from the start'. You can find it in your heart to say that? My wife has loved you as her own son. You wretched man! Everything she has said about you, or what has been reported to you as said, I have said; therefore she has only ever quoted me. I offered in my April letter from Berlin to tell you myself verbally what I have said about you when I was displeased. You avoided this in Vienna in May; we were interrupted whilst I was trying to do so when you last arrived in Bayreuth, and since then you have avoided me completely and shunned my house. What was uttered in beer cellars (where you were Niemann's companion) we are now discovering from disgraceful newspaper reports. Your whole letter is proof of where these dirty tricks emanated from. As I become more difficult to attack, so my wife now becomes the target, presumably because from the outset she was the only one who strove to mediate and also took care to keep trouble away from me. What was your source?? Let's see: 'One said that . . .' and 'again I was told that . . .', and with such slanders you defile the countenance of a woman who possesses the one thing you all lack—truthfulness!

At first I wanted to write to you calmly and as 'Master', which is still how you address me. As such I had so much to say to you. Much too about the 'ruin' which, in your opinion, you have decided my wife has brought you. But you have neither the judgement nor the will to understand what I would tell you. So go with your self-inflicted absurd madness to wherever it leads you! I have no more time to waste on such stupidities. Tell Director Jauner that I am not coming to Vienna, because I do not wish to contribute further to your ruin. Goodbye![17]

In her diary Cosima describes Richter's letter as 'foolish and crude',[18] though she only discovered it when she questioned her husband about the reason for his restless night. She herself decided to write to Richter as well 'with a full heart',[19] and reminded him that his primary loyalty was to her husband, not to her, and that without it he could not be a participating servant in the great Wagnerian adventure which lay ahead. She began by stating her total ignorance of the offending newspaper article and the ensuing correspondence between the two men. She then reminded Richter of his sojourns at Tribschen and the way he had naturally integrated into the family there, and of her vigorous defence of him when he stood accused by critics of ignoring Liszt's works while in his post at Pest. Turning to more recent events Cosima listed complaints that Wagner had made against him, his lack of preparation for the most recently composed episode of the *Ring*, *Götterdämmerung*, his withdrawal from the Vienna concert (which Wagner then had to conduct), and the organizational details that had gone awry and created difficulties. As for those notorious cymbal clashes in the Overture and second act of *Fliegende Holländer*, Wagner called out in her presence, 'Oho! Richter is helping my orchestration!' and expressed the opinion that, rather than being invited to witness such occasions, he should be encouraged to stay away. The catalogue continued with his late arrival at the summer rehearsals for the *Ring* and the various rumours and misunderstandings which gradually led to a cessation of communication between Richter and the Wagners (hence the accusation that he avoided going to Wahnfried). With an apparently genuine wish on Cosima's part that bygones should be bygones, she went on: 'Now I beg you most urgently and for the sake of old times at Tribschen, when you knew so precisely what we meant to you, make up for everything by telegraphing my husband as soon as you receive the letter which he wrote last night and express your heartfelt regrets that you listened to such nonsense, and promise to return to us and be the Hanns [*sic*] Richter of old.'[20] She concluded by forgiving Richter and reminding him that he would be as defensive about his wife as Wagner had been about her, but also alluded to the 'special' circumstances of how she and Wagner were united. She was well schooled by her husband (deliberately or not on his part) in twisting words, playing on emotions, and evoking memories. She recalled the happy days of the *Idyll* and protested in all innocence, 'if I have hurt you personally, I regret it from my heart, but when was this?'[21] Richter's response was immediate. On the day he received both letters, he dispatched a contrite telegram, 'Deeply revered Master. Forgiveness for the penitent. Please write to me all you wished to say, so that I can acknowledge my wrong completely and repent. Regretting my absence, assure you never to have wavered in my loyalty. Your life-long devoted Hans Richter.'[22]

From Wagner's telegraphed reply, it appeared that a letter was required from Richter as a reassurance, and because the composer sensed that the 'cramp which displaced your feelings has passed, and all further declarations will be welcomed with an open heart. . . . You have been obscured in a cloud of wrong impressions, and unfortunately I must assume that the smoke in a public house at eleven o'clock at night contributed to the misinformed perception which you formed.'[23] Wagner reminded Richter that he had highly recommended him to Vienna on his arrival from Pest earlier that year, and of the significant part he had played in his appointment in the spring. His opinion of Richter had in fact been so unqualified in its praise that it was now a matter of some embarrassment to him with Viennese musicians and officials that newspaper reports of Richter's supposed doubts about the viability of his projects had appeared. It had even been suggested to him by Jauner and Scaria that he might add Ernst von Schuch from Dresden (a proven Wagnerian) to work alongside Richter, but Wagner still had confidence in his original choice. He alluded to a statement he had made earlier that year (6 May in Vienna) at Dr Standhartner's house that he wished to stage his own operas himself (the forthcoming Vienna *Lohengrin* was meant here) because no-one knew how to do so except he himself. This also included Richter's interpretation of *Meistersinger*, which was essentially influenced by their close contact at Tribschen and Munich in the years 1867 and 1868, when the work was being copied and then performed. Wagner only intended to placate his critics and pacify his enemies with such statements; they were not intended to insult his supporters. The composer's letter closed the affair and normal relations were restored.

It must, however, be doubted if their relationship ever achieved the closeness of former years. Wagner's opinion of Richter fluctuated greatly throughout their acquaintance according to the composer's own volatile moods. He appeared to consider him a highly gifted musician with a natural and instinctive talent, but with a limited intellect. 'Richter is and remains a country bumpkin. In Tribschen he would sit all day in the domestic quarters and tell them of Schopenhauer's writings without understanding a word of them himself. He's nothing more than a bungling artisan. As a man he's always been a burden to me.'[24] By no longer being Wagner's constant companion and minion, Richter was not on the spot to please his master; others were doing that. Wagner, the great communicator, was often liable to misunderstand matters communicated to him (and more was probably reported to him out of mischief than otherwise). Cosima soaked up her husband's views by natural osmosis even if she subsequently misrepresented them, and after 1876 it was, significantly, twelve years before she invited Richter to return to Bayreuth to conduct. Although as the two of them got older they became closer, Richter never treated her with the reverence that he gave without hesitation to her husband. With Richter she never got anywhere when it came to artistic domination or interference in the performance of Wagner's music-dramas. She could practise that on any other conductor in her employ (or even elsewhere) but not upon Hans Richter.

Back in Vienna Richter returned to duty at the Opera and awaited the return from the country of his wife. Vienna, at the end of August, was enduring a heatwave and he told her to stay away until it was over. The refreshing rains occurred at the beginning of September. Letters survive between them dated from 28 August to 7 September (the date of her return), but nowhere is any mention made of the troubles between him and the Wagners; his concern for her well-being included sparing her any worry over that issue. Between performances of *Norma*, *Dom Sébastien*, *Zauberflöte*, *Oberon*, *Don Giovanni*, *Lohengrin*, *L'Africaine*, *William Tell*, and *Fliegende Holländer* he had rehearsals for a forthcoming première which elicited this understatement. 'A new opera is being rehearsed, *Carmen* by G. Bizet, a French composer who died in Paris only recently. I hope this new work, although not of significance, will nevertheless be entertaining.'[25] Richter's opinion of the opera may well have been influenced by Berta Ehnn's portrayal of the heroine, seen for what it was only when Pauline Lucca took over shortly thereafter in a memorably fiery interpretation. The première took place on 23 October and was its first performance outside France. Ernest Giraud had been specially commissioned to write recitatives for use instead of spoken dialogue, but as he was dilatory in sending his work (it did not arrive until late September), the version used (and which would remain in use until Mahler's arrival in Vienna at the end of the century) was a mixture of recitative and dialogue, the latter predominating in all the personal and comic scenes.

If Richter underestimated the opera he was currently preparing, he made no such mistake with his important and growing concert work.

Today we had a very interesting meeting on Philharmonic concert business. Herbeck was very keen that he should get the opera orchestra for the Gesellschaft concerts. He counted on it, partly because he was once its Director and established a lot of good for the orchestra and put them in his debt, and partly because he could promise them good financial rewards. At first the votes were divided. But then I intervened and said that the Philharmonic would be very unwise if they created competition for themselves. The artists' pride would also be humbled if making money made them faithless to their flag. If, however, they were to play in both concerts it would no longer be a competition between both orchestras, but rather between the conductors. They should feel themselves too proud to be handled by either like goods. When I finished there was universal jubilation. Not one person said that he wished to play under Herbeck. That was a downright victory![26]

The dual function of the Vienna Philharmonic orchestra was considered sacrosanct, but several attempts had been made over the years to combine its operatic work, where its members were employees of the Court, with its concert work, where it was autonomous in directing its affairs and choosing its conductor. It was Otto Nicolai who, on 28 March 1842 (Easter Monday), conducted the orchestra's first concert at the Redoutensaal, opening with Beethoven's Seventh Symphony. Nicolai's premature death in 1849 provoked a serious crisis in the concerts' affairs. His successors, Georg Hellmesberger, Wilhelm Reuling, and Heinrich Proch, were dull conductors with lack-lustre personalities who could never compete with virtuosi such as Paganini, Liszt, and Jenny Lind. They dominated European musical life and transformed salons into concert halls and concert halls into salons during the middle years of the nineteenth century. It took the likes of Hans von Bülow to establish the authoritative imprint of the conductor as a careerist in the 1860s. The social upheavals of the post-1848 revolutionary years forced concert plans to be shelved and, during the 1850s, there were only spasmodic attempts to mount various events, two or three concerts each season and always subject to the availability of the conductor from his operatic duties. In the latter years of the decade this was Karl Eckert, who was devoted to the Philharmonic and its concerts. It provided a welcome relief from the struggles at the Opera which he undertook on Wagner's behalf against the city's critics, the Court censors, and the anti-Wagner faction amongst the public, but before he finally gave up the fight in September 1860 he had established the subscription system for the orchestra's concerts which persists today. He also had to compete with concerts under Herbeck put on by the Gesellschaft der Musikfreunde.

Eckert's successor for fifteen years (1860–75) was the 24-year-old Otto Dessoff, born in Leipzig and appointed a Kapellmeister at the Vienna Court Opera in 1860. This year is considered today to be the starting-point of the Philharmonic concerts. Besides programming a regular concert series the orchestra established its autonomy, and could now choose its own man to lead it; in effect it built itself an island republic, surrounded on all sides by the Imperial might. At the same time it resisted any temptation to combine with the Gesellschaft der Musikfreunde, although both organizations shared the new Musikverein building as a venue for their concerts. The writer John Ella visited Vienna in 1866 and recorded his observations of a Philharmonic Society concert on 11 November.

The band, mustering about sixty musicians, was well conducted by Dessoff, and every composition was executed with the most scrupulous attention to details which often escape the vigilance of our London conductors with their one hurried rehearsal. I could not but remark the respectful address of the maestro on entering the orchestra with a *bâton* about half the size of those distracting white wands used at London concerts. On mounting the dais, he salutes, in silence and with solemn reverence, his critical audience. At the end of the composition, if satisfactorily executed, enthusiastic applause follows. The conductor then makes a respectful obeisance, and at the third round of applause, the whole band rise from their seats and salute the audience. This custom is far more respectful than the foolish habit in London of the players usurping the privilege of the public in applauding their conductor *before* the music is played.[27]

Dessoff was a supporter of Wagner, Liszt, Bruckner, and Brahms, whose First Piano Concerto and Haydn Variations he premièred in 1871 and 1873 respectively. He reshaped the Philharmonic's programmes by eliminating the solo items (in which the orchestra would leave the soloist to perform either unaccompanied or with piano accompaniment). After his performance of Beethoven's Ninth Symphony on 11 April 1875, he exchanged Vienna for Karlsruhe. It was the exact moment that Hans Richter arrived in the capital. The modern era (many prefer the term 'golden era') began when he, originally one of their own number from the Kärntnerthortheater days, was appointed as Dessoff's successor after his impressive début at the Wagnerverein concert on 24 January 1875. Apart from the 1882–3 season (when Richter resigned after a dispute), he remained Director of the Philharmonic concerts until 1898. Like his predecessor, Richter restricted the number of works presented in each programme and, even if he was more circumspect before making decisions on new works and composers, he nevertheless presented many first performances along with the established compositions. The programme for his first concert, on 7 November 1875, was Wagner's Faust Overture, Bach's Third 'Brandenburg' Concerto, and Beethoven's 'Eroica' Symphony (this midday concert was followed by an evening performances of Verdi's Requiem, an example of the sort of gruelling schedule that awaited him in his new post). The essential ingredient of the programmes of the Vienna Philharmonic from its very first days under Nicolai has been the symphonies of the first Viennese school. Dessoff introduced the works of Wagner and Liszt; Richter added those of Brahms, Bruckner, Dvořák, Tchaikovsky, and a host of others, but the common thread was always and remains even today the symphonies of Haydn, Mozart, and Beethoven. Critical reaction was mostly favourable; Theodor Helm reported cries of 'Richter! Richter!' from the audience and, in the 'Eroica', pronounced himself satisfied with the conductor's 'brilliant triumph. We cannot remember such an enthusiastic reception for this gigantic work.'[28] A review also appeared in the *Oesterreichische Musiker Zeitung*, this time by Joseph Scheu, who, after his observation that many people had to be turned away because the event was oversubscribed, wrote that

Richter is indisputably one of the most significant conductors of the present day, and all friends of the Philharmonic concerts who had doubts about their continued prosperity after Dessoff's departure may look to the future with confidence. Richter conducted the whole concert from memory, and his cast-iron stillness was eloquent proof of the absolute confidence with which he sets about his task.[29]

This Philharmonic concert was given in the presence of Wagner and his wife ('first movement of the "Eroica" dragged, the last in which R. had been of help to him, goes quite well', Cosima wrote in her diary[30]). The couple had arrived in Vienna on 1 November as Wagner wished to rehearse *Tannhäuser* and *Lohengrin* for performance under Richter three and six weeks later respectively. On the day of their arrival Richter was conducting the first of two performances of Verdi's Requiem followed by *Carmen*. The Wagners heard both works and Cosima wrote: 'In the evening Verdi's Requiem, about which it would be best to say nothing,'[31] and the next day, 'Good relations with the Richters. In the evening *Carmen*, a new French work, interesting for the glaringness of the modern French manner.'[32] Wagner was reputedly more enthusiastic ('Here, thank God, at last for a change is someone with ideas in his head'[33]), Hanslick was predictably restrained, but the Viennese public took the work to their hearts, none more so than Brahms, who heard it time and again.

Richter's diary recorded his reunion with the Wagners, the first since the heated correspondence of the late summer. 'On 1 November the Master came to Vienna with his wife and children. Nothing was mentioned of the past. The Master and his wife were very charming with us all the whole time. I am pleased and happy from the bottom of my heart that, without any resentment, I am free to come and go with the greatest Master.'[34] The seal was set on their reconciliation on 14 December 1875 when Hans and Marie's first child was baptized Richardis Cosima Eva at her parents' flat at Margarethenstrasse 7/1 in Vienna's 4th district. The Wagners needed no further proof of the sincerity of Richter's regret. Richardis was born on 9 December, and the next day Wagner agreed that he and Cosima would be the child's godparents. 'We did not expect you to do anything else,' Wagner told Richter.[35] That evening a dinner was given by Count

Hohenlohe and his wife attended by Richter, the Wagners, Franz Jauner, Gottfried Semper, Baron Leopold Hofmann (an important civil servant in the Foreign Ministry and at Court who had once been the official censor and who would, after a period as Minister of Finance, become President of the Gesellschaft der Musikfreunde in 1877 and General Intendant of the Hof Theater in 1880), Franz Dingelstedt (Director of the Hofburg Theatre), Josef Hellmesberger (Director of the Vienna Conservatoire), Count Karl Seilern von Aspang (a Court adviser), and Salomon Mosenthal (a civil servant but also librettist to Nicolai, Flotow, and Marschner). Two days later the Wagners left Vienna.

On 22 November Richter had conducted *Tannhäuser* under Wagner's watchful eye. The production was notable for its use of the 1861 Paris version in which the Overture leads directly into the Venusberg scene. Wagner had agreed in October to come to Vienna when it was confirmed that 'the Vienna performance is serious',[36] but he was worried about some members of the cast. Amalie Materna sang Venus and Berta Ehnn (fresh from her portrayal of Carmen) took the part of Elisabeth after Marie Wilt had been dropped. Wilt, a member of the company from 1867 until 1878, was not considered suited to the role by Wagner, who was ruthless and candid in stating his preferences or demands regarding his singers' physical suitability. In casting the role of Tannhäuser he had less success. In his letter to Richter he already expressed doubts about Leonhard Labatt. 'Do not forget to tell our honoured Swede [Labatt was born in Stockholm] that he must sing the second Finale complete. There'll be trouble with the A in the Adagio ("erbarm' dich mein!"), and also with the passionate section in the final Allegro. I must state here and now that I will allow nothing to be cut and in case Labatt declares that he cannot, I must for my part declare that for that reason I would be unable to be content with him in a model performance of *Tannhäuser*. Therefore I fear there are some hard nuts to crack!'[37] Wagner was right. Labatt (whom he called 'the one-eyed Swedish Jew'[38]) was not up to the role and there were many problems with both him and Louis Bignio as Wolfram during the preparations. Nevertheless the performance was a success, 'good beyond all expectations', Cosima noted,[39] although as soon as Wagner had gone Vienna reverted to the more standard (Dresden) version of *Tannhäuser* with no Venusberg scene.

The occasion was marred, however, by the composer's speech before the curtain at the end of the evening. He said, 'In May [1876] it will be fifteen years since I heard my *Lohengrin* for the first time, here with you in Vienna. You accompanied my endeavours at the time in friendly fashion, and today it appears you have wanted to do so once again, whilst I try (as much as the resources available to me will allow me) to make my works clearer to you. Please accept my heartfelt thanks for this encouragement.'[40] This statement was immediately construed by the press (in spite of Wagner's protests) as criticism of Labatt's limitations and they savaged the composer for his apparent ingratitude. Wagner had as always been extremely demanding upon his singers, who had to attend all rehearsals even if they were not being used. Although he raised the weakest of them to higher standards, with others he met resistance and stubbornness he could not overcome as well as a limited ability to cope with an uncut performance, and Labatt was one such singer. Wagner called the company together three days after the première and insisted that 'resources' meant not people but the technical facilities available at the theatre. He told them that as a sign of his sincerity Director Jauner was at liberty to report what he was saying to them to the press (he could not and would not do so himself) but that if he did, then he would have to leave Vienna the next day. Needless to say no one called his bluff, although reports did appear in the papers of his speech to the company with tabloid style headlines such as 'I hate the press'. His quarrel with Richter was buried in public when, after the dress rehearsal for *Tannhäuser*, 'Richter thanks him on behalf of the orchestra, kisses Wagner's hand, and thanks him again personally for his consideration toward himself'.[41]

Instead of leaving Vienna, Wagner began immediately with rehearsals for *Lohengrin* with Emil Scaria (Heinrich), Georg Müller (Lohengrin), Mila Kupfer-Berger (Elsa), Georg Nollet (Telramund), and Amalie Materna (Ortrud). This work was also given without cuts, and the composer was once again the first victim of his own short temper and demanding rehearsal technique (he nearly left the city in disgust when Labatt pleaded ill health to the management and the performance of *Tannhäuser* on 30 November was replaced by Gounod's *Roméo et Juliette*). Despite the gloomy entries in Cosima's diary she described the opening night as a 'wonderful performance'.[42] Wagner had scored a particular success with the chorus, working with them in detail and providing some of them with little cameos to act out in various scenes. His good working relationship with them and their high regard for him led to one more visit to the Austrian capital three months later.

By the end of 1875 Hans Richter was well established in Vienna. He was regularly to be seen at the Opera, mainly in the works of Wagner, Weber, and Mozart, but by the New Year he had also conducted his fourth Philharmonic concert (on 26 December with Friedrich Grützmacher as soloist in Raff's Cello Concerto and Josef Hellmesberger taking the viola solo in Berlioz's *Harold in Italy*). One of Richter's players at this time was Arthur Nikisch, who sat among the second violins, but by 1878 he was off to Leipzig (at Dessoff's recommendation) as chorus master and second Kapellmeister assisting Josef Sucher. The beginnings of Nikisch's career were in many ways similar to those of his older compatriot Richter. Another orchestral player, who remained throughout Richter's period in Vienna, was the cellist Joseph Sulzer, who recorded several interesting observations of the conductor's work both from the concert platform and from the orchestra pit in the Opera.

A lot of nonsense has been written about Richter's alleged distaste for Mendelssohn and Schumann. . . . The truth is that the works of [these composers] were never performed more brilliantly or stirringly than under Hans Richter, in whose eyes I have seen tears in the Andante of Mendelssohn's A minor ['Scottish'] Symphony. On the other hand Richter made absolutely no secret of his antipathy towards certain composers, especially those who represent the soft and sweet approach with Gounod in their forefront. Richter's disposition and his musical make-up was founded in the Wagnerian school, so he found such music repugnant. It was highly amusing for those in the know to observe how he conducted the pompous phrases in *Roméo et Juliette* with seeming seriousness and broadly swinging arm movements. In the overture to this opera there is a cello solo which frankly I loved to play, because it is especially grateful and one can really 'get inside it'. I did so and Richter, for his part, regularly gave me a sarcastic glance which clearly and reproachfully asked the question, 'How can anyone like such stuff?' However Richter did not always use this pantomime method of expression. When he had to conduct an opera which either did not interest him or was simply loathsome, one could quite understand how he used funny asides in an effort to make his unpleasant job easier. When Germont first enters in *La traviata*, he is indignant at first with his son's sweetheart, but then, when she shows him the contract in which she makes over all her goods and chattels to Alfredo, he suddenly calms down and, full of honourable grief, cries out, 'Why must your past bring shame upon you!' [Ah, il passato perchè, perchè v'accusa!], which Hans Richter paraphrased with the words: 'Aha, now he changes his tune because his son gets the furniture.'

In *Carmen*, after the first bars of the third entr'acte, Richter put down his baton and let the splendidly melodious piece play by itself to the end without conducting. When the applause followed, Richter asked with comic seriousness of those players seated nearest to him, 'Now then, is a conductor really necessary?' *Carmen* was one of Richter's favourite operas and it was admirable how this Master of every style conducted this refined work, and brought to light its secret treasures. A sad incident sticks in my memory of one of the first rehearsals of *Carmen*. At the end of the rehearsal Director Jauner informed the orchestra that he had just received a telegram informing him of Bizet's death. Deeply moved by this sad news, we spontaneously got to our feet in honour of the late Master, whose work had just now been charming us.[43]

Richter's regard for Verdi was limited and, although *La traviata* was in his repertoire, Siegfried Wagner recalled that 'there was once a great mess in one of the choruses in a performance. The Intendant went up to Richter afterwards and said angrily, "Herr Kapellmeister, such things must not be permitted to happen here in the Court Opera," to which Richter replied coolly, "Have I not always said that the chorus should be cut?"'[44]

8

1876

Bayreuth

BIRTHDAY greetings from Richter to Cosima Wagner at Christmas 1875 and the naming of his daughter Richardis put the finishing touches to healing the rift, or, as Cosima herself described it, the 'passing illness'[1] of the previous summer. Richter was once again the 'alte Geselle', the disciple of old. The Wagners were to pay another visit to Vienna before concentrating entirely on Bayreuth during the momentous year of 1876, and it proved to be Wagner's last visit to the city. In order to show how pleased he had been with the chorus in *Lohengrin*, he offered to return in March 1876 to conduct a benefit performance for them of the work (Richter had been conducting it in the meantime). Jauner feared that Wagner would demand extra rehearsals but to everyone's surprise none took place. He arrived during the evening of 1 March, conducted the opera on the following evening, and, by the next day, was gone for good. Although Richter recorded in his diary that there was no rehearsal prior to the performance, Max Morold contradicts this in his *Wagner in Wien* (Vienna, 1930) with an account of a short rehearsal (a 'top-and-tail' session where difficult corners were tackled). This was to be the first and last time that Wagner ever conducted *Lohengrin*. According to Joseph Sulzer's account, Richter's role was as follows:

Wagner in no way handled a baton like a professional conductor. In some places he would conduct extremely precisely and rhythmically, at others, when either tired or wrapped in his own thoughts, nonchalantly or not at all; as a result an unholy confusion would sometimes have occurred were it not for a rescuing *deus ex machina* in the shape of Hans Richter who intervened at such moments. In fact Richter had foreseen such potential calamities, and in honour of the Master had undertaken the job of timpanist, from which seat he conducted with his sticks without Wagner noticing.[2]

Richter himself said he was elsewhere.

Apart from a few moments of uncertainty, the performance was excellent. To achieve a greater security the chorus master [Karl] Pfeffer and I conducted from the wings. [Wagner] received a storm of applause each time he appeared in the pit. The Master refused any reimbursement of costs or even his fee, a magnanimous act which was the greater as he himself had, during his recent November/December visit to Vienna, offered to conduct for the benefit of the chorus, who had given him so much pleasure. Throughout his stay the Master was in the best of moods. On 3 March [he] visited us and commented favourably on the appearance of my little daughter. At 4 p.m. supper at the Hotel Sacher organized by Director Jauner who had every reason to be grateful to the Master for his words after the performance. At 9 p.m. that evening we both went to the North-West station for the Berlin train [Wagner went from Vienna to the Prussian capital to stage *Tristan*]. A large part of the chorus had gathered at the station. At the Master's wish we sang the 'Wach auf!' chorus [from *Meistersinger*]. It sounded superb in the high-vaulted waiting hall. The Master was moved to tears. A heartfelt farewell. Each member of the chorus received 35 florins as share of the benefit.[3]

Sulzer described Wagner's instructions to the orchestra during *Lohengrin*. In bar 54 of the Overture, for example, the cymbals should not be damped, and the tempo of the D major postlude to the Bridal Chorus in the third act was significantly slower than they had known it hitherto. The composer apparently warned against producing any resemblance here to a Song without Words by Mendelssohn, though Sulzer admitted that, because Wagner's words tailed off into an incomprehensible mutter, he could only speculate on what the comparison was that the Master actually made. This description of Wagner's last visit to Vienna ended with an account of the casting of the third nobleman of Brabant with whom the composer had an altercation. This was Angelo Neumann, near the end of his short singing career before becoming an impresario and stage director. As such he mounted the first touring production of the *Ring* before directing the Opera in Prague. The dispute with Wagner concerned his refusal to be one of those designated to carry away the corpse of Telramund. More detailed records of Wagner's instructions for *Lohengrin* were taken down by the young Felix Mottl, currently a junior member of the music staff at the Court Opera in Vienna and a member of the Nibelungen Chancellery.[4]

Highlights of the rest of the season in Vienna for Richter included a performance of Weber's *Der Freischütz* on 26 February 1876, notable for

Josef Drexler as Caspar in his final operatic appearance after a career spanning thirty-seven years. The Philharmonic concert on 7 March concluded with the première of Karl Goldmark's 'Rustic Wedding' Symphony, a 'successful' performance which Richter conducted 'with the score'.[5] A concert on 16 April included an orchestrated version by Richter of a Largo by Handel for violin and viola, accompanied by winds alone. This was the period for reworking such baroque pieces; Hellmesberger had provided the missing slow movement of Bach's Third 'Brandenburg' Concerto with his orchestration of an Adagio from a violin sonata by the same composer at Richter's début concert with the Philharmonic Society at the beginning of the 1875–6 season. In April he made a short trip to Wiesbaden to hear the opera *Melusine* by Karl Grammann conducted by Jauner's future successor in Vienna, Wilhelm Jahn. He then returned to Vienna via Salzburg, where he met Hermann Zumpe of the Nibelungen Chancellery, and once back in the Austrian capital completed his operatic and concert duties before leaving on 29 May for Bayreuth and the first Festival.

Richter and Marie lunched at Wahnfried with the Wagners on 30 May and then moved into their quarters at 21 Ludwigstrasse, where they would spend the duration of the Festival. Rehearsals were delayed until 3 June because celebrations in Berlin for Botho von Hülsen's twenty-five years as Intendant at the Opera there prevented orchestral players from departing as planned. The orchestral rehearsals for the first Bayreuth Festival began with two-hour sectionals in the morning from nine to eleven o'clock with the wind and brass followed by strings from eleven to one o'clock. They were devoted to the first scene from *Rheingold*, and Richter noted that 'the tuba players and bassoonists had made significant progress during the past year. At the string rehearsal the Master came for a few minutes and expressed his pleasure with both the sound and the orchestra's performance.'[6] The full orchestra combined that evening from five to seven o'clock for a rehearsal of the same scene. Rehearsals for *Rheingold* were daily from 3 to 11 June; from 7 June they were all full orchestral stage rehearsals, although an orchestra-alone session had to be called for the morning after the final scene had been rehearsed with stage. Another ten days were given to *Walküre*, with the same basic pattern of act-by-act sectionals with strings, winds, or brass, followed by the full orchestra before combining with the singers on stage. Badly copied parts caused delays which disrupted the schedule and Wagner had dental problems during the *Walküre* sessions, so Richter had to supervise piano-production rehearsals as well as his orchestral sessions. Wagner tended to change his mind from one day to the next about moves and interpretation.

Betz and Niemann were quite happy that the Master was absent from rehearsals for several days. They're short-sighted and conceited if they believe that things go better without him. As happy as I am that he is absent from my part-correcting and special study sessions, I still realize how vital his presence is at the full orchestral and stage rehearsals. No one else knows the true way in which it's all to be staged. Common theatrical routine and actors' tricks aren't enough! In musical terms too, after the technical difficulties have been ironed out through detailed study in the primary rehearsals, his presence regarding tempos and interpretation is both very important and desirable to me.[7]

Problems arose when the boilers providing steam on stage leaked into the pit, putting a harp out of tune and ruining a pair of drums. The Rhinemaidens had their own difficulties. Each was strapped high up into a cradle atop a frame and had to simulate swimming. They had three men apiece, two to guide their stage placing and movement, the other a stage conductor to guide them through their music, Anton Seidl for Lilli Lehmann as Woglinde, Franz Fischer for her sister Marie as Wellgunde, and Felix Mottl for Minna Lammert as Flosshilde. Meanwhile the orchestral players sweated and toiled down below. They complained about heat and draughts so Wagner came round to placate them. 'Not only did I compose the operas, I also have to come and shut the windows for you,' he teased.[8] Tensions were released by frequent visits to the Angermann inn to make merry with improvised cabaret turns. On 11 July, a rehearsal day for Act III of *Götterdämmerung*, Richter wrote, 'At six o'clock this morning Herr Gustav Richter died suddenly of a heart attack. He was a viola player from Berlin, the first death amongst the Nibelungen company. An old acquaintance of the Master, he played under him in the first Berlin performance of *Fliegende Holländer* in 1844.'[9]

Throughout June and until 12 July the *Ring* was slowly rehearsed day by day, act by act. On one occasion Richter jokingly asked Felix Mottl whether he wished to conduct a rehearsal. 'My heart stood still at such a thought,' recorded Mottl.[10] Richter's main complaint concerned the condition of the parts, but he often recorded that Wagner was satisfied. On one occasion, however, he was not; this was after a rehearsal of Act II of *Walküre* when the composer wrote criticizing his conductor's failure to base his tempos on dramatic rather than musical considerations. In his view the roles of stage and musical director were one and the same; a conductor with no sense of drama had no insight into his music.

My friend, it is essential that you attend the piano rehearsals, else you will not get to know my tempos. It would be very trying to have to make up for this in the orchestral rehearsals, where I do not like to discuss matters of tempo with you for the first time. Yesterday we hardly ever refrained from dragging, especially Betz, whom I have allowed fiery tempos at the piano rehearsals, and also with Materna. Even the Valkyrie were held back in some of the heated moments in the ensembles with Wotan. I really believe that throughout you are bound too much to beating crotchets, which always hinders a tempo, particularly as the long notes dominate when Wotan is angry. In my view one should beat quavers where one needs to be precise, but you cannot maintain the mood of a lively allegro by beating crotchets.

Wagner then goes into technical detail with a description of how to conduct the music from Wotan's entry on his hunt for the errant Brünnhilde. Initially he must conduct 'only in an energetic four, both to fire up the orchestra and to deal with any inattentiveness on their part', but having established this supremacy (at least by the point where Wotan sings 'Wo ist Brünnhild?', 'from which point on there are no more melodic quavers in the vocal line'), the tempo must be at a speed where he can go into two or *alla breve*, and it must be maintained when the Valkyrie sing 'schrecklich ertos't dein Wetter' (in the score this word is 'Toben') but not hurried. 'Stop conducting crotchets! Away with them!', he wrote. Likewise at 'Zu uns floh' die Verfolgte' ('*never* would I have beaten crotchets here, they constrict you and your freedom is impeded'). When Brünnhilde appears from the protective ring of her sisters he must go 'slower, yes, but not too slow, so that when the somewhat faster speed appears we have a less moderate tempo than the first impassioned one'. He should beat crotchets here to keep control of the complicated syncopations in the orchestral part and thereby prevent any hurrying by the players, though 'I would achieve that by beating in two'. Wagner concluded:

My dear Hans! The services you have rendered my orchestra are so great, and my recognition of your quite unique abilities and achievements is so meaningful that I fear we have given vent to our pleasure and left undiscussed—and with a certain timidity—the most important issues, as if to avoid giving the impression of disagreement or unhappiness. But by so doing we would be wickedly injurious to both our souls and the truth, and that is not our way!

As I say, I attribute the reasons for our differences simply to your being *on the whole* still much too much unable to assist at the piano rehearsals; it is, however, at these where I seek to the utmost to establish my will and to engage with the singers. Be so good and, like me, consider it impossible for any offence to be taken between us. With all my heart, your old Rich. Wagner.[11]

More than twenty years later Richter wrote in his diary, 'I remembered the Master's letter about beating 4/4. He was right! At the time there were many players from Meiningen in the orchestra who were inexperienced in opera, and not one musician from an orchestra with a modern conductor from whom they were used to seeing the *alla breve* style. Those conductors mainly came later, and predominantly from Vienna.'[12] On his famous 'Final Request' to his artists which was pinned to the company notice board on the opening day of the Festival, 13 August 1876, Wagner wrote, 'The long notes will take care of themselves; the small notes and their text are what matters.' It is appropriate here, in view of accusations of Richter's both dragging and hurrying, to record the performance timings of his first *Ring*. Figures such as these do not always convey either the swiftness or the slowness of a performance, for a quick two-in-a-bar can seem like a slow one-in-a-bar. Of the timings, which are listed in greater detail in Egon Voss's *100 Jahre Bayreuther Festspiele* (Regensburg, 1976), which covers the period between 1876 and 1970, it is interesting to note that variations occur either side of Richter's timings. Table 1 shows, for comparison, three of his professional descendants. Furtwängler in 1936 is both faster and slower, Knappertsbusch in 1951 is much slower, whilst Karajan (also in 1951) is faster.

On 2 July there was a brief interruption of rehearsals for the *Ring* when a play-through of Wagner's Centennial March, written for a commission fee of 5,000 thalers from Philadelphia for the occasion of the centenary of American Independence, was organized for eleven o'clock that morning. Unfortunately a wind rehearsal of the first act of *Götterdämmerung* overran by half an hour, and when Wagner arrived punctually to conduct his march he found no musicians in the pit because all the string-players, hearing the winds at rehearsal after the appointed hour, had remained outside the theatre until they were sure that the rehearsal would begin, demonstrating once again the reluctance of the musicians to be in the pit any longer than necessary. Wagner stormed out to sulk with Cosima in one of the boxes.

TABLE 1. *Performance timings for the* Ring *operas* (hrs. and mins.)

	Das Rheingold	Die Walküre	Siegfried	Götterdämmerung
Richter	2.31	3.39	4.00	4.19
Furtwängler	2.36	3.38	3.58	4.14
Knappertsbusch	2.42	3.53	4.05	4.40
Karajan	2.25	3.36	3.53	4.20

1876: Bayreuth

FIG. 4 *Illustrated Sporting and Dramatic News*, 26 August 1876.

Richter meanwhile had finished his rehearsal and, breaking the embarrassed silence in the theatre, appealed to the composer for understanding. 'I must rehearse, Master,' he said, 'otherwise we shall not be ready.'[13]

There was a late crisis when 'that lout Scaria'[14] demanded a high fee to sing Hagen. Wagner could not meet his demands and Richter suggested Gustav Siehr from Wiesbaden, who was drafted in on 15 July. Comments in his diary about Unger at rehearsals of *Siegfried* show concern for his musical reliability, which was later justified. Towards the end of July there were cleaning-up sessions, but a slightly less taxing schedule allowed trips to the surrounding countryside in Wagner's company. First dress rehearsals began at the end of July on alternate days until 4 August, the final dress rehearsals beginning in the presence of King Ludwig two days later. They were on successive days, and an hour before *Götterdämmerung* on 9 August there was an orchestral scene-change rehearsal of Siegfried's Funeral March at which Richter read out a letter of appreciation of the players' contribution to the Festival from the king. The dress rehearsal of *Götterdämmerung* lasted from 5 p.m. to 11.15 p.m., which elicited the comment 'Great exhaustion' from the conductor.[15]

Rheingold opened at 7 o'clock on 13 August. The fanfares summoning the audience to take their seats for the 1876 Festival were selected by Richter, Donner's 'Heda, heda, hedo' for *Rheingold*, the Sword motif for *Walküre*, Siegfried's motif for *Siegfried*, and the Valhalla motif for *Götterdämmerung*. Richter's diary entries for the performance days of the first Festival, when compared with Cosima's diaries or Richard Fricke's account, beg the question whether they were all participating in the same event. 'A splendid performance, only a few little mishaps on the part of the stage machinery. Great jubilation at the end, but the Master did not appear. Loge (Vogl) and Alberich (Hill) were applauded.'[16] No mention here of Wotan mislaying the Ring as Alberich places his curse upon it, nor of the backcloth suddenly being raised to reveal shirt-sleeved stage hands. Neither is there any reference to the neck-less dragon (whose head had already mistakenly travelled to Bayreuth via Beirut) in *Siegfried*. Richter did, however, note comments made to him by friends in the audience:

The orchestra does not make itself fully felt; that miserable canopy over the violins robs the strings of too much sheen. 14 August: 4 o'clock: *Walküre* Niemann as Siegmund was unsurpassable. A success on the whole; only Betz was hoarse towards the end and Frau Materna went wrong with the text so she had to wait a few bars until she got back in. The orchestra was splendid. Unfortunately complaints once again from all sides about the covered orchestra. 15 August: *Siegfried* had to be postponed because Betz was unwell. 16 August: 4 o'clock: *Siegfried* huge and unexpected success. 17 August: 4 o'clock: *Götterdämmerung*. Tremendous demonstrations at the end. The Master appeared at last and spoke a few words which were maliciously interpreted.[17]

Wagner was reported as saying that hitherto Germany had had no art, whereas what he actually said was that in post-1870 Bismarckian Germany the country now *had* a national art, particularly in the field of music-drama with the staging of his tetralogy. On the evening of 18 August, the day following the end of the first cycle, there was a banquet given by Wagner for the Festival company, patrons, and other invited guests, a total of 700 people. The Richters were given places of honour, Hans opposite Wagner and Marie to the composer's right. The excitement and tension of the past few days were too much for the conductor, who escaped to the neighbouring town of Bamberg on the following day, but only for a few hours. He was drawn back to Bayreuth by midday and that evening had a piano rehearsal with the tenor Engelhardt as substitute for Georg Unger as Froh,

1876: Bayreuth

who was saving himself for his performances of Siegfried. Other substitutions occurred in the second cycle, which, Richter noted, went better than the first. As well as Marianne Brandt coming to the rescue of the ailing Luise Jaide as Waltraute, Hedwig Reicher-Kindermann took over the same artist's role of Erda. The third cycle 'was the most perfect. I was really affected when I walked into the pit to conduct this great work for the last time.'[18] At its conclusion Wagner addressed the audience and then called for the curtain to be raised to reveal the whole company, singers and orchestra alike (he had ordered no curtain calls during the cycle). At their centre stood Hans Richter, the only occasion on which he was seen on stage at Bayreuth by the public. This was not false modesty on the conductor's part. What he offered Wagner, apart from his musical talents, was an unswerving loyalty to his cause. He never accepted a penny from the composer or his widow for working at Bayreuth, nor, unlike most other conductors, did he ever go up on stage after a performance to acknowledge applause from the audience (his presence on stage in 1876 was as part of the assembled company and not an individual curtain call). Due to the covered orchestral pit they never once had a glimpse of Richter from within the auditorium. He maintained throughout his professional life that all credit for the performance was due to Wagner even after his death, and that he would not be responsible for deflecting any on to himself. Even at his last appearance at Bayreuth in 1912 with *Meistersinger*, and despite impassioned appeals and cries from orchestra and audience alike to appear before the curtain to acknowledge his ovation, Richter quietly left the pit forever and walked out of the theatre into retirement.

Wagner's nature prevented him from doing anything but taking such loyalty for granted. He simply expected it from anyone involved in his work, whether they were singer, conductor, designer, or stage carpenter. He was often too swift to condemn and when he did so he was often wrong. Letters to Richter during the spring and early summer of 1876 criticise his artists, among them Materna, Unger, and Schlosser. The composer completely misjudged the two tenors at one point. 'Schlosser is worrying me', he wrote 'and I am thinking a lot once again of your Schmidt. If he is hopeful of getting away, go through Mime with him once. If he is good, we can take the matter seriously for Mr [*sic*] Schlosser appears to have gone to the dogs. P.S. Unger is progressing well; with this one I hope to win my case. There is nothing the matter with him.'[19] Events would prove that his impatience with Schlosser and his confidence in Unger were entirely misplaced. Possibly Wagner remembered that Schlosser had created the role of Mime in the 1869 Munich *Rheingold* of which he so strongly disapproved at the time. There is an interesting postscript to this incident, for when Schlosser died in 1916 Richter (who himself had only two months to live) wrote an appreciation.

Max Schlosser, the first David (1868) and Mime (1876), has died of old age. Perhaps it would interest those many friends of this revered artist how Schlosser came to sing David. Rehearsals for *Meistersinger* were already underway, but no apprentice had been found. Various attempts with different singers came to nothing. Then the actor Rüthling told me that he knew a singer from one of his previous engagements whom he had seen and heard perform marvellously in the *buffo* tenor roles of Lortzing, and who would be very suitable for David. At the time he was a baker in Augsburg, but he still had a longing for the theatre. I travelled to Augsburg at once and met a charming man with a splendid beard. He sang 'Die Weisen' [from Act I of *Meistersinger*] at sight; I saw and heard that he was the one. I left him with the role of David to study. After a few days Schlosser came to me at Munich and delighted me with what he had achieved. We decided to go at once to the Master. But there was still a drawback to overcome—that beautifully cared for beard! I knew how to overcome Schlosser's misgivings. With such a fine beard one could play the role of the venerable father in *Norma*, but never the fun-loving apprentice-cobbler from Nuremberg. As soon as the barber had removed this obstacle, we hurried off to the Master. He was highly pleased and Schlosser was engaged. There was total unanimity when recognizing his achievements. His musical security pleased conductors and his flawless acting pleased directors. Honour the memory of this outstanding artist and splendid man![20]

Among those who attended the Festival were Richter's mother and friends Camillo Sitte (who, as an architect, was probably responsible for Richter's diary criticisms of the acoustic shell covering the orchestra pit), Josef Sucher, and Reinhard Schäfer. Richter was often called upon by Wagner to be at his side as he welcomed his guests of honour such as King Ludwig of Bavaria, Don Pedro the emperor of Brazil, or the German Emperor Wilhelm I. The latter was taken on a tour of the theatre by Wagner on 14 August, during a heatwave. When they came to the orchestra pit the composer said to the emperor, 'This is where my musicians have to sweat.'[21] Richter received recognition from the aristocracy on 17 August at the conclusion of the first cycle, when the first of many orders he was to receive throughout his life was conferred upon him by the Grand Duke of Mecklenburg-Schwerin, Friedrich Franz II. A week later he received an order from the Grand Duke of Meiningen followed by one from the Grand Duke of Weimar. When King Ludwig of Bavaria returned to

Bayreuth for the third cycle he conferred the Order of St Michael on the singers Niemann and Betz, and upon Wilhelmj and Richter. Several artists and critics, even some of the performers, wrote their impressions of the first Bayreuth Festival and some provided their memories of Hans Richter. Lilli Lehmann sang the Rhinemaiden Woglinde and the Valkyrie Helmwige in the *Ring*.

The singers saw almost nothing of the conductor. A black cloth was nailed behind him against the wall of the pit so that Hans Richter and his white shirt-sleeves could be seen, for he conducted in his shirt-sleeves, and whenever he had the chance he drove up to the rehearsals at the theatre, driving an ox-wagon in the shimmering heat. Everything was new, the immense distance between the conductor and the stage, and the lack of a prompter.

... Hans Richter, who achieved a purely unbelievable task by his equal devotion to Wagner, his works, his success, and his family through a continually renewed enjoyment and love for his work for which he could never do enough.[22]

Tchaikovsky was one of several composers present at the Festival. His only glimpse of Richter was leading the musicians in the procession which accompanied the German emperor from the station to the Theatre on 12 August. Bruckner was present but wrote nothing, but Grieg called Richter 'the conductor of genius',[23] and Sir Charles Villiers Stanford described him as 'a fair-haired Viking'.[24] Wilhelm Ganz was also a visitor and wrote admiringly:

He performed a great feat in conducting [the *Ring*] without having the score before him, entirely from memory, such a thing never having been done before in the musical world. I have already mentioned that my father conducted the classical operas by heart, but this was child's play compared with Dr Richter's accomplishment of conducting the difficult and complicated music, vocally and instrumentally, of the *Ring*, and in those days it was extraordinary that a work so intricate and difficult should be memorised by one man. Dr Richter, like the members of the orchestra, was in his shirt-sleeves as the heat was so great.[25]

Whether Richter really conducted without a score from the very first performances of the *Ring* is unlikely. Bechstein's drawings of him in the pit at Bayreuth show a score on the stand; admittedly these were done in rehearsal, but with the scenic mishaps of *Rheingold* and Materna forgetting her text in the performance of *Walküre* Richter would probably have wanted the score to hand. He was also under the critical eye of his conducting colleagues such as Mottl, Fischer, and Zumpe, but in addition many came from other cities and towns to assess the *Ring* for staging in their own houses. Among them were Hermann Levi, Leopold Damrosch (from America), and Franz Wüllner, the first conductor of *Rheingold* and *Walküre* who had dared to take Richter's place in Munich in 1869, but who was now reconciled with Wagner and a keen Wagnerite himself. Also present at the 1876 Festival was an army of music critics from all parts of Europe and elsewhere. Most of them spent a great deal of their reports dealing with either the story of the *Ring*, or a description of the new Festival theatre, or just the social whirl and tittle-tattle associated with such an occasion, leaving them relatively little space for a serious assessment of the Festival's actual achievement. It was generally the case at this time that the conductor barely received more than a passing mention in the course of a review (those of Richter's Vienna concerts and opera performances omitted his name once he was established in the city). Joseph Bennett made an exception in his report for the *Musical Times*.

Herr Richter of Vienna filled the all-important post of conductor in a manner absolutely beyond reproach. I may say this the more emphatically, because I have never seen Herr Richter. He is known to me only 'by his fruits', and assuredly never did music so exacting receive such ample justice. It may be urged that the completeness of the performance arose from a multitude of rehearsals. Of course it did—otherwise Herr Richter and his men would have wrought a miracle.... Enough if I invite the reader to imagine all he can in the way of merit, and then believe that he has not done justice to Herr Richter's wonderful band.[26]

There remains the question of Wagner's ultimate feelings about Richter and his performances at the 1876 Bayreuth Festival. In 1878 he complained, 'I do not leave one person behind who understands my tempo,'[27] but as far back as 1865 he had told King Ludwig that 'apart from me no one [other than von Bülow] understands how to conduct'.[28] From this statement it is fair to deduce that, but for the obstacles between them of a domestic nature, von Bülow would have been the composer's first choice to conduct the *Ring*. Once it became obvious that the cycle would not soon be restaged in Bayreuth the question arose of a conductor for *Parsifal* in 1882. The solution then lay in the hiring *en masse*, for reasons of economy, of the opera orchestra from Munich. Ludwig took the choice of conductor out of Wagner's hands by insisting on Hermann Levi (he assisted Wagner in the rehearsal year of 1875 and the performances in 1876, and conducted the first *Ring* in Munich in 1878). In spite of Cosima's comment 'Richter not sure of a single tempo',[29] made when she and Wagner reflected on the summer of 1876, it was to Richter that she turned when the *Ring* was staged again twenty years later in 1896 and for six Festivals thereafter. Her differences with Richter and her knowledge that he would never subordinate himself to her nor tolerate any interference with his musical interpretation were all overruled by her strong desire by then to have around her anyone who had worked for her late husband.

Richter was, above all else, a musician whose musical backbone was strengthened by his phenomenal ear and his incredible practical understanding of all the orchestral instruments. There would be no more Franz Strausses to complain of unplayable passages. Wagner knew this, and his son Siegfried later said that 'he was no conductor in the manner of von Bülow. He always remained a musician.'[30] By that he meant what his father had always asserted, that a grasp of the drama was as vital as that of the music, and at this stage of his life Richter lacked that ingredient. If Siegfried disparaged the craft of musicianship he probably took his cue from his father, who, in the year after the first Bayreuth Festival, set his sights on another project, the creation of a new Royal Music School in the town of Würzburg. Writing to King Ludwig on the subject he added, 'as if we did not have enough "musicians" already!' In the same letter he advocated linkage between the new school and his Bayreuth Festival, in order to create a new performer specifically for his own purposes, the product of 'what I shall describe as the only true School of the new German music-drama';[31] their study of drama would have to be an integral part of their study of music. In 1879 he again told the king:

I am completely lacking a *musician* and a *dramatist*. I know of no conductor whom I could trust to perform my music correctly nor any singing actor whom I could count upon to give a correct performance of my dramatic characters, unless I first went through the part with him, bar by bar and phrase by phrase. The bungling incompetence that is to be found in every aspect of German art is without equal, and every compromise which I have occasionally sought to enter into with it has led me to the point at which my exalted lord and dearly beloved friend found me on the evening of the final performance of *Götterdämmerung* in Bayreuth, when I sat behind him, starting up violently on several occasions, so that my most precious friend was moved by his concern for me to ask me what was the matter? But it was too humiliating for me at that very moment to admit what it was that had reduced me to such despair, and to explain that it was my horror at realizing that my conductor—in spite of the fact that I consider him the best I know—was not able to maintain the correct tempo, however often he got it right, because—he was incapable of *knowing why* the music had to be interpreted in one way and not another. For this is the heart of the matter: anyone may succeed *by chance* at least once, but he is not aware of what he is doing, for I *alone* could have justified it by means of what I call *my* school.[32]

When Wagner wrote his review of the first Bayreuth Festival in 1878, he singled out the male singers, in particular Gustav Siehr as Hagen and Franz Betz as Wotan, for special praise. No mention of the ladies, and the petulant antics of Betz, who did not take kindly to the ban on curtain calls, were forgotten. He mentions Richter almost in passing, typical in itself but recalling his cry, 'Oh Richter! I had almost forgotten you,' at the 1875 rehearsals. It occurs when he describes the erection of a marble tablet in the theatre foyer on which are inscribed the names of the participants in the first Festival (apart from the orchestral players, who were highly offended when they discovered this tactless omission). 'I named . . . the conductor of the orchestra, my Hans Richter, who achieved the impossible, was much put to the test, and accepted responsibility for everything.'[33] Wagner would praise Richter to the heavens or damn him to hell, but whatever his moods may have led him to write or say, it is true that Richter's current lack of dramatic awareness was the reason for the composer's dissatisfaction with him, not his musicianship; it is a charge which can only be levelled at Richter at the age of 33. He had several hundred performances of operas by Wagner ahead of him in which to mature.

Three of these occurred after his return to Vienna on 3 September. Within a fortnight he had conducted *Fliegende Holländer*, *Tannhäuser*, and *Lohengrin*. For the rest of 1876 he continued his operatic duties including the first Viennese performance of Ignaz Brüll's *Das goldene Kreuz* and his own first interpretation of Meyerbeer's *L'Étoile du Nord*. The first four Philharmonic concerts of the season also took place, the first, on 12 November, including two re-orchestrations of baroque works (by Bach and Boccherini) and the second, on 26 November, notable for the first Viennese performance of Tchaikovsky's Fantasy Overture *Romeo and Juliet* and Brahms's participation as conductor of his own Haydn Variations. Jules de Swert was the soloist in his own new cello concerto at the concert on 10 December, which also premièred one of Robert Fuch's string serenades (Fuchs was highly thought of as a composer by Richter). The final concert of the year began with the first Viennese performance of Méhul's Overture *Horatius Cocles*, followed by Anton Rubinstein's Third Piano Concerto played by his pupil Vera Timanova, and ended with Schubert's 'Great' C major Symphony, Richter's first performance of this masterpiece. *Tannhäuser*, *L'elisir d'amore*, and *William Tell* (Rossini's opera was currently the work he had performed most since his career began in 1865) completed the year 1876, an *annus mirabilis* to which he bade farewell at Bayreuth on a two-day social call after Christmas.

9

1877

London

'NEXT year we will do it all differently,'[1] said Richard Wagner to Richard Fricke after the 1876 Bayreuth Festival, in spite of the fact that the composer had originally intended the cycle to be staged just once in a temporary theatre on the banks of the Rhine, after which both score and theatre would be burnt. Hans Richter had addressed the orchestra when they parted on 30 August 1876 with the words, 'he who is sincere will come back again next year!',[2] and he was greeted by a chorus of approval. When Richter left Bayreuth on 2 September 1876 for Vienna he recorded that 'the Master was very moved'.[3] Little did he know that it would be another twelve years, and five after the composer's death, before he entered the orchestral pit in Bayreuth again. In response to a request from the Leipzig Intendant Dr August Förster, who was eager to stage the *Ring*, Wagner had replied in September 1876, 'give me time to present my work once more in a carefully corrected form next year here in Bayreuth'.[4] The reason that the Festival theatre would not reopen its doors for another six years was the deficit of 148,000 marks (1½ million in today's money) incurred by the events of 1876. Wagner was now forced to consider any means by which he could recoup his losses. This included granting performing rights to other theatres, whose managers sensed they had him at their mercy. Jauner in Vienna, mindful of having released Richter and Materna to Wagner in 1876, secured the city's first *Walküre* in a series of nine performances conducted by Richter between 5 March and 12 April 1877. This was the first occasion, thanks to the introduction to the building of electricity, on which the start of the opera and the end of the intervals were announced by the sounding of an electric bell in the theatre. Leonhard Labatt sang Siegmund, Berta Ehnn Sieglinde, Emil Scaria was Wotan, and Amalie Materna Brünnhilde. The city's native son

Josef Hoffmann had made the designs for Bayreuth, and these were used again in Vienna. The production, by Jauner himself, was a triumph; even Hanslick's review was milder when he conceded defeat in the wave of the 'thunderous applause' and acknowledged Wagner as 'the darling of the public'.[5] Jauner employed the Empress Elizabeth's own riding master, who brought eight horses and their Polish riders to double as the eight Valkyrie. The distance reserved for them to gallop to Valhalla with the dummy dead heroes astride their mounts stretched from the exit to the Kärtnerstrasse to the exit to the Operngasse. After a few performances, and due to a variety of mishaps, the horses, apart from Brünnhilde's Grane, were withdrawn from the production. Jauner drove his deal home with Wagner by securing the complete *Ring* and giving the composer 10 per cent of the proceeds. Nine more performances of *Walküre* took place before *Rheingold* was staged in January 1878.

Edward Dannreuther, founder in 1872 of the Wagner Society in England, had also been active on Wagner's behalf and proposed a series of concerts in London in May 1877 at the Royal Albert Hall. After some considerable hesitation (due to his own bad experiences there in 1855 and reports of von Bülow's financial losses incurred by touring England) Wagner agreed, and having taken the decision to go ahead, he did so with his customary enthusiasm. Richter and Materna were enlisted from Vienna. The concert agents Hodge and Essex originally proposed twenty concerts, but this proved too ambitious and the number was cut to six. 'It is essential that you are with me in carrying out this project,' Wagner wrote from Bayreuth. 'Yes, without your help I could not consider staging these concerts. I hope this request, which I am also sending to the administration of the Court Opera, will meet with a favourable response, and I also ask you to obtain the parts for my earlier operas if they are not being used.'[6]

Richter was briefly in Bayreuth on 27 March and wrote in his diary, 'on Wednesday morning I was with the Master (Liszt was also present). I was sad when I saw how easily the Master was prepared to surrender *Rheingold*, *Siegfried*, and *Götterdämmerung* to the stage. With that act the great and pure artistic concept of the Bayreuth Festivals could remain dead for a long time.'[7] On 17 April he left for London, leaving Marie behind with Richardis and their new daughter, Ludovika Viktoria, born just five days before the Vienna première of *Walküre*. Hans was accompanied by Josef Medez, a singing pupil of his mother's, who travelled with him to London to assist in the administration of artistic matters (he later became an opera singer in Pest). After an overnight stay at the Charing Cross Hotel, the two men took lodgings at 46 Petersburgh Place in the Bayswater Road.

I have not really regained my senses yet, but I will try and give you an account of my journey and my short stay to date in London. We had quite a good journey as we had a coupé throughout just for the two of us. The journey along the banks of the Rhine as far as Cologne was wonderful. There we had a three-hour wait, which I used to seek out an old friend, the art dealer Lesimple, who founded the Wagner Society in Cologne. He took us to a fashionable gentlemen's club, where we were sumptuously entertained. At night we travelled on. At the station I saw my picture on sale, together with those of Niemann and Materna. In Brussels, where we arrived on Thursday at five in the morning, we had over two hours to look round the city in the beautiful morning weather; I also went to the theatre and wallowed in memories of seven years ago when I first conducted *Lohengrin* there. At 10.30 we were in Ostend. We had to wait a long time for the tide before the ship [the *Louise Marie* with fifty-three passengers] could sail. At first I felt quite well, but as the waves grew higher I was forced to give up my breakfast, together with everything else in my stomach, to the sea. Soon I was better. Some waves even came on board ship but I stuck bravely to the deck, although I was soaked through, for down below it was too steamy and revolting on account of others who were seasick. From Dover we dashed up to London, where we arrived at seven o'clock in the evening; I had regained my appetite once again. Later that night we went to our friend Dannreuther to gather information.

. . . I spent the whole of Saturday getting an insight into the huge bustle of this city. Yes, but how to find the words to describe it! We live in the vicinity of Hyde Park, which lies in the middle of the city and is roughly the size of the Prater in Vienna. There are several such parks in the city, and only due to them, in spite of the fog and smoke, is it possible to stay healthy. Eating and drinking is first-rate. Never have I seen nor eaten such lovely meat. The English way of life pleases me very much. Today, Sunday, all is very quiet in the city. We travelled to the Zoo by underground railway, which criss-crosses the whole of London and even goes under our lodgings. I was astonished by the good order and good manners of the public. I could soon get used to being here. Everything exists for the family, not the pub or the coffee house.

A start has already been made to promise me an engagement; but tell *no-one* about it. . . . Wagner arrives on 1 May. I live very cheaply here, only travelling by cab is expensive, something I could not deny myself during the first days as I wanted to see everything. . . . These few weeks will soon pass; I think it will work to my advantage in every respect.[8]

Richter suffered terribly from seasickness throughout his life and came to dread crossing the Channel. It was reputedly the reason that he never made longer voyages, such as to America, and it was even rumoured to be one of the reasons that he refused the appointment as conductor to the Boston Symphony Orchestra in 1893 as successor to Arthur Nikisch. Meanwhile Wagner sent a letter addressed to 'Mr John Richter Esqu.' which began,

'Oh, Richter! I hear you are in London? What are you doing there? Oh yes! I forgot! Yes, yes! I know now!' before becoming more serious.

Save what you can of *Tristan* to retain the totality of the programme. Unfortunately Seidl has gone off with the full and vocal scores, so I cannot really say how the matter is best resolved. I would have liked to do Brangäne's A flat ¾ aria, but an ending must be worked out after it. Perhaps you can do it; also the finale of the act from the entry of Marke onwards (Hill will be very good in this) must be done by the orchestra. Then the Prelude and Finale 1 (Act III). *Bon*! I'd have thought?? I swear this is the last time I have anything to do with singers and musicians.... *Adieu* old friend! Very best greetings to Wilhelmj, Dannreuther, and the two sons of the Rhine.'[9]

Rehearsals for the concerts began on the morning of 20 April with a wind sectional followed by the full orchestra in the first act of *Walküre*. 'The orchestra cheered me most heartily after the rehearsal.'[10] There were 169 players; of these, according to the *Athenaeum*, 'the wood, brass, and percussion included fifty-seven players (the wood trebled and the brass doubled) beyond the ordinary complement. To counterbalance this sonorous phalanx, there were 102 [*sic*, though in view of the following breakdown this figure should be 112] strings, divided into twenty-four first violins, twenty-four second ditto, fifteen violas, twenty violoncellos, twenty-two double basses, and seven harps.'[11] Dannreuther and Wilhelmj, meanwhile, were intent on getting Richter a concert with the Philharmonic Society during his London stay, but after he had first made an impression with the Wagner concerts. He was also introduced to the influential critic of *The Times* James Davison, whose reaction to Wagner was respectful but cool (Richter described him as 'the English Hanslick'[12]). Davison had been present at Bayreuth in 1876 and, according to Richter, 'received me most flatteringly and introduced me to an English lady as "the foremost conductor not only on earth but under the earth", which was a reference to conducting at Bayreuth'.[13] Negotiations for a Philharmonic Society concert continued (a fee of 200 thalers was even proposed) but eventually came to nothing. Richter never conducted for the Society, though he was given honorary membership in July 1905. He accepted two concerts for the 1908 season, but pulled out late in the day, which caused bad feeling. Myles Birkett Foster, a Director of the Society, compiled the records for the centenary in 1912 in which Richter appeared as a footnote. 'The writer had the honour of taking part in that [1877 Wagner] Festival, and well remembers the all-controlling power of Hans Richter, hidden behind Wagner's conducting desk, but really conducting everything; for Wagner, in the enjoyment of his own splendid creation, frequently forgot the baton altogether.'[14]

By the time Wagner arrived on 1 May (greeted at Charing Cross station by a deputation from the orchestra headed by Richter and Wilhelmj), the singers had also gathered in London. They included Josef Chandon (from Vienna), Materna, Sadler-Grün, Unger, Hill, and Schlosser, collectively described by Hans as 'a little Bayreuth'.[15] Richter's relationship with the orchestra grew warmer and closer. 'I have no trouble,' he wrote home. 'On the contrary I receive a lot of pleasure; they like me very much and I am greeted with applause at every rehearsal and cheered at the end. They all express the wish that I stay here.'[16] His photograph was sold to the public during the intervals of the concerts, and he was fêted at the German Club in London. He continued to report favourably on his impressions of London and the English way of life. 'It is not cold here, but always foggy; I have not actually seen the sun clearly yet. You have to wash every hour because your face and hands get blackened by the coal dust. But you easily get used to the English way of life, it is much steadier than ours. Children and adults look splendid here in spite of the fog and dust, the large gardens and broad streets let in a lot of fresh air to prevent the concentration of the smoke. And still that good nourishing meat!'[17] Rehearsals were also held for the chorus of the Deutsche Gesangsverein (German Choral Union) for *Fliegende Holländer*, whilst those for the vocal soloists took place at Dannreuther's house (12 Orme Square, Bayswater, where Wagner lodged). Richter's preparation of the orchestra had been exemplary. 'The Master is exceptionally pleased with everything; at today's rehearsal he gave me a kiss as a sign of his satisfaction in front of everyone and to the accompaniment of the cheering applause of the whole orchestra.'[18] That evening Richter joined the Wagners at the home of the Hungarian journalist Maximilian Schlesinger, who handled the fiscal affairs of the concerts together with Dannreuther, Wilhelmj, the painter Rudolf Lehmann, and the poet Robert Browning. The composer Hubert Parry, then a piano student of Dannreuther's, also kept a diary, and met Wagner at his teacher's house on 2 May. He attended several rehearsals and all the concerts, but his assessment of the composer as conductor differs from the views of other observers at the time.

May 4th: All the morning at the rehearsal at the Albert Hall. The hero was there and in good humour, and pleased with the band. They chiefly practised *Siegfried* second act and finale. Richter conducted wonderfully and drilled the incompetents with vigour.

May 5th: (Evening) . . . then to Dannreuther's, where there was a goodly company of artist folk to see Wagner, who was in great fettle and talked to an open-mouthed group in brilliant fashion. He talks so fast that I could catch but very little of what he said.

May 7th: All the morning at the rehearsal at the Albert Hall. Wagner conducting is quite marvellous; he seems to transform all he touches, he knows precisely what he wants, and does it to a certainty. The *Kaisermarsch* became quite new under his influence, and supremely magnificent. I was so wild with excitement after it that I did not recover all the afternoon. The concert in the evening was very successful and the Meister was received with prolonged applause, but many people found the *Rheingold* selection too hard for them.

[May 8th]: Tuesday rehearsal, part of the morning. It did not go well.

May 9th: Rehearsal in the morning, not very satisfactory as far as I heard it. . . . The concert in the evening was very successful and the Meister was well received. Hill singing as the Holländer was superb. The long first act of *Walküre* was a severe test on the public, without the assistance of the scenery or dramatic action, and many went out; but the applause at the end was great nevertheless.

May 12th: Went to the *Valkyrie*, it was a triumphant success. The great last act of *Walküre* was overwhelming and very few went out before the end. And the cheering and clapping was prolonged and enthusiastic. The *Walkürenritt* was encored bodily. That last scene is soul subduing and many people that I saw afterwards were as moved at it as I could have wished.

May 14th: We had the great Telramund and Ortrud scene from *Lohengrin*, in which Hill let himself out and surpassed everything I ever heard before for dramatic singing. . . . He half-acted it throughout and made me quite wild with wonder. He is a real genius—Materna also worked up at the end and gave us some fine bits of singing and warmth. Then Unger and Sadler-Grün followed with the lovely scene 'Das süsse Lied' and failed rather, Unger seeming quite out of voice and shambling. We had also the opening scenes of *Götterdämmerung* which were splendid. Wagner has compressed it for the occasion, curtailing the Norns' bit and cutting out Tagesgrauen and fitting it so into one scene, which is very effective. We had the introduction to *Lohengrin* which he takes very slow and quite teased the band. At the concert in the evening Unger shortly showed that his voice had utterly collapsed. The beautiful scene in *Lohengrin* was quite painful, and at the end of the first part a general change of the programme was determined on—all *Siegfried* had to be missed out; and the short second part consisted only of the *Walkürenritt*, which was again encored, and the opening scenes of *Götterdämmerung* as in the morning, in which Unger cut a very poor figure—after that Wagner cut short the proceedings by taking Materna by the arm and walking straight out.

May 15th: To rehearsal. The last scenes of *Götterdämmerung*. Didn't go well and the wind had to be drilled alone. Unger did not appear at all.

May 16th: A day to have lived for. The concert was a perfect triumph. They were in great difficulties about it as Unger and Hill were both unable to sing from hoarseness, so the programme presented difficulties, but the result of the arrangement is the best concert by far we have had. A great deal of *Meistersinger*; of which the lovely opening to the third act was encored and Brünnhilde's great scene at the end of *Götterdämmerung* which was quite splendid, and best of all Siegfried's Tod, which seems to me the greatest thing in the world and made me quite cold with ecstasy. The applause was tremendous for nearly a quarter of an hour after the concert, shouting and clapping renewed again and again.

May 18th: [*Tristan* rehearsal] Wagner got into a charmingly unsophisticated rage with some of the band for beginning badly, threw down his baton and seized his coat and comforter and put them on . . . and walked up and down the platform in front of the orchestra till time and the appeals of those of the orchestra more in favour had cooled him down a bit.

[Concert] There was tremendous enthusiasm and prolonged cheering at the end, and addresses were read to Wagner, Richter, and Wilhelmj. Wagner was crowned and Richter received an ivory baton [Wilhelmj was presented with a violin bow] and the great trio embraced amid renewed cheers.[19]

Richter's notes in his diary fill in some of the details and take a different view from Parry. In the second concert on 9 May in the first act of *Fliegende Holländer* he described 'great uncertainties; the musicians often did not understand the Master's beat. Hill complained bitterly about the slow tempos and angrily wanted to leave; only with great effort could Wilhelmj and I persuade him to sing. The impression made upon me by the Master's conducting was of complete insecurity and—unbelievably—ignorance, or even more, total obliviousness of his own works. Thereafter the Master handed the baton over to me, as he became aware of the friction amongst the performers. Half jokingly he said to Wilhelmj, "It seems I have upset you nicely!" '[20] The *Musical Times*, in common with most reviews, noted Wagner's discomfort on the podium. '[Wagner came] to personally conduct his own music, a task for which he knew himself, as all of us now know him, to be unfitted. . . . The master, great as he is in other respects, is a poor conductor, equally lacking spirit and the power of control . . . but now is the time to acknowledge the very valuable services of Herr Richter, the Wagnerian conductor *par excellence*. Whenever the baton fell from the nerveless hand of the master, Herr Richter took it up to retrieve the fortunes of the day. And right well he did this. New life appeared to animate the orchestra, every man of whom seemed to be in a measure inspired.'[21]

On the same day George Bernard Shaw wrote: 'Herr Wagner, as a conductor, must be very unsatisfactory to an orchestra unused to his peculiarities. He does not . . . lack vigour, but his beat is nervous and abrupt; the player's intuition is of no avail to warn him when it will come; and the tempo is capriciously hurried or retarded without any apparent reason. Herr Richter, whose assumption of the baton was hailed by the band on each occasion with a relief rather unbecomingly expressed, is an excellent conductor, his beat being most intelligible in its method, and withal sufficiently spirited.'[22] Another observer of the Wagner Festival was Hermann Klein, who wrote about Wagner's attempt at the *Fliegende Holländer* Overture which had so surprised Richter.

I was introduced to [Hans Richter] by Franke, and his hearty handshake and the open, fearless expression of his eyes prepossessed me in his favour. Thick-set and broad-shouldered, slightly below medium height, his beard square-cut and of a golden-brown tinge, he wore spectacles and spoke with a boisterous vivaciousness in a broad Viennese dialect. He had then just turned thirty-four. This was the man who was to save the Wagner Festival from disaster, to create for us a new orchestral standard, to give us fresh ideas on orchestral conducting, and, generally speaking, to change the course of musical taste in this country. . . . I had never seen Wagner conduct, and was anticipating it with the liveliest curiosity. Would the greatest living authority on the art of conducting prove himself an equally great exponent thereof? . . . Why, then did he hesitate to raise his baton? Yet hesitate he did; or rather he raised and lowered it twice or three times before he actually gave the signal for the start. . . . Twice a fresh start had to be made. Wagner grew more nervous and flurried every instant. . . . His next gesture was to beckon to Richter who was standing near. . . . He then went down from the platform to the arena and sat there facing the orchestra, looking very glum and dejected, whilst Richter continued the rehearsal. The change was magical. Under that strong, clear, unhesitating beat the very tone of the instruments seemed to take on a different hue; the animation of the players revived as their spirits rose, and the music went splendidly.[23]

The *Athenaeum* was not well disposed towards Wagner. Of the first concert it wrote: 'Whether the composer was fatigued, or felt he had no longer the sympathy and support of his previously rapturous listeners, we cannot say, but after Loge, the cynical God of the Walhalla, had his solo, the baton was taken by Herr Richter, the capital conductor of the Imperial Opera house at Vienna.'[24] The paper's opinion of Wagner's conducting grew more critical a week later. 'At the present period of this so-called "festival" . . . it is quite evident that the heart of the composer is not in his work as conductor of these concerts; he has grown more listless and apathetic apparently at each successive performance, and when Herr Richter takes the baton, a new spirit actuates the band, and the compositions obtain really animated interpretations.'[25]

Some years later Shaw had occasion to reflect upon the Festival and its main protagonists Wagner and Richter.

Herr Richter's popularity as an orchestral conductor began not in the auditorium, but in the orchestra. It dates from his first visit here in 1877 to conduct the Wagner festivals at the Albert Hall. At these concerts there was a large and somewhat clumsy band of about 170 players not well accustomed to the music, and not at all accustomed to the composer, who had contracted to heighten the sensation by conducting a portion of each concert. It is not easy to make an English orchestra nervous, but Wagner's tense neuralgic glare at the players as they waited for the beat with their bows poised above the strings was hard upon the sympathetic men, whilst the intolerable length of the pause exasperated the tougher spirits. When all were effectually disconcerted, the composer's baton was suddenly jerked upwards, as if by a sharp twinge of gout in his elbow; and, after a moment of confusion, a scrambling start was made. During the performance Wagner's glare never relaxed; he never looked pleased. When he wanted more emphasis he stamped; when the division into bars was merely conventional he disdained counting, and looked daggers—spoke them too sometimes—at innocent instrumentalists who were enjoying the last few bars of their rest without any suspicion that the impatient composer had just discounted half a stave or so and was angrily waiting for them. When he laid down the baton it was with the air of a man who hoped he might never be condemned to listen to such a performance again.

Herr Richter let slip the secret that the scores of Wagner were not to be taken too literally. 'How', exclaimed the average violinist in anger and despair, 'is a man to be expected to play this reiterated motive, or this complicated figuration, in demisemiquavers at the rate of sixteen in a second? What can he do but go a-swishing up and down as best he can?' 'What indeed?' replied Herr Richter encouragingly. 'That is precisely what is intended by the composer.'[26]

Towards the end of the series it became apparent that financially all was not well. The inexperienced agents Hodge and Essex (whose main claim to fame was the sale of American Estey organs) had failed to spot that as many as 2,000 of the seats or boxes in the Albert Hall were privately owned, which meant that no income from their sale would reach Wagner. This eradicated a potential £12,000 and reduced the net income of the venture to £700, despite an extension of the Festival by popular demand with two extra concerts on 28 and 29 May. Richter sensed the unease felt by the orchestral players when, instead of cash, they began to receive money-orders or bank drafts from Hodge and Essex. Their restiveness in turn unsettled Wagner, who snapped and lost his temper at the slightest provocation, and, having handed the baton to Richter at some point during each concert, would sit in an armchair facing the audience and glower at them, as if daring any brave soul to criticise what he was hearing. Richter, on the other hand, kept a cool head and distanced himself from events, thus endearing himself even more to one and all, and Wagner, to his credit, covered the singers' fees of £1,200 from his own pocket. Richter was paid £100 by Hodge and Essex for the Wagner Festival, of which he took £51 back home with him. The rest was spent in London on trips and social engagements.

I got to know a marvellous old gentleman, a Rhinelander by the name of Schwaben. I have never eaten so well as at this man's house. Fresh cherries, grapes, heavenly peaches, incomparable asparagus, pineapples, the most superb wines. On Tuesday the choral society celebrated Wagner's 64th birthday with a banquet. It was wonderful. Curiously enough, ladies were not permitted into the hall, but had to observe our enjoyment from the gallery. Even Frau Wagner was up there. After the Master, I was toasted with thunderous applause by the full hall and made a fitting reply. The next day I was once again the guest of Herr Schwaben, who took me to the London docks. Ships arrive there from all parts of the world to load and unload their wares. In the underground wine cellars we tasted the best wines, we inspected a large ship which was off-loading lead and at the same time taking gigantic cannon shells for Gibraltar on board. Then we drove among the alley-ways leading from the banks, which was extremely fascinating. The largest riff-raff in the world live here, the dross of humanity, drunken women, rank thieves and robbers; as we drove through, a young man was stabbed in the throat and the assailant arrested by a policeman. It would be a rash act to go into this area in daylight without a police escort. Then we travelled on the Thames down to Greenwich to eat a fish supper. This is a speciality of London. Only fish is served, but in an incredible variety of sorts and preparation. Late at night we travelled home by train over the roofs of London. Everything was free, for Herr Schwaben would not permit us to pay for anything. It must have all cost at least 150 florins. It was a real life of idle luxury. Nevertheless I am looking forward even more to my rice soup in Vienna, for then I will be back again with you all![27]

On 30 May Richter left London for Vienna, where he conducted several performances at the Opera throughout June, including *Oberon* and *The Merry Wives of Windsor*. On 17 July he travelled with his friend Viktor Schembera, a Viennese poet and journalist, to Switzerland. An unexpected pleasure awaited the pair in Lucerne, as he told Marie:

Surprise! Surprise! We arrived at 7.05 on Wednesday evening in Lucerne. I immediately looked the people up who used to run Tribschen for the Master, and discovered from them that Wagner was arriving in Lucerne at 8.35 that very evening. Naturally we hurried off to the station. The first person Wagner saw when he stepped down from the carriage was me. Enormous pleasure on his part and no less so from his children and wife. He told me he had been thinking of me the whole day, and how nice it would be if we found one another together in Lucerne. He wanted to telegraph me from Ems [where Wagner had been staying since his departure from London], but did not know my summer address. You can imagine everyone's joy! He said, 'Na, Richter. God meant us for one another!' . . . We spent the whole of Thursday in his company. At midday we ate together with him and his family (except Loldi) and in the afternoon we all travelled to Tribschen, and visited the house in which *Die Meistersinger*, *Siegfried*, and *Götterdämmerung* were composed. [In his diary Richter added, 'In the afternoon we went by boat to Tribschen, which was occupied by a French family. Oh how different it all looked there! A vocal score of *Les Huguenots* lay on the piano.'[28]] He and Frau Wagner were in a charming mood. We left late in the evening. Early today, Friday, the Master together with his family left for Munich at six o'clock. Naturally we accompanied him to the station. Once again they were wonderful hours that we spent with him. I was most pleased for Schembera, whom he got to know, and also with the quite different impression I had of Frau Wagner. You would be amazed![29]

From Lucerne Richter and Schembera were joined by another former school friend in Vienna named Doleschal, now a bookseller in Lucerne. Together they travelled to Lake Maggiore in Italy via the St Gotthard Pass, where they saw the railway bridge (known as the Devil's Bridge) and tunnel currently under construction to provide a direct rail link between Switzerland and Italy. On their early morning upward climb the three men had to make do with a post coach drawn by six horses, with conditions harsh enough to force the passengers to walk alongside the coach. Once over the top the team of horses was reduced to two and they galloped down the other side. The journey was completed by train and ship to the island of Isola Bella. The three companions were so happy here that they postponed plans to travel on to Genoa, and instead spent two more days sightseeing. Hans found an excellent organ in a tiny church and was soon playing it to the delight of the 150 island inhabitants, mainly poor fishermen. Two days were spent in the port of Genoa, where a swimming expedition took them so far out to sea that his two friends became seasick in the swell of the waves (Richter for once was not). That evening they visited the town's opera-house and heard Ricci's *Crispino e la Comare*, but were unimpressed. Hans returned to Vienna at the end of July, and by mid-August he was hard at work with performances at the Opera.

Richter never worked for Wagner again during the composer's lifetime.

The two men met just once a year during the four years 1878, 1879, 1881, and 1882, when Richter paid brief visits to Bayreuth. The few letters from Wagner during these years were merely notes which mostly berated him for not writing, or were complaints of press rumours which upset him. Richter, on the other hand, was hard at work not only consolidating his position in Vienna, but also in London, where his impact during the spring of 1877 had been enormous and comparable to the one he had made in the Austrian capital in 1875. The result of the Viennese concert for the Wagnerverein had been his move to the city and a lifelong career there; the outcome of the Wagner Festival in London would bear fruit two years later with the first of the annual Richter Concerts in 1879.

10

1878–1879

Vienna

AFTER the London Wagner Festival of 1877 Hans Richter concentrated his musical activities in Vienna, adding London after 1879 and Bayreuth after 1888 (with occasional forays to such places as Birmingham and the German venues of the Lower Rhine Music Festival). Vienna was his domicile, where his family was based and his children born and educated. It was here that his living had to be earned and the mouths of six children (all born between 1875 and 1882) filled. His post of Hofopernkapellmeister at the Opera was not well paid, and to increase his income Richter gradually acquired other posts in the city. The first of these was Vize-Hofkapellmeister (or Deputy Conductor to the Court Chapel) at the end of 1877, when a vacancy occurred upon the death of Johann Herbeck.

Here was another example of Richter returning to his musical roots. Already he was conductor (Hofoperntheater-Kapellmeister) to the opera orchestra in which he had begun his musical career as a horn-player at the Kärntnerthortheater, now he was rejoining the Court Chapel where he had been a choirboy twenty-three years earlier. Herbeck had also held a multitude of posts in Vienna, among them Vize-Hofkapellmeister from 1863 and Kapellmeister upon his promotion three years later. The domino effect of his death on 28 October 1877 was that Josef Hellmesberger sen. succeeded him as Kapellmeister, thus vacating his own post of Vize-Kapellmeister, for which written applications were received from the Court organists Anton Bruckner and Rudolf Bibl, from the Director of the Singakademie Rudolf Weinwurm, and from Ludwig von Brenner, a native of Vienna but currently Music Director in Berlin. Verbal applications (made after discreet soundings had been taken by Court officials) were received from Hans Richter, Pius Richter, and Josef Hellmesberger jun. for the two posts. A letter of recommendation from Prince Konstantin zu Hohenlohe-Schillingsfürst was sent to Emperor Franz Joseph, as a result of which Hellmesberger sen.'s promotion was formally ratified on 11 November 1877, and Hans Richter's appointment followed on 19 November. Pius Richter, a Court organist and no relation of Hans, was awarded an honorary Vize-Kapellmeistership. Ironically Richter, in his capacity as Hofoperntheater-Kapellmeister, conducted a chorus assembled to sing at Herbeck's funeral in St Peter's church in Vienna on 30 October 1877. Richter's conducting duties at the Court Chapel commenced on the third Sunday in Advent, 16 December 1877, with Haydn's *Missa brevis* in B flat, the Gradual 'Ora pro nobis' by Ludwig Rotter, and Mozart's 'Sancta Maria' as the Offertory. The choir consisted as always of boys from the Court Chapel augmented by men from the Opera. Vocal soloists and orchestral players came from the Opera. For the next twenty-three years, Sundays would often consist of conducting the music for the service at the Court Chapel until midday, followed either by a Philharmonic concert or an evening performance at the Opera, on occasion even all three. On 20 March 1893 Richter in turn succeeded Hellmesberger, and remained in post as Hofkapellmeister until 24 March 1900, his salary set at 1,800 florins with an increase to 4,000 florins in 1900. He also received a subsistence allowance of 500 florins which was doubled in 1900, and a five-yearly bonus of 200 florins. In 1898 he received 150 florins as an allowance for a special uniform for mourning the death of the Empress Elizabeth.

Besides organizing, rehearsing, and conducting the music for the service each Sunday in the Court Chapel, the Hofkapellmeister was responsible for organization and discipline among the choristers and the orchestral players, and was often obliged to explain the behaviour of his musicians to the Court officialdom. On one occasion two violinists, Arnold Rosé and Josef Hellmesberger jun., absented themselves from a service. It was Richter's responsibility to explain their actions and to carry out the judicial verdict of the administration. Rosé had taken umbrage because he had not been given a solo that day and Hellmesberger refused to sit next to Jacob Grün (currently leader of the Philharmonic orchestra). Richter pleaded on their behalf for a written reprimand and warning that such behaviour would meet with instant dismissal in the future. Although this was acceded to, Richter's own actions were not viewed sympathetically by the authorities, who considered he had no understanding of the gravity of the original misdemeanours. The habit of players and choristers arriving up in the choir balcony during the service after the sermon was stamped out by Richter.

Everyone had to be in position, their music on the stands, candles lit, before the service began, and everyone had to wait for members of the Court to depart before they themselves left. Today, despite the absence of royalty, services are still held in the Chapel, a tall narrow building of several storeys and with boxes inset into the walls. The conductor now stands at the edge of the choir balcony with his back to the altar; in Richter's day this was not so. He sat facing his choir and players who were arranged to either side of him, with the singers placed to the front up to the balustrade of the balcony and the orchestral players alongside or behind him on both sides. His chair was raised on a podium and he looked straight ahead at the altar. The organist sat behind him at the instrument, which is set against the back wall. The player (often Bruckner) depended upon mirror contact with conductor and priest.

The writer John Ella described his impressions of the Imperial Chapel choir after a visit to the city in 1866.

The band, led by Hellmesberger, including the *élite* of the profession, consists of twelve violins, four violas, three violincellos, three double-basses, and a complete set of wind instruments, with drums. Never overpowering the voices in accompaniment, yet this little orchestra, with the aid of an organ most skilfully played, in the choruses, is quite powerful enough for the size of the chapel. Of the fourteen masses I heard, on Sundays and Saints' days, at this chapel, by Salieri, Haydn, Mozart, Cherubini, Beethoven etc. in the winter of 1866–67, the Requiem of Cherubini and Beethoven's Mass in C made the deepest impression on my feelings.[1]

There was a problem concerning the respective singing styles of soloists at the Opera who also sang in the Chapel. Richter's belief that singers should earn their place in the Chapel choir only through merit came to a head in 1896 when the singers Hermann Winkelmann, Friedrich Schrödter, and Franz von Reichenberg were denied confirmation of their trial status on Richter's recommendation because they could not master differences between the operatic and church styles, 'which must remain separate'.[2] Curiously the opera administration took the view that the three singers would revenge themselves on Richter by reneging on their opera contracts (their value as Wagnerian singers was highly prized), and opposed his action, but Richter insisted and threatened the Chapel administration with his own resignation over the affair. 'What can a general do with soldiers who cannot handle weapons?', he asked.[3] The emperor himself was embroiled in the affair, which by now had caught the attention of the Viennese press, but, after about a year, Richter eventually won the day. On another occasion Richter banned the bass Ludwig Weiglein from solo appearances with the Chapel choir after several incidents of drunkenness.

All musicians' appointments were confirmed only after a lengthy trial period during which they were called *Exspectanten*. This system was threatened with abolition in 1898, and Richter came vigorously to its defence. He argued that a consequence of change would be 'that a well-trained ensemble would be destroyed and in its place would arise a continually changing, unreliable collection of choristers and orchestral players, who with time-consuming and costly rehearsals would not be well placed to fulfil their Imperial duties'.[4] As an example he went on to cite the demands made upon the members of the Hofkapelle in Easter week and its associated services. 'Nothing would be achieved by rehearsals in this case, as the music for these services is not to be found in scores or parts, but from liturgical books, a knowledge of which is only acquired through time and experience.'[5] Richter was a strict disciplinarian, both revered and feared by the choirboys, adult singers, and players alike. Yet he was not humourless despite a gruff manner and his outwardly intimidating appearance, created by his Barbarossa-red beard and clear blue eyes behind thin, oval-shaped glasses. Spotting Brahms and Bruckner among the congregation below, though seated far apart from each other on either side of the Chapel, Richter whispered to the choir, 'Look down there lads, those two would much rather gobble each other up.'[6]

In 1875, at the time of Herbeck's resignation from the Opera and Richter's arrival, another move was played in Vienna's game of musical chairs when Brahms resigned from his Directorship of the Gesellschaft der Musikfreunde. His last appearance was on 18 April that year when he conducted a cut version of Max Bruch's secular oratorio *Odysseus*. The Directors of the Society immediately turned to Richter and offered him the vacant post. Richter telegraphed from Budapest on 30 March 'on acceptable terms, prepared to accept honourable post', which he followed up the next day with 'free in the autumn, possibly earlier if necessary. Pest contract not renewed from my side. Request written detailed account.' The Society responded on 1 April and asked that he should keep himself free on any account, only to follow up a week later with a complete volte-face. During the course of that week Herbeck had changed the situation. He had been their Director from 1859 until 1870, but now offered himself again after his resignation from the Opera. The Committee felt obliged to offer him his former post, and shamefacedly wrote to Richter withdrawing their offer to him but hoping for continued good relations. Richter, for his part, acted with full honour and magnanimity. Despite further attempts to lure him after Herbeck's death in 1877, it was November 1884 before he took up the appointment in succession to Wilhelm Gericke, and held it until April 1890. The concerts (on average four each season with occasional additions) were centred on the Singverein, the choral association operated by the Society, and with which he was now committed to a weekly rehearsal. This had not always been the case; Brahms had initially refused his appointment because the orchestral and choral activities were not unified. Richter received a concert fee of 500 gulden which included his weekly sessions with the choir. The repertoire for the Gesellschaft concerts during those years makes fascinating reading.

Richter performed the choral classics of his age, from the works of Bach, Handel, Haydn, and Beethoven, but also premièred the newest works of Dvořák (*Stabat mater*), Bruckner (Te Deum), Robert Franz (Psalm 117), Arnold Krug (*Die Maikönigin*), and Friedrich Kiel ('Es gibt so bangen Zeiten'). A startling omission is Mozart, just the Offertorium 'Venite populi' for double choir. For the bicentenary in 1885 of the birth of Bach he gave the first complete performance in Vienna of the Mass in B minor on 31 March. This was preceded on 26 February by a performance of *Saul* to celebrate Handel's bicentenary in the same year. In other years there was also Handel's *Joshua*, *Theodora*, and *Ode to St Cecilia*, Bach's Magnificat and *St Matthew Passion*, the first complete *Christmas Oratorio*, and some of his motets and cantatas. Another first Vienna performance was Berlioz's Te Deum at Richter's second concert on 14 December 1884. Haydn was represented in Richter's period by *The Creation* and *The Seasons*, Mendelssohn by Psalm 115, Psalm 43, *St Paul*, Schumann by his *Scenes from Goethe's Faust* and *Paradise and the Peri*, Schubert by his *War Song of Miriam* and excerpts from his opera *Fierrabras*, and Berlioz by the Requiem and *The Damnation of Faust*. His opening concert on 23 November 1884 included not only the première of Kiel's work, but also the newly discovered *Cantata on the Death of Josef II* by Beethoven, whose *Missa solemnis* received two hearings besides a complete performance of *The Ruins of Athens*. In addition to celebrating the births of Bach and Handel, the society also marked the centenary of Weber's birth in 1886 with a concert on 21 November consisting of the *Konzertstück* for piano and orchestra and the choral pieces *Hymne* and *Kampf und Sieg* (Battle and Victory), written to celebrate the Battle of Waterloo in 1815. The tercentenary of the birth of Heinrich Schütz was the reason for including his *Seven Last Words* on 10 January 1886, and in the same year Liszt's death in the summer was mourned in a memorial concert on 12 December with *Die Ideale*, the Second Piano Concerto, Psalm 13, and the Fourth Hungarian

Rhapsody. Brahms also featured in Richter's programmes with the Alto Rhapsody, German Requiem, and *Triumphlied*. Among orchestral works were Joachim's *Overture in Memory of Heinrich von Kleist*, the première of Joachim Raff's Second Violin Concerto, and the first Viennese performance of Liszt's Second Hungarian Rhapsody.

By 1885 Hans Richter was thus predominant in the four organizations which formed the pillars of Vienna's musical life. Two were Court-based, the Opera and the Court Chapel, two were civil-based, the Philharmonic concerts and those promoted by the Gesellschaft der Musikfreunde; they were four autonomous musical institutions with a good number of the singers and players participating in more than one organization (for example the opera orchestra then, as now, gave the Philharmonic concerts). His predecessors Dessoff, Herbeck, and Hellmesberger sen. had all adopted a similar pluralistic approach to their careers. Richter, however, did not include teaching at the Vienna Conservatoire as part of his musical activities, as they had. The Conservatoire, founded by Salieri, was housed together with the Gesellschaft der Musikfreunde (of which it was an offshoot) in the Musikverein building. Richter had one tangential contact with the educational institution when he was appointed to the jury of the Beethoven prize, a sum of 5,000 gulden awarded to a student composer. The jury members consisted of the Director of the Conservatoire (Hellmesberger sen. from 1851 until 1893), a conductor from the Opera and one from the Court Chapel, the Director of the Gesellschaft concerts, and three composers. These often included Brahms and Goldmark, as, for example, when Richter joined the committee on 15 December 1881. Among the unsuccessful applicants that day was Gustav Mahler, then a 21-year-old graduate of the Conservatoire studying at the University. The work he submitted was *Das klagende Lied*.

The names of Brahms and Bruckner have now appeared. Their presence in the city created a polarity of opinion which caused much dissension, discussion, and disunity. Wagner was, predictably, the reason for the controversy. Little of the fight was personally waged by either composer, despite an unattractive streak in Brahms's character. This manifested itself in the Hans Rott affair of 1880, when an already unstable young composer was actively discouraged from his chosen career by Brahms, to whom he had turned for advice and encouragement. Further deranged, Rott was placed in an asylum where he died four years later. Bruckner never forgave Brahms for his part in Rott's decline. Brahms, a north German uneasily domiciled in Austria, always considered attack the best form of defence. Although he never lost his respect for Wagner, it was the blind loyalty of his adherents that he could not abide, and it was they whom he attacked. Among them were to be found the youthful composers Mahler, Rott, and their third companion Hugo Wolf, all of whom were active in Vienna's Wagner Society founded in 1872 by the 16-year-old Felix Mottl. Their activities, their loyalties, and above all their compositions were equally frowned upon by Hellmesberger, the ultra-conservative Director of the Conservatoire, and by several of the teachers of composition there such as Johann Nepomuk Fuchs. On the other hand Bruckner, teaching at the University, was a devoted admirer of Wagner and attracted the forward-thinkers among the new generation of composers and musicians. Besides those reactionaries amongst the musical institutions in Vienna there was also the press, led by the arch conservative of them all, Eduard Hanslick. He loathed Wagner, derided Bruckner, and considered Brahms to be the torch-bearer of Classical Romanticism as derived from Beethoven. The Music of the Future, or the New German School, had to be resisted at all costs.

Hans Richter stepped into this maelstrom and dominated the musical scene of Vienna. He was a man who was closer to Wagner in musical terms than anyone save Cosima, and yet he was accepted by the conservative musical institutions of the city. In the spring of 1877 he was conducting the Wagner Festival in London; at the end of the same year he was presenting Brahms's new Second Symphony to the world for the first time. He was walking a tightrope of diplomacy, trying to keep everyone happy and maintain his neutrality in musical affairs. Although he would never work for Wagner again, he was to prove his most ardent ambassador in England, by including, or even filling his concert programmes with, Wagner's works suitably adapted for performance away from the opera-house. In Vienna, however, he was more discreet and took an impartial line which is apparent from his concert programmes. In his first decade as conductor of the Philharmonic concerts (1875–85) Wagner's name appears only four times on Richter's programmes (though he was very active on his behalf in the city's opera-house). He would prove equally loyal to Brahms, and even more so to Dvořák, who needed to establish himself from Prague. Brahms's Second Symphony was given for the first time on 30 December 1877 at the fourth Philharmonic concert of the season, and was played after Mendelssohn's Overture *Ruy Blas*, and three movements from Mozart's Serenade for thirteen wind instruments, the programme concluding with an orchestral arrangement of a prelude, chorale, and fugue by Bach. The symphony was given a triumphant reception, the Scherzo having to be repeated by public demand. Brahms had not had such a success with his dramatic First Symphony, which was performed a year earlier in Vienna, but the contrast in moods between the two works was instantly grasped by the public, who responded in typical Viennese fashion to the sunnier mood of the more pastoral Second Symphony. Hanslick used the occasion to show the Wagnerites that the symphony as a musical form was not dead. He applauded Richter, 'who conducted the symphony at the express wish of the composer. He had studied the work with loving care and performed it to perfection, which does him full honour.'[7]

Richter's views and activities on Bruckner's behalf can be described as more equivocal. Eventually he came to admire his works, but initially he was sensitive to Hanslick's disapproval and gave a wide berth to the composer's symphonies. With Herbeck's death Bruckner had lost the only true champion in Vienna who could do anything for him at that time. Richter probably met Bruckner for the first time in 1875. The composer first mentions his name in a letter to Moritz von Mayfeld. 'My Fourth Symphony is finished. The "Wagner" Symphony (D minor) is significantly improved. The Wagner conductor "Hans Richter" was in Vienna, and is reported in various circles to have said how glowingly Wagner speaks of it. . . . Richter is rumoured to have said that he wishes to perform the D minor Symphony in Pest.'[8] At the time (December 1877) when Richter was about to première Brahms's new Second Symphony, Bruckner expressed his astonishment at the conductor's capacity to ingratiate himself with Wagner's strongest opponents. He voiced his complaint in a letter to the conductor Wilhelm Tappert in Berlin dated 12 October 1877, little knowing that by the end of the same month his greatest champion in Vienna, Herbeck, would be dead. The implications of Herbeck's death were inextricably linked to the 'Wagner' Symphony. On 16 December, just two weeks before Brahms was to enjoy his triumph, poor Bruckner experienced his own disastrous first performance of the revised version of his Third Symphony. After each movement the hall emptied more and more until only twenty-five people (seven of those in the stalls) remained in the hall. They were mainly his loyal composition students and included Mahler and the Schalk brothers. It was a disgraceful débâcle, with members of the orchestra deliberately playing wrong notes either to show their own opinion of the music or to register their protest at the composer's inept conducting ability. This concert was promoted by the Gesellschaft der Musikfreunde and Herbeck would have conducted the work, but with his death Bruckner was forced to do it himself. Hellmesberger conducted the first half, but there was no question of his doing Bruckner's work. He could not abide the composer's symphonies, although he later encouraged a performance after a gap of thirteen years of the Mass in D minor and secretly, for fear of incurring Hanslick's wrath, expressed his admiration for the F major String Quintet. Richter had good reason to avoid offending both Hanslick and his former teacher Hellmesberger.

The viola player Sigmund Bachrich drew attention to the diplomatic manner Richter adopted in the chauvinistic atmosphere of musical Vienna. The way in which the public could best demonstrate their various allegiances was manifested in the applause (or booing) which broke out at the end of performances of new works, an important adjunct to which was the appearance on stage of the composer. Bachrich compared the three composers continually at the centre of public interest, Brahms, Goldmark, and Bruckner.

One could take for granted that, at the performance of a new work by Brahms, the Protestant community would set to work, whereas with Bruckner the so-called Christian socialists would break out in thunderous applause. I do not wish to imply that political beliefs had crept into the concert hall, but it could not be overlooked that each party chose their leader in defiance of the other. After the last chord of a newly performed work the public manifested its desire to see the composer onstage with persistent applause. Having got him there each party sought more. With Brahms this was no easy matter. He heard his new works from some hidden nook in the hall, and before the end crept out. Only the vigilance of the orchestra's stewards ensured he was restrained from leaving. Richter had to get hold of him and drag him gently but firmly on to the podium. Things went more smoothly with Goldmark, although he too was reluctant until he appeared before the public. Only good old Bruckner was always ready and in position. No sooner had the noise erupted than he was there, acknowledging it unendingly in his childlike, naïve manner. We in the Philharmonic had already packed away our instruments, but Bruckner was still being lauded by his supporters. Hans Richter, however, maintained his nonalignment to any party. All that mattered to him was the work, whose beauty he had the sensitivity to understand and in whose performance he knew how to achieve a perceptual plasticity. I have played under many conductors, and also seen many modern [1914] ones on the podium but I must confess that—with the exception of Wagner—none has made such a truly great and artistically serious impression upon me as Hans Richter. If I were to try to recall all his great musical deeds here I would know with which I would begin, but I would truly not know with which I could end.[9]

In 1878 Richter's operatic repertoire consisted mainly of Meyerbeer, Weber, Boïeldieu, Auber, Bizet, Gounod, Rossini, Ignaz Brüll (*Das goldene Kreuz*), Nicolai, Cherubini, and the Wagner operas (Vienna's first *Ring* proceeding apace with *Siegfried* added to the list on 9 November). In spite of a few cuts, *Siegfried* took five hours to perform largely on account of Jauner's insistence upon intervals in which hot food was available. When *Götterdämmerung* followed on 14 February 1879 it too was cut, Wagner having sanctioned the cuts and provided them himself for Richter in a realistic approach to performances away from Bayreuth. The works of Wagner could be said to have arrived in Vienna when Richter conducted the final scene from *Meistersinger* at the command of Emperor Franz Josef to celebrate his silver wedding anniversary on 24 April 1879 (as an Imperial Chapel choirboy Richter had sung at that wedding in 1854). If the operatic and orchestral repertoire of 1878 was not particularly innovative, the records Richter kept of the artists who sang or played for him were of greater interest. There was a Verdi Requiem with the Swedish Christine Nilsson, Zélia Trebelli, Angelo Masini, and Karl Mayerhofer on 13 April, Pauline Lucca sang in *L'Africaine* two days later, and Amalie Materna (Brünnhilde), Emil Scaria (Wotan), and Ferdinand Jäger (a new and impressive Siegfried) were the main protagonists in the last two operas of Wagner's tetralogy. Richter shared the podium with Luigi Arditi, who conducted the Italian repertory (including Adelina Patti in *La traviata*). He also encountered Delibes, who came to conduct his ballet *Sylvia*, and Anton Rubinstein, who conducted his opera *Die Makkabaër*. Another visiting composer was Saint-Saëns, who played Beethoven's Fourth Piano Concerto under Richter on 3 March 1879. Two months later the conductor had given the first of three orchestral concerts in London, designed to capitalize on and extend his success of two years earlier.

11

1879–1880

Friends and Enemies

HERMANN FRANKE and the impresario brothers Schultz-Curtius were the organizers of the Orchestral Festival Concerts as they were known. Franke, a violinist pupil of Joachim and well connected in London's musical circles, had shared the first desk of the first violins with the leader August Wilhelmj in Wagner's London Festival orchestra. Hermann Klein takes up the story.

[After the Wagner Festival] for a while life jogged along in the regions of opera and orchestra as though nothing had happened that was worth remembering. Then Hermann Franke's brain began to get to work again. The unpractical, well-intentioned muddler who was primarily to be credited with an ill-organized undertaking that was nevertheless an achievement of historic value, suddenly bethought him once more of Hans Richter. . . . In the spring of 1879 he told me that there was a scheme on foot . . . for giving a series of Orchestral Festival Concerts, conducted by Richter, at St. James's Hall in the following May and June. The position of leading violin was to be shared by Franke and Schiever, and the band was to be a picked one, consisting of about half English, half foreign players.[1]

The result justified his [Franke's] expectation; two years later [in 1879] at St. James's Hall, Richter's feat of conducting not only Wagnerian fragments but Beethoven symphonies entirely from memory furnished an absolute novelty and created quite a sensation. Thenceforth, Hans Richter's popularity in England was assured, and his concerts, given once, and sometimes twice, every year, became a regular feature in the economy of London musical life.[2]

Richter made notes in his diary on his fortnight in London. He arrived on 1 May and stayed with Franke at 11 Bentinck Street off Cavendish Square (the home of Hermann Klein's parents). At ten o'clock the following morning he met his orchestra for the first time. 'For the most part it consisted of acquaintances (1877) who greeted me heartily.'³ His soloists included George Henschel and Clementine Schuch-Proska, wife of the Dresden-based conductor Ernst von Schuch. The preponderance of the works of Wagner and Beethoven was symbolized by the presence on the platform in front of the orchestra of the busts of both composers. The first concert on 5 May opened with Wagner's *Kaisermarsch* and continued with operatic arias and duets, the Prelude to the third act of *Meistersinger*, Schumann's *Manfred* Overture, and Beethoven's Seventh Symphony. The string strength was seventy players (18.18.12.12.10). To begin with attendance was poor, but, as word spread on the excellence of performance standards, so the audiences grew. The second concert, during the afternoon of 7 May, was a mixture of Wagner, Beethoven, Gluck, and Liszt, while at the final concert on 12 May Richter played works by Berlioz, Handel, and Beethoven. A seal of approval was conferred upon the Festival with the presence of royalty on this occasion. 'The Prince of Wales summoned me and appeared to be really pleased.'⁴ Richter was fêted throughout his stay in London. He visited the Tower, the Zoo, and the port of London, where he boarded a three-master. He made new friends and cemented friendships formed two years earlier, and names of artists and patrons of the arts in London make the first of many appearances in his diary; some of these friendships would last until the end of his life (Joshua, Cyriax, Chappell, Alma Tadema, Armbruster, and Dannreuther). The concerts would probably have been financially more successful if Franke and Messrs Schultz-Curtius had had more practical experience. Richter observed as much in his diary, but added that 'Sir Julius Benedict agitated against the enterprise in an underhand manner. When, however, artistic success was already assured at the first concert, he slyly wanted to be a part of it by pushing himself forward as Frau Schuch's accompanist in the Lieder; but his offer was rejected soundly and in manly fashion by Franke.'⁵

Parry met Richter on 4 May. 'After lunch spent some time with Dannreuther. Richter was there for a time—a jolly, burly, hearty animal talking strong Viennese dialect which to me was unintelligible.' His comments on the performances included, '*Manfred* overture powerfully done. [Beethoven's Seventh] A major symphony in most respects better than I ever heard it before. . . . *Meistersinger* overture and *Cellini* very fine. Henschel sang Hans Sachs's monologue very well. The first movement of the *Eroica* was much too fast, the others fine. Richter had an ovation afterwards.'⁶ Richter's timings were recorded in the diary of the young Herbert Thompson, later music critic of the *Yorkshire Post*:

Monday May 5th 1879: St James' Hall for first 'orchestral festival' concert, Richter conductor, *Kaisermarsch* Wagner (10′), Introduction to Act 3 of *Meistersinger* (6½′), Wolfram's scene from *Tannhäuser* (Henschel), duet from *Dutchman* (Henschel & Frau Schuch-Proska), Overture to *Manfred*, Schumann (10½′), Beethoven 7th Symphony (46½′). Splendid, perfect band of 110. *Richter conducts without music*! Also aria from *Entführung*, Mozart.

Wednesday May 7th 1879: Second Richter concert at 3 p.m. Wagner's *Faust* Overture (11½′), Introduction & closing scene from *Tristan* (14½′), *Walkürenritt* (5′), Wotan's farewell etc. (Henschel), Liszt's symphonic poem *Les Préludes* No. 3 (16½′), Airs by Mozart & Gluck, Schuch-Proska doing the Mozart, Radecke doing the Gluck, Beethoven C minor Symphony (35′). Concert over at 5.30 precisely.

Monday May 12th 1879: Third concert at 8 p.m. Overture to *Meistersinger* (9′), Scene from *Tannhäuser* (Wolfram) (Henschel), Siegfried's Totentrauer symphony from *Götterdämmerung* (6′), Berlioz overture *Benvenuto Cellini* (11½′), *Eroica* symphony (53½′). Prince & Princess of Wales present.[7]

Sir Charles Villiers Stanford was also involved in setting up the Richter Concerts when Franke travelled to Cambridge to consult him.

Franke's enthusiasms and ambitions were far more enjoyment to him than playing the violin, and they speedily overshadowed his instrumental powers. Like the rest of the orchestra, he was fascinated by the personality and dominant force of Hans Richter, and laid plans to secure him for further concerts where he could show the British public his mastery of the works of other composers besides Wagner.... The way seemed clear enough for such an undertaking. The Philharmonic was in somewhat feeble hands... Manns, practically the only metropolitan conductor of merit, confined his energies to the Crystal Palace, Hallé to Manchester. The outcome of our conversations was the establishing of the Richter Concerts, in which I was able, thanks to my personal acquaintance with several enthusiastic amateurs of means, to assist by building up a guarantee fund. The first series took place in May 1879, and consisted of three orchestral and one chamber concert. The third, fifth, and seventh symphonies of Beethoven were given with a perfection which was nothing less than a revelation to the public, too long accustomed to persistent *mezzofortes* and humdrum phrasing. Richter's popularity with the band was increased by his quaint efforts to express himself in English. Le Bon, the oboist, who played an A natural instead of an A flat was so startled by hearing Richter call out 'As' (the German for A flat) that he began to pack up his instrument and take up his hat, until a German neighbour assured him that there was no allusion to long ears. A *pizzicato* which gave the impression of being produced by nail power, he corrected by the request to play 'not with the horns but with the meat' [presumably the cornified part of the finger was confused for the former, and flesh = *Fleisch* = meat for the latter].... In the course of time Franke found it impossible to carry the whole weight of responsibility on his own shoulders, and posterity has done him the usual kindness of forgetting the fact that the inception of the whole scheme was his, and his alone. I once heard Richter indignantly condemn this injustice by saying, 'Er hat's gewagt' (he dared to do it).[8]

Apocryphal stories of Richter's English became legion over the years, but he did seriously intend to master the language. When it looked as though trips to England were to be a regular fixture in his calendar, he wrote, 'Start of English language studies with Fräulein Martin.'[9] Turning to the press and their reception of Richter as man and artist, the *Musical Times* summarized the Festival (a word the critic considered unsuited to the event) with an all-embracing review.

As to Herr Richter's conducting. Whatever may be said in favour of invisible orchestras, we are of the opinion that the audience assembled during the memorable Bayreuth performances have lost a considerable element of assistance in the appreciation of the music of the Tetralogy in not seeing Hans Richter conduct. His bâton speaks the intentions of the composer whose work he is interpreting with an eloquence which at once attracts and fascinates to the end, both executive artists and audience alike. His individual reading, for the impress of his individuality upon his orchestra, as may be inferred, is most marked, has in it nothing eccentric or obtrusive, while his manner is entirely free from the ecstatic and ostentatious ways of some modern conductors. The result is invariably a performance harmonious in all its parts, a fact which, if examples are to be quoted, was especially noticeable in the three symphonies, the splendid rendering of which no one present will easily forget.[10]

The *Athenaeum* noted that 'the cordial reception given to Herr Hans Richter . . . was not more than is due to an orchestral chief of the first class and to a musician who is doing good work in his own country. . . . Herr Richter did not fail to give some delicate Viennese touches in the colouring of [Beethoven's Seventh] symphony, which must have surprised his London hearers. . . . The players seemed quite at home with their conductor, who, although new to them, had them well in hand; he conducted from memory. The concert was a representative one, and that of the 12th should be supported by all amateurs and artists who are not wedded to routine in art matters.'[11] A week later the same reviewer concluded a favourable report with the words, 'If the coming of Herr Richter has no other artistic result than a reform in the method of conducting that prevails in this country it will have been most beneficial.'[12] Louis Engel, writing in the *World*, also greeted Richter's arrival in England with great enthusiasm.

'Was it not wonderful, splendid, phenomenal, gigantic?' and so on. Such were the expressions we heard on leaving St. James's Hall after the first Orchestral Concert. The concerts led by Hans Richter are the most perfect ideal of orchestral leading, and, to a certain extent, of orchestral playing. The public generally cannot be aware of the reason why so extraordinary a leader as Hans Richter is and must be so great a rarity. In addition to the thorough knowledge of the score before him, the knowledge of the capabilities of each instrument is of course imparted to any musician who knows instrumentation. But to know how to play every instrument in the band, to take the clarionet, the horn, the fiddle out of the performer's hand and to show him how to play it, that is given, among all the conductors we know, only to Hans Richter. He led, moreover, all these scores by heart, never erring as to the entry of any instrument. This extraordinarily great feat of memory and ability fills the band with that respect so necessary to produce discipline; it is the combination of his grand conception of the masters, his indomitable energy which carries the band with him, and his unshakeable calm when he wishes to check their *entrain*; it is the conviction that he is their master . . . which makes the orchestra attend to every movement and every degree of movement of Hans Richter; because firm and short as is his beat in ordinary progress of movement, it lengthens in crescendos and slackens in diminuendos, his left hand continually indicating smoothing passages, or with an eloquence of rare ability the way of attacking the instrument to obtain the effect at the moment desired. Hans Richter plays the orchestra as a pianist sits down and plays his instrument, or, like a violinist, he places his left hand on the fingerboard called the band, and, with his bâton as bow, overcomes all difficulties, sings all cantilenas, accelerating or retarding, increasing or diminishing the force from the utmost sonority to a whisper, the like of which was never heard in St. James's Hall. He plays as only a great and gifted master can play, and his leading both of Wagner and Beethoven is a never-to-be-forgotten treat. It must not escape notice that what Richter accomplished, he did with an orchestra collected at random, where some unequalled performers play by the side of quite indifferent ones; but as Napoleon I said, 'Soldiers are made by the general'; and thus every man in the band rivalled in obeying the leader's will. He increases movements suddenly; marks out one single note, as he did with the C sharp on the harp or the low C on the double bass; and his *troupes* follow the leader's command, and leave him the responsibility, sure to be led to victory; and the unbounded enthusiasm of one of the most musical and critical audiences of London proved that their trust was not misplaced.

We had it on our heart to give the great man the great recognition he deserves, because we have made it our duty to speak out the truth where favourable or unfavourable. If we meant to be hypercritical, we would ask Mr Richter—because from him we expect perfection in the full sense of the word—why he did not make the bassoon mark its plaintive two notes F flat and E flat in the 45th and 46th bar of the Andante of the Fifth Symphony, and why he did not subdue the full chords in the same Andante from bar 114–120, which prevented the motif on the violoncellos and basses from being distinctly heard? But it would be a superhuman performance that left not the possibility of a remark.[13]

According to his biographer, Harry Plunket Greene, Stanford thought orchestral concerts were generally in the doldrums at this time, August Manns at Crystal Palace providing much the best. He also pointed out that it was not opportunism that engendered Richter's success. 'Anyone who has played or sung under Richter will recall the spell which that great man threw upon those who followed his beat. Massive, leonine, tranquil, he held them with the fire of his eye. These were the days before conductors had acquired *prima donna* status; to Richter his music was all in all. Manns, fiery, with gloved hands and slightly effeminate beat useless to those unaccustomed to it, and Hallé, scholarly, gentle, and uninspiring, were supposed to be his only rivals. But Richter stood unrivalled. His very reserve breathed confidence and filled the air with inspiration.'[14] The Czech-born Wilhelm Kuhe, a pianist and teacher domiciled in London who was used to organizing festivals, also had little good to say about standards of concerts before Richter's arrival.

But enough is it for me to lay emphasis on the wondrous change wrought in England of recent years . . . thanks to such men as Hans Richter. For whereas in former years no sort of concert in London attracted a large gathering unless it brought to the platform a vocal or instrumental 'star' distinguished in the musical firmament, we now behold the spectacle of an audience, crowded, alert, and expectant, drawn only by the magic name of a Richter or a Mottl. In the old days the poor conductor, who had to work so hard before he got his forces to a sufficiently high level of excellence to ensure a finished performance, was a mere harmless and necessary figure in a scheme of attractions in which his 'drawing' capacity was not reckoned. Now he is a veritable power in the land of music, and his name is printed in the type formerly accorded to none save a Patti, a Jenny Lind, a Liszt, or a Rubinstein.[15]

Back in Vienna by 18 May for another *William Tell*, Richter resumed all his duties, including services at the Court Chapel, until the summer break. Notable in this period was the opera orchestra's second visit to Salzburg for two concerts in the Aula Academica on 17 and 18 July promoted by the International Mozart Foundation. The orchestra's first visit had been two years earlier in 1877 under Otto Dessoff, who returned from his new post in Karlsruhe to conduct it as Music Director of the Salzburg Festival, and the occasion was notable for being the orchestra's first concert tour away from Vienna (a trip to London had been mooted in 1865). At the time of the 1877 tour Richter, as described in the previous chapter, was holidaying in Switzerland, but in 1879 the orchestra, by then firmly wedded to its own conductor, insisted on his presence. The *Salzburger Volksblatt* described how the players had travelled in two trains and were met at the station by a welcoming committee. Bands played and guns fired salutes from the hills high above the city. 'Hofkapellmeister Hanns Richter, with his artistic appearance and engaging manner, arrived earlier and now also awaited his trusty orchestra.'[16] The format of the three-day Festival was unchanged from 1877, an orchestral concert on each of the first two days followed by a matinée of chamber music on the third.

The first concert opened with Mozart's Overture to *Zauberflöte*, after which Hellmesberger jun. played a violin concerto by Bach and Clementine Schuch-Proska sang an aria from Mozart's *Idomeneo* to celebrate the centenary of the opera's composition. The concert also included Schumann's Manfred Overture, Schubert's Unfinished Symphony, an aria from *Zauberflöte*, and Beethoven's Seventh Symphony. On the following evening the programme featured three composers, Beethoven (Overture Leonore No. 3 and the Violin Concerto played by the leader Jacob Grün), Mozart (his Double Piano Concerto, an aria from *Figaro*, and his Symphony No. 39), and Wagner (the third act Prelude and Hans Sachs's Monologue from *Meistersinger*, soloist Dr Krauss). Richter had to fight for a place on the programme for Mozart and his E flat Symphony. 'Now, despite Salzburg's own wishes, I return to the subject of Mozart's E flat major Symphony,' he told the Festival director. 'The great and immortal son of Salzburg is really only modestly represented on the Festival programme.... Therefore I entreat you to give way to my request and to advertise the Mozart E flat major Symphony instead of the one by Mendelssohn.'[17] Wagner's *Meistersinger* Overture, on the other hand, replaced a Serenade by Volkmann at the Festival's instigation. They did not want to lose the chance of hearing Wagner's music conducted by the next best man to the composer himself.

In Vienna on 2 November 1879 Richter conducted Mendelssohn's Overture *Athalie* together with Beethoven's 'Eroica' Symphony, after which Brahms conducted his own German Requiem. Brahms was in the middle of a 'golden era' at this time; his first two symphonies had appeared, the Violin Concerto, Tragic and Academic Festival Overtures, and the Second Piano Concerto were all products of the five fruitful years 1876–81. Richter's name would be linked with early performances of all of them. Von Bülow was Brahms's greatest champion from 1877, having switched his allegiance from Wagner. He established a base with the Meiningen orchestra with which he toured and gave fine performances. Having cemented his place in Vienna's musical life it is not surprising that Brahms was able to help another composer at this time, Anton Dvořák. Brahms recommended his colleague to the music publisher Fritz Simrock, to Vienna's Opera Director Franz Jauner, and to Hans Richter. As it happened, things were going well for Dvořák too at this time; he was courted by publishers, was prolific in his output, and received several performances throughout Germany in the 1879–80 season. One work in particular was popular with the public, his Third Slavonic Rhapsody in A flat, Op. 45. Wilhelm Taubert gave its première in Berlin in September 1879, and Wiesbaden, Karlsruhe, and Budapest performed it even before Richter could do so in Vienna on 16 November in the first Philharmonic concert of the season. After the concert Dvořák wrote enthusiastically to his friend Alois Göbl.

I was sitting beside Brahms at the organ in the orchestra and Richter pulled me out. I *had* to come out. I must tell you that I won the sympathy of the whole orchestra at a stroke and that of all the novelties they tried over, and there were sixty as Richter told me, my Rhapsody was best liked. Richter actually embraced me on the spot and was very happy, as he said, to know me and promised that the Rhapsody would be repeated at an extra concert at the Opera.... I had to assure the Philharmonic that I would send them a symphony for next season. The day after the concert Richter gave a banquet in his house, in my honour so to speak, to which he invited all the Czech members of the orchestra. It was a grand evening which I shall not easily forget as long as I live.[18]

Dvořák declared himself well satisfied with Richter ('who is renowned as a very pronounced Wagnerian') in a letter to Simrock written a few days earlier. 'I have nothing but the highest praise for the performance of the work. It was played with incomparable beauty, and the impression upon me of the Rhapsody, which I had not yet heard, was overpowering. Everyone played with enthusiasm.'[19] Most reviews, including Hanslick's, were favourable and Dvořák's immediate musical future looked secure, in total contrast to the home-grown Bruckner, whose time had not yet arrived. Dvořák's promised symphony, completed a year later in October 1880, was his Sixth in D major and the dedicatee Hans Richter. There were, however, many obstacles ahead which would prevent Richter from conducting the work until he finally did in London in May 1882. Meanwhile the remainder of the 1879–80 season was notable only for some of the soloists with whom he worked, some of them composers performing their own works. There was Xaver Scharwenka with his Piano Concerto (14 December), David Popper with his Cello Concerto (28 December) together with the second performance of Brahms's Second Symphony, Sarasate in

Mendelssohn's Violin Concerto (6 January 1880), and Leschetizky in the Fourth Piano Concerto by Saint-Saëns (21 March). Other works receiving their first Viennese performances that season were Heuberger's *Orchestral Variations on a Theme by Schubert* and three movements from Mozart's Posthorn Serenade. Mozart featured prominently in Jauner's years at the head of the Opera. Of the cycle of operas he staged, Richter conducted the comparatively rare *La clemenza di Tito* during this season. It opened on 27 January 1880, the composer's birthday. It was to be Jauner's last season as head of the Opera, for he took offence at the appointment without consultation of an Intendant, Baron von Hofmann, and resigned his post in the summer after a five-year tenure.

Meanwhile Richter returned to London for a season of concerts. These were now rechristened the 'Richter Concerts' and took their customary place in London's season during May and June. There were ten and they featured such works as Parry's Piano Concerto (Dannreuther), Robert Fuchs's Serenade, Scharwenka's Piano Concerto (the composer as soloist), Dvořák's Third Slavonic Rhapsody, Beethoven's Fourth Piano Concerto (Hallé), Saint-Saëns's Fourth Piano Concerto (the composer as soloist), and (at the final concert) Beethoven's 'Choral' Symphony with the Richter Choir of 180 voices, a work whose performance was to become an annual finale to the concerts.

Parry recorded a few comments on the performance of his Piano Concerto in F sharp major which August Manns had already premièred at the Crystal Palace on 3 April.

May 8th: To rehearsal by 10.30 at St James's Hall. Had to wait a long while during the rehearsal of Schumann's symphony [No. 4]—so careful is Richter. Then they had a patient and laborious grind over the six sharps business in the *tutti* at the beginning. Had to settle to have it transposed. June 1st: In the evening the Richter banquet. June 5th: In the evening to Dannreuther's to dine with him, with Richter, Franke, Jameson etc. The talk was astounding, an incessant flow for about six hours. June 14th: The performance of the things from *Tristan* [Prelude and 'Liebestod'] was intoxicating at the concert in the evening, and the performance of the Ninth as near perfection as could be imagined. All except [Thekla] Friedländer [soprano] who sang abominably. Richter had an ovation after and justly.[20]

Herbert Thompson attended once again and made several detailed observations of the 1880 season.

May 10th 1880: First Richter concert at 8 p.m. Overture to *Meistersinger* (9½'), Beethoven's first Symphony (26'), Parry's manuscript piano concerto (33') played by Dannreuther, Schumann's fourth Symphony (28'), numerous and attentive audience. *Capital* audience! Conducted by Richter in his magnificent style.

May 20th 1880: Second Richter concert at 8 p.m. *Anacreon* (9½'), *Siegfried Idyll* (14½'), Spohr's *Dramatic* Concerto (Neruda) (20½'), Fuchs' Serenade for string orchestra No. 2 in C (17'), Beethoven's 2nd Symphony (42½'). Concert over at 10.30 p.m.

May 24th 1880: Third Richter concert at 8 p.m. *Italian* Symphony (32'), Scharwenka's Piano Concerto (29½'), Beethoven's *Eroica* (51½').

May 27th 1880: Fourth Richter concert at 8 p.m. Wagner's *Faust* overture (11½'), Beethoven's 4th Symphony (35½'), Dvořák's *Slavonic Rhapsody* (13½'), Beethoven's fourth piano concerto (Hallé) (29'), Schubert *Great* C major (51'). Concert over at 10.55 p.m.

May 31st 1880: Fifth Richter concert. *Euryanthe* overture (9'), Haydn's D major *Salomon* Symphony (31½'), Volkmann's cello concerto (20'), Beethoven's 5th (37').

June 3rd 1880: Sixth Richter concert. *Kaisermarsch* (10'), *Hunnenschlacht* (14½'), Beethoven *Emperor* concerto (32') (played by Barth splendidly), *Tannhäuser* overture (13'), Beethoven's 6th Symphony (41').

June 4th 1880: *Lohengrin* at 8 p.m. Nilsson/Elsa, Tremelli/Ortrud, Candidus/Lohengrin, Galassi/Telramund, Behrens/King, Monti/Herald. Richter conductor. Orchestra and singing of principals (excepting Nilsson who was occasionally out of tune) good. Stage management disgraceful and ludicrous. Chorus bad. Cornets substituted for trumpets.

June 7th 1880: Seventh Richter concert. *Roman Carnival* (9'), Schubert *Unfinished* (22½'), Saint-Saëns Fourth piano concerto (22½'), Introduction and closing scene from *Tristan* (15'), Beethoven 7th Symphony (43').

June 10th 1880: Eighth Richter concert. Brahms Symphony No. 2 (41½'), Bach Double Violin Concerto (14') (Franke & Schiever), Henschel Concert Overture (MS) (10½'), Beethoven Symphony No. 8 (27').

June 11th 1880: Extra Richter concert (Franke's benefit). Liszt's *Faust* Symphony (60'), Scene from *Meistersinger*, Beethoven unfinished violin concerto (15'), Wagner's *Siegfried Idyll* (14½'), Introduction to Act 3 of *Meistersinger* (6'), Beethoven *Leonore* No. 3 (12').

June 14th 1880: Ninth and last Richter concert at 8 p.m. Mozart's Symphony No. 40 (27'), Introduction and closing scene from *Tristan* (14½'), Beethoven *Choral* Symphony (81'). Splendid performance. Solo vocalists Friedlaender, Hohenschild, Candidus & Henschel the weakest feature. Large and enthusiastic audience. 25 minutes getting out.[21]

Thompson's entry for 4 June mentions the second of three performances of Wagner's *Lohengrin* given at Her Majesty's Theatre (the third, on 12 June, was Richter's fiftieth performance of the opera). His contract had only been drawn up with the impresario Colonel James Mapleson as late as

11 May but rehearsals with soloists and chorus began in earnest straight away and were fitted around Richter's concert commitments. Earlier in the spring Carl Rosa had given the first performances in English of the opera at the same theatre and Richter inherited the orchestral parts used by his conductor, Sir Michael Costa. According to Mapleson, '[Richter], after some fifteen rehearsals, declared the work ready for presentation. He at the same time informed me that on looking through the orchestral parts he had discovered no less than 430 mistakes which had been passed over by his predecessor, and which he had corrected.'[22] Richter noted that 'the many mistakes in the parts, which have often been used for years, are both annoying and amazing in the extreme',[23] but the conductor's diary reports a total of a dozen rehearsals, including all piano sessions with soloists and chorus. If Mapleson remembered fifteen orchestral rehearsals there were simply not enough hours in Richter's days at this period. On occasion (e.g. 24 May) he would even conduct an afternoon rehearsal of the opera (Act II from 2 to 5 p.m.) between a morning dress rehearsal and the corresponding evening concert. He dismissed his players from the afternoon rehearsal on 19 May because Mapleson's promise that a full orchestra would be present was not honoured.

In his diary Richter listed the cast and his opinion of some of them: 'Lohengrin—Candidus, Elsa—Nilsson (awfully bad), Ortrud—Tremelli, Telramund—Galassi (very good), the King—Behrens (Well! well!), Herald—Monti (a brute!). Orchestra really fine, chorus full of good intentions. Chorus master Smithen a first-rate musician and fine chap. I had much success but was not overjoyed after the performance. Rotten!'[24] The *Musical Times* was pleased to note that the public's attention had shifted away from 'petted vocalists'[25] to focus upon the conductor and his interpretation of the opera and how it would stand up to performances hitherto given by Italians. It had already dubbed Richter 'an accomplished musician, a born leader, and a conscientious worker',[26] but results even exceeded expectation. 'It may safely be said that never was the meaning of Wagner so clearly revealed in the choral portions of *Lohengrin* as on this occasion, and never were the hearers so deeply impressed with its poetical significance.... We have nothing but praise for the orchestra... every shade of colour was so minutely attended to that it appeared as if we were listening to a new work.... We have the satisfaction of knowing that Wagner's own intentions were in every case fully realised, and cannot but be gratified that Herr Richter was resolved to assert his independence.... Herr Richter, both on entering and quitting the orchestra, was overwhelmed with applause.'[27]

Louis Engel, writing in the *World*, confirmed Richter's view that Christine Nilsson was past her best, but that she also 'seemed to take more interest in staring at the Princess of Wales, who sat in her box, the most charming picture, with one of her little sailors on each side, than in looking at Mr Candidus, with his red cheek and white beard. That certainly is very excusable, but should not Elsa listen to and look at Lohengrin in preference to the audience?'[28] A more serious disaster was the non-appearance in the second act of the Herald 'for some minutes after he had been announced, [which], but for Herr Richter's resolution to continue the accompaniment, in spite of there being no singer, might have proved seriously detrimental to the dramatic action'.[29] Engel had earlier drawn attention to the problems Richter had had in the preparation of the opera and then went on to give a vivid sketch of the conductor at work together with a consideration of his merits.

As to Richter, people never weary looking at him and wondering how every coming passage casts its shadow beforehand—by his arm, his head, his hand—for he has a whole dictionary of phrases to communicate to his orchestra without ever speaking. It is a sort of deaf-and-dumb language discernible only by the eye, and most interesting it is; and those who understand what it means to know a score or a part of a score by heart are amazed at the infallible certainty with which he indicates everything that is to come. The oftener he does so, the less should it astonish. You should get reconciled to the fact that this man can do what so many others never would try to do; but when you see him lead one score after the other, and never make a mistake, never even hesitate, never even move more than absolutely necessary—as if it was a thing of everyday occurrence to know a score of scores by heart—you cannot help feeling that you are in the presence of one of those rare organisations to whom the impossible is possible by dint of extraordinary gift, study, and will.... People often, when Richter leads, pay all attention to the phenomenal concentration of power in his own hands—those hands that give the whole palette of tints, from *pianissimo* to *fortissimo*, so clearly, that the public learn to understand it as well as the band. As the right arm indicates the increase of force, so the left hand indicates the decrease; if sudden, with the outstretched arm, as if to check the impulse; if gradually, the fingers bend and unbend; and he actually plays the orchestra with his hand, as a man would play a stringed instrument. One reason of his greatness is a quality which is so rare in life: he knows exactly what he wants.... He no more allows his orchestra to carry him away than he lets them drag behind. He is as undemonstrative as ever you saw a conductor, but just therefore every little extra movement tells, and has instantly the desired effect. Determined and varied as his beat is, it never fights through the air; it is precisely what it ought to be. Like every man of solid capacity there is no fuss about him, deeds not words!

The orchestra, recognising the merits of the conductor, obeyed with that docile readiness which only a great conductor can command, and were simply perfect. It was clear from the first moment through the manner of their playing the introduction, how they would go through the work. Richter takes some movements slower than is the custom in this country; and not only must this be accepted, because better than anybody else does he knows Wagner's intentions, but it has of late become the fashion in our operatic orchestras to rush through the tempi at the cost of clearness and correctness, and it is very desirable that moderation should take the place of hurry.[30]

Engel was obsessed by Richter's ability to memorize scores, though he did warn his readers against being distracted by the conductor's *tour de force*, and urged them to take a score of the works to the concerts and follow it without looking at the conductor. He added, 'Were [Richter] perpetually [in London], he could form his orchestra as he understands it, drill them, and go with them through a number of scores. There would then be a hope to have model performances as a lesson for those provincial conductors who need this. That all provincial conductors do not stand in need of the lesson, Mr Hallé has triumphantly proved.'[31] The prophetic Engel did not realize that within twenty years Richter would live in England and conduct Hallé's own orchestra in Manchester.

The *Athenaeum* confined its reviews during that early summer of 1880 to Richter's orchestral concerts. After the first one on 10 May he was praised in particular for the clarity and dynamic control brought out and exerted in the Schumann and Beethoven symphonies. The inclusion of Parry's Piano Concerto was praised though it was noted that it was the only English work to be played in the season, and worthy though it might have been for inclusion, there were other (unnamed) native composers who also justified consideration. Fuchs's Serenade, performed at the second concert, was not thought to be worth its place. 'Novelty is not a necessary consideration at the Richter concerts,' wrote the reviewer, 'but if it be included, works of genuine interest should be brought forward, or, in default of them, some slight attention might be bestowed on English music.'[32] Dvořák, currently enjoying popularity in London with performances of his String Sextet and Piano Trio, Op. 26, received praise for the Third Rhapsody; Volkmann's Cello Concerto was dismissed as 'unsympathetic and not likely to endure'.[33] The wisdom of placing Schubert's Ninth Symphony, a fifty-minute work, at the end of a programme which (as Thompson recorded) had already contained an hour and a half of music was questioned. Hallé's contribution to that programme as soloist in Beethoven's Fourth Piano Concerto underlined the abilities of this great pianist, and Richter's now famed interpretation of the same composer's Fifth Symphony at the next concert was praised. 'For the first time in these symphonies, Herr Richter allowed the full power of the orchestra to be heard, and the effect of the entire mass of instrumentalists, playing with faultless precision, yet with the utmost vigour, was almost overwhelming.'[34] An 'electrical' *Tannhäuser* Overture was the highlight of the concert on 3 June, and Richter's faster tempos in the first movement and rustic trio of the third were noted in the 'Pastoral' Symphony. The piano was considered flat in the 'Emperor' Concerto, and Liszt's *Hunnenschlacht* dismissed as 'vague and indefinite'. After a successful *Carnaval romain* (7 June), Richter was criticized for his reading of Schubert's Unfinished Symphony. 'With all diffidence we venture to disagree with the conductor's reading of the *andante con moto*, which appeared to us a shade too fast, whereby much of the poetry was lost.'[35] Unfavourable comparison was also made here with the superior quality of Manns's woodwind-players at the Crystal Palace.

The last three concerts were reviewed in one issue. There was praise for the performance of Brahms's Second Symphony, which revealed 'beauties hitherto concealed or at any rate only half suggested. The dreamy *adagio* was taken at a slower pace than at previous performances, and the *nuances* observed with the utmost delicacy and care. Again the spirited finale went admirably, the fine *crescendo* near the close being given with especially imposing effect. Herr Richter is usually deaf to the demand for encores, but on this occasion he granted the unanimously expressed wish for a repetition of the quaint and fanciful *scherzo*.' Richter yielded to the same demand at his final concert, though this time the work was the 'Liebestod' from *Tristan*. This was played after a well-received performance of Mozart's penultimate Symphony in G minor (using clarinets according to the composer's revised version) and before a rendition of Beethoven's Ninth.

He takes the trio of the *scherzo* considerably faster than we have ever heard it before. If the metronome marks given in the score are by Beethoven himself, which we believe to be the fact, there can be no doubt that Herr Richter is correct in his *tempo*. The conductor's remarkable power of bringing out the individual parts in his orchestra caused many details to be heard which are generally imperceptible; and in spite of the complexity of the music, the general effect was clearer than we ever remember to have heard it. [Despite Beethoven's] reckless regard of the capabilities of the average human voice, the performance was certainly as good a one as we have heard, or probably are likely to hear.... Herr Richter has once more proved himself the greatest of living conductors, and with a second-rate band has achieved results far superior to those obtained by many conductors even with the best possible material under their hands.[36]

What the press did not know were the frustrations which bedevilled Richter in the second of those final three concerts, in which he performed Liszt's *Faust* Symphony. As his diary records, 'In the morning a heavy rehearsal for the evening benefit concert for Franke. Angry about the cymbal-player's antics, came in late and bungled it. The whole thing a risky undertaking with only one rehearsal for this difficult work. After the *Meistersinger* Prelude [to Act III] I received a bouquet which I then gave to Miss Bailey.'[37] The *Musical Times* printed a retrospective view of Richter's 1880 concert series. All the Beethoven symphonies had been given in numerical order in just over a month, and the success of the venture had been, according to the paper's critic, variable. Opinions had been aired about his tempos, in particular a slower than usual 'allegro', but the detail thereby revealed had been a remarkable feature of the performances. Those given of the Fifth, Seventh, and Ninth Symphonies 'were splendid, more than worthy of Herr Richter's reputation, and absolutely astonishing, having regard to the quality of the orchestra. . . . Herr Richter produced with second-rate means an article of, in many respects, first-rate excellence. The performance of the 'Choral' Symphony was naturally looked upon as a supreme test of the conductor's greatness, and on its account such a crowd assembled that the seating capacity of St James's Hall proved inadequate to the necessary accommodation. But it was worthwhile standing to hear the orchestral movements of Beethoven's greatest work played with so much perception and such clear expression of their meaning. . . . It was marked by many high qualities that helped largely to raise the standard by which hereafter the rendering of these works will be judged.'[38] This tribute to Richter is one of the most significant made in this country, coming as it did only three years after he first appeared on a London concert platform. Standards of music-making in England would never be the same again as musicians and public alike absorbed the impact he made on their practice and enjoyment of music.

12

1880–1881

London and Vienna

FROM the moment Hans Richter first worked in London with Wagner in May 1877 he was the talk of the town and on everyone's invitation list. By the time he had established his own series of Richter Concerts in 1880 he had no difficulty in filling his engagement diary with social calls, luncheons, dinners, or tourist trips in the company of new-found friends. At first he was courted by Edward Dannreuther, Karl Armbruster, and other admirers of Wagner who lived in London and championed his cause there; but Richter was monopolized by no one and soon enlarged his circle to include those who were more catholic in their taste or wished to see native composers advanced. Many of those he met were of German parentage or Germans living in London, and much of the social activity such as chamber music and dinners took place at the German Athenaeum Club in the capital. This was also true of cities elsewhere in England, particularly in Liverpool, where Max Bruch was in charge of the Philharmonic Society from 1880 to 1883, and in Manchester, where Hallé ruled the musical life until his death in 1895. By 1880 and his third visit to London certain names were recurring in Richter's diary. There was the painter Lawrence Alma Tadema, the music critics James Davison, Francis Hueffer, and Joseph Bennett, the singer George Henschel, the impresario Hermann Franke, the pianist Walter Bache, the violinist Ernst Schiever, but there was also a name which was to recur until the end of Richter's life, Marie Joshua.

Mrs Joshua was a music lover and patroness of the arts who had a wide range of artists and musicians in her circle. Richter was soon among them, and was a welcome guest at her weekly open Sunday lunches or at special gatherings which she organized either at her London residence in Westbourne Terrace, Hyde Park, or at the family's country house at Felixstowe in Suffolk. Later (thanks to Richter) she became a fervent supporter of Elgar and it was to her that the composer wished to dedicate his Violin Sonata. Though honoured by his request she gently declined the offer because she felt herself unworthy to join the ranks of other dedicatees such as Hans Richter and the unnamed recipient of the Violin Concerto. Sadly she died on 10 September 1918 before being able to reply to his offer (dated four days earlier), though her family agreed with Elgar's next suggestion that a dedication using only her initials MJ might appear. Early editions of the work carry them, but later they were withdrawn. Mrs Joshua's granddaughter, the late Mrs Winifred Christie, described her grandmother to the author as 'a strange character; a mass of contradictions. She was tremendously opinionated, yet completely humble in her own conceit.' She was a German Jewess and her sister was married to Sir George Lewis. Although Marie held the opinion that music could only come from Germany, she admired Elgar because his works related so closely to music from that country.

Richter arrived in London on 4 May 1880 and took a flat in Cavendish Square, where he found 'a charming basket of flowers and a cake as a welcome from Mrs Joshua'.[1] During his six weeks in London a lot of his social engagements were centred around the artists with whom he was working (such as Saint-Saëns) or who were currently in the capital (such as Minnie Hauk, who was singing Aida). He dined at the Denmark Hill (south London) home of the writer Edward Speyer, who became a close friend and admirer and who devoted a chapter to Richter in his memoirs.[2] For the following season in 1881 Richter took a flat in Vere Street, and within a week was lunching with the Joshuas (her husband is rarely mentioned by Richter or others). He made his annual visit to the Zoo, and gathered his friends and acquaintances around him in cosy pubs, with Sir George Grove and Charles Villiers Stanford now added to the company. On 17 May 1881 he visited Jenny Lind at her London house and found her 'exceptionally friendly',[3] and that evening attended upon the widow of Ignaz Moscheles. After rehearsing with Ludwig Straus and Sir Charles Hallé on 22 May he lunched with Joachim, who was visiting London. Two days later a trip to another pub, the Old Welsh Harp, was undertaken, and on 25 May he received honorary membership of the German Athenaeum. The occasion was a concert at which mostly chamber music was performed (including Brahms's F minor Quintet and a ballad by Sullivan), but Richter conducted the *Siegfried Idyll*, after which Hallé gave a speech in his honour. Richter replied with a speech of thanks not only to the Club but to Hallé for the contribution he was making to English musical life.

On 26 May he visited Charles Darwin's daughter, Mrs Litchfield, and on the following day was taken to lunch with the great man himself at his house in Downe, Kent. Darwin was then 72 years old and had only a year to live. Richter was received in friendly fashion and after lunch he played to his host, 'mainly at the instructions of Mrs Darwin who is very musical. A son is at present in Strasbourg and plays the bassoon very well. At half-past three Franke and I took a beautiful walk back to the station. This day remains unforgettable for me. What kindness lay in the eyes of the great man! "I think you have several instruments in your pocket," he said after I had played a few pieces by Beethoven, Wagner, and Mozart.'[4] The artist Alma Tadema (for whom Richter sat at this period) was one of several painters with whom the conductor became acquainted; another was the Pre-Raphaelite William Holman Hunt. Stanford was becoming a close friend, and his Psalm 46 was performed at Richter's fifth concert on 30 May 1881, at which 'all the ladies of the choir wore black and yellow! After the *Tannhäuser* Overture there was enthusiastic cheering. I received a baton twined in yellow silk.'[5] His former mentor von Bülow attended the next concert on 2 June, which included 'a very good performance of Brahms's C minor [First] Symphony. Bülow appeared enthusiastically happy.'[6] Appearances seem to have been deceptive according to a letter which von Bülow wrote in quite different terms to his mother, although he was not in the best of moods, declaring himself little pleased with London. He also complained about Anton Rubinstein, currently playing in the capital, that 'his financial success is greater than his artistic one, his playing, as always, very uneven. You would not have liked his playing of Bach and Mozart, but he remains a powerful individual nevertheless and a great painter of colours. In any case it has been a useful study to hear him. I can say the same only to a certain extent in the case of evaluating Richter as a conductor. He is still causing a sensation here, but his casualness is finding for him his proper place amongst the ranks of the decreasing greats. His direction of a Mozart symphony was good, that of one by Brahms inadequate.'[7] Von Bülow's conducting technique was as different from Richter's as Richter's would prove to be from Mahler's. Von Bülow would never have been sympathetic to his younger colleague's more restrained approach to music-making. He seemed more interested in the dog show at Crystal Palace, and regretted that his travels prevented him from owning a four-legged creature who would not cause him as much pain and anxiety as his fellow two-legged variety.

On 8 June Richter visited Cambridge as a guest of Stanford, and the next day his host took him around the University city and showed him manuscripts by Handel and William Byrd. Their relationship was growing: the first surviving letter from Stanford to Richter dates from the autumn of 1881 and is written in German using the formal 'Sie'. In it Stanford thanked Richter for returning his Cambridge hospitality by putting him up in Vienna, where he and his wife arrived on 15 September. The point of the letter, however, was to get the conductor to show Stanford's opera *The Veiled Prophet* to Director Wilhelm Jahn (who succeeded Jauner that summer) with a view to getting it staged in the Austrian capital (it had received its première in February that year in Hanover).

I think the opera would have a good reception in Vienna. Here [Hanover], as I understand it, the public response was warmer at each performance, especially in the third and fourth galleries (where the best public is to be found everywhere), and all this has given me the courage to have the cheek to write you this letter. It would be to the greatest benefit of us English composers [wrote the Irishman!] if an English opera were performed in Vienna, and if my opera proves worthy, then this success would have as great an effect throughout Germany as it already has in Hanover. . . . The first step is taken; if you could help me to take the second you know already how I and other serious-minded composers, who live and work for our art, would thank you. It's out! Forgive me for my vain plotting, and believe me that I appreciate all your goodness towards me.'[8]

On 13 June Richter's concert programme included Beethoven's Seventh Symphony. 'I had to stop the Andante of the A major Symphony twice after the opening bars,' he recorded, 'as sounds of a trumpet were coming from the street; the public was grateful for that!'[9] Between rehearsals for Beethoven's *Missa solemnis* he continued to sit for Alma Tadema and, on the evening of 16 June, was taken by Franke to meet the 38-year-old Adelina Patti and her future husband, the tenor Ernest Nicolini, at the home of a mutual acquaintance. She was 'exceptionally kind. I played some things. She was quite delighted and promised to come to Bayreuth to visit the Master after her American tour, for I suggested that she should conclude her artistic career with Wagner. Nicolini was also there; it was a very merry evening.'[10] Patti did not take up Richter's suggestion. On 21 June he went to Alma Tadema for a final sitting and visited Speyer at Denmark Hill; the Princess of Wales (Princess Alexandra) arrived at St James's Hall in time for the end of the rehearsal next morning of the *Missa solemnis*, and that evening Richter was entertained at Mrs Joshua's home. He considered the performance on 23 June of Beethoven's masterpiece 'for the most part first-rate'.[11] The soloists were Louise Pyk, Ellen Orridge, William Shakespeare, and George Henschel. Stanford was at the organ, and a chorus of 200 took part. The work was repeated four days later when the conductor thought it 'quite excellent, much calmer than the first'.[12] The *Musical Times* concluded that 'a profound impression was made by the work, so profound that the audience largely refrained from applause at the end, and went away silently, as though from a place of worship', but it regretted that the chorus were not 'the strong-lunged, deep-chested men and women who triennially astound musical visitors to Leeds'.[13]

Parry attended several of the concerts and met Richter socially on occasion.

19 May: Went off to Richter Concert to hear Cowen's Symphony again [Scandinavian] and was as much pleased as ever with the slow movement and the scherzo, the latter is quite astonishing.

23 May: Dined with the Hickens and went to the Richter with them. Was more impressed with the *Tragic* of Brahms than at the C.P. [Crystal Palace]. I think Richter gets a better total impression out of a work than Manns does. The C minor [Beethoven Fifth] superb of course.

30 May: In the evening Richter concert. Absurd programme (Stanford's *Psalm*, Liszt concerto, Haydn symphony, *Tannhäuser* overture). Parts of the *Psalm* are fine and the setting good. Some parts dull and wanting in vitality; even Mendelssohnian—Dann. in great form with the concerto and had a succès d'énthusiasme [*sic*] at which I was quite delighted. Haydn mostly quite charming and naïve, overture splendid and as exciting as if one had never heard it before. But what an incongruous bundle!

18 June: Got off early to dine with the Ionides (Constantines) and meet Richter. It was the most sumptuous dinner I ever ate and the company was delightful. Richter was very vocal and bellowed at dinner like a wild bull of Basham. It was a pretty late affair and I walked back with Dannreuther and his Mrs. after.

20 June: In the evening to Richter concert. Very fine performance of *Coriolan*, the new Venusberg music for *Tannhäuser*, Pogner's *Ausrede* etc. The *Eroica* was superb of course.

23 June: Then to Richter's performance of the great Mass [*Missa solemnis*]. It was a fine performance altogether, though Pyk was not quite up to the mark and the chorus were overpowered. He made his brass play very loud. The work is more and more awe-inspiring. I never was more moved or rather quite mastered.[14]

With that triumph Richter concluded his summer season in 1881 and returned to Vienna via Bayreuth, where plans for *Parsifal* were well advanced. Marianne Brandt was also visiting and Wagner offered her the role of Kundry during her stay. Cosima, who was suffering from toothache, wrote in her diary, 'I would like to stay in bed, but there is Hans Richter with all his sunny good humour to receive. . . . Lunch much enlivened by Richter's countless anecdotes. . . . Memories of Tribschen the happiest of all!'[15] These were two happy days, and Richter was, as ever, quite at ease with the Wagner children together with Marie and their oldest daughter Richardis, who joined him there for his stay. Wagner gave his former deputy a warm welcome, and together with Brandt the Richters set off for Vienna on 2 July. The two men were to meet just once more during the *Parsifal* performances the following year. By now Wagner had let his protégé go his own way and no longer expected servile attendance or continual letter-writing. At the end of one of his last letters to Richter the composer joked, 'You do not trouble yourself over me any longer, do you? Well, you can look forward to Hell!'[16]

The Vienna season 1880–1 had some noteworthy landmarks under Richter's baton. The Opera staged the first performance there of Cherubini's *Medea* (26 November 1880), the rare *Preziosa* by Weber, and Gluck's one-act opera *Le Cadi dupé* coupled with Schubert's ballet music from *Rosamunde* (22 March 1881). There were also numerous performances of *Carmen* (with Pauline Lucca), the standard fare of Wagner, Beethoven, Mozart, Halévy, Gounod, Marschner, Meyerbeer, and Ignaz Brüll, as well as a rarity for Richter, Verdi's *Rigoletto*. On the concert platform at the Philharmonic concerts there were first performances of works which have since become standard repertoire together with those whose concert 'shelf-life' has been brief (such as Robert Fuchs's Piano Concerto and Goldmark's Overture *Penthesilea*). More enduring have been Brahms's Tragic Overture (first performed 26 December 1880), and the first Viennese performances of Mozart's Symphony No. 31 (K. 297) 'Paris' (5 December 1880), Haydn's Symphony No. 87 in A (6 March 1881), Berlioz's Overture Les Francs-Juges (21 October 1880), and Brahms's Academic Festival Overture (20 March 1881). For the concert on 6 January 1881 the committee of the orchestra received a request for a pair of tickets from Andreas Schubert to hear the Ninth Symphony composed by his brother, who had died over half a century earlier. The concert was sold out and the committee could therefore not oblige Herr Schubert. The soloist in Goldmark's Violin Concerto at a special Philharmonic concert for the pension fund of the Opera on 10 April 1881 was Arnold Rosé. He was appointed the orchestra's leader that year and held the post until 1938, an incredible fifty-seven years, which would have been longer had it not been for the German annexation of Austria which forced him into exile.

It was typical of Richter to present Rosé at a charity concert, for such events were to become part of the conductor's life throughout his career. It was at a similar event, on 20 February 1881 for the Association of German Schools, that he gave another world première, that of Bruckner's Symphony

No. 4 in E flat, now known as the 'Romantic'. This occasion was to prove the watershed in Bruckner's career. The concert was an unmitigated triumph which transformed his reputation amongst the Viennese public, if not all the critics. After he had been disparaged hitherto as an eccentric composer of incomprehensible music, the reputation of the 57-year-old Bruckner was now rehabilitated, and he was recognized both as a genius in their midst, and as the leader of progressive music, though neither the Philharmonic concerts nor the Gesellschaft der Musikfreunde had taken a risk by promoting the concert; it was left to the Wagner Society in Vienna, which had made the composer an honorary member just three weeks earlier on 27 January. The Vienna Philharmonic had already rejected the first version of the symphony, deeming only the first movement playable and the rest mad. It took another fifteen years, until 1896, the year of the composer's death, before they dared to perform it under their own auspices. It was as the Court Opera orchestra that they played on 20 February 1881, and for Bruckner the success of this concert wiped out memories of the fiasco of the performance of the Third Symphony four years earlier in 1877.

Hans Paumgartner wrote in the *Wiener Abendpost* that 'the public received the symphony with undivided enthusiasm, which it expressed by stormy and triumphant applause. After each movement Bruckner had to appear four or five times. In a word, Bruckner came through it all brilliantly. Since last Sunday he belongs to our most distinguished composers and, as an artist, he has become public property.'[17] Even Hanslick gave Bruckner grudging praise in the face of such positive public acclaim, though he could bring himself only to describe it as an 'unusual success', and did not profess to understand the symphony.[18] Bruckner's reputation as a musician had been mainly earned by his outstanding ability as an organist, his name as a successful composer being largely limited to his choral output. Now he was clearly identified by many as Vienna's natural successor to Schubert as a symphonist, even if his name was inextricably linked to that of Wagner; yet even here Eduard Kremser in *Das Vaterland* had the foresight to qualify the comparison by writing, 'Bruckner is a Wagnerian in so far as Wagner is a Beethovenian, and Beethoven in turn a Mozartian, but in no other sense.'[19] The success of the occasion had its repercussions, particularly after Wagner's death just over a year later, for with Wagner gone Bruckner now became the target of Brahms's supporters in Vienna. In the aftermath of his triumph, however, Bruckner was blessed with an unusually happy period in his life. It is significant that he immediately set about composing his Sixth Symphony (a positive and joyous work in A major), completing two movements before interrupting it to compose his even happier Te Deum.

The programme of the concert on 21 February also included Beethoven's Overture *King Stephen* and his Fourth Piano Concerto. The soloist was Hans von Bülow, who just ten days earlier had astounded Vienna's public by playing the last five Beethoven piano sonatas in one recital. On this occasion, however, though successful as solo pianist in the concerto, von Bülow suffered an embarrassing failure as composer when he conducted his own orchestral ballad entitled *Der Sängers Fluch*. Its reception was respectful but no more, but unfortunately for von Bülow it was followed immediately by Bruckner's triumphant success. This did not endear him to the temperamental von Bülow, who neither forgot the occasion nor forgave the man. He never conducted a note of Bruckner's music and only grudgingly recognized the composer's genius after a Berlin performance of the Te Deum in 1891. In the meantime he had become an ardent admirer and fervent champion of Brahms. Seeing a printed score of Bruckner's Fourth Symphony lying on the counter of the publisher Albert Gutmann, he dryly asked, 'Is that supposed to be German music?',[20] but then he is also reputed to have said that the best German music was being written by the Russians. When Theodor Helm sent von Bülow a copy of his new book on the Beethoven string quartets, von Bülow, writing in acknowledgement, asked to be forgiven for ordering his local bookbinder to reduce it to 307 pages and then rebind it. Page 308 began with a study of the string quartet after Beethoven, including reference to Bruckner's new F major String Quintet. Bruckner did not reciprocate such petty bad feeling; his regard for von Bülow the conductor remained high (even if he found his pianism somewhat cold), and he always attended concerts given by the visiting Meiningen orchestra, often in order to hear and understand Brahms's music, which von Bülow programmed.

There were happier incidents to report in the preparations leading up to the first performance of Bruckner's symphony under Richter. It was Ferdinand Löwe, a prominent member of the Wagner Society, who first brought the work to the conductor's attention when a play-through on two pianos was organized in the presence of the composer, Löwe being one of the pianists. Bruckner later told August Göllerich that Richter said to him afterwards that ' "nothing like this has been written since Beethoven. Bruckner, you have been misjudged!", whereupon tears came to his eyes'.[21] The rehearsals were to prove a pure joy for the composer. At one point where a printing discrepancy between score and parts had been discovered Richter asked him which note was correct. Bruckner responded with typical deference, 'Whichever you wish, Herr Hofkapellmeister.'[22] After one of the final rehearsals an incident occurred which touched Richter deeply.

This thaler is a souvenir of a wonderful day. . . . When the symphony had ended Bruckner came up to me, beaming with enthusiasm and happiness. I felt him press something into my hand. 'Take this', he said, 'and drink my health with a glass of beer.' I took the thaler, had it set, and always carry it on my watch chain in remembrance of a capital man and the tears which flowed gratefully from the old musician, and which was so typical of his touching naïvety.[23]

Richter's schedule for the day of the première of Bruckner's symphony was daunting. The concert took place during the day, and later that evening he conducted *Don Giovanni* at the Opera. Three days later, on 23 February, he took the orchestra on one of its rare excursions away from Vienna to Graz to perform a totally different programme with works by Mozart, Beethoven, Weber, Berlioz, and Wagner. Meanwhile he was also in correspondence with Anton Dvořák. The Czech composer had promised Richter and his orchestra a symphony (his Sixth in D major to be dedicated to the conductor) for Vienna after the triumphant reception there (and in London) of his Third Slavonic Rhapsody in the spring of 1880. The first sketch was completed in September that year, with the instrumentation completed by the end of the following month. Dvořák was anxious to hold Richter to his promise of a performance in the 1880–1 season as none of his orchestral works apart from the Slavonic Dances and Rhapsodies had been performed outside of his home country. By the end of November the work had been played through by Richter and the composer. Richter was extremely enthusiastic and kissed Dvořák after each movement, and plans were made for the work's première.

The first date chosen was 26 December, but Richter withdrew it, giving as his reason that the orchestra was so overworked that he could only programme repertoire works (Beethoven's Eighth was the symphony, though this was also the concert at which Brahms's Tragic Overture received its première). Instead he offered to perform it on 6 March 1881. Dvořák readily agreed and asked Richter to play it through with the orchestra so that they too might gain a favourable impression of the symphony. Unfortunately for Dvořák, Richter was beset by problems at home when two of his children contracted diphtheria. No sooner had the other two healthy children been sent to his mother when she herself went down with the disease, and her grandchildren had to be evacuated again, this time to a hotel. To add to his problems Marie was expecting another child. Richter was nervously awaiting the outcome of this crisis and could not possibly undertake 'such a serious enterprise in these sad days'.[24] Dvořák's reply was completely sympathetic, reminding the conductor that he himself had lost two children, his 1-year-old daughter Ruzena and 3-year-old son Otakar, within a month in the autumn of 1877. Yet a letter sent to Richter a week after the proposed, but now cancelled, date for performance implies that Dvořák discovered other reasons for the delay. 'So it did turn out as I had suspected. Well, I must not lose heart because your final words offer me boundless consolation and renewed courage to write new works.'[25] John Clapham, in his biography of Dvořák, says that the composer 'found out that some members of the orchestra, who were consulted when programmes were being arranged, objected to playing music by a new Czech composer in two successive seasons. Richter had repeated the third *Slavonic Rhapsody* on 29 March 1880.'[26] Otakar Sourek, on the other hand, thought that the orchestra might have been offended when the dedication was not granted to the Philharmonic Society, for which it was especially composed. Whatever the reason it quickly found a première in Prague on 25 March 1881. Richter never conducted it in Vienna, but gave the second English performance on 15 May 1882 (Manns did the first at Crystal Palace on 22 April). Wilhelm Gericke gave the first Viennese performance for the Gesellschaft der Musikfreunde on 18 February 1883, but the Philharmonic concerts waited until 1910 (six years after Dvořák's death) before programming it, and by then Richter was resident in Manchester.

After his London season of 1881 Richter returned via Bayreuth to Vienna, where he spent the quiet summer months of July and August, mainly confining his duties to the Court Chapel and the opening performances of the opera season in August and September. He was grateful to be with his family after the harrowing experience of their ill health during the previous winter and was also enjoying the newest addition to his family, Mathilde Leonhardine, born on 21 January that year, the fifth child and youngest daughter, destined to live longest of all his children. She died in 1978 at the age of 97. In July Cosima Wagner wrote to him to say that a second copy of the role of Kundry in *Parsifal* would be sent to him by their assistant Engelbert Humperdinck. In the absence of any firm commitment by Marianne Brandt to accept the role (despite going through it with Wagner earlier that month), Richter was asked to show it to Amalie Materna and assess her as a possible alternative. What with also coaching Brandt in the role in Vienna until she made her decision, 'you will not be short of seducers!'[27] Visitors came from London, Hermann Franke and George Henschel in mid-August followed by the Stanfords in mid-September and the violinist Ludwig Straus a few days later. On 8 October the Richters moved house to Sternwartestrasse 36 in the Cottage district of

Währing, the suburb of Vienna where they would remain for the rest of their time in the Austrian capital. Then on 17 October Richter left for London for an autumn season of two concerts (a provincial tour having been considered but prudently abandoned).

The concerts took place on 24 and 29 October, the second consisting of several orchestral extracts from Wagner's operas and Beethoven's 'Eroica' Symphony. The first concert concluded with a repeat performance of Beethoven's 'Choral' Symphony but was preceded by Wagner's *Meistersinger* Overture, Berlioz's *Nuits d'été*, and a new piano concerto by the young Eugen d'Albert who was also the soloist. Parry was at the concert: 'To Richter concert where young d'Albert made his appearance with his new concerto for pianoforte and orchestra. He is a young marvel. His style of writing is already masterly; he handles his orchestra and his pianoforte superbly; expresses his ideas clearly and fluently; and the tone and quality of the thing is always full, rich and effective and he is yet only seventeen. I never saw or heard such gifts. If the man only develops in his inside he will be far and away the first composer ever sprung in this country. I went round for a moment and said "How d'ye do" to Richter & Co. and a word to d'Albert.'[28]

13

1881–1882

Richter and d'Albert

D'ALBERT, whose mother came from Newcastle, and whose father Charles (a composer and ballet master born in Hamburg) was French, was then aged 17. He was a pupil of Ernst Pauer and had received his musical education at the National Training School for Music in London (forerunner of the present Royal College of Music), whose Principal was Sir Arthur Sullivan. He was also under the patronage of Marie Joshua. Pauer and Mrs Joshua brought him to the attention of Richter, who decided that his future as a concert pianist was assured and decided to take him back with him to Vienna and to effect an introduction to Liszt. The story runs in the Richter family that with six mouths to feed a seventh would make no difference, but letters from the young man to his father infer that a financial arrangement was made with Richter which also provided Eugen with pocket money. Several letters were written by d'Albert over a six-month period to Mrs Joshua, first from Richter's Vienna home and subsequently from Weimar after he had become Liszt's pupil. His Anglophobia and love for things German meant that he never lived in this country again once his studies had started with Liszt. His father had arranged for him to study with Leschetizky, but Richter advised against this and undertook to tutor him himself. He considered that Leschetizky would completely change d'Albert's hand position and technique, and also held the view that he was too interested in his female pupils for d'Albert's own good. The letters to Mrs Joshua provide a fascinating account of life in Vienna, where the young man spent six months under the roof of its foremost conductor. He wrote in English with an occasional use of a German word or phrase. Some small corrections have been made to grammar or punctuation.

Richter set off for home with d'Albert on 30 October 1881 and within a week the young man had dispatched his first impressions to Mrs Joshua. In many ways d'Albert's place in the domestic household bears an uncanny resemblance to Richter's own youthful experience at Tribschen. D'Albert's immediate popularity with the Richter children reflects exactly the sympathetic relationship Richter had had with Wagner's children and stepchildren fifteen years earlier.

After many accidents and adventures, I have at last reached Vienna under Hans Richter's wonderfully kind care, quite safe and well. The morning on which we started was very cold, but fine with no wind, and we never expected that the sea would be so rough as we found it on arriving at Dover. The Herr Hofkapellmeister is always very ill on the sea; even the very smell of the salt water affects him, and consequently he was very sorry to have to cross in such bad weather. We were three hours in crossing, as we had to wait an hour outside the harbour, because the sea was so heavy. I was not at all ill. When we arrived at Calais, Herr Richter was so anxious to leave the sea that he got into the first train we came to without asking whether it went to Cologne or not. After we had been travelling an hour or so, the guard came into the carriage and the Hofkapellmeister asked him whether it was the right train for Cologne. He said that it was not; it was the Paris train, and all we could do was to get out at Boulogne and go back again to Calais. This was very troublesome, but we had to do it, losing thereby two hours and a half.

They had to wait ten hours in Calais, so it was not until four o'clock the next morning that they set off. After a two-hour delay in Lille they then had to change three times before reaching Cologne. 'Herr Richter took me all over the town in the evening, showing me both the inside and outside of the beautiful Cathedral, and the theatre where they played Schumann's *Genoveva*, and introduced me to the director of the Wagnerverein there, with whom we dined.' The pair travelled on via Würzburg, where Richter met Kapellmeister Fritz Steinbach, then on to Nuremberg, and down the Danube to Vienna.

We reached Vienna on Wednesday at six o'clock in the morning, after travelling three nights and three days instead of two nights and two days. I like Vienna very much indeed; everybody seems to understand music here and to enjoy it. Frau Richter is very kind to me, and very nice, only rather delicate, and I am very fond of the five beautiful, dear little children who sing songs from Wagner's *Ring des Nibelungen* quite splendidly, and can 'beat' all the measures perfectly. Here everything is Wagneriana. . . . Herr Richter is very fond of his children, and is never tired of playing with them.

I like the people and living here very much, they are so homely and natural. In the morning I generally go with Herr Richter to the rehearsal if there is one; if there is not I stay at home and work, and in the evening if there is no opera performance I go with him for a walk, generally to Pötzleinsdorf, where we go to a country inn, and Herr Richter talks and drinks beer with the peasants; I played there one evening in the Schoolroom, and they were so enthusiastic. Afterwards two students at Richter's request played to me some *Wienertänze* on the zither and guitar.

Herr Richter is very kind to me; he says that he would like me to stay always with him here in his house, as long as I remain in Vienna, and not to go anywhere else to stay, if the noise of the children and their always wanting to play with me does not stop me working.

He will not let me have any lessons from Leschetitsky, as he says that I do not need it; what advice I may need he will give me. He plays Bach with me every evening. On Sunday morning I was at the first Philharmonic Concert—they played the *Meistersinger* Vorspiel, a new concerto for strings of Bach, which is one of the finest things I have ever heard, and the *Pastoral* Symphony. [This was the first Vienna performance of Bach's 'Brandenburg' Concerto No. 6.] The orchestra here is on the whole very much better than the London one, and they seem to understand Richter better. On Monday evening I was at the Opera House for the first time.... They played *Lohengrin*, and in every way it was a splendid performance; only the Lohengrin, Herr Labatt, the same who played it at Covent Garden, was a failure, but Frau Materna as Ortrud was very grand indeed. Her reading of the part was truly wonderful. Of course the orchestra was splendid. Herr Richter has spoken to Brahms about me, and he will see me and hear my compositions as soon as he comes back from Pesth. Was it not kind of Richter to speak to him? Brahms has written a new pianoforte concerto, which he says is the longest work he has ever composed; it is in four movements, and takes an hour and a quarter in performance.[1]

D'Albert seized every opportunity to hear Richter conduct at the Opera, six operas in nine days between 7 and 16 November.

I have been six times to the opera. I have seen *Lohengrin, Fidelio, Don Giovanni, Romeo and Juliet, Barber of Seville*, and last but *not* least *Die Walküre*. It is indeed a great advantage for me to be able to hear so much music, which is good with very few exceptions. I have never before heard so much music; every day there is something. The Saal Bösendorfer especially has the first place as the stronghold of chamber music, pianoforte recitals etc. Herr Richter says I must give a concert there later in the winter.

The performance of *Fidelio* was absolutely perfect in every respect; Richter himself said so, and the marvellous beauties of that truly grand work appeared to me more fully than before. Frau Materna as Fidelio was really the true artist, not like the Italian singers with their runs and trills, but as a musician whose soul is not in the amount of applause he receives, but in the interpretation of the composer's spirit. *Don Juan* was not so well played although a star played the part of Donna Anna—Madame Pauline Lucca; but although the Viennese run after her, and when she sings the seats are all sold out, not a bar can she sing in tune, and Mozart's beautiful music was completely spoilt.

Die Walküre was to be played on Saturday, but owing to the illness of one of the singers, had to be changed for a very different sort of opera, *Romeo and Juliet* by Gounod. I think it was dreadful to see Richter conducting such a work, because it was quite bad enough to have to listen to it; it is just a weak, ineffective reproduction of his *Faust* without any of the redeeming points of that work; nothing but sweet-sounding duets, without even any musician-like workmanship, everything done to gain applause no matter by what means, and that means nearly always the Cymbals and Big Drum. In the last scene Juliet awakes before Romeo dies, so as to give opportunity for another duet with scales and brilliant runs, although they are supposed to be dying. The *Barber* was much better and Mademoiselle Bianca [*sic*], the chief singer at the Vienna Opera House, sang the part of Rosina, in a very brilliant manner, and earned a great, well deserved success. But the greatest work of all was the *Walküre*, and I was so delighted to be able to hear it at last, after having waited for so many years. I don't think, of course excepting the original Bayreuth performance, that I could have heard it better performed; only the stage management was not so good compared with the orchestra and the singers. For instance in the second act, for some reason or other Brünnhilde appeared without the 'feierlicher Ross' Grane, which was an omission that appeared all the more strange as the other eight Walküren were provided with horses, although in this case they might certainly have been left out as they did at Bayreuth because (accompanied by the wild, wonderfully exciting 'Rittmusik') on Wednesday night the horses were supposed to gallop wildly across the back of the stage amid thunder and lightning, but unfortunately some of them appeared to have wooden legs and nearly stood still in the middle of the stage. But the Feuerzauber was splendidly executed—the whole stage seemed to be on fire, and Richter says they manage it better here in Vienna than anywhere else.

Sunday I went with Mrs Richter to hear *The Creation* in the Musikverein under the direction of Hofkapellmeister Wilhelm Gericke. This was the first concert of the Gesellschaft der Musikfreunde. Johannes Brahms was there in the gallery, and I saw him for the first time. I have not yet been introduced to him as he has just come back from Pesth, where he gave a concert entirely composed of his own music, including his own new concerto played by himself, and which I hear the band say was 'boring'. As he is now in Vienna Richter will take me to him as soon as possible, most likely next week as this week Richter is far too busy to be able to trouble himself about anything else but his own business. He has a rehearsal today and tomorrow for the second Philharmonic [concert] on Sunday next and Saturday morning a rehearsal and Saturday evening a performance of *Die Meistersinger*. It shows how much they like Wagner here that they play three of his operas in one week, viz. *Tannhäuser*, *Walküre* and *Meistersinger*.

... I speak German with everyone here except Richter himself, who does not want me to forget my English, and wishes to exercise his own; he takes *three* lessons a week and studies very hard at it every day.[2]

D'Albert was now well and truly part of the family. He was taken everywhere and met Vienna's resident musicians. The Philharmonic concert series for 1881–2 was now well under way and he freely gave his opinion of the musicians he heard there and elsewhere.

I have been again to a great many concerts, both interesting and uninteresting. The best were those of the Gesellschaft der Musikfreunde, of Hellmesberger (Quartet) and the Philharmonic. The last concert was especially fine with Gluck's overture to *Iphigenia* (with Wagner's concert ending) and Beethoven's G major concerto played by Mr [Heinrich] Barth. I was at the rehearsal as well as the concert, and at both he played very finely and clear. He had a great success, being called forward three times. Mr Richter introduced me to him after the rehearsal and he was *very* nice to me; he knew about me from Joachim in Berlin and he not only wished to hear me play, but wished to hear all that I intended to do. He particularly desired me to play in Berlin. We had luncheon with Mr Barth, Concertmeister Grün, and Herr Bösendorfer after the rehearsal in a Kaffeehaus near the Opera, which was very pleasant. One learns so much from the conversation of great musicians that cannot be read in books.

I have been twice to the concerts of Hellmesberger which are the best quartet concerts in Vienna, and perhaps in the world. They play many new compositions. Mr Richter said that if I had written a quartet, he would have played it as he takes an interest in rising composers. Last night they played a new octet by [Hermann] Grädener, which was very clever indeed, though not very original, being very much like Brahms. In the quartet in B flat by Saint-Saëns a new pianist played— Herr B Schönberger. He plays really very well and with great taste. He must be very young indeed, being still a student and under the guidance of Anton Door who turned [the pages] for him. [Benno Schönberger was in fact a year older than d'Albert.]

There was a very bad performance of *Die Meistersinger* given at the Opera [19 November]. Richter was very angry. All the singers except Madame Ehnn, who always sings well, sang badly, and one under the influence, as he said, of severe hoarseness could not sing at all in the last act, so they had to leave the greater part of it out and commenced with the last scene. This singer, Herr Scaria, always disappoints the public. In *Die Walküre*, where he plays Wotan, he has twice been unable to sing at all, the performances having to be postponed. Richter says that there is nothing whatever the matter with him, only as Director Jahn resigns at Christmas and they have to appoint a new Director, it is his ambition to become Director, which, if it happened, Richter would at once resign. The affairs are in great confusion at the opera house.

Last week I was at the Wiener Stadt-Theater with Mr and Mrs Richter; they played *Der Sklave*. It was very amusing. But what I enjoy as well as anything is going with Mr Richter either to Pötzleinsdorf, which was the favourite resort of Schubert and Beethoven and now of Rubinstein, Brahms and Mosenthal, or to *Die Höhle* where there [are] always many musicians at supper. I have been there very often and the conversation is principally about Wagner, which I like very much. The composers I principally study are Bach, Beethoven, Liszt and Wagner; the others Richter does not think much of. I also read much Schopenhauer. Mr Richter's mother comes very often to see us; she was principal singer at the opera [houses] in Russia, and still retains her voice wonderfully for her age, she sings beautifully Sieglinde in *Die Walküre*. . . . I asked Mr Richter whether I should accept the Mendelssohn Scholarship, and he said 'ja, natürlich', and so of course I was obliged to do so.

I take long walks every day, and it is very nice to walk along the very paths that Beethoven has traversed before. I am helping Mrs Richter to arrange a scene out of *Götterdämmerung* for the children to play on Christmas Eve; it is quite a secret from Mr Richter, I think it will surprise him. The children are very funny, they either call me Herr Albert or Kaiserlicher Rath [Imperial Counsellor].[3]

The next Philharmonic concert took place on 4 December and, among standard works by Mendelssohn, Mozart, and Beethoven, another was revealed to the world for the first time, Tchaikovsky's Violin Concerto. The soloist was Adolph Brodsky, a fellow student with Richter at the Vienna Conservatoire and a life-long friend. He had been a professor of violin at the Moscow Conservatoire and went on to a similar post in Leipzig, but eventually preceded Richter to Manchester, where Brodsky succeeded Hallé as Principal of the Royal Manchester College of Music. Brodsky's wife Anna wrote in her memoirs:

Soon after our arrival [in Vienna] A. B. went to see his former colleague Hans Richter. Richter had been in his last year at the Conservatoire when A. B. entered. . . . Many of the brightest memories of A. B.'s student life are connected with this extraordinarily gifted friend. Richter's principal instrument was the horn; he was also a fine pianist. He could play the flute and clarinet, as well as all the string instruments in the orchestra down to the double bass. On account of these manifold accomplishments he had earned in the Conservatoire the surname of *Nothnagel* (Hope in Extremity), and in whatever difficulties the orchestra might be placed by the absence of some member, Richter promptly came to the rescue. Once, at a public concert in the Conservatoire when the *Tannhäuser* overture was being given, he performed on three instruments at once. Besides his horn, he played the cymbals, which he fastened to his knees, and the triangle. He hung the latter on the music stand and struck it whenever he could free his right hand from the horn. Another pleasant recollection is connected with a rendering of Beethoven's septet at an open practice. Richter gave a magnificent interpretation of the horn part, A. B. was leader in the first movement, and Risegari led the adagio.[4]

Brodsky brought Tchaikovsky's concerto with him, and after going through the score Richter decided to play it through with the orchestra at one of the sessions used to try out so-called novelties. The concerto (completed in March 1878) had originally been written for and dedicated to Leopold Auer, who declared it a radical work and technically so difficult that it was unplayable. According to Mrs Brodsky, her husband impressed the orchestra enormously and was offered an engagement; on the other hand Tchaikovsky's concerto was rejected. Brodsky insisted upon playing it and he was eventually allowed his way, although both he and his wife became very nervous as the day of the première approached.

I sat in the front row. Brodsky appeared before the large orchestra ready to begin.... I was appalled at the greatness of his task. Looking at the hundreds and hundreds of people who filled the hall, I realised what a daring thing it was to play this extremely difficult concerto for the first time before such an audience and my heart beat violently. Then I became all attention. The first few notes showed some trace of A. B.'s nervousness, but then the music he loved took possession of him, and he forgot everything else. His face grew composed and happy, he played his very best.... I never saw an audience more attentive; there was a wonderful stillness during the whole performance. After the first movement the applause was unanimous and prolonged. Then came the dreamy poetic second movement, which passes into a finale full of energy and fire, original and free alike in its conception and in its form. After the finale enthusiastic applause filled the hall. This must have been too much for the conservative portion of the audience. They wished to check it by signs of protest, and for some seconds unmistakeable hisses mingled with the applause, but this seemed only to emphasise the success, for people stood on their feet to shout 'Bravo!' and the opposition was soon overcome. Again and again Brodsky had to appear, and bow his acknowledgments to the excited audience.[5]

Coincidentally Tchaikovsky had arrived in Vienna on 25 November and spent two nights there before going on to Rome, apparently without knowing that his concerto would see the light of day within a week. He would have hated such an occasion and must have been grateful for his narrow escape. Press reaction was mostly negative with none so vehement as Hanslick's in the *Neue freie Presse*. 'For a while it advances in customary fashion, is musical and not uninspired.... Tchaikovsky's Violin Concerto brings us face to face for the first time with the revolting thought—may there not also exist musical compositions that we can hear stink?'[6] Time (and a few judicious cuts in the last movement) has proved the conservative members of that 1881 audience and the Viennese critics wrong. Brodsky earned the composer's gratitude and the dedication of the concerto when it was withdrawn from Auer. Later he was completely won over by the work and championed its cause, particularly as a teacher by instructing a whole generation of young violinists in the concerto.

D'Albert had nothing to say about the concert in his next letter to Mrs Joshua. She had written to him anxiously when news broke in the press of a disastrous fire with heavy loss of life at the Ringtheater in Vienna on 8 December. He replied to reassure her of his safety.

I am sure that in London no-one can have any idea of the extent of the loss of life or of the terribleness of the event. Such a night has not been known in Vienna for many years. The loss of life is now estimated at 1,000! Every person who took a ticket for the gallery met with death. Not one escaped! In the second gallery alone 600 tickets were issued, and being a public holiday, the theatre was unusually full so much so that the ticket-seller was just going to put up the 'sold-out' [notice] when the fire broke out. People were there that perhaps hardly ever go to the theatre. In the gallery the bodies [were] found four and five rows deep. In some instances the father and mother have gone to the theatre and perished, leaving their entire family of little children at home. There is not one family in Vienna that has not lost a relation or an acquaintance. Mr Richter knows many that have perished in the flames. One player in the orchestra was already safe in the street when he remembered that he had left his hat in the theatre, and foolishly going back for it, was also burnt.

The Ring Theatre was externally one of the handsomest theatres in Vienna, and a great ornament to the Ringstrasse; at present there remain only the four walls still smoking. It will not be re-erected as a theatre. The fear that has come over the people of Vienna is terrible; in the Burgtheater—the next best Vienna theatre—on the following evening there was not one seat sold, and all the theatres are more or less empty, so that they have to issue placards stating that firemen will be stationed outside the theatre during the performance etc. I believe that in the forthcoming performance of the *Ring des Nibelungen* the fire will be left out in the last scene of the *Walküre*.

On the evening of the fire Mr Richter was at Pötzleindorf and I was at home where Mrs Richter had invited a few little friends of [their eldest daughter] Richardis, as it was her birthday and she was six years old. At seven o'clock the coachman who had called for some of the visitors said that it 'burned in the Ringstrasse'—he did not say that the Ring theatre was burning, but we had not to wait long before we knew it; soon in every street there were people running in the direction of the Ringstrasse crying 'the Ring theatre is burning' and then the glare of the fire lit up the night so that it seemed like broad daylight, even as far as here in Währing. The crowd before the theatre was enormous, I believe, although I did not see it until the next morning when the theatre was still burning. One person was trodden to death in the crowd. The sight must have been dreadful and yet wonderfully grand; when the gasometer exploded, it was [feared] that the walls would fall on the people, which happily they did not.

Everything concerned with this fire was caused by stupidity. First that the spirit lamp was not properly guarded and then the blunder of turning out the gas. This was the greatest misfortune. The people, trampling and fighting, scrambled to try and get out, but in the total darkness none could find the way. Many in the gallery, where none escaped, leapt over into the [stalls] in sheer despair. Then all the doors not usually in use were locked and were utterly useless. In a quarter of an hour the entire theatre was on fire, and when the flames reached the gallery the cry of terror that came from the poor helpless people huddled up there was such as had never been heard before. Kapellmeister Hellmesberger, as well as his son, wife and daughter were there, but I did not know that his wife had fallen a victim to the flames. I heard she had escaped, and that only his daughter-in-law had her foot sprained. Mr Richter's mother intended to go that evening, as one of her friends was playing in the piece, and she only stayed away because of Richardis' birthday.

... As yet I think nobody in Vienna can think or talk about anything else. Director Jauner was reported to have shot himself and as yet I do not know if it was true [ironically he did shoot himself in 1900]. Mr Richter thinks that all theatres should be closed, and only concert halls left open; all the theatres in Vienna are very dangerous, indeed only the Opera House has taken precautions against fire.[7]

Bruckner had wished to go the Ringtheater that evening (it was right next door to the building where he had a top-floor flat) but chose not to when he saw that Offenbach's opera was being played. When he returned from a walk, he was horrified to see the conflagration and rushed up flight after flight to rescue his compositions from the threat of the fire next door. His arrival was timely for the window sills of his flat were rapidly being scorched. Richter had a very busy few days at the Opera at this time, conducting *Rheingold*, *Carmen*, *Walküre*, and *L'Étoile du Nord* on consecutive days (13–16 December) followed by *Siegfried* four days later. On Christmas Day Pauline Lucca organized a memorial concert for the victims of the fire. The orchestra agreed to her request that they should participate at the matinée, and several soloists also offered their services. After solo and chamber works by Liszt, Brahms, Schumann, Goldmark, and Mozart, the orchestra played the Funeral March from *Götterdämmerung*. Richter conducted, having already officiated at the Court Chapel earlier that morning and with a performance of Gounod's *Roméo et Juliette* at the Opera later that evening. A happier occasion took place the next day when Brahms was soloist in the first Viennese performance of his new Second Piano Concerto, premièred only six weeks earlier on 9 November in Budapest. D'Albert described these Christmas events to Mrs Joshua.

While the presents were being laid out, Mrs Richter was having great trouble with the scene from *Götterdämmerung* for the children. The rehearsals for the musical part had been begun already a month before, in consequence of which the scenic portion had been a little neglected. What the children sang was the song (eight bars of it) of the Rhinemaidens in the *Rheingold*, as the music in *Götterdämmerung* was too difficult for them; the scene from *Götterdämmerung* was selected to give the little Hans [as] Siegfried something to do. Not until a week before did Mrs Richter begin to have the dresses made, and not until the very same day was the Siegfried costume bought. The dresses of the little girls were two green and one red with garlands of flowers turned round them; they looked very pretty. The Siegfried costume was the most difficult to manage, especially in such a hurry, but was at last arranged. Mrs Richter borrowed from the theatre Lohengrin's costume (a smaller pattern) and bought a toy sword and a helmet which had 'Fireman' written on it. But the most difficult of all was the [scenery]. This consisted of a back-[cloth] scene of a terrace overlooking a garden . . . and a side scene representing a door. They were the only available pieces of scenery that could be procured at the time. The Rhine was a piece of cloth arranged in folds. There was great difficulty in keeping Mr Richter out of the room in which the preparations were carried on. He at last became quite angry because he said he was not a child, that they should hide anything from him that he was to receive. But when at last after much trouble the doors were opened he was very pleased with the sight, especially with his little Hans, who made a very good and, luckily for once, quiet Siegfried. Afterwards Mr Richter drank my father and mother's health because he said he liked them so much. Then he played with me some Fugues of Bach, and I alone played to them for the first time the new *Parsifal* and die Zaubermädchen of Wagner, which is very beautiful. The next day (Christmas Day) was very warm, no sign of winter whatever; at five o'clock in the morning Mr Richter had to conduct in the Imperial Chapel a Mass of Hummel, and at twelve we went to the concert of Pauline Lucca to aid the funds of the Ring Theatre disaster.

The next day was the fourth Philharmonic concert, the chief item in which was the new piano concerto by Brahms, played by himself—I had already heard it twice in the rehearsals. It really is very fine and very difficult, the most beautiful movement is the slow movement. Brahms was called forward twice, which was not so much as I expected, although Richter says it is more than he has ever received here before. It would perhaps have been even more successful if someone else had played the piano part. . . . Mr Richter leaves on Saturday for London. I play here as soon as he returns. He is so glad he is going to London. He is not contented with the Philharmonic Concerts where he has so many rehearsals and so many concerts and only 700 gulden (£70) a year.[8]

14

1882

Richter and d'Albert

THE reason for Richter's trip to London at the beginning of 1882 was to conduct another charity concert whose receipts would be donated to the survivors and families of victims of the Ringtheater fire. It took place on the morning of 7 January at the Royal Albert Hall and was devoted to works by Beethoven and Wagner. According to Richter's diary it was a very solemn occasion, the full house standing for the Austrian national anthem. Besides conducting the six orchestral items, he also accompanied Marie Roze at the piano in Agathe's aria from Weber's *Der Freischütz*. D'Albert reported to Mrs Joshua what he had heard from Richter after a rehearsal for the concert.

He wrote to me that 'das Concert wird grossartig' [the concert will be splendid], and to Mrs Richter that he was astonished at the choir of 300 ladies, all of whom he said were from the best families, and who brought their relations to the rehearsals to the number of two thousand. I wish that he had not gone to London simply for this one concert, for in his absence were given the only two performances of *Götterdämmerung*. Although a better and more artistic performance could hardly be possible, still I should have preferred to have seen Richter at the desk instead of Kapellmeister Fuchs, who although a very good general conductor has not studied the *Ring des Nibelungen* so thoroughly. [Johann Fuchs was a staff conductor at the Opera and Director of the Conservatoire at the time.]

On Friday I was with Mrs Richter in the third concert of the Gesellschaft der Musikfreunde. In the programme were included the new work for chorus and orchestra by Brahms, *Nänie* (Schiller), a new scherzo by Hugo Reinhold, a resident composer of not very remarkable ability, and the F minor concerto of Chopin performed by Frau Essipoff. The new work of Brahms was a comparative failure. It met with hardly any applause, partly owing to its excessive length and to the absolute lack of all melody, and to the monotonous repetition of the words. The scherzo, though more fit for the ballet than the Musikverein, was, on the contrary, very well received, which shows the taste of the Viennese public. Essipoff's playing was simply perfect![1]

Richter travelled on to Hamburg from London. There he stayed with his old friend Josef Sucher, and attended a performance of Smetana's opera *The Two Widows* and a Beethoven concert conducted by Hans von Bülow on 10 January. Two days later he was back in Vienna in time for the next Philharmonic concert on the 15th, which featured another Viennese première, this time the 'Scandinavian' Symphony by the English composer Frederic Cowen (Richter had conducted it in London on 19 May the previous year), followed that evening by *Lohengrin* at the Opera. Cowen later recalled his visit to Vienna.

I arrived the day before the rehearsal, and being a little tired after the long journey, I was sleeping peacefully at nine o'clock when a messenger came to the hotel from Richter to say that he was waiting for me. I had no idea that the Viennese people commenced work with the dawn, but of course I jumped out of bed at once, dressed as quickly as I could, and went over to the hall. I missed the first movement, but I was still in time to make a little speech to the orchestra in my best German and to listen to the rest of the work. The performance took place on the next Sunday. I believe it was a very fine one, as it was bound to be under Richter, and with an orchestra considered to be one of the best in the world; but I confess I did not hear it, for this being practically my début as a composer on the continent, I was too nervous to sit among the audience and remained in the artists' room at the back. I judged, though, that it was successful from the fact that I had to come forward and bow many times, particularly after the adagio and at the close of the work. . . . A further proof of this [success] was that it was at once accepted for publication by a Viennese firm, and very soon found its way to all the chief musical centres of Europe.[2]

D'Albert was very unenthusiastic about the symphony; he was also patronizing about the Viennese public, whose taste in music he considered to be superficial.

I suppose you will be also interested to hear about Mr Cowen's symphony, and I will try to tell you about it, but you know I do not like his compositions, even this his best work, and it is not prejudice, for I have unfortunately heard it three times here, and have tried to find something nice in it, although I certainly did not succeed. It is true what Professor Hanslick said about it, 'dass es hübsch ist', but 'hübsch' and 'schön' are very different expressions and the prettiness of the themes is not equalled in the beauty of the construction. A true symphony must improve on hearing, but Mr Cowen's symphony, as even Mr Richter says, disproves the former favourable impression. The first three movements pleased very well here because they skim lightly over the frivolous surface of the Vienna public without in any way searching into the depths. Mr Richter is glad of the success of the work, not on account of Mr Cowen, but on account of the English nation.

... The other evening Graf [Geza] Zichy played in the Bösendorfer Saal. He lost his right hand hunting and plays with the left as if with both. Although not very artistic, his playing is wonderful. He is a pupil of Liszt.

Mr Richter is the only real, true, pure German character that I have met with since I have been here, for the Viennese are rather of the Slavonic than of the German stock, and their character is lighter, more akin to the French than the true German. I know too, through Mrs Richter, such a great many Hungarians, who though a very characteristic and musical [nation], is not at all a true or deep nation, and above all they dislike the Germans in which opinion, to my great distress, Mrs Richter coincides. Mrs Richter (née von Szitányi) is in every way very 'geistreich' [clever] and nice (excepting her German prejudices). Her family had at first great objection to her marrying Herr Richter, as her family is very proud, and had great aversion to the idea of her marrying an artist. For a year they prevented Mr Richter seeing her, when Mr Richter used, through bribery of a servant, to send her letters enclosed in musical scores; one was enclosed in *Lohengrin* at the time of its first production in Pest, wherein he said that *she* was the 'Gralstaube' [dove of the Grail] which would give him courage to further works. The only composition of Mr Richter is a very nice Lied called *Marie* (that is Mrs Richter's name, or in Hungarian 'Mariska'). The accompaniment of this song is purposely made very difficult because at that time in Pest Mrs Richter was taking singing lessons from Herr Richter under whose care her guardian had placed her as he thought him (Mr Richter) very solid and like himself very strict. These lessons should have lasted [half] an hour each, but Mr Richter always made them last an hour. Mrs Richter used to sing much in the evening when Counts and gentlemen of the Hungarian aristocracy accompanied her on the piano. This made Mr Richter very jealous so that he wrote this song and made it so very difficult that nobody might accompany her but himself. In the singing lessons one song that they always kept to the last, and which Mrs Richter never could find easy, was by Liszt, *Ich liebe Dich*.

Wagner must have a beautiful house in Bayreuth. Mrs Richter says that everything in the house is perfect in construction. Every chair is differently designed. The walls are decorated in fresco, with paintings of Wagner's different dramatic scenes. There are two children, Eva and Siegfried and they must be very nice, although not at all talented. Wagner is very fond of them, and when there is a good pudding he will rather not take any himself than that they should not have any. Wagner must be passionate. Nobody dare praise any other person in his presence. Once somebody said that Richter was a good Kapellmeister to him, and he was furious. Once a very tall singer came to see him, and rather than have to look up to him, he jumped on a chair and so conversed with him.

I think all musicians must be a very sad and bitter set of people. Hardly one has led a happy life, and not one bears a friendly feeling to the other; they are all jealous, ungenerous, and illiberal. . . . The only composer who one can in every way love and respect, and who, while blending almost Beethovenish genius with wonderful self-consciousness, never is ungenerous or illiberal, and who nobly withstands all the mocking of raving Wagnerianers is that great man Brahms; as a man one can say nothing against him, and only those who are blinded by prejudice could say anything against him as a composer. Richter says he is the only musician that I may know well in Vienna.

Musical matter must be very tame in England, very different from here. All the papers speak favourably of Mr Cowen's symphony. They mention him as 'der feine Gentleman'. He is staying at the Grand Hotel on the Ringstrasse and has been once to see Mr Richter.[3]

D'Albert made his début with the Philharmonic orchestra on 26 January 1882, when he played the first movement from his own A minor Piano Concerto (Richter rejected the other movements as too unoriginal for public performance). The programme also included the first performance at the concerts of Mozart's 'Linz' Symphony and Berlioz's *Harold in Italy*, in which Adolph Brodsky played the viola solo. D'Albert told his father that the occasion was awesome. 'At last the attendant came and said that the symphony was over and I had to go on. Richter was waiting for me at the conductor's podium, and no sound came from the public because I was a foreigner. But, dear Papa, I will never forget the applause when the concerto was over. I was called forward three times. Richter had tears in his eyes when he pressed my hand, and then I had to appear a fourth and fifth time.'[4] He also sent a short note to Mrs Joshua together with a few reviews which appeared in the press.

I must tell you that Richter wishes me to say that it was a *very* great success, the greatest I have ever had, and that I was five times called forward. The orchestral players were so nice; many people came to see me. Mrs Richter was so 'aufgeregt' [excited] that she wept during the concert as did Dr Schäfer, Mr Richter's mother, and Herr Bösendorfer. At home there were people invited to dinner, and the children, at least the oldest, gave me a beautiful wreath on which was written 'Richardis, 26 Feb. 1882', and which, as my first, I shall ever prize. Mr Richter made a speech in which he said (from *Die Meistersinger*) 'Eine Meisterweise ist gelungen, von Junker Walther gedichtet und gesungen etc.' [sung by Hans Sachs shortly before the quintet in Act III], and even the little baby had with me 'angestossen' [clinked glasses].[5]

Hanslick had some barbed comments to make. He erred in calling d'Albert 'a pretty young man of about fifteen', and preferred to call the work 'one movement from a piano concerto' rather than use the description in the programme 'a piano concerto in one movement'. He dismissed it as an immature piece which owed much to this or that work and whose sparse ideas were drowned in virtuosic flourishes, though he failed to doubt that the talented d'Albert would compose better things in years to come.

But we can wait, and so should the Philharmonic. This famous concert organization should limit its ambitions to the grove of the Hesperides [daughters of Atlas and Hesperus whose duty it was to guard the tree bearing golden apples] of music, and not a kindergarten. Just as we have protested against the fifteen-year-old Mozart appearing in the programmes of the Philharmonic, so we may well protest against the fifteen-year-old d'Albert. The reception of the latter on the part of the audience was of course the friendliest and warmest imaginable. One need never count in vain upon the well-meaning kindness of the Viennese public towards foreigners and children. Naturally Herr Hofkapellmeister Richter knew that when he recently performed the symphony of Mr Cowen in a Philharmonic concert and now the young d'Albert. Notwithstanding all the relative appreciation of both these Richter protégés, we would not wish to see our Philharmonic concerts develop into a little English colony.[6]

This is a significant review not so much for d'Albert as for Richter and indicates the niggling way in which a campaign was beginning to take shape in the press which resulted in his withdrawing from the Philharmonic concerts at the end of the 1881–2 season. The more that comments such as Hanslick's about Richter showing favouritism towards English music and musicians began to appear in the press, the more reluctant he became to promote performances abroad of new English works. Evidence that he wanted to despite such opposition lies in his reply to Stanford's letter of 3 October 1881 with reference to the opera *The Veiled Prophet* (quoted on p. 159, Chapter 12, text attached to n. 8). Written from London to Stanford in Cambridge, it is quoted unaltered to show the state of Richter's English at this period. He had now—unwisely it would appear —stopped receiving language tuition.

Verzeihen Sie mir dass ich Ihnen nicht sofort von Wien antwortete [forgive me for not answering your letter immediately from Vienna], but I had to much to do during the changement of my dwelling house. After my returning to Vienna it will be my first matter to promote your *verschleierter Profet*, and I hope the best. We have so few novitates, dass wir uns gratuliren können, wenn so gute Künstler, wie Sie sind, uns new works send. [that we can be pleased if such good artists as you send us new works]. I would come to you but all my fry time I must spend to our opera.[7]

The *Wiener Abendpost's* critic ended his review of d'Albert's appearance with a warning against just such a negative campaign. He for one realized

Richter's worth and began, 'Hans Richter is the most significant of living conductors and the Philharmonic can count themselves lucky to play under him. In spite of this some organized agitation on the part of just a few musicians has been noticeable recently.' The reviewer accused this anti-Richter faction of trying to provoke the Director of the Opera Wilhelm Jahn into taking over Richter's post, but pointed out that he (Jahn) was already overworked and in no position to take on more responsibilities, nor did he wish to. 'The overwhelming majority of members of the orchestra are enthusiastically well disposed towards Hans Richter, and this agitation will soon find an end,' it concluded.[8] Underneath this notice d'Albert scrawled, 'Yesterday in the rehearsal Brahms told Richter that he was present at the concert and thought the pianoforte a very bad one.' The pianist confirmed in a later letter that Bösendorfer only possessed one concert grand piano and that was currently in Prague, so d'Albert had to make do with a salon piano. The same letter reports Richter's fury at the reviews, 'Mr Richter was terribly enraged about it; he said it was out of enmity towards him, as every journalist in Vienna is his enemy, except two, even Hanslick lately; and they all know that I am staying with him.'[9]

D'Albert's biographer Wilhelm Raupp points to frustration and boredom eventually setting in by the spring of 1882 when the young man went his own way to Weimar. Certainly d'Albert was fretting at not having enough solitude to compose when he told his parents in January that he had only composed thirty-two bars of a symphony which Richter had promised to conduct in London in May. He also felt that he could not devote enough time to piano practice; Richter was constantly urging him to give recitals of his own music at social functions, whereas he wanted to study the basic repertoire for his chosen career as a concert pianist. The conductor was very strict with d'Albert, taking the role of surrogate father and mentor very much to heart. Before d'Albert's first public performance with the orchestra Richter forbade him to leave the house for a week; instead he had to concentrate his mind solely upon the occasion ahead of him. A glimmer of d'Albert's frustrations is apparent from the next letter to Mrs Joshua. It describes in detail a walk in the Vienna woods which took him to Heiligenstadt and the house once occupied by Beethoven, not a museum in those days. He then wrote about another of von Bülow's monumental piano recitals, this time devoted to the music of Brahms. The composer was present, and, having waited three months, d'Albert was at last introduced to him by Richter.

It is a very pretty country village not far from Währing, and surrounded by the Wienerwald. I saw the house in which Beethoven composed the *Pastoral*

1882: Richter and d'Albert

symphony, a small, at the present time very dirty peasant's house with only two windows, quite small in the chief room. In the summer it is let as lodgings where last year a person had lived who played waltzes and operatic airs every evening in the same room as Beethoven composed his immortal work! . . . By the side of the path runs a little brook which comes down from the heights above; but of all beautiful spots the *holiest* is a little way further on where the path descends into the valley. Here is a secluded spot surrounded by trees and bushes, and here it was that the great man loved to sit. There is erected here a monument by the Gesellschaft der Musikfreunde.

A little while ago I was at a musical evening given by Dr Standhartner who is very celebrated here. . . . I played nearly all the evening and I will only say that Mr Richter was *very* pleased and that I was very happy. Other people also wished me to go to them but Mr Richter does not allow me to go out at all in the evening, and is very particular, which is very good of him.

A great event here was the Brahms concert given by Dr Hans von Bülow. The programme consisted entirely of Brahms' compositions, and that even the most enthusiastic Brahmsverehrer [worshipper] admitted was a mistake. After four pieces it became dreadfully langweilig [boring], which is never the case with Wagner. The finest number was the Sonata Op. 5. What made it even more tedious was that Bülow repeated so many of the numbers simply that the people might hear them twice to better appreciate them. After the concert I went with Mr Richter and Frau to the supper in the 'blauer Igel' inn [its real name was 'rother Igel' but 'blau' is slang for drunk]. The guests were Professors Epstein and Door and their wives, Dr Standhartner, Ludwig Bösendorfer, Hans von Bülow, and Brahms. It was so nice to be introduced to so many great men, but of the greatest worth was my introduction to Brahms, who said he would have me at his house when he returned to Vienna for good. Although I admire Brahms so very much as a composer, please do not think for a moment that I prefer him to Wagner, or that my admiration of the latter is in the least degree abated, for although Brahms could make me think and coldly philosophize, only Wagner could truly inspire me, and he is my greatest inspiration when I am not happy. . . . P. S. Richter says that next season he will go out *nowhere* in the evening, and that nothing will induce him to see anything.[10]

Richter conducted two more Philharmonic concerts in the Vienna season; the first (on 12 March) included Brahms's Violin Concerto with Hugo Heermann as soloist, the second (on 26 March) featured Frau von Stepanoff as soloist in the first Viennese performance of Grieg's Piano Concerto. On 2 April he conducted Beethoven's 'Choral' Symphony in aid of the operahouse pension fund, and two days later took the orchestra to Graz for a programme of Wagner, Beethoven, Weber, and Volkmann's Serenade. He took d'Albert along to the Styrian capital. During their six-hour journey the orchestra of 108 men travelled by train over the spectacular Semmering Pass (a two-hour climb to complete twenty-six miles). Once there d'Albert played for the local dignitaries and was booked immediately for a recital the following winter. Meanwhile he had at last had a lengthy meeting with Brahms, arranged during Richter's rehearsals for the concert on 12 March.

On Monday I was with Brahms for an hour! Mr Richter had asked him in the rehearsal to hear me play his concerto and other pieces, but he said he disliked hearing his own compositions, and would prefer to hear something else of mine; and so he appointed me to go to him on Monday at ten o'clock. I had to go alone as Richter received at the last moment a telegram that he must conduct in the rehearsal, and so he could not go with me as he intended to do. I had more anxious feelings about going to Brahms than in playing in the Philharmonic concert, but as soon as he had spoken with me I felt at ease. He was so kind and so friendly, quite different to what I had expected. He was disappointed that I had not brought the [music] with me, as he said he liked so much to read music. He would not hear me play any of his own compositions, but gave me little hints about the Rhapsody which I have to play here. He then heard my Suite, and it pleased him very much, especially the Trio of the Gavotte. I had studied so much Bach. He asked me what other compositions I had written and said he would like to see them, and asked what I was writing at present; when I said a symphony, he wished to hear the first subject and other bits out of it. I only regret that it is not finished as it was Mr Richter's greatest wish to play it in London, but playing in this concert has distracted me; indeed I think one cannot compose when one lives in a family. One must live alone to have the requisite quiet, although not for the world would I have left this house where I am so happy. Brahms says he has no prejudice against the English, as people say, and the only reason that he does not go to London is because of the long journey and he hates travelling. I translated for him a letter from Mr Stanford, asking permission to dedicate his new edition of fifty Irish melodies to him. Brahms is glad I was not with Leschetitsky as he cannot bear the lady-pianists. These females, he said, have no artistic feeling and practise away with the window wide open like machines. Most of Leschetitsky's pupils are young ladies.

He said I ought to live in Germany, it would be better for me. He lives very simply and homely and has three small rooms with a Streicher piano. When I went in he was reading the score of the *Meistersinger*, so that even he, himself so great, so famed, feels there is a greater.[11]

A month later he told Mrs Joshua, 'I was again with Brahms for more than an hour, and he was even nicer to me than before; he showed me many things and spoke a great deal. He told Mr Richter he was pleased with me, and is coming out here before we leave.'[12] Richter's sixth and last child, his younger son Edgar Ludwig, was born on 18 April. A week later, on 26 April, Richter left Vienna for a two-month London season of Wagner operas at the Theatre Royal, Drury Lane, and his own Richter Concerts.

Two days before his departure he was visited in Vienna by Augustus Harris, impresario and lessee of the Theatre Royal, to conclude business arrangements for what was to be an epidemic of Wagner operas in London (described in the next chapter). D'Albert played Anton Rubinstein's Fourth Piano Concerto in Richter's first London concert (3 May), Adolph Brodsky had a triumphant reception with Tchaikovsky's Violin Concerto (8 May), and Dvořák's Sixth Symphony was at last conducted by its dedicatee (15 May). 'This morning', wrote Richter to the composer, 'was the first rehearsal of your fine work. I am proud of the dedication. The orchestra was quite enthusiastic. The performance is on the 15th. I am certain of a great success. It has also been rehearsed with love.'[13]

By this time d'Albert had returned to Vienna for his first encounter with Liszt: 'Last week I played and had a lesson from Liszt, and he was so kind and so nice.'[14] By the end of May he had left Richter's home for good and moved to Weimar. They next performed together in Vienna on 4 November 1886, when d'Albert played Brahms's Second Piano Concerto.

On Saturday morning I went to see Liszt and he was very nice to me. He said I shall only stay a month but we shall work hard together, and he kissed me; but I think he kisses a little too much. Then he asked his housekeeper Pauline to find me a room; he wished to have it near Gutmann, the music publisher here and told her to recommend me. . . . In the afternoon I was at the first lesson with Liszt. He introduced me to all the other people and was very kind. All I played him was the cadenza to his piece, which he liked so much I had to play it twice over. He has about fourteen pupils and a great many play very indifferently. He is, however, not so easy with them as I expected. He was very particular about my being comfortable in my house and said he would come and visit me soon as he lives quite close by.[15]

D'Albert's last surviving letter to Mrs Joshua again provides a unique insight into Liszt's piano classes.

On Saturday last I had a great success when I played for the first time in the lessons. Before I had only played the little cadenza, on Saturday I played my Suite. I don't like to tell you, but it was really beautiful. Liszt was so charmed; he said repeatedly, 'It is as if Tausig had returned', and then again said to the others that I played like Tausig when he first went to Liszt. He kissed me often and invited me to go to him the next day when there was a matinée, and he played a solo, a Liebestraum by himself. It is enough to come to Weimar simply to hear him play, because he still plays simply wonderfully. The technique is perhaps not any more so good, but the Adagios are beautiful. In the lessons he also plays very often. On Sunday he invited me to dine with him the next day, and so on Monday I was there at two o'clock. There were eight persons to dinner, but none of his other pupils. The dinner lasted two hours and Liszt was very witty indeed the whole time. After dinner he told the others to ask me to play to them and he went into his room to sleep. I then played a great deal, perhaps for two hours, until he awoke, when he came out of his room and said he had heard the Berceuse and Polonaise and was very pleased. Then he went a little way with us in the park and then returned home.

Today was again a lesson with him and he wished me to play the Berceuse and Polonaise over again in both of which he gave me several hints. Last Tuesday there was a chamber music soirée at his house led by the fine violinist Concertmeister Koempel, who is really a splendid artist and plays beautifully. They played a quartet by Tchaikovsky and the seldom played posthumous Fugue by Beethoven. I never knew anyone so loving as Liszt, although he can also be very grob [rude]. There is no doubt I hope that I shall learn a great deal from him while I am here.[16]

On the journey to London with Richter, they stopped in Nuremberg and d'Albert sent Mrs Joshua a postcard. 'I write principally to know if you have heard that Richter has given up the conductorship of the Philharmonic concerts, and perhaps also the Opera. I think it dreadful, but I dare say he may go back to Vienna, but not to conduct.'[17] Matters were not quite as serious as d'Albert described. Richter would return to Vienna, he would not leave the Opera, but he had resigned the conductorship of the Philharmonic concerts. It would appear from the minutes of the Vienna Philharmonic Orchestra that there was a certain dissatisfaction with Brahms among some of the players. Richter would have none of this, particularly when he began to be attacked in the press, and so he simply walked away from trouble. Behind him he left a whirl of speculation, rumour, and gossip as he headed north to London to conduct, among other things, the British premières of two of the greatest Wagner operas, *Tristan* and *Meistersinger*.

15

1882–1883

The Master's Death

It must have been with some relief that Richter shook off the dust of Vienna with its carping critics in the press when he set off for London, though no relaxation of his busy schedule lay before him. There was intense interest and fierce competition in the air when he arrived in the English capital with Wagner once again the focus of attention. Angelo Neumann was presenting three cycles of the *Ring* (followed by additional performances of *Walküre* and *Götterdämmerung*) throughout May, the first to be staged in London. This first cycle was already over when Richter arrived, the other two competed with the season of German opera promoted by Hermann Franke and Bernhard Pollini which he was contracted to conduct. Neumann was Director of the Leipzig opera-house, whilst Pollini held the same post in Hamburg. What with Ernest Gye's concurrent season including singers of the calibre of Patti, Sembrich, and his wife Albani, and his two successes *Aida* and *Carmen* (with Lucca in the title-role), London's opera-goers and critics were treated to a surfeit of choice. Neumann had a star-studded cast including several singers from the Bayreuth Festival of 1876, among them Niemann, the Vogls, Lilli Lehmann, and Schlosser. Emil Scaria, Theodor Reichmann, and Hedwig Reicher-Kindermann were also in the company. *Siegfried* and *Götterdämmerung* were subjected to several unsatisfactory cuts, and the scenery was deplorable. The conductor was Anton Seidl, whom Richter had brought to Bayreuth into the Nibelungen Chancellery. He made a good impression upon public and critics alike.

The Drury Lane German opera season, on the other hand, had Hans Richter as its leading attraction. The operas to be presented were (in order of opening) *Lohengrin* (18, 24, 25 May), *Fliegende Holländer* (20 May, 8 June), *Tannhäuser* (23 May, 1, 7, 14, 16 June), *Fidelio* (24, 31 May, 9, 21,

27 June), *Meistersinger* (30 May, 3, 6, 10, 15, 17, 22, 23, 29, 30 June), *Euryanthe* (13, 28 June), and *Tristan* (20, 24 June); thirty performances of opera with six Richter Concerts interspersed among them (including Brahms's Requiem, and Beethoven's *Missa solemnis* and 'Choral' Symphony), a taxing schedule of staggering proportions for a conductor who memorized his scores. His singers, as eminent as those at Her Majesty's Theatre, were Rosa Sucher, Therese Malten, Marianne Brandt, Josephine Schefsky, Hermann Winkelmann, Franz Nachbaur, Emil Kraus, Eugen Gura, and Josef Ritter. Karl Armbruster was in charge of the 100-strong chorus imported from various German opera-houses, and the orchestra was that of the London Richter Concerts. Richter's supreme achievement was to weld together an ensemble of singers and players from disparate sources within two weeks (his first orchestral rehearsal for the operas was on 5 May and was devoted to *Tristan*). Hermann Klein summarized the achievements of this enterprise.

They created a new standard, a new mental perspective, not only for the rising generation of opera-goers, but for those critics whose insular experiences had been confined exclusively to the lyric art of the Italian and French schools. Henceforward we were to understand what was signified by Wagnerian declamation and diction superimposed upon a correct vocal method, as distinguished from mere shouting and a persistent sacrifice either of the word to the tone or of the tone to the word.... Imagine the advantage of hearing *Tristan und Isolde* and *Die Meistersinger* for the first time with such a noble singer and actress as Rosa Sucher as Isolde and Eva; with such a glorious Tristan and Walther as Winkelmann; with the famous Marianne Brandt as Brangäne; with that fine baritone Gura as King Mark and Hans Sachs. Those artists were in their prime and sang their music as few German singers have sung it since.[1]

Lohengrin was sung for the first time in German; hitherto it had been either in Italian or, when Carl Rosa's company performed it, in English. Unfortunately Richter hardly merited a mention in the reviews, for by now it was taken for granted that he would give an exemplary performance and inspire everyone to do their best. As Louis Engel put it in the *World*, 'To praise Richter as *chef d'orchestre* would be carrying coals to Newcastle.'[2] Only *Tristan* aroused any negative criticism in the press, this from the *Illustrated Sporting and Dramatic Chronicle* being typical.

The two chief personages forfeit our sympathy, and the story is repulsive.... The orchestral music, almost entirely built up of the *leitmotifs* which abound in the work, is often superb, but much more often distressing to the listener, owing to harsh progressions, incessant changes of key, and the absence of 'full closes'. . . . Herr Wagner's supporters assert that posterity will universally admire his opera-dramas. We venture to assert that posterity will do nothing of the kind so long as music is regarded as a source of enjoyment rather than a laborious exercise of the mind.[3]

The *Musical Times*, which described Richter as 'a tower of strength to the project', questioned only the standard of *Euryanthe* which, it felt, had been neglected in favour of the Wagner operas. *Meistersinger*, on the other hand, was recognized for the success of its ensemble work and earned the accolade that 'if the German Opera existed for this completeness only, it would not have been set on foot in vain; for so long have we been accustomed to the slovenly habits, perfunctoriness, and individual self-seeking of the Italian stage, that we needed an example of what can be done when all are in earnest and ready to subordinate themselves to the general good'.[4]

The whole venture eventually proved so tiring for Richter that Parry's First Symphony, scheduled for its first performance on 19 June, was cancelled. The number of rehearsals required for *Tristan* would have left too few for this new work, although another reason given by Richter in his diary (13 June) for this cancellation was the poor quality of the parts. Though disappointed, Parry still attended several of the operas.

7 May: In the morning to Richter by appointment for my Symphony but he was too busy.

9 May: In morning by appointment again to Richter, who after long pottering about at last went through the Symphony with me, seemed to like the slow movement and scherzo and promised to do it. Rehearsal to be early in June.

18 May: Then to *Lohengrin* under Richter. The singing was not first-rate but the orchestra, ensemble and chorus and acting generally was good and the total result was most enjoyable. An enthusiastic audience containing many singers and all sorts of curiosities.

20 May: After dinner went to *Fliegende Holländer* under Richter. Performance not on the whole good—no better than Carl Rosa's. Gura as the Dutchman was disappointing. Sang untidily and did the acting too much. [Rosa] Sucher as Senta was very good. The chorus singing was bad, unsteady. Acting fair. It did not make the impression on me I know it should.

23 May: In the evening to *Tannhäuser* under Richter, a really fine performance. The first time that it has really gone home to me in its entirety. The house was full and enthusiastic.

30 May: In the evening to *Meistersinger*. I went with extreme anticipation of delight, and was far more delighted than my utmost expectations could rise to. I think I never enjoyed any performance in my life so much. Beckmesser (Ehrke) was splendid and so was David. Winkelmann (Stolzing) was, I should say, good without being remarkable. His voice told well and kept up right through, though

it was a little anxious work in the last act. Gura (Sachs) was said to be very good, but at present I am not powerfully in love with Gura; he seems to me to be praised too highly on the grounds of his reputation; Stanford particularly was extravagantly laudatory but it seemed to me that the performance, though very fine, was not of the very highest order. A little heavy and consciously self-satisfied. The house was perfectly crowded and there was tremendous enthusiasm.

2 June: Had to go to Vere Street in the morning, and was kept waiting for an hour and a half as Franke was too busy. All that came of it was that Franke couldn't tell me anything, though the rehearsal was fixed for tomorrow; only I had better bring the parts to Richter. Did so in the afternoon and saw Richter, who said he was very tired and had too much to do—couldn't have rehearsed tomorrow, but would try to find an opportunity before long, any spare hour in the theatre. Don't expect we will come off and shall not believe in any performance at all till the full rehearsals are over and it is definitely announced. They always seem to want to back gracefully out of it.

5 June: To Richter concert where we had much that was interesting. Liszt *Hungarian Rhapsody* for orchestra perfectly superb and such a performance as no one but Richter can get. [Josef] Sucher's Cantata struck me much, beautiful writing for the voices and very fine scoring, the general conception good too. The performance of the C minor was the best I ever heard. Richter has got his band into most wonderful order; the things sound quite unsurpassable. I went down to see him for a few minutes in the interval and was introduced to Sucher, a curious fat, thick-fibred, ponderous animal, but with a pleasant bright expression in his unhandsome countenance.

13 June: To a rehearsal of my symphony at 9.30. Unfortunately a great part of the band did not come. No first clarinet, no third horn and a great part of the strings absent. They struggled on roughly for a long while against heaps of mistakes in the parts and it finally became evident that with this one rehearsal which only is possible Richter could not make it go and so it was given up. A good deal of it sounded well, but even the men who were there were tired and not up to the mark and shirked their work. In the evening to *Euryanthe*; parts of it beautiful but the posturing and posing on the stage and the preposterous plot to me insupportable. The orchestra sounds often poor and raw especially the trombones which he uses in the conventional ways.

20 June: To *Tristan and Isolde* at Drury Lane; at which I was much too tired to enjoy it to the full—but what I did not enjoy I at all events perceived. The performance was on the whole remarkably good. Winkelmann made me rather nervous at times as his voice showed signs of giving way, but [Rosa] Sucher was better than ever as Isolde, Gura's Marke was also first rate and Kraus's Kurvenal the same.

24 June: To *Tristan and Isolde* at Drury Lane, Sucher was quite magnificent as Isolde, far finer than on Tuesday especially in acting.

25 June: Then I went to lunch at the Joshua's to meet Richter & Co. A very swell and elaborate feast. Richter was very friendly.

26 June: Richter concert. Dann[reuther] played the Liszt A [major] concerto and after there was the Ninth Symphony. Richter must evidently be under a strain of nerves—every single movement was too fast, restless and hurried.

27 June: *Fidelio* under Richter. Not quite a first-rate performance, but interesting.[5]

Two of the singers from the company recalled their impressions of London and events surrounding the 1882 German opera season. One of them was Rosa Sucher, who gave a graphic account of the contrast between affluence and poverty evident on the streets of London. They stayed in the Covent Garden Hotel, but often Richter would take them to his favourite haunts to eat and drink, in particular to restaurants with a German kitchen. Karl Armbruster was their self-appointed guardian and urged them to stay away from certain areas of London, where, he assured them, their property and even their lives could be in danger. 'When we left the Opera at night for the hotel we saw crumpled figures, mostly women with faces white with drunkenness, sitting in almost every doorway. In daylight we would see them sitting and eating at the kerbside. In the dressing rooms one could always hear the cries of the drunks from the streets.' Ladies riding in Hyde Park, on the other hand, gave a completely different picture of London. Sucher continued, 'The *Minstrels*, reputedly an all-negro choir, were much the talk in those days. They were singing at St James' Hall, where the Richter concerts were being given. One evening we were going to the the artists' room to see Richter, when we were followed in by one such negro. I got such a shock when he called out in a broad Viennese dialect, "Greetings, Herr Hofkapellmeister!" It was a former chorister from the Vienna Opera.' Sucher described her preference for, and greater success as, Senta in *Fliegende Holländer* rather than Elsa in *Lohengrin*, and recalled an alarming incident in a performance of the former. 'As I hurried up the rock in the third act to jump off it into the sea, I checked myself at the last moment and saw that the stage was completely opened up, down to the deepest cellar. At that moment I did not know what to do. Richter was gesticulating angrily up at me from the conductor's desk—I nearly jumped down out of sheer duty, but I pulled myself together and ran off into the wings. Richter came angrily on stage and then he saw the whole mess. The stage hands had not understood the director's orders and closed the stage too late. No bone in my body would have remained unbroken had I jumped. I felt too faint to go before the curtain.'[6]

Hans von Bülow offered to coach Sucher in Hamburg in the part of Isolde when he heard of her participation in the English première. 'It would certainly have been wonderful to study the whole role with him,' she recalled, but she never got beyond the first act because of the amount of time she needed for the other operas also being performed.

I shall never forget how he did not get further than the point 'Mein Herr und Ohm' [Isolde quoting Tristan in her narration to Brangäne in the third scene of Act I]. My God, I had to take it all in on the wing, for he was so idealistic in his thinking; but I am proud to this day that he took an interest in me.... The day of the performance arrived. I was somewhat hoarse from the strain of it all, but sang nevertheless, and after a few bars excitement and enthusiasm conquered my hoarseness. I was in very good Isolde-voice.... When the first act ended there was a small pause, and I wanted to say to Winkelmann (Tristan), 'I think we are...', but I got no further for such a storm of roaring, shouting, and cheering broke out; though stunned, we took at least ten curtain calls—nothing in the world can compare with this moment.... The critics were enthusiastic about my Isolde, and Patti stayed the whole evening and waved her greetings to me. Richter was pleased with me. We were together a long time after the opera, and in the flush of success he suggested we should use the familiar 'Du' form of address.[7]

Eugen Gura was the other company singer who later recalled the events of the season. This took the form of a letter to his sons dated 2 June 1882, published in his memoirs.[8] The highlight of the tour for him was a State Concert at Buckingham Palace hosted by the Princess of Wales in Queen Victoria's absence. The soloists were not only from the German opera company; singers such as Albani and Nilsson also took part, but Richter was not the conductor. He is mentioned in the *Meistersinger* triumph in which Gura sang Sachs, and also as conductor of the 'Choral' Symphony in which he sang the bass solo at the concert on 2 June.

When news of plans by the two companies from Hamburg and Leipzig to perform so many Wagner operas in London virtually at the same time reached the composer, he became concerned that the victim of any rivalry between Pollini and Neumann and their ventures would be his operas. In response to Richter's annual birthday letter on Christmas Day 1881, Cosima replied from Palermo where the family was wintering. The letter was extremely friendly and warm. Time and separation from the *Tribschener Kind* (child of Tribschen), as she called him, were creating a bond of friendship between them, which put aside all previous aggravation. Wagner, she said, was concerned on two counts. Neumann had laid his plans for the first London *Ring* long before Franke approached the composer with his own project (and Neumann was proving a loyal propagator of Wagner's works), but he was also concerned lest any animosity should develop between the two conductors, Richter and Seidl.[9] During the course of the season word got back to Wagner from Richter and others of its success. Cosima replied, 'Your letter came right *a tempo* to cheer my husband up. *Meistersinger* has, as you know, also become especially close to his heart, and it has really delighted him that you have found an impartially pleased public for it. . . . We can imagine how overworked you are. Soon we too shall be; in the meantime the splendid scenery [for *Parsifal*] is nearly ready, and your invention, concerning the bells, has proved excellent. I still find it somewhat strange that you are not here working with us, for I associate the production of most of the works with you.'[10] The *Musical Times* had this to say about Richter's ideas for the Temple bells in the new opera.

We read in the *Allgemeine Deutsche Musik Zeitung*: 'Among the new and most remarkable effects included in the forthcoming performances of *Parsifal* may be mentioned a bell-instrument, manufactured by Steingräber of Bayreuth after a design made by Hofkapellmeister Hans Richter. The mechanism of this instrument consists in a keyboard of four keys, some six centimetres wide, each striking upon six pianoforte bass strings, whereby the sound of four distinct bells is produced. In connection with four gongs (manufactured in England) of corresponding tonality, the peal of bells is so exactly imitated that we seem to hear four mighty brass tongues speaking down from the giddy heights of a cathedral spire.'[11]

In the extremely limited periods when he was not performing, rehearsing, or correcting parts (those for *Lohengrin* were particularly bad), Richter led his usual gregarious social life. Franz Betz (the Wotan of 1876 and Richter's soloist for the concert on 8 May) was in town when he arrived at the end of April, and on 1 May he sang for Richter and his friends at the conductor's lodgings at 11 Bentinck Street, off Manchester Square. Brodsky also played, and the assembled company included Parry, Stanford, Sir George Grove, Charles Ainslie Barry, the tenor William Shakespeare, and the pianists Walter Bache, Oskar Beringer, and Fritz Hartvigson. On 4 May he 'went to St Paul's in the morning; [John] Stainer played the organ quite wonderfully'.[12] Two days later he 'visited Crystal Palace in the afternoon. Gypsies. An electric exhibition.'[13] On 21 June he began another series of sittings, this time for Hubert Herkomer, who became a lifelong friend; Richter received his completed portrait from Alma Tadema five days later. By the end of his stay it was becoming apparent that Franke was in serious financial trouble with the German season. Richter had already recorded a debt of £580 owed to him after the final performance. Others, particularly the orchestral players, found themselves in worse circumstances. Richter, with his unique loyalty to his men, returned later in the year to give two concerts entirely for their benefit, and it says much for the respect that he continually earned from players that Louis Engel was able to write in the *World*: 'I call Richter an autocrat in recognition of his common sense, because in an orchestra, as he the other day quite correctly remarked, there obtains no Republican principle; one man is responsible and he can be so only when his will is implicitly obeyed.'[14]

This close bond between conductor and players is further illustrated by a letter he received in London from his men back in Vienna during the German season. As d'Albert had reported to Mrs Joshua, Richter simply walked away from the press campaign that had been building up against him during 1882 (one of their complaints was his agreement with the management to perform Wagner's operas with cuts. The composer, whilst opposing such directives, understood and forgave him). Once in London he was able to obtain a clearer perspective of his options, and he resolved to stick to his decision to resign. Already in March speculation in the press had been rife. In view of Richter's resignation the concerts were to be offered to his Director at the Opera, Wilhelm Jahn, but Jahn protested that as long as Richter was even in Vienna he could not accept them. The papers went on to propose such alternatives as von Bülow, Dessoff, or Hermann Levi. The influential critic Ludwig Speidel attacked Richter to promote Jahn as, years before, he had intrigued against Dessoff to ensure his replacement by Herbeck as conductor of the Philharmonic concerts. Denials and counter-denials flew back and forth between the press and the orchestra's committee. Finally they sent a deputation to Richter on 23 April, shortly before his departure for London, begging him to reconsider his decision. Richter promised no more than to make his mind up and send his answer before the orchestra began its annual leave on 15 June. On 8 June he sent his reply couched in the warmest terms. He wrote that he had proved himself loyal to the orchestra during his seven-year tenure. He was 'the most enthusiastic Philharmoniker' and had done nothing to harm the ideals he had inherited from his predecessors. 'My most wonderful achievement has been your choosing me as conductor.' All this had brought him to the conclusion that he could no longer conduct their concerts. 'Let me withdraw from the Philharmonic concerts with proud conviction.... There is one other matter I ask you to consider. The opinion has been expressed by several people, even those who are well disposed towards me, that the reason for my withdrawal lies in differences between me and my faithful orchestra. I authorize you to dispel such rumours.' The orchestra had plenty of choice before them, and his name was not 'to be seen to be insulted by the well-known gutter press'.[15]

Wilhelm Jahn (Director of the Opera since the summer of 1880) agreed to conduct the 1882–3 season, but only until a permanent successor to Richter could be found and on condition that he received no fee for doing so. Jahn and Richter were good friends, and musically they complemented one another so perfectly that it became a relatively tranquil time for the Vienna Opera, giving rise to the so-called golden era. Where Richter excelled in the German repertoire, it was Jahn who enjoyed conducting the Italian and French operas, though it should not be forgotten that he conducted an uncut *Meistersinger* at Wiesbaden, much to Wagner's approval. Jahn had begun his career as a singer in Hungary and also mastered several orchestral instruments. His career as a Kapellmeister had taken him to Amsterdam, Prague, and Wiesbaden before he came to Vienna. Like Herbeck (appointed Director in December 1870), Jahn combined the administrative work with the post of Kapellmeister, a life-style which Richter vowed never to repeat after his years in Pest. Jahn, who was also a very capable stage director, remained in post for seventeen years until 1897, when he was succeeded by Mahler. Because of their friendship Richter was able to influence decisions on repertoire and casting without the responsibilities associated with the top post. Like Richter, Jahn was a well-built man, with bespectacled blue eyes and a beard, though his was more refined in cut than his colleague's. Jahn, an extremely likeable man who made few enemies, had no ambitions as a concert conductor, and the eight programmes he devised for the 1882–3 season contain little of interest apart from the first performances of Dvořák's *Legends* (26 November) and two movements, the Adagio and Scherzo, from Bruckner's new Sixth Symphony (11 February). Jahn's strengths lay in recognizing his own limitations and deferring to Richter to conduct the post-*Tristan* Wagner operas (which he agreed must be staged in Vienna), and in concentrating on his own tastes. His operatic successes were first performances of Massenet's *Manon* and *Werther*, Mascagni's *Cavalleria rusticana*, Leoncavallo's *Pagliacci*, Humperdinck's *Hänsel und Gretel*, Smetana's *The Bartered Bride*, and Verdi's *Otello*.

Together Jahn and Richter built an ensemble of singers that became the envy of the operatic world. Wagner's operas alone had Materna, Winkelmann, Reichmann, and (until 1886) Scaria. The only area in which the two men failed to make any headway, because the public or the critics were not yet ready for it in Vienna, was the question of cuts. In the 1880s *Meistersinger* was performed under Richter without the 'Aufzählung der Weisen' or Hans Sachs's 'Dichterbelehrung'; even more incredible was *Götterdämmerung* without the Norns, Waltraute, or Alberich. The best description of Vienna's stagings of Wagner's operas comes from the house logbooks a decade later. In the year of Jahn's departure (1897) both *Tannhäuser* and *Lohengrin* lasted three hours twenty-six minutes. *Götterdämmerung* on the other hand took only *four* minutes longer and *Meistersinger* just *twenty-four* minutes longer. Within three years Mahler had extended *Lohengrin* to exactly four hours, and *Götterdämmerung* and *Meistersinger* to four hours and fifty minutes. When *Tristan* was first staged under Richter on 4 October 1883 a fifth of the work (600 bars) was cut, but Mahler restored enough to leave this masterpiece less disfigured at four hours and ten minutes.

July 1882 was a month of unaccustomed rest for Richter, with just his weekly duties at the Hofkapelle to fulfil. It is easy to imagine him as d'Albert described him in a letter to his father soon after his arrival the previous autumn. 'Herr Richter cannot go out without someone doffing his hat to him by way of greeting. Nevertheless he is quite at home eating roast chestnuts in the street.'[16] At Bayreuth *Parsifal* was given its first performance on 26 July, but the Richters did not travel there until 22 August. He had 'the warmest reception by the Master; once again the children were especially devoted and sweet'. Three days later the couple attended the civil wedding of Blandine von Bülow to Count Gravina (the bride's grandfather, Liszt, was also present), and that afternoon went to a performance of the new opera. Richter wrote just two words in his diary about the new work: 'great emotion'. That night, at eleven o'clock, they returned to Vienna. Richter never saw Wagner alive again.

Two additions to his operatic repertoire in Vienna occurred in September 1882, *Le Roi l'a dit* by Delibes and Adam's *Le Postillon de Lonjumeau*, an operetta which he had conducted years earlier in Pest. In London for the German opera season orchestral benefit concerts (9 and 14 November), he conducted Stanford's new G major Serenade (first heard at Birmingham), and gave an early performance in London of the Prelude to *Parsifal*. At the second concert Dannreuther played Brahms's Second Piano Concerto, which Parry heard. 'Dann[reuther] got through the Brahms concerto very well. It was an extraordinary difference from the performance at the Crystal Palace. He and Richter between them made musical sense and poetry out of it, what at the C.P. was altogether wanting. . . . Richter's first word was "Oh! The symphony! I will play it the first thing in the summer next year!" He was nice and cordial, his reception was splendid and well did he deserve it.'[17] Richter's altruism and humanity towards his players was noticed by the press. 'At both concerts', said the *Musical Times*, 'Herr Richter was loudly applauded, the added emphasis of his reception arising no doubt from a sense of the noble generosity which made him prefer others to himself, and add to his own sacrifices that those of his subordinates might be lessened.'[18] Louis Engel was an enthusiastic supporter of Richter, though it is doubtful if the conductor would have bothered to heed his advice.

It is evident that Richter has the speciality of leading Wagner and Beethoven although very different opinions are entertained about the movements he takes in the latter composer's works. Anyway I think that Mr Richter might be persuaded to give up his position in Vienna for a well-secured position in England; and although we are not exactly without good conductors, we could not but be pleased to have so clever a *chef d'orchestre* in our midst. Only Mr Richter must understand one thing. He has made a great many enemies in Vienna, and although his capacities are not so much made of there as they are here, yet he is recognized as an excellent conductor. But what did harm him there will do him harm here. A man may make a position, but to keep and to strengthen it does not merely depend on his artistic merit. . . . Mr Richter must understand that in this country, more perhaps than anywhere else, a man is responsible for the choice of his friends.[19]

Once back in Vienna Richter embarked on a series of Wagner operas before Christmas, largely to present Hermann Winkelmann to the public in *Lohengrin*, *Meistersinger*, *Siegfried*, and *Götterdämmerung*. Sandwiched incongruously between the last two (on 18 December) was a revival of Grisar's one-act *Bonsoir Monsieur Pantalon*, which Richter had conducted on 3 April that year, its first staging in the Court Opera. On 3 January he travelled to Budapest and gave a concert for the widows and orphans of its opera orchestra's players, the programme of which included Beethoven's 'Choral' Symphony. The next day he was back in Vienna conducting the first performance there of Leschetizky's only opera, the one-act comedy *Die erste Falte* (The First Wrinkle). Charity concerts seemed in abundance at this time for on 6 January he conducted another, this time for flood victims in the Tyrol. In the first six weeks of 1883 he conducted *Aida*, *Lohengrin*, *Carmen*, *L'Étoile du Nord* (Meyerbeer), *La traviata*, *The Barber of Seville*, *Walküre*, and Gounod's new opera *Le Tribut de Zamora* (the first German performance on 30 January).

He was conducting *Der Widerspänstigen Zähmung* by Hermann Goetz on 13 February 1883. In his diary he wrote, 'After the performance I heard rumours of Richard Wagner's death; I could not believe them, but I could not get any proof at the telegraph office.'[20] He sent a telegram to Siegfried Wagner: 'The shocking news unbelievable. How is the honoured Master? Reply paid Hofoper Vienna. Hans Richter.'[21] By the next day the awful truth was confirmed and Richter set off at once for Venice, where the composer had died, arriving in the early afternoon of 15 February. The next day the funeral cortège left on its sad journey by train for Bayreuth, Richter sending telegrams ahead to organize the solemn ceremonial reception of the coffin. The small funeral party, which, besides the Wagner family, included Paul Joukowsky (designer of *Parsifal*), Adolf von Gross, Dr Friedrich Keppler (Wagner's doctor), Wilhelm Kienzl (music critic and composer), and Richter, arrived just before midnight on 17 February; *en route* Hermann Levi had joined it in Munich. At four in the afternoon of Sunday 18 February Richter was one of the pall bearers at Wagner's funeral. After the coffin was placed in its last resting place in the garden at Wahnfried, everyone dispersed to allow Cosima to come from the house and be alone at the graveside. Meanwhile a party of about thirty had collected on the stage of Wagner's opera-house on the hill. Richter led the speeches after which a collective resolution was taken by all present to preserve and further the Master's cultural legacy. From Bayreuth Richter sent a telegram to Marie in Vienna, 'the children are already more composed. Cosima inconsolable.'[22] Wagner's last letter to him, written from Venice two weeks before his death, was largely about the baritone Karl Sommer who wished to come to Bayreuth to study the roles of Klingsor and Amfortas. Wagner welcomed the idea and suggested the singer should arrive by 20 June and observe the rehearsals and performances scheduled for July. He was always 'delighted to meet new young talent and to develop them in my style.... Adieu! Good Hans! Greetings from the heart! Your good old Rich. Wagner.'[23]

Richter had no time to mourn Wagner, the man whose music shaped the pattern of his career and his life. A memorial concert was given under his direction on 1 March in the large Musikverein hall, in which he conducted Siegfried's Funeral March from *Götterdämmerung* and music from the first act of *Parsifal*. Apart from the current operatic fare (which also included the first of several performances of *Muzzedin* by the principal viola of the orchestra Sigmund Bachrich on the following day), there were two concerts in April. The first, concluding a cycle of Mozart's operas mounted by Jahn, included the Requiem, the second consisted of just Beethoven's 'Choral' Symphony and was in aid of the Bayreuth Festival. The nine (mostly weekly) concerts given that year in London took place between 7 May and 2 July. The first was in Wagner's memory and the last concluded traditionally with Beethoven's Ninth Symphony; of interest in the otherwise standard programmes were Raff's Third Symphony 'Im Walde' (10 May) and the Second Rhapsody, entitled 'Burns', by Alexander Mackenzie (21 May). Mackenzie had first written to Richter two months earlier, after he heard from Narciso Vertigliano (or N. Vert as he styled himself), the conductor's new manager since 14 November 1882, that he was interested in performing the composer's symphony. Preoccupied with the première under Carl Rosa of his new opera *Colomba* (9 April 1883), Mackenzie suggested his 'Burns' Rhapsody to Richter instead. Manns had given the work its first performance in March 1881 'as the tailpiece of a concert at the Crystal Palace, and no-one has heard it. No. 1 is never played except on weekdays at the Palace, and although both works are often mentioned in the press they would be positive novelties for London. Allow me to tell you that it has pleased me very much that you are prepared not only to look through our English products but also to play them.'[24] Mackenzie the Scot and Stanford the Irishman thought it more prudent to present themselves to the musical world as English in order to get on. 'The Rhapsody', remarked the *Musical Times*, 'deserves unqualified admiration,'[25] and the successful performance elicited a grateful letter from the composer 'for the trouble you took, and with such good humour, over my work. It was a pleasure for me to hear it played in such a splendid manner, that you can believe of me. I hope you were as happy with the success as I was, and that you would consider it worth doing in Vienna.'[26] Louis Engel was effusive in welcoming Richter back for the 1883 concerts.

The great conductor has returned, his first concert has taken place and no need to say how it succeeded. Richter is himself a guarantee for the success, whatever may be the pieces which compose the programme. Everything is foreseen. The rehearsals are such that no hesitation, no doubt remains; the performance therefore comes as near perfection as human forethought can make it. I said some time ago, after Richter's last concert, that he plays the orchestra as if it were a piano or a violin, and I am happy to find the very words repeated in a great daily [paper]. . . . To speak again about Richter's quite inconceivable memory, to which again and again new proofs are added, might at last become monotonous if it was not something so gigantic.[27]

Richter travelled alone to London and the Bentinck Street apartment off Manchester Square, whilst Marie stayed behind with the six children. Their life in Vienna was idyllic in their home situated away from the busy city centre, though it must have been tiring to have to travel back and forth by carriage (a half-hour journey) on performance days. On Sundays when he had a morning service, followed by a Philharmonic concert and then an opera in the evening, he might spend the afternoon resting with friends or in his room at the Opera. A glimpse into his domestic life in 1883 is provided by letters from Bentinck Street back home to 'my dear little wife'. Broadwood had supplied him with a piano for the duration of his stay, and

Mrs Joshua had once again placed a floral welcome in his apartment. 'You can send me a good Debrecziner bacon in the tin box in which nails and tools are now kept; also a little pot with white ointment for all purposes, but it must be soon.'[28] He was very specific about which newspapers must be sent, so that he could keep up with events back in Vienna: Friday afternoon's *Figaro* with its gossip column 'Wiener Luft', the *Fliegende Blätter*, and the *Mercur*, all to be bought at Gerold's shop on the Stefansplatz and dispatched every week.

Whilst in London Richter wrote to Jahn about operas he had seen, a duty he often undertook to report new ones which Vienna should stage, or to warn against those to be avoided. In June he heard the first staging in London of Ponchielli's *La gioconda*, and it was performed in Vienna the following year. He also gave his opinion of the scores he had perused of *Il Guarany* by Antonio Gomes (1870) and Saint-Saëns's new opera *Henry VIII* (1883), neither of which he found satisfactory. Of the former he wrote, 'although it contains a few well-constructed ensembles, some nice ballet music at times, and some "grateful" numbers for the singers, I cannot recommend the performance of this opera at the Court Opera. The melodies are so ordinary, if not downright shamelessly Italian.' The latter 'contains decent, indeed often fine, music, and one would not expect anything else from so genuine a musician; only I could not find a single number which made a dramatic impression upon me.... Given a quite outstanding cast—as for example with [Gounod's] *Le Tribut de Zamora*—it would be possible to give this opera a short life, but under no circumstances do I seriously expect it to have a lasting success.'[29] Meanwhile the long-suffering Parry was still having little luck getting his First Symphony performed by Richter. It had already been postponed from the previous season because of the *Tristan* rehearsals, and he was having no luck a year later. 'Made it up with Richter and had a hug by way of reconciliation in the artists' room in St James' Hall and he was very affectionate. Too fascinating to be reasoned with. Had luncheon with him another day, but he was seedy and out of spirits.'[30]

Interspersed among the concerts were day trips with friends to Richmond, Hampton Court, Windsor, Woking, Virginia Water, and a picnic at Sevenoaks. On 18 May, and despite a severe toothache, Richter visited the Royal Normal College and Academy of Music for the Blind, opened in 1877 in Westow Street, Upper Norwood, on the southern border of London. It left a lasting impression. 'The Director, Mr Campbell, is himself blind; the musical performances were perfect, an organist [Bach], the choruses [from Mendelssohn's *Lobgesang*], and a pianist [playing

Schumann's concerto] in particular were excellent. Their gymnastics and the way they ran about freely in the garden were astonishing.'[31] The pianist, a pupil of Fritz Hartvigson, was Alfred Hollins, whose published memories include an account of the facilities such as the gymnasium at Norwood. He does not mention Richter's visit in 1883 (instead he described the school concert that summer at nearby Crystal Palace when he played Beethoven's 'Emperor' Concerto under Manns), but Hollins heard *Meistersinger* under him at Bayreuth in 1889. His observation then reflects the heightened senses of the blind listener.

Richter conducted the *Meistersinger*.... No doubt it was presumption for an inexperienced youngster like myself to dare criticise Richter, one of the greatest of conductors, but although the performance was magnificent, I thought there were flaws in it. The chorus and orchestra were not always together, and I felt throughout that the orchestra did not accompany so sympathetically under Richter as under Mottl. I think Richter was more at home in a purely orchestral piece, and I have not heard anyone else give so satisfying an interpretation of the *Meistersinger* overture as he did.[32]

The concert season ended as usual with Beethoven's 'Choral' Symphony (2 July). Among the soloists was the mezzo-soprano Ellen Orridge who, since 1881, had been a regular performer as a soloist in choral works for Richter. This was sadly the last time they worked together. That summer she visited her parents at Guernsey, where, at only 27, she contracted typhoid and died on 16 September. Coincidentally Richter's two sons Hans and Edgar also contracted typhoid in the summer and became dangerously ill. Marie took the healthy children off into the country, leaving Hans and his mother to look after the sick boys, who had the strength to pull through. Throughout this crisis Richter was hard at work in Vienna. His 1883–4 season at the Opera, apart from single performances in August of *Mignon* and *Il trovatore*, began with his normal repertoire. The most significant event that autumn was the city's first staging of Wagner's *Tristan* with Hermann Winkelmann and Amalie Materna in the title-roles. It was also the first occasion on which the name of the producer (the newly appointed *Oberspielleiter* or resident director Karl Tetzlaff) was advertised on the poster. The press had a field day, most of them unable to keep awake during even a cut version of the opera, but the public was enthusiastic. Richter was particularly pleased with the orchestra, 'which fulfilled its duty with thrilling enthusiasm. Of the singers Emil Scaria as Marke was especially moving, for I knew what importance the great Master laid in particular on this figure and his scenes, and ensured that Scaria did not make the cuts which are usual in Leipzig and Berlin. I should also stress at this point that the cuts that were made *were* the same as those sanctioned (or at least tolerated) by the Master in his day for Berlin and Leipzig.'[33]

At the end of October Richter returned to England for three London concerts and a trip to Manchester with his orchestra on 7 November. Wagner and Beethoven predominated in the programmes. Richter had travelled via Hamburg (where he visited his old school friend, the conductor Josef Sucher, who was recovering from an operation on his tongue) to London, where anarchists had been active. He wrote a letter reassuring Marie 'of my well-being if you have read or heard about the dynamite explosion in the underground railway: I was not there. . . . I have just returned from buying shoes for the children. I hope they fit. I have written the names of each child on its corresponding shoe, so that the customs officials can see that they are not intended for resale, and perhaps they will demand less duty. We are having warm weather here, so my winter coat is only necessary after the concerts. What is Pylot up to? Is he behaving himself? Don't forget to give him water.'[34] Pylot was the newly acquired family dog, large, black, and full of character; named after the Steuermann in either *Fliegende Holländer* or *Tristan*, the creature became an important member of the family. Richter's granddaughter Eleonore recounts an incident when Pylot was brought to the Opera to be used onstage to pull a small wagon. When the dog caught sight of his master seated at the conductor's desk in the orchestra pit, it leapt off the stage with a howl of recognition and on to Richter's music desk. Fortunately this incident occurred during a rehearsal, and the cart was not attached at the time.

The Manchester concert (consisting of Wagner orchestral extracts and Beethoven's 'Eroica') gave Richter particular pleasure because, as he recorded in his diary, there was 'great intrigue against our concert, as a result of which there were gaps in the better seats, but the gallery and elsewhere were full. Petty artistic dealings with Forsyth (acting on Hallé's behalf).'[35] Writing to Marie, Richter said, 'the success of the Manchester concert was colossal; even I did not expect it. We can be especially proud of conquering this city, for the Manchester public consider themselves the best and most critical public in England. All the local experts did everything they could to bring our enterprise down, but yesterday's success brought disgrace upon all their chicanery. That was no ordinary applause, the public joyously shouted their hurrahs, clapping their hands and stamping their feet.'[36] Always careful in matters of money, Richter, in his last letter home before returning to Vienna, listed the serial numbers of the banknotes he had received as a fee (one fifty-pound note and thirty in tens), in case something should happen to him. Arriving at Vienna's Westbahnhof at ten o'clock in the morning, he had barely an hour before the rehearsal for his first Philharmonic concert, followed that evening by a performance of *Tristan*.

Jahn stepped aside with relief when Richter agreed to resume the Philharmonic concerts, but the conductor's programmes for 1883–4 contained little new music apart from the concert on 2 December. It consisted of Mendelssohn's Fingal's Cave Overture, Dvořák's new Violin Concerto (first performed in its final version two months earlier in Prague) with Frantisek Ondricek as soloist, and ended with Brahms's new symphony, the Third in F major, which Richter dubbed his 'Eroica'. It is unfortunate that no interesting correspondence exists between Brahms (or Bruckner) and Richter, because all three men lived in the same city. Despite efforts by the anti-Brahms and pro-Wagner faction in the audience, the new work was a great success, and continued on its triumphant way throughout Germany. Ondricek had a concert of his own on 11 December when he repeated Dvořák's work and also played the Violin Concerto by Heinrich Ernst. Vienna's first *Ring* cycle began the next day with three days between each opera, as part of a presentation of all Wagner's operas from *Rienzi* to the *Ring*, which Jahn staged in the first three weeks of December. Richter also conducted *Lohengrin*, *Tristan*, and *Meistersinger* as part of the event, and 1883 ended appropriately on a lighter note with *Bonsoir Monsieur Pantalon* on New Year's Eve.

16

1884

More Opera in London

DESPITE all six children contracting measles one after another from the end of November 1883, Hans and Marie spent Christmas *en famille*. There were Court Chapel duties on Christmas Day and a midday concert the following day, but the couple went alone to Hans's home town of Raab for a short break before *Don Giovanni* on 29 December. Concerts in Vienna between the New Year of 1884 and the end of the season celebrated or featured musicians of lowlier rank than Brahms. The Second Serenade by Robert Fuchs formed the centrepiece of the concert on 6 January, and was followed by a new symphony by Giovanni Sgambati (the eminent Italian pianist and pupil of Liszt). Ignaz Brüll was the soloist in his own Second Piano Concerto on 20 January. Robert Volkmann had died in October 1883, and Rosa Papier sang 'An die Nacht' on 16 December in his memory, but a more substantial tribute was paid to the composer when his Second Symphony was played on 9 March. Stanford's Serenade ended the same concert, whilst the one on 23 March began with Mendelssohn's rarely played Overture to his opera *Camacho's Wedding*. The final concert of the season (6 April) contained two other comparatively rare overtures, Spohr's *Jessonda* and Volkmann's *Richard III*. Volkmann was not the only colleague who died at that time; another was Gustav Hölzl, the first Beckmesser in *Meistersinger* at Munich in 1868. In February 1884 Gustav Weber also died; it was he who had bedecked the podium in Pest for Richter's final appearances there. 'An honourable man and a warm, artistic friend. Wagner received from him the wild vines which decorate Wahnfried. When I left Pest I received a picture from him which showed all the dates of concerts I had given in Pest. We last saw one another at the Vienna performance of *Tristan*.'[1] Another friend whose death he mourned was Louis Brassin (at St Petersburg on 17 May), to whom Richter was grateful for securing the Brussels *Lohengrin* engagement of 1870, when the young conductor's fortunes were low after the *Rhinegold* affair.

When Richter set off for London on 15 April it was (as it had been in 1882) for a lengthy season of concerts and German opera, but this time at the Theatre Royal, Covent Garden. 'After another very bad and stormy crossing', he wrote to Marie, 'I am happily arrived in London and today have my first rehearsal already behind me. The orchestra welcomed me, as always, with sincere happiness and enthusiasm. All our friends send you greetings.' London was still under threat from anarchists at the time, several bomb attacks having already occurred, and so the conductor's luggage, sent on ahead whilst he stayed one night in Brussels, was searched by the railway police. 'Every suitcase is opened within half an hour of its arrival at the station. They have to take this precaution since the last dynamite explosion. . . . It was a very good thing that you packed my winter coat because it is very cold and stormy here. I shall write today to Weber, who should bring my tails, which I forgot. You also forgot to pack my slippers; I shall buy some new ones today. The English *cuisine* has cured my upset stomach, I am quite well again. . . . Do not move the piano about for the feet could break.'[2] The first concert was in London on 21 April, followed by another trip north for two in Manchester and one in Liverpool (where Hallé had now succeeded Bruch) before returning to London; the programmes for all five concerts were mainly Wagner and Beethoven, and interestingly no soloists were engaged. From comments in his diary ('orchestra generally very intelligent' in Manchester and 'thanked the orchestra for their performance' in Liverpool) it was not his London band which travelled north with him. He had two rehearsals with the northern-based players in Manchester, followed the next day by one for the Liverpool concert (the same programme plus the *Kaisermarsch*). Sandwiched between Wagner and Schumann on 5 May were Jules de Swert's Cello Concerto (with the composer as soloist), Mackenzie's orchestral ballad *La Dame sans merci*, and the first performance in England of Brahms's *Gesang der Parzen* (Song of the Fates, to words by Goethe), a six-part choral work written in 1882. A week later Brahms's Third Symphony was heard for the first time in England, and repeated a fortnight later. Concerts in June, interspersed amongst the operatic performances, included the first English performances of Liszt's Third Hungarian Rhapsody and Raff's Overture to *Romeo and Juliet*, and a repeat of Parry's Piano Concerto with Dannreuther as soloist. The Richter Choir ended the season on 16 June with Brahms's *Song of Destiny* and Beethoven's Ninth Symphony. The concert agent Pedro Tillett recalled:

The Richter Choir, trained by Theodore Frantzen, held their rehearsals at Store Street Hall owned by the pianoforte firm of Wornum & Sons. It was rather a forlorn place, but central, being just off Tottenham Court Road and handy for the male chorus, who came on from their business. I used to attend these rehearsals to check up the attendance and when they had finished, not having had any dinner, one or two of us used to repair to the Bedford Head Hotel nearby, where they specialised in suppers, the favourite ones being tripe and onions and Irish stew washed down with a tankard of Reid's stout—some beer in those days. The Doctor [Richter] used to come for final rehearsals.[3]

Karl Armbruster was chorus master for the 100-strong imported German chorus for the opera season which ran for five weeks from 4 June to 11 July. The operas performed were Wagner's *Meistersinger*, *Lohengrin*, *Fliegende Holländer*, *Tannhäuser*, and *Tristan*, Weber's *Der Freischütz*, Beethoven's *Fidelio*, and the first staging in England of Stanford's *Savonarola*. A performance on 5 July of Liszt's oratorio *St Elizabeth* was planned but cancelled when problems with Stanford's opera arose. The month of May became a very arduous one for Richter, who was not pleased with the calibre of all his singers, nor with the orchestra from the Royal Italian Opera which was 'not of the same high standard as my concert orchestra two years ago. The chorus was very good.'[4] His singers included Clementine Schuch-Proska (Eva and Aennchen), Emma Albani (Senta and Elsa), Lilli Lehmann (Isolde and Venus), Heinrich Gudehus (Max and Walther), Theodor Reichmann (Dutchman, Sachs, and Telramund), and Karl Scheidemantel (Pogner, Kurwenal, Wolfram, and Rucello in *Savonarola*). By the middle of June, and after variable performances of the operas, it became clear that Franke was in financial trouble once again. On 15 June Richter noted 'very serious conversations' with the impresario in his diary. Apart from the chorus, Albani seems to have been the only one who kept up his artistic spirits during the season. On 11 June he conducted the 'best *Lohengrin* until now with Albani as Elsa, who was wonderful'. On 20 June, in *Fliegende Holländer*, her Senta was 'quite excellent; above all an artistic relief from this calamity'.

Stanford had conducted his opera *The Canterbury Pilgrims* at its première on 28 April at Drury Lane by Carl Rosa's company (Richter attended on 3 May and found it 'a quite splendid work' according to his diary), but *Savonarola* did not enjoy such success. Rehearsals began on 22 June, though it had originally been the intention to open on 18 June. The preparations were beset by problems. The opera had already been performed in Hamburg, with the soprano Rosa Sucher in the double role of Clarice/Francesca, but in London the part was assigned to Biro de Marion of the

Royal Italian Opera. Richter described her performance as Agathe in *Der Freischütz* as 'tolerably good, but as Elisabeth she was bad; she began the role of Clarice/Francesca but then gave it up. A Frau Waldmann-Leideritz took it over, but she seems to me to be quite inadequate. Vederemo! [We shall see!]'[5] Three days later matters were desperate and telegrams were sent to Sucher, holidaying in Austria. 'It came to nothing. Now *Tristan* is just around the corner with grossly inadequate rehearsals,' wrote Richter.[6] '*Fidelio* on 25 June and *Tannhäuser* on 27 June were wretched performances! Then a ray of hope! In spite of unsatisfactory rehearsals a relatively good performance of *Tristan* on 2 July. Lilli Lehmann—Isolde excellent, Gudehus—Tristan very good, Luger—Brangäne, Wiegand—Marke, Scheidemantel—Kurwenal did their best. The orchestra better than ever. The enthusiasm of the public was at its highest right to the end.'[7]

Richter's opinion of Frau Waldmann-Leideritz proved accurate, for by the beginning of July she too had withdrawn and the role was now assigned to Fräulein Schaernack who had a week in which to learn it. According to Richter she did so, and correctly too, but it was not enough to save *Savonarola*, which was given 'at last after a few bad rehearsals'[8] on 9 July in German 'with maimed rights, a bearded hero, insufficient rehearsals, and incompetent stage management. In addition the owner of the libretto having refused to allow it to be sold in the theatre, the public knew

FIG. 5 Hans Richter returning to Vienna with the financial rewards of his first visit to Covent Garden. Cartoon by Hans Schliessmann.

nothing of its meaning in a foreign tongue, and I scarcely recognised the opera I had seen a few weeks before.'[9] Parry was in the audience.

[*Savonarola*] came off at last, the unfortunate Wachmann decidedly having been supplanted and a fair prima donna in the shape of Schaernack provided in her place. Considering the short time she had to learn it, she did very well. Stritt, who did Savonarola, was very much out of voice and his rendering of the part was stagy and bad. The performance altogether was rough and unsatisfactory and I was very much disappointed with the piece. It seems very badly constructed for the stage, poorly conceived and the music, though clean and well-managed, is not striking or dramatic. The claque, which was tremendously strong, overdid their business, and by being too eager to bring C.V.S. before the curtain as early as the Prologue, caused a sensation which prevented them from being able to make any effort over the later act. At the end there was very little applause except from his personal supporters, but Richter brought him before the curtain instantly without waiting for any demonstration and this made sense of an appearance of a good reception, but I felt impressed with the fact that so far the thing was a decided failure in every way.[10]

Louis Engel was far harder on Stanford's new opera.

The only seeming guarantee for the success was the name of Hans Richter; but as I have pointed out before, Richter is one of the greatest conductors but his business capacities want very strongly a better guide than his own tact. Never have I seen him so demonstratively swing his arms through the air, exerting himself frantically to keep the band and the chorus (and what a chorus!) and the soloists (and Heaven preserve us from hearing them again!) together in some decent manner. But bad as the executants undoubtedly are, if Rubini and Malibran, Lablache and Tamburini came out of their graves, they could not make this unmitigated rubbish succeed.[11]

The opera season ended on 11 July with *Lohengrin*, but still Richter was in conflict with those running the financial affairs of the venture. Lilli Lehmann, writing about her first Isolde under Richter in London in 1884, recalled how some of her colleagues dealt with the same officials.

Hans Richter, who worked with the orchestra day and night, was as tireless as ever when it came to breaking a lance for Wagner. A large cut was to have been made in the philosophical conversation about day and night, for at that time no one dared to give the English public a fully uncut *Tristan*. It became clear at the first orchestral rehearsal, however, that more than half of the great narration in act one was cut, as well as other places, following the parts and version used in Vienna. According to Richter they were not copied at all, whereupon I absolutely refused to sing the role. The next day the notes were forthcoming (perhaps they *were* either in the parts or they were pasted over) but the first act was done without cuts. The excellent production justified the brilliant success for which Richter got great credit.... Even before my arrival in London I had heard of unpleasant financial differences between the artists and the impresario, who at each occasion took shelter behind a guarantee fund. Until that time nothing was owing the principal singers, but the chorus and orchestra, who had already given serious warnings, went on strike at the first *Tristan* rehearsal, and only played on when paid. A further nice surprise awaited me in *Tristan*, when Gudehus whispered to me after the love-duet, that he would not sing further, because he had not received his fee after the first act. I pleaded with him not to upset the performance, not to debase our love for Wagner and his work, but the interval between the second and third acts lasted interminably because Gudehus really would only continue to sing after he was paid. He was actually right, for why should artists suffer losses because of mismanagement of badly funded enterprises?[12]

Richter was not short of critical support in England; he had a far easier time of it in London than in Vienna, where he would be the means with which the critics could attack or support composers. A loyal supporter in London was Hermann Klein, who thought that the second (1884) season of German opera was not as financially successful as the first (1882) because 'the time when German opera should take abiding root in the affections of the London public was yet to come'.[13] He and Richter became close friends during the earlier season, and even in 1925 he considered that the conductor's 1882 premières of *Tristan* and *Meistersinger* had never been surpassed in both singing and orchestral playing. Richter visited Klein for tea after a *Tristan* and the conversation turned to the change of heart that Klein had undergone towards the work as a result of his experience of a staged performance.

I had made my public *apologia* to Wagner in the previous issue of the *Sunday Times*. Evidently Richter had read it, for he began: 'I am glad to see you have changed your mind about *Tristan*. You are not the first, but you are younger as a critic than my friend Eduard Hanslick (Wagner's doughty Viennese opponent). He has also altered his views about certain works some time ago, but has not yet overcome his obstinacy sufficiently to retract them in print. One must never have fixed prejudices where music is concerned; no not even when one is a great critic!'

I remarked that we still had a few elderly examples of the sort here in London; but as to the others, if they could not change their attitude it was from sheer conviction that it would be equivalent to high treason to do so. 'Then they are just as bad as Hanslick', replied Richter. 'To me it is unbelievable (*unglaublich*) that intelligent musicians should fail, much less refuse, to recognize how beautiful, how almost supernaturally beautiful, are these later works of Wagner. Of course you have got to know them and know them through and through (*durch und durch*) before you can perceive all the beauties that lie hidden in them. But whereabouts is the "ugliness" that one hears of? I cannot find it. Besides what is "ugliness"? Do you say that Beethoven wrote "ugly" music because he wrote discords like the strident one which announces the singer in the last movement of the ninth symphony?' 'Of course not', I said. 'Only you must give us time. Remember it is only five years since you first came and conducted at the Albert Hall [1877], and until last week we really knew nothing of *Tristan* but the Vorspiel and the Liebestod.' 'I agree', said Richter. 'Perhaps I am a little impatient with your old colleagues; besides the audiences here are wonderful. So is my London orchestra—*ausgezeichnet* [excellent]. And have you not got a splendid tenor, Edward Lloyd, who sings the *Preislied* better than anyone I know? Well directly we finish at Drury Lane I go off to Bayreuth to hear the first performance of *Parsifal*. No, I don't conduct that; Hermann Levi is doing it. That is a great work if you like, and perhaps next season London shall hear the Vorspiel.'[14]

Richter's tolerance with critics comes across in that conversation with Klein, but after his 1884 German opera season in London it was to be another nineteen years (in 1903) before London heard him conduct opera again. His patience with critics and the public was counterbalanced by a marked impatience with incompetent impresarios. The miseries of the 1884 opera season had been alleviated by excursions elsewhere to the Manchester Zoo, the Health Exhibition, the Court Theatre to see *My Milliner's Bill*, to be photographed by Elliott and Fry and painted by Gustav Gaupp, to Cambridge on 10 June where he conducted Beethoven's Seventh Symphony in a concert given by the University orchestra at the Guildhall, and to Birmingham, though it was not as a tourist that he visited this last city. The Birmingham Triennial Festival lost its conductor when Sir Michael Costa died on 28 April, but he had resigned his post as death approached. On 26 April Richter met Messrs Johnstone and Milward to discuss the 1885 Festival. His appointment as Festival conductor was ratified on 10 May, and on 5 July he visited Birmingham where 'Spencer showed us the Town Hall and the Free Library. In the afternoon to Milward's house in the country to meet the chief members of the Festival Committee; very nice family.'[15]

Richter's appointment was viewed in some quarters with alarm, but in one particular case with considerable anger. Joseph Barnby, Frederic Cowen, Charles Villiers Stanford, and Sir Arthur Sullivan were all leading contenders whose names were bandied about in the press. Reviewing Costa's successors in his posts at Birmingham (Richter), Leeds (Sullivan), the Sacred Harmonic Society (Hallé), and the Handel Festival (Manns), the *Musical Times* noted at the end of Costa's obituary that 'instead of one naturalised Englishman we have three foreigners and an Englishman born'. Its report of Richter's appointment to Birmingham was well reasoned. It gave the Festival Committee its 'hearty and unreserved approval. Possibly, and indeed probably, there may be some heartburning over the fact that a foreigner is to be entrusted with the reins on an occasion so essentially English, but those who cavil at the choice of Birmingham will do well to remember that art knows no frontiers.'[16] Sullivan was furious, though, as Joseph Bennett quite correctly pointed out to him, as conductor of the Leeds Festival he could hardly have expected the appointment to a rival organization. Sullivan wrote to everyone of influence in the musical press. 'Why do you encourage these blooming Germans so much?' he asked Davison of *The Times*. 'I think the whole business of those Franke/Richter concerts is an insult to us.' Hermann Klein of the *Sunday Times* was told that 'foreigners are thrust in everywhere, and the press supports this injustice',[17] and to Bennett, editor of the *Lute*, he called it 'an affront, a bitter humiliation for all us English. . . . [Others] would have done the work well—a hundred times better than a German [*sic*] who cannot speak the language, who had never had any experience in dealing with English choruses, and who knows none of the traditions of those choral works which form a large element in the Festival.'[18] These works were presumably *Elijah* and *Messiah*, written by the Germans Mendelssohn and Handel!

Stanford, one who could well have been a contender for the post, was emphatic in his approval of Richter's appointment.

The crying need for reform necessitated the leadership of an exceptionally strong man, who could speak with European authority; for the Birmingham Festival was as important in its way as the Festival of the Lower Rhine, and as such affected our musical position amongst other nations. Apart from his great gifts as a conductor, his power of getting the best out of his orchestral players, of saving time, and of minimizing grumbles was invaluable at such a moment; he was too international in his tastes and policy to justify any permanent feeling of grievance on the ground of patriotism. The appointment at this crisis turned out to be no hindrance but rather a great help to English music and English artists alike.[19]

Louis Engel also had something to say on the Birmingham appointment as part of a glowing review of a Richter concert on 19 May 1884.

A more magnificent orchestral performance than that of Wagner's *Siegfried's Rhine Journey*, *Trauermarsch*, and *Walkürenritt* at the fifth Richter concert I never heard, and it once more shows what Napoleon said to be perfectly correct, 'A general can make the best army out of any soldiers'. What was the orchestra when Richter took it in hand? What is it today? Equal to any demand, but mind you well, under him only. Hanns [*sic*] Richter's conducting is so perpetual, he does not merely superintend the time, he plays the piece with his head, with his left hand; the way he bends forward more or less, the immense, the perplexing knowledge of each entry which his memory never misses; the great respect and unbounded confidence of all the performers, who, with the music before them, trust to him without the music, render him a phenomenon at the conductor's desk, and although I am sorry to see that the great, the important Birmingham Festival Committee saw fit to pass over the claims of any Englishman to succeed one who was also a foreigner—Sir Michael Costa—it is impossible to blame them for selecting the one man who has given such frequent and ample proof of his being capable of doing what until now no Englishman has proved himself to be able to do. Do not let it be said again that conducting by heart is a secondary accomplishment. Before all let those who see Dr von Bülow or Richter conduct by heart and feel inclined to pooh-pooh what they cannot do, try a Wagner score. Beethoven's symphonies many people know by heart. But then see the perpetual *rapport* of Richter with his orchestra; his eye is everywhere and it could not be everywhere and on the paper too. He can carry or steady his band just as he pleases and what a force of will that requires, only those who understand the affair thoroughly are able to say.... Richter combines all [the numerous requirements of which I have spoken so often as being necessary to make a first-rate conductor] and therefore he has been chosen over the head of a number of others, some of them undoubtedly very meritorious.[20]

Engel gave further insight into Richter's conducting technique when reporting the concert on 9 June 1884.

I have often been asked about Richter's long beat. I fully admit that a short beat seems by far preferable, but Richter himself uses sometimes a short beat; yet he is so eloquent with every movement he makes and his orchestra seems so well to understand him that the usual objection to a long beat, the disrupted chords, cannot be urged in his case. I must say however that on two or three occasions such chords did happen on Thursday night, and three times I saw what does not usually occur, that he put up his left hand—the accustomed sign for *piano* playing—without the immediate effect it used to have as a rule.[21]

The *World* could also be highly critical of Richter's business acumen, and summarized his 1884 season as 'not so successful as many of his friends would have wished.... If Herr Richter had kept strictly to his London concerts with undivided attention, I believe he might have done much better. His provincial concerts... were inconceivable failures financially speaking.' Beethoven's Ninth, it decided, was 'partly splendid and partly not so. In several instances in this... he wanted to keep the band down and lifted up his left hand, which used to be of an instantaneous effect, but it was not always so in these latter concerts.' Having chastised Richter for a fast tempo which blurred the double dotted figuration in the violins in the *Tannhäuser* Overture, it then suggested that the Finale of Beethoven's last symphony should be transposed down a tone to help the overtaxed chorus. 'These reserves in all truth and justice made, it is impossible not to recognize, as I have often done, the undeniably superior skill of Hans Richter in conducting the greatest masterpiece of the greatest master.'[22]

The *Musical Times* for May had carried another significant paragraph reporting that 'Herr Hans Richter, the Capellmeister *par excellence* has been appointed successor to Herr Gericke, the late director of the concerts of the Gesellschaft der Musikfreunde at Vienna, an important post in the musical world, as all amateurs know.'[23] A summary of Richter's programmes for the Society until his resignation in 1890 was given in Chapter 7. Meanwhile Italian operas were included among his duties at the Vienna Court Opera during August 1884; there were three Verdi operas, *Il trovatore*, *Aida*, and *La traviata*, and two by Rossini, *Barber of Seville* and *William Tell*. The operas new to Richter's repertoire in the 1884–5 season were Auber's *Le Maçon* (The Builder), the composer's first successful work written in 1825, and Ponchielli's *La gioconda*, which the conductor had recommended to his Director Jahn the previous year when he heard it at Covent Garden.

Three concerts took place in London in the autumn of 1884 (the only work new to London was Liszt's Fourth Hungarian Rhapsody on 4 November). Parry attended the second (on 11 November) and wrote to his wife:

I went round to see Richter too. He looked rather played out and had got a bad cold. The place was perfectly crammed. When we got there there was not a single vacant place in the hall, but by luck while we were discussing with the ticket man what was to be done, a man brought back a couple of tickets he didn't want and so we got in among the scornful down below. King tried to sing the Fire Scene when Wotan bids adieu to Brünnhilde but his voice is gone; poor little soul, I saw him after but I could say nothing. The thing was ever so much too big for him.[24]

The final concert was the first performance in English of Beethoven's Ninth Symphony, 'the purely orchestral movements of which were rendered to absolute perfection by the splendid body of instrumentalists inspired by that most sensitive and communicative of all batons wielded by Herr Hans Richter'.[25] Amy Sherwin, Isabel Fassett, Edward Lloyd, and Frederic King were the soloists. Engel wrote, 'the first season (1879) of the Richter Concerts was a clear loss and an important one because the stuff that conductor was made of was not known. Patiently and hopefully was that loss borne by a gentleman whose confidence in the ultimate success of Richter as a conductor was so brilliantly justified.'[26] Pedro Tillett discovered an old ledger in which gross receipts (but not expenditures) were listed as follows:

1883	Summer	£3808.8.9	Autumn	£1624.5.2
1884	do.	£3644.0.0	do.	£1582.0.0
1885	do.	£4027.0.0	do.	£1400.0.0
1886	do.	£3645.0.0		

He considered the figures 'of interest as evidently a loss must have been incurred on the autumn concerts',[27] despite a packed house at all three concerts.

Richter arrived back in Vienna on 14 November in time for his first Philharmonic concert two days later (Rosé playing the first Viennese performance of Bach's E major Violin Concerto), and his first Gesellschaft der Musikfreunde concert a week later. Moritz Rosenthal was the soloist in Liszt's First Piano Concerto at the Philharmonic concert on 30 November, which also included the première of Robert Fuchs's First Symphony. December was a very busy month, consisting of much Wagner (one full *Ring* cycle, *Meistersinger*, and *Rienzi*), and a *Walküre* after a midday Gesellschaft concert in which two first Viennese performances were given, Mottl's orchestration of Schubert's Wanderer Fantasy and Berlioz's Te Deum. Richter was happy that 'on 29 November Isolde and Eva [Wagner] came to the Wagner cycle in Vienna. They stayed with us, and I do believe that they really felt at home with us. They travelled home on 22 December.'[28] Despite Cosima's grief-stricken withdrawal from contact with most of the outside world, her children had not forgotten their older playmate and friend, the 'child of Tribschen'.

17

1885–1886

Vienna, London, and Birmingham

INTEREST in the year 1885 is focused on England, though at this time Richter served four organizations in Vienna. The five remaining concerts of the Philharmonic series between January and April 1885 produced only one work new to the city, Dvořák's Slavonic Rhapsody in G minor, Op. 45. Though the composer was grateful when he saw his name in the programmes for the 1884–5 season, he nevertheless had doubts which he aired.

I have some misgivings on the grounds that the Viennese public has shown a certain prejudice against compositions with a Slavonic flavour; a feeling which must lessen the success the work would have in other circumstances. In London or Berlin it might be alright but unfortunately our national and political relations being as they are, it will not do in Vienna, therefore I would ask you, dear friend, if you would not choose another of my compositions.[1]

Though Dvořák suggested his Hussite Overture (which could be renamed Dramatic Overture to avoid a similar Czech association), the Scherzo capriccioso, or the Sixth Symphony, the conductor left the programme as it stood for 1 March 1885. Dvořák wrote again after the concert to remind him of the two new works as well as asking for a performance of his *Stabat mater* in Vienna (it was performed in April 1886 by the Wiener Singakademie, but not under Richter). Dvořák also hoped they would meet at Birmingham later in the year, when he would conduct his own newly commissioned *Spectre's Bride*.

Richter engaged d'Albert as soloist in Beethoven's 'Emperor' Piano Concerto for the concert for the Gesellschaft der Musikfreunde on 4 January 1885. D'Albert had now Germanicized his Christian name from Eugène to

Eugen and alienated his English friends by claiming he had never learnt a thing in the land of his birth. This behaviour distressed such supporters as Mrs Joshua, despite reassurances from her friend Sir George Grove that the young man was only going through a *Sturm und Drang* period in his life. He reminded her that it was only because d'Albert was so good when she introduced him to Richter that the famous conductor took him to Vienna and, within a very short time, presented him at a Philharmonic concert, all of which could have only reflected well upon his training in London. Grove was correct here, though his dismissal of d'Albert's behaviour as a consequence of youth was proved wrong when his Anglophobia intensified throughout his life.

Lilli Lehmann repeated her London triumph as Isolde in Vienna. She also sang Leonore in *Fidelio* and Constanze in Mozart's *Die Entführung aus dem Serail*, all under Richter. Lucca, meanwhile, was not only a stunning Carmen, but also a fine Gioconda. The violinist Marie Soldat played the Brahms concerto on 8 March 1885, and the season ended with a triumphant performance of Beethoven's Ninth Symphony to celebrate twenty-five years of Philharmonic concerts since they were started by Otto Nicolai. The concert began with his own *Kirchliche Festouvertüre* (Ecclesiastical Overture) and had, unusually, five star soloists (baritone and bass) for the symphony, Marie Lehmann, Rosa Papier, Hermann Winkelmann, Theodor Reichmann, and Karl Mayerhofer. That same evening Richter conducted Bellini's *L'Étoile du Nord* at the Opera, and two days later (14 April) headed for London. Some of his orchestral personnel here were new, but more importantly the touring itinerary was extended, with his first concert taking place in the provinces. The programme, a mixture of Wagner, Mozart, Beethoven, Liszt, and Gluck, with soprano soloist Lena Little, was given on consecutive days in Nottingham, Liverpool, Leeds, Manchester, Sheffield, and Oxford. The reason for visiting this last city on 25 April was for Richter to receive an honorary doctorate in music. His diary records that he travelled there in the morning for the ceremony, and at 1.30 p.m. Vice-Chancellor Jowett conferred it upon him. 'The whole orchestra was present at the ceremony. I was led in procession to the concert hall by the Vice-Chancellor. The concert, at two o'clock, was splendid. In the evening a Richter dinner. I conducted the concert in my doctoral robes (red and white silk) and in my doctoral hat. Apparently it looked festive.'[2] There was a report in *The Times*.

In a Convocation held for the purpose this afternoon, the degree of Doctor of Music, *honoris causa*, was conferred on Herr Hans Richter in the presence of a large concourse of spectators. The Rector of Lincoln, who as Public Orator presented Herr Richter, referred to his wonderful powers as a conductor and to the extraordinary memory that enabled him to direct the most elaborate performance without book. Allusion was also made to the part played by Herr Richter as an interpreter of Wagner's works. Almost immediately after the Convocation a performance of remarkable excellence was given under Dr Richter's direction in the Sheldonian Theatre, which was densely crowded. In the evening the University Musical Union entertained Dr Richter at dinner. Invitations were issued to the leading College Organists, and the gathering was of the most representative character. The chairman, in proposing the toast of the evening, remarked that it was nearly a century since the University had conferred the degree of Doctor of Music on a foreigner, and that by a curious coincidence, in 1885 as in 1791, the distinction was accorded to a musician from Vienna, concluding by welcoming Dr Hans Richter as a worthy successor to Dr Joseph Haydn. In reply Dr Richter declared that two days in his life would be especially impressed upon his memory—one was the day on which he first met his beloved master and friend Richard Wagner, the other was the day on which he had received the honour of recognition by the ancient University of Oxford, a distinction that he should do his utmost to vindicate.[3]

Louis Engel used the occasion to attack a protective ring which, in his opinion, had been building up around Richter, preventing access to the great man.

The Oxford University has caused the small press of this good city of London great agony by offering the degree of Doctor *honoris causa* to the famous conductor Hanns Richter. One of their great grievances is that a number of English musicians should have been made Doctors in preference, not one but all of them. . . . I have more than one bone to pick with the newly created Doctor. He presses upon the public a set of people who exist only by their so monopolising him, that his friends have no choice left but to bear up with them or do without him. I for one chose the latter course. . . . If a conductor is not to be got at except through a double hedge of unbearableness, I can do without him; but if his chums are not to everybody's taste, his conducting stands high above the ordinary level. . . . Richter has done wonders with a number of players who are by no means the best you can get here *moyennant finance*, nor are their instruments such as to create the great tone of the Philharmonic band. Yet he has made them do what the audience admires, because he had the talent, unavoidably necessary for a conductor, to make them obey him with a preciseness and an instantaneous rapidity very difficult to be found in a band, and which it is one of the extraordinary qualities of an extraordinary conductor to create in such a degree. . . . Let all those who think the University was wrong write to Oxford, but not to me, who am of an opinion that the University honoured herself by honouring a distinguished musician.[4]

Elsewhere in the press Richter was beginning to receive notices reflecting a weariness with Wagner and Beethoven, though the public was acknowledged to be flocking to the concerts. As long as the audiences were satisfied neither Franke nor Richter intended to rock the boat, although the season did not prove to be a financial success. There were some new works that spring; an occasional Haydn symphony, Glinka's *Kamarinskaya*, Brahms's Alto Rhapsody, yet another new (the Fifth) Hungarian Rhapsody by Liszt and two extracts from his oratorio *Christus*, d'Albert's new Overture *Hyperion*, Berlioz's *Symphonie funèbre et triomphale*, Robert Fuchs's C major Symphony, and Stanford's new *Elegiac Ode*. Liszt's music invariably got dismissed in the press as bombastic and frivolous, d'Albert's work was described as 'hopeless . . . [with] its inordinate length, excruciating harmonies, and want of cohesion and intelligibleness. . . . We are glad to say that Mr d'Albert's overture met with the fate it deserved. Faint applause and unmistakable sibillation combined to extinguish it beyond hope of revival in this country.'[5] It was now open warfare between the musical press and the young pianist/composer. Berlioz's work was received without enthusiasm by the public, and Fuchs's symphony regarded by the press as commonplace.

Parry attended a concert and remarked, 'they had Beethoven's No. 2 and I was surprised how much was still left to do in the way of expressing it. Scherzo and last movement much too fast and rhythms not nearly incisive enough. Saw Richter for a few minutes. He is not well, suffering from a bad leg.'[6] A few days later he 'saw Richter again, his leg still very bad. Inflammation of the knee joint.'[7] The ailment in his right knee had been troubling Richter for more than a week when Parry saw him for the second time, because the conductor noted that 'Dr Vragassy massaged me, which appeared to do no good for the knee swelled up severely. Dr Semon brought Sir William MacCormack to me, and he ordered cold compresses.'[8] At the concert Parry attended on 18 May, Richter conducted sitting down. Walter Damrosch also came to see him to negotiate the possibility of a trip to America to conduct a German opera season, but nothing came of it. 'On the same evening', he noted, 'I learned that that grand old bassoonist Raspi—80 years old—who could still play first bassoon wonderfully on the provincial tour had died in Manchester of a throat illness. A splendid old eccentric fellow!'[9] Raspi had been a member of Hallé's orchestra since its inaugural concert in 1858. Another musician who died at this time was Sir Julius Benedict, about whom Richter was not so charitable. 'I had not seen the old sinner since the summer of 1883; on that occasion he said something in praise of Brahms, but added as an afterthought "but he does not come anywhere near the one [Wagner] *we* lost in the New Year". Shortly before or after saying this, he said the meanest and most commonplace things about Wagner to an American interviewer. [Benedict was] a musician who played for tips.'[10]

Although the Birmingham Triennial Festival was not due to open until the end of August, Richter (due to arrive only eleven days before the first concert) travelled to the city during his London season in the spring for some preparatory rehearsals with the chorus. On 29 May he had his first encounter with them in Stanford's new oratorio, *The Three Holy Children*, and Beethoven's 'Choral' Symphony. It was 'a surprisingly good choir, well appointed and with an able choirmaster, Stockley. Very warmly welcomed by the chorus.'[11] On 5 June he returned to rehearse *Elijah*, 'which went well'. Stanford joined him a week later to be present at a rehearsal of his own new work, and *Messiah* was rehearsed on 17 June. However familiar the music of Mendelssohn and Handel might have been to the Birmingham chorus, they were not so to Richter. In the case of *Elijah*, he had the daunting task of conducting his first performance in the city which had commissioned it and which was the scene of its unveiling barely forty years earlier under the composer himself. There were many in Birmingham in 1885 who had been present on that auspicious occasion, including the original chorus master, now the Festival organist, James Stimpson. Richter was therefore under many a watchful eye and had to justify his appointment to prove his critics wrong.

On 23 June he travelled to Switzerland, where he holidayed with his family, who had come to meet him from Vienna. The month from 12 July to 12 August was spent at home with his duties at the Hofkapelle and Opera, and on the 11th he returned to Birmingham with Marie. Rehearsals were concentrated and intense until the Festival opened on the morning of 25 August with *Elijah* (with soloists Emma Albani, Zélia Trebelli, Anna Williams, Janet Patey, Edward Lloyd, and Charles Santley). The Festival had an extremely taxing schedule with two concerts each day, one in the morning and the other in the evening. A large choral work took up most of the programme, but there was also room for both orchestral items and concertos, with the soloists often adding a short item unaccompanied or with piano. According to Stanford, Richter 'remodelled the orchestra, rectified the balance of strings and wind, and made the programmes of the evening concerts, which had mostly consisted of a farrago of operatic airs and selections, as artistically interesting as those of the morning. This Festival set an example of sufficient rehearsal and preparation, and of the selection throughout of worthy music, which has since been followed by all other gatherings of the kind.'[12] The conductor's orchestra at its fullest strength consisted of twenty each of first and second violins, sixteen each of violas and cellos, fourteen basses, quadruple woodwind plus two piccolos, a fifth clarinet, a cor anglais, and a double bassoon, six each of horns and trumpets, four trombones, two tubas, four percussionists, and six harps.

He was given sterling support by both his chorus master William Cole Stockley, and the organist James Stimpson. Richter's work-load in 1885, compared with later Festivals, was comparatively light. As well as *Elijah* he twice conducted Gounod's new oratorio *Mors et vita* (commissioned after the triumphant unveiling of his *Redemption* at the 1882 Festival) on 26 and 28 August. There was also *Messiah* in Robert Franz's version on 27 August, and *The Three Holy Children* by Stanford programmed with Beethoven's 'Choral' Symphony on the morning of the final day, 28 August. Richter's first Festival was remarkable for the number of new works performed, for in addition to his own conducting of music by Gounod and Stanford, Cowen came to conduct his *Sleeping Beauty*, Mackenzie his specially commissioned Violin Concerto (with Sarasate as soloist), the local amateur composer Thomas Anderton conducted his cantata *Yuletide*, Dvořák his *Spectre's Bride*, and Ebenezer Prout his Third Symphony, the first time that a symphony had been commissioned by the Festival. Richter remarked in his diary at the conclusion of the celebrations, 'Chorus splendid, orchestra likewise. The soloists mainly excellent. After and *along with* Bayreuth, the finest thing I have experienced.'[13]

The public and press were also happy. Having outlined Richter's task in equalling Mendelssohn's precedent in 1846, the *Musical Times* wrote:

Herr Richter's entrance into the orchestra was greeted by an outburst of enthusiastic applause.... The choral singing throughout the work was fully up to the highest standard of the Birmingham choir—indeed the leads under the unerring beat of Herr Richter—were simply perfect, the great chorus *Thanks be to God* producing a thrilling effect, partially marred however by the usual rustling of dresses accompanying the movement of those who, in their hurry for lunch, forgot the respect due to Mendelssohn and Mendelssohn lovers. A thoroughly appreciative audience was attracted by the first performance of Gounod's new Oratorio *Mors et Vita*.... The absolutely perfect manner in which [it] was rendered [makes] it difficult to select special pieces for commendation.... To the labours of Herr Richter, who worked hard to ensure the triumphant success achieved, the warmest thanks are due, and we heartily congratulate him upon this gratifying result of his efforts.

The choruses [in *Messiah*] were admirably given throughout ... and the steady and intelligent beat of Herr Richter was sensibly felt by the band, chorus, principal singers, and we may say also, by the auditors.[14]

A memorable performance of Beethoven's *Choral* symphony.... Although doubt may exist as to the time in which he takes some of the movements, there can be no two opinions as to his power of ensuring a rendering ... rarely heard in this country. The manner in which every point ... seemed instinct with a new life under the conductor's baton will not be easily forgotten by the spellbound listeners. . . . A success which must certainly be recorded as one of the most decisive and important of the Festival.

. . . Herr Richter, whose personal as well as artistic qualities have gained him the esteem of all with whom he has been brought into contact in Birmingham, was called forward and positively overwhelmed with applause, the members of the chorus presenting him with a large lyre formed of flowers.[15]

On the way back to Vienna (and after yet another dreadful sea crossing) Richter called in at Bayreuth. Cosima received him, the first time they had met since her husband's death two and a half years earlier. A note in his diary reads '1886 Festival',[16] but no details survive of any discussions between the two regarding a return to Bayreuth. There were no Festivals in 1885 and 1887; Cosima took them over officially in 1886 with *Parsifal* and *Tristan*, although as far back as the autumn of 1883, the year of Wagner's death, she had sketched plans to the end of the decade. Richter first appears in these in 1889, entrusted with the *Ring*; the most work was intended for Hans von Bülow, down to conduct all the operas from *Lohengrin* to *Meistersinger*, with *Parsifal* reserved for Levi. For von Bülow, however, the turbulent events surrounding his divorce precluded any contact with his former wife. Richter was not yet able to return to Bayreuth with so many commitments in both London and Vienna, and so Felix Mottl joined Levi in the summer of 1886 to conduct Bayreuth's first *Tristan*.

Richter spent six weeks in Vienna before returning to England for another tour. His operatic diet was solely Mozart and Wagner during that time and, apart from duties at the Hofkapelle each Sunday, his concert life in the capital had not yet started. In England he began with a London concert and then set off to conduct one each in Newcastle, Glasgow, Edinburgh, Dundee, and then back to London via Glasgow and Edinburgh again. There was little variation in the programmes from the usual diet of Wagner, Beethoven, Mozart, Liszt, and Schumann, but then the provinces (apart from Manchester and Liverpool) were getting their first taste of the famed conductor, whose reputation had spread north from London. Richter was enormously impressed by the Scottish countryside (and Edinburgh's city organ, which he was invited to try on 26 October). Glasgow was August Manns's territory, and Richter was convinced that his rival was making mischief before he arrived in the city; but despite this the concert was a triumph. He was duly elected an honorary member of the Glasgow Society of Musicians.

Back in Vienna (on 14 November) he gave one concert each with the two orchestral societies which he now headed. Raff's Second Violin Concerto, with the Weimar-based Carl Halir as soloist, and Bach's Magnificat in Robert Franz's edition were new to Vienna in the programme given by the Gesellschaft der Musikfreunde on 22 November. Another novelty was included in the first Philharmonic concert of 1886 (3 January) when Richter uncharacteristically conducted a French work, Massenet's *Scènes pittoresques*. January was an auspicious month for new works; at the Gesellschaft concert on 10 January Bruckner's Te Deum was heard for the first time, and a week later, in the Philharmonic concert, Hermann Grädener's new Comedy Overture was overshadowed by the first Viennese performance of yet another new Brahms symphony, his fourth and last.

Bruckner was currently basking in his own golden era. On 10 March 1885 he enjoyed what was probably the greatest triumph of his life, the Munich première of his Seventh Symphony under Hermann Levi. The conductor of *Parsifal* was convinced of the greatness of the new work, though he summoned the composer to Munich for his advice concerning the last movement, which he did not entirely understand. Bruckner came and attended the orchestral rehearsals. August Göllerich maintains that whereas Levi was only too glad when Bruckner interrupted the music to express his opinion on the playing, Richter in Vienna would shut him up with an abrupt 'I know better than you'.[17] In the following months Bruckner's stock rose considerably, not only throughout Germany, but also in Vienna, where more serious attention was now being paid to his music. On 23 April 1885 Ferdinand Löwe and Josef Schalk played the First Symphony and the first movement of the Third in a two-piano recital at Bösendorfer's hall. Theodor Helm, in the *Deutsche Zeitung*, challenged Richter to include Bruckner among the new works in the 1885–6 season:

One can believe that the gentlemen of the Philharmonic must be overcome by feelings of shame when they read the latest reports from Leipzig, the Hague, and most recently from Munich of the enthusiastic reception of the symphonies of Bruckner, which they stubbornly withhold from the Viennese public. We do not give up hope that our leader of the Philharmonic, known to be progressive, will discharge part of this guilt towards Bruckner next season, for time marches on and this debt, which has been accumulating for years, must finally be settled.[18]

On 2 May 1885 the Te Deum was performed under Bruckner's own direction, but accompanied again by only two pianos. This time, however, Richter needed no further prompting and he programmed the work for performance with the Gesellschaft der Musikfreunde in the next season. He also approached Bruckner with a view to playing the Seventh Symphony in the 1885–6 concerts with the Philharmonic. The composer reported this to Levi, adding, 'I told him it should be performed in Vienna only after it had been published. The Herr Hofkapellmeister should not let the work be ruined for me by Herr Hanslick etc. In the meantime he could perform another symphony which already had been ruined.'[19] Later in the autumn of the same year Bruckner replied to yet another request from Richter, 'I cannot allow any works to be performed in Vienna because Hanslick and his associates pull them to pieces so much that I can no longer find a publisher.'[20] The Philharmonic decided to press ahead with its plans for the symphony, despite very serious reservations expressed by the composer to its committee and his own friends. He even withheld the manuscript parts from the orchestra, and it was only because they were published early in 1886 that the work was able to be performed on 21 March. Meanwhile the Te Deum had a predictable success from the public and a majority of the critics, but an equally predictable drubbing from Hanslick. The Seventh Symphony, meanwhile, was getting further performances, at Hamburg on 19 February, and its first in Austria. This took place at Graz on 14 March (one week before Richter's in Vienna) under the 27-year-old Karl Muck, who found 100 mistakes in the parts during his fourteen rehearsals. The Wagner tubas were played by members of the Vienna Philharmonic.

The Philharmonic concert on 21 March 1886 was the first occasion when a complete symphony by Bruckner was played at a concert under its own auspices, seventeen years after the composer first arrived in Vienna. Despite several members of the audience taking to their heels after each movement, the occasion was an undoubted success. The composer was called onstage several times, starting from the end of the first movement, which was always a good sign. With his customary modesty Bruckner deflected the applause, with a sign, to both Richter and his orchestra, and was seen to mouth 'Küss d'Hand, küss d'Hand' to the cheering audience. For him, however, the greatest compliment came from another Viennese colleague who managed to remain aloof from the city's musical politics, Johann Strauss. He sent Bruckner a telegram, which awaited the composer even before he returned from the concert hall. 'Am quite shattered—it was one of the greatest experiences of my life.'[21] That evening there was a banquet given for Bruckner by the Wagner Society of Vienna, at which Richter made a speech.

Gentlemen, you would not believe how, to begin with, my musicians were so quick to mistrust and reject this Bruckner symphony. But you also have no idea how quickly this mistrust became the most glowing enthusiasm as the

Philharmonic, with ever growing interest and eagerness, rose to the tremendously difficult task of playing it; nor how in the end, on the day of the concert, each and every musician from the leader of the violins to the timpanist was absolutely determined to summon up his best for the work, which they now recognized in all its profundity and power. A radical transformation had taken place in the minds of all the members of the Philharmonic regarding Bruckner, and you may rest assured, gentlemen, that from now on our brilliant compatriot will never again find it necessary to make a detour via a host of foreign musical cities in order to present his works to the Viennese public. No, in future every new Bruckner symphony will first be changed from the pages of the score into sound in Vienna itself, in the concerts of the Philharmonic.[22]

Bruckner was delighted with Richter's performance of the Seventh Symphony, but dreaded Hanslick's response. He happened to meet the critic's cook whilst out shopping, and asked her to put in a good word for him with her master! Hanslick's review was slightly milder than usual, for although he attacked the work for all its adherence to Wagner, he was forced to acknowledge its acceptance by the public. More serious, however, was Richter's speech that evening at the banquet. It fell like a bomb into the Tonkünstlerverein, or Society of Composers, whose members were mainly followers of Brahms. Max Kalbeck, writing in the *Wiener Presse*, threw down a challenge to Richter if his praise of the composer was to be taken seriously. 'There is nothing left for him to do except to play the earlier six symphonies one after the other.'[23] This sarcasm belied a genuine

FIG. 6 Brahms, Johann Strauss, and Richter playing cards. Silhouette by Otto Böhler.

fear that Bruckner was in the ascendancy and Brahms in decline. It was, however, unfounded, for any new work from Brahms's pen tended to be announced in forthcoming programmes even before conductors had seen the score. This was particularly true of the Fourth Symphony, completed in October 1885 and performed for the first time within a matter of days. Brahms and von Bülow took it on a tour of Germany and Holland in October and November, and Richter secured it for Vienna on 17 January 1886. The composer was present and happy with the performance, despite his worries that the last movement, with its thirty variants of the opening eight bars, might prove incomprehensible to both players and public. Hanslick praised the new symphony, of course, though a little warily when it came to the 'novel' last movement.

The early months of 1886 added a new opera to Richter's repertoire, Viktor Nessler's *Der Trompeter von Säkkingen* (25 February). Guest singers in the first four months of 1886 included Emil Götze as Lohengrin and Walther von Stolzing, Marie Wilt as Norma, and Pauline Lucca one again as Carmen. Berlioz's *Faust* was sung in the first of two extra Gesellschaft concerts on 17 March with Carl Hill in the title-role and Rosa Papier as Gretchen. She also sang Brahms's Alto Rhapsody in the fourth Philharmonic concert on 4 April, two weeks after the triumphant Viennese première of Bruckner's Seventh Symphony. The second extra concert for the Gesellschaft der Musikfreunde concluded the season on 20 April with Beethoven's *Missa solemnis* (Wilt, Papier, Gustav Walter, and Hellmesberger jun. as solo violinist). Richter then left for England with Marie and began his season there with several concerts in the north. The London ones in May were of more interest. They included the first performance in England of Brahms's new symphony (10 May), Stanford's music to the *Eumenides* (with a student chorus from Cambridge University) a week later, and, on 24 May, the first hearing of d'Albert's new F major Symphony. Louis Engel wrote of Richter's part in this, 'the execution of the symphony was worthy of such an orchestra under such a conductor, and it is the noblest answer Dr Richter can give to those who decry him as a foreigner that, whenever and wherever he can, he brings forward an English composition which deserves conscientious recommendation and bestows on it all the care and attention he can to help the work.'[24] Parry heard the concerts.

May 10th: Richter concert in the evening. The new Brahms symphony [No. 4]. Fine of course, and tone noble and rich. The last movement quite a new experiment and some of it very harsh and all extremely abstruse. If it were not Brahms no audience would listen to it. It is quite a new departure as a Chaconne stands as a sort of extra after the usual last movement, the Scherzo for once being left out. I don't quite agree with Joachim about the slow movement being Beethovenish. It seems to me not at all that. The first movement is very noble. Saw Richter and he was not at all cordial.

May 17th: In the evening to the Richter concert where we had a lot of the Eumenides. It was very successful and impresses me with Stanford's ability greatly. It is his best thing so far.[25]

Richter had promised Bruckner a London performance of the Seventh Symphony in his 1886 season, but Göllerich says that Brahms and Hanslick dissuaded the conductor from doing so, despite a plea from Levi to Richter on Bruckner's behalf. C. A. Barry wrote an article on the composer in the June issue of the *Musical Times*, in which he stated that 'a performance of Herr Bruckner's seventh symphony . . . is promised at a forthcoming Richter concert'.[26] It did not happen that year; Bruckner told Moritz von Mayfeld, 'Herr Richter performed nothing. Of course Hanslick was in London. In high places I am told that behind Hanslick stands Brahms.'[27] Göllerich tells how Richter informed Bruckner that he was ill for the first rehearsal of his work; from Richter's diary the only illness mentioned is 'a dreadful cold' on 11 June when he paid a short visit to Birmingham to discuss Festival business for 1888, but the entry also continues with the words 'at this time Hanslick and Gericke were in London'.[28] A more significant reason for not programming the work could have been Franke's increasingly precarious business situation. He was nearly bankrupt (a Grand Wagner Operatic concert had to be given for his benefit at the Royal Albert Hall on 16 June with a 150-piece orchestra) and matters came to a head with Richter later in the year. A Bruckner symphony under those conditions would probably have been far too great a risk. In an article which began, 'Immense is Richter and immense is his obstinacy,' Louis Engel concluded, 'At the same time I will not for one moment dispute that his obstinately forcing down the throat of his audience such a quantity of Wagner, which . . . must at least result in the audience dwindling down into a nucleus of enthusiasts, may cause the popularity of the conductor and the success of the concerts seriously to suffer.'[29]

Richter's London concerts in the spring of 1886 included the new Liverpool Exhibition Overture by Cowen, after which Hallé played the 'Emperor' Concerto on 31 May, and, on 7 and 10 June, a concert performance of the second act of *Tristan* followed by the third act of *Siegfried* in which Therese Malten and Heinrich Gudehus were the principal singers. At the final concert on 28 June members of the Leeds Festival Chorus joined the Richter Choir in Beethoven's *Missa solemnis*. Doubtless the presence of their conductor Stanford at the organ helped to bring about the innovation of combining forces. Richter's non-musical activities that spring and summer included a visit to the Epsom races on Derby day, to the theatre to see Henry Irving in *Faust*, and a social occasion at which a fellow guest was Adelina Patti. There was also a similar even on 30 April when the names Behrens and Rodewald appear for the first time, names which will feature more prominently a decade later when Richter was being sought as a replacement for the late Sir Charles Hallé in Manchester. Parry wrote of Richter's June concerts:

June 10th: To the Richter where we had the second act of *Tristan* and the last act of *Siegfried*—after this I feel as if it was no good to try and do anything more in music. It is the most mighty and comprehensive expression of dramatic music possible.

June 23rd: Paid Richter a visit. Found him with a very bad cold. His wife a tidy sort of person and communicative. They discoursed on d'Albert. Richter says he is not on the right road in composition; 'What does he want with writing such melancholy music at his age? It is all Schopenhauer. I like Schopenhauer well enough, and read him often, but I don't like him in music'. He thinks the amount of playing he does is not good for his music and says his nerve is gone from the hard strain. I don't believe much of it, but think d'Albert will do well. Richter asked me to come again and bring the Shirley Ode—F Symphony.

June 24th: I played him the Ode and he said he liked it. Then we went at the symphony together. He read it amazingly, always seeing where the instruments were which were prominent, and never making a mistake in transposition, but reading it right off—horns, clarinets and all; and all I had to do was to fill in and play bass, which I did abominably. He offered to try it in the autumn and thinks he will play it next season. Of course I know it won't come off. But he was pleasant.[30]

A highly interesting entry in Richter's diary occurs on 17 May with reference to a letter to Jahn in Vienna 'regarding Bayreuth; great upset over refusal of leave'.[31] Richter was the first musician to be allowed to take leave from the Hofkapelle for engagements elsewhere rather than for a genuine holiday, but with the Opera he had to negotiate each request. There are unfortunately no letters from Cosima to Richter between December 1882 and July 1888, but in July 1887 she offered to write to Count Hohenlohe on his behalf with regard to the 1888 Festival, 'it goes without saying that it will be a pleasure to do everything to secure your leave for us'.[32] The question therefore arises, was Hans Richter Cosima Wagner's first choice as conductor of *Tristan* in 1886? The Festivals of 1883 and 1884 had been no more than a summer extension of the Munich Court Opera, with the whole orchestra moving from the Bavarian capital to Bayreuth (as in 1882). Then two things transformed the Festival; the first was Cosima's sudden decision, taken in 1885, to emerge from her mourning and step in to assume the mantle of her dead husband, the other was the suicide on 13 June 1886 of King Ludwig of Bavaria. It occurred a matter of weeks before the Festival and all Munich's plans to transfer its players to Bayreuth were abandoned. This forced Cosima to revert to the methods used a decade earlier when the best players from Europe were recruited individually. With Richter unavailable, she turned to Felix Mottl, who would not only have conducted the opera, but also raised and lowered the curtain for her if she had asked, such was his complete devotion.

Richter did not mourn Ludwig's passing, but two other deaths that summer touched him. On 26 July Emil Scaria 'was buried in Frankfurt am Main. An irreplaceable loss. His masterly accomplishments in the roles of Marke, Wotan etc. made one forget the bad impression he made as a person. Whatever the case, he was the one singer who learnt most from the Master. He annoyed me much with his "cuts"; a scourge of répétiteurs and prompters.'[33] A few days later, on 3 August, he attended Liszt's funeral in Bayreuth, but he thought it a shabby affair and 'not worthy of the great dead man'.[34] Little did Richter know that, when he stood at Liszt's graveside, he stood only a matter of feet from where his own ashes would be laid to rest thirty years later.

July and August 1886 were spent in Vienna, the opera *Undine* by Lortzing being added on his repertoire on 25 July. On 13 September the Richter family moved from No. 36 to No. 56 Sternwartestrasse, a larger house which he bought for 29,500 florins that summer. The opera *Marffa* by Johannes Hager (pseudonym for Johann Halslinger, a Foreign Office official and amateur composer) was premièred by Richter on 14 October, and five days later he set off for his autumn season in London. Before leaving, however, there was an orchestral session on 15 October set aside for trying out new works, which might, having been approved by the orchestral players and conductor, be programmed during the following season. One of the works submitted on that day was by Hugo Wolf, and the episode was to prove an unhappy one for all concerned.

Wolf had first met Richter in 1875 when the conductor was newly arrived in the city from Budapest. At the time the 15-year-old Wolf was a recalcitrant student at the Vienna Conservatoire. He had sought Richter out with a view to taking lessons from him, saying that he could learn

1885–1886: Vienna, London, Birmingham 229

nothing from 'that ass Hellmesberger'.[35] Richter was not impressed, and rebuked Wolf for insulting a man with whom not only he, but virtually every other significant musician in Vienna, had studied. Richter did, nevertheless, manage to get the youth into Bayreuth, where he could partake of the music of his beloved Wagner. Wolf also harboured a hatred of Brahms which was as fervent as his worship of Wagner was intense. He had consulted Brahms in his early years as a composer, but the older man could only suggest that he take further lessons in counterpoint from Gustav Nottebohm, not advice that Wolf (who was eventually expelled from the Conservatoire in 1877) wanted to hear. Then in 1884, at the age of nearly 24, he was appointed music critic to the weekly *Wiener Salonblatt*, and from that vantage point was able to avenge himself in print on Brahms for a period of three and a half years. From early youth Wolf had displayed an unstable disposition, and in later life this mental instability was exacerbated by the effects of syphilis contracted in Vienna at the age of 17. Until the autumn of 1885 Wolf held a high regard for Richter as a conductor, though he was less complimentary about his selection of programmes. When Richter took over Gericke's direction of the Gesellschaft der Musikfreunde in 1884, Wolf pointed to a lack of judgement in selecting novelties at the Philharmonic, singling out too many performances of older compositions (by this he meant the many suites and concertos by Bach in rather spurious editions by Robert Franz or Sigmund Bachrich).

For two years Wolf took many opportunities to laud Richter, but it is worth noting the sudden change of attitude at the end of 1886.

Hans Richter gave an utterly admirable account of Mendelssohn's music to a *Midsummer Night's Dream* [on 20 January 1884], a performance indeed beyond all praise. [27 January 1884]

The gentlemen of the Philharmonic will surely not flatter themselves that the public streams to their concerts in expectation of artistic delights. We know that Herr Kapellmeister Richter is anything but malicious, but the devil himself could not have behaved more maliciously than our Herr Kapellmeister in the choice of such a programme as . . . Gade, Dvořák, Molique, and, as an act of mercy—what a stupendous undertaking—a symphony by Mozart. Bravo Herr Kapellmeister! You display taste, goodwill, industry, dedication, earnestness, stamina, and a goodly portion of ambition. Where will it all lead you? You will not, for God's sake, aspire to the dizzy heights of performing Haydn's children's symphonies? Beware the strain of such works. . . . No, Herr Kapellmeister, you must take care of yourself. . . . Spare us your precious life . . . for the battle against the Philistines. . . . These [works] may be agreeable to you, given your pronounced love of rehearsing, but to us they are a horror . . . and we shall forever be denied the prospect of hearing just once under your direction Czerny's *School of Velocity*, whose orchestration Herr Bachrich would doubtless undertake as a favour, and for a commensurate fee. [8 March 1885]

We think far too highly of this conductor, risen to eminence under the precepts of Richard Wagner, to believe in the sincerity of his attitudes towards Brahms's manner of composition. [22 March 1885]

It was the orchestra under Hans Richter that provided the high point of the evening. For us this most recent orchestral achievement in the final scene of *Götterdämmerung* has never unrolled so graphically, and with such crushing and exalting effect. [14 March 1886]

The uniquely difficult [*Missa solemnis*], making hair-raising demands on the chorus and soloists especially, was precisely played under the sure hand of Hans Richter. [25 April 1886]

Kapellmeister Richter, however, is on leave, sits comfortably in London, and makes the most of his talent [*Pfund* means both talent and pound sterling], happily adding to both his renown and his weight. That is only just, but it hardly does us any honour that Richter's merits are rated more highly in every respect in England than here, and it is greatly to be regretted that we should be deprived of his capacities at a time when his absence is more keenly felt than ever. And so we have had a very inadequate, listless production of the Nibelung trilogy [under Johann Nepomuk Fuchs]. [23 May 1886]

Herr Dr Johannes Brahms . . . reckons that he owes it to his fame to impregnate every form of instrumental and vocal music with that precious sensation [boredom], on which account he has recently taken to composing symphonies, to the great satisfaction of our worthy Hofkapellmeister Dr Hans Richter. [5 December 1886]

The memorial for Franz Liszt came off well, even peaceably, without the originally threatened execution of *Les Préludes*. And yet it would have been better if our jovial Hans Richter had stuck to 'his' *Les Préludes*, in which there is nothing left to destroy, rather than murder 'our' *Ideale*. We will gladly do without any Liszt if he is to be played with so little spirit and feeling as was the case with *Ideale*. We assume that Herr Richter first learned this piece at the dress rehearsal. . . . Herr Richter would be well to memorize thoroughly the Preface to Liszt's symphonic poems, but he should do it before the dress rehearsal. [19 December 1886]

It is to be hoped, in the interests of [Wagner's *Siegfried Idyll*], that the Philharmonic will not make it a part of their repertoire, at least as long as Herr Richter remains the conductor. Is it indeed necessary that our worthy Kapellmeister demonstrates his friendly disposition toward anything related to Wagner's concepts by laying his hands on Wagner's works and mutilating them when even our revered Herr Hanslick is content with the sacrifices laid out on the altar of the goddess of impotence in honour of her chosen favourite, Johannes Brahms? . . . Our worthy Kapellmeister . . . is not concerned with such childish matters as principles, views, confessions of faith, etc. That kind of trivia is not for him, and whenever he senses the stirring of something like scruple or doubt, he takes his baton and smashes the offender to bits and pieces. Unfortunately this impenitent time-beater remains untouched by scruple even when confronted with a musical work... such as the 'Faust' Symphony, whose second movement Herr Richter simply chopped into mincemeat.... Under such circumstances it would be better to play nothing but Brahms symphonies. In them at least there is nothing to be spoiled. [27 March 1887][36]

During 1885 Hugo Wolf completed the score of his new symphonic poem *Penthesilea* and decided to show it to Richter, which proved no easy task as the conductor was very difficult to track down at the Opera. Instead Wolf resolved to leave it at Richter's house in Währing. 'To my surprise Richter himself rolled forth out of the background and let me in. He was somewhat astonished over my visit and looked at me in great distress when he became aware of the score. I assured him that... he should look over the notes for their appearance and not for their content. My modest request relieved him.... Finally he asked for my score for more detailed examination and promised me a performance, in spite of Bachrich and his associates, if he should find the work a good one.... God only grant that *Penthesilea* may please Richter, for it would be devilish if nothing were to come of it.'[37] Wolf was further encouraged by reports from friends and colleagues in Vienna that Richter was impressed by the work and considered it important. Clearly his hopes were high for a trial with the orchestra, but a series of unexplained postponements and bitter disappointments occurred over the next year. There had also been the matter of his String Quartet in D minor which he had been trying to have performed by one of the many chamber groups in the capital. One was led by Arnold Rosé, and its viola player was Sigmund Bachrich, to whom Wolf had not been kind in print. The quartet was left at the stage door of the Opera for Wolf to collect along with a note expressing the unanimous refusal of the players to play it. Both Rosé and Bachrich were also prominent members of the Philharmonic orchestra and Wolf's action in placing himself at the mercy of people to whom he had been less than polite, and whose personal and collective support for Brahms was well known, is difficult to understand. He was clearly asking for trouble.

The 1885–6 season came and went and still *Penthesilea* remained untried, until the beginning of the following season, when Richter scheduled it for 15 October. Wolf immediately asked the conductor for permission to be present, something which was never allowed. Richter himself takes up the story.

I wanted to fulfil Wolf's wish to play the work through. He could give old Obenaus [the orchestral steward] a few coins to get himself into the organ gallery from where he would not be seen. 'But', I added, 'officially I know nothing about your being present at the rehearsal.' And that is how it happened. When the piece was over, I said to the musicians, knowing full well that Wolf was present, 'You see, gentlemen, how this man, who disparages Brahms every week in the *Salonblatt*, can compose such impossible stuff.'[38]

Wolf never forgave Richter for this comment to his players, nor did he ever forget the guffaws and laughter which his music elicited from the orchestra. To try and understand what might have caused this (and though their behaviour was inexcusable, no one, apart from Richter, knew that the composer was hidden in the hall), mention must be made of what was demanded of them at that morning rehearsal. They had *eight* new works before them, beginning with Glazunov's Overture on Three Greek Themes and concluding with Wolf's piece. In between were works by Prince Henry of Reuss-Koestritz, Finck, Degner, Hofmann, Herzfeld, and Krinninger totalling two symphonies, two serenades, an overture, and a suite by men whose names and music are unknown today (Prince Henry eventually conducted his symphony in Hamburg in December 1887). One can well imagine the state of the players' minds when confronted with a complicated and difficult work at the end of a morning spent sight-reading mediocrities. A rule at such occasions was that no rehearsal should take place; whatever mistakes occurred, the conductor would keep going in the hope that eventually the players would find their place, but meanwhile a veritable cacophony might ensue. None of the eight works played that day received a majority vote from the eighty-three players present. Glazunov mustered the highest total of nine and Wolf received none.

The affair plagued Richter on and off for years, particularly after the composer's death in 1903, when the Hugo Wolf Association in Vienna took up the dead man's cause. Some Wolf biographers have been quick to condemn the conductor, and it is true that he does not emerge blameless from the incident. Frank Walker adopts the fairest attitude towards both men in his biography of Wolf. He concludes that Richter's opinion of the work changed between his first sight of it in 1885 and the play-through in October 1886, probably basing his criticisms on its scoring, which was far too thick. In an open letter to *Die Musik* in 1906 Richter wrote that he warned Wolf of his reservations about the orchestration, but that the composer wanted the trial to go ahead as he had never heard a note of his music played by an orchestra. Richter's responsibility here was not properly discharged. He should have refused to play *Penthesilea* through unless it was revised, or he should have called on the composer's friends to bring pressure to bear.

Despite denials, Richter must have been aware of Wolf's attacks on Brahms in the press, such as this one which appeared on 24 January 1886 after the Vienna première of the Fourth Symphony. 'What an original, profound artist Herr Brahms must be when he can compose symphonies not only in C, D, and F, as Beethoven could, but even, unprecedentedly, in E minor! Heavens! I begin to stand in awe of Herr Brahms's uncanny genius. . . . May Herr Brahms be content to have found in his E minor Symphony . . . the language of the most intensive musical impotence.'[39] Walker concludes that Richter did not intend to wound or ridicule Wolf; he wished to do no more than teach him a lesson, but matters got out of hand and he seriously underestimated Wolf's temperament. It was not often that Richter misjudged a work, even if he did misjudge the composer in this case, and Walker's assertion that in *Penthesilea* 'we possess one of the grandest romantic conceptions of the nineteenth century'[40] must remain a matter of opinion. In a letter to Ludwig Karpath after the composer's death, Richter candidly acknowledged that 'members of the Philharmonic and Kapellmeisters can be wrong',[41] but he was also correct in pointing out that the *Penthesilea* being played then (1904) was not the same work he had conducted in 1886. Walker supports this, for Josef Hellmesberger jun. and Ferdinand Löwe revised it in 1903 to the extent of removing 168 bars as well as carrying out a thorough revision of the orchestration.

There was another composer who was also having little success in winning Richter's support at this time. This was Ferruccio Busoni. In 1883 (at the age of 17) he had arrived in Vienna and took an orchestral suite to the publisher Albert Gutmann. He was given an introduction by both Gutmann and Hanslick to Richter, who responded favourably to the youth's playing of his own composition, and once again a trial was promised for the work with the Philharmonic. A series of postponements then ensued which lasted into the beginning of the next season, but it was finally tried out on 4 October 1884. Again the rule governing the presence of the composer within the hall seems to have been broken (two years before the Wolf incident) and Busoni secreted himself in the gallery to hear his work played. Despite Richter's favourable opinion the orchestra had the last word, and by just one vote the suite was rejected. Though bitterly disappointed, Busoni, unlike Wolf, got on with his life, though he did leave Vienna.

Richter left Vienna for England after the Wolf débâcle and gave concerts in London, Birmingham, Brighton (at the Dome), Newcastle, Dundee (where he observed the whaling ships), Glasgow, Edinburgh, and Nottingham. Apart from Scotland's first hearing of Brahms's Fourth Symphony and the performance of Dvořák's *Spectre's Bride* with the Sacred Harmonic Society in Nottingham's Albert Hall, there was nothing new to Richter's programmes on this visit. On 3 November he wrote to Mrs Joshua on the touchy subject of changing the management of his concerts, for in his opinion Franke was no longer competent. He now wished to explore further an approach by Chappell's made to him a few years earlier through Mrs Joshua, though he swore her to secrecy 'until the concert on 9 November is over and the orchestra has been paid. . . . On the 10th Franke will be told everything, but before that my worthy musicians must be paid.'[42] According to Richter's diary for the day on which he effected the transfer of the running of his concerts from Franke to Chappell, his musicians were not paid. This, together with public weariness of an unchanging diet of the same programmes, was the cause of the reorganization of his English concert life. Arthur Chappell was considered the possessor of a more eclectic and imaginative spirit.

Richter arrived back in Vienna in time to bid farewell to his dying mother-in-law and attend her funeral on 15 November (with an inappropriately timed performance of the operetta *Der Trompeter von Säkkingen* later the same day). His first Philharmonic concert of the 1886–7 season had taken place the day before with d'Albert at last playing Brahms's Second Piano Concerto. The young pianist was becoming firmly established on the concert circuit, with triumphs in Berlin and Leipzig already behind him. The year ended with tributes to composers, Weber, whose birth had occurred exactly a century earlier, and Liszt, who had died earlier in the summer. Works new to Richter's programmes in Vienna in concerts in the last two months of 1886 were Mackenzie's Violin Concerto with Rosé (28 November), Dvořák's Scherzo capriccioso (5 December), Liszt's Fourth Hungarian Rhapsody (12 December), and the F major Symphony by one of the principal choral conductors in Vienna, Richard Heuberger (19 December).

18

1887–1888

Return to Bayreuth

THE concert on 19 December 1886, with its programme drawn appropriately from the Vienna-based composers Beethoven, Heuberger, and Fuchs, was Hans Richter's 100th Philharmonic concert. Just before it took place he wrote the orchestra a letter which reveals the excellent relationship between the conductor and his men. He referred to the famous Viennese brothers Johann and Josef Schrammel, who by 1886 had a tavern quartet of two violins, guitar, and clarinet and had added the word *Schrammelmusik* to Vienna's musical language. They had begun as a trio in 1878 and in 1884 wrote a Hans Richter March. They mainly played old Austrian folk music at beer gardens, wine festivals, and inns, and among their keenest admirers were Richter, Brahms, Johann Strauss, and Emperor Franz Josef. Although these groups dated back to the Middle Ages when they were used to spread music and news up and down the banks of the Danube, by the late nineteenth century they were called Schrammel quartets. The clarinet was replaced in 1893 by the accordion.

From the newspaper I understand that you wish to celebrate my 100th Philharmonic concert. It gives me great pleasure and pride to know that you deem my achievement so worthy, but I feel I am still too young to celebrate a jubilee even though the top of my head already shows signs of several thoughtful general pauses! If I should be granted a 200th Philharmonic concert to conduct, then in God's name celebrate; I'd then be old enough to be forgiven the pleasure of such vanity. So, I ask you to observe Sunday as an ordinary concert day, i.e. with fullest enthusiasm for the job in hand and otherwise to make no fuss. However I shall gladly take this opportunity once again to take it easy with you afterwards; after so much serious and often difficult work, real jollity can only do us all good. So let's fix a definite place to meet—just amongst ourselves—on Sunday. There we shall hear how superbly a splendid tavern band [*Schrammeln*] play those incomparable waltzes by Lanner. I can offer you nothing better.[1]

Whereas Richter was now being criticised in London for unimaginative programming, no such criticisms were levelled at him in Vienna. The Philharmonic programmes for the rest of the 1886–7 season (to 20 March) contained some first performances in the city, even a symphony by Haydn (No. 100, the 'Military') together with Bizet's Suite *L'Arlésienne* on 2 January, and Dvořák's new Symphony No. 7 in D minor a week later. He was also innovative at the Gesellschaft der Musikfreunde in the same period, conducting Handel's *Ode to St Cecilia* with Marie Lehmann and Bach's *St Matthew Passion*. At the Opera there was just one new work, *Harold* by Karl Pfeffer (3 April), but the performance of *Walküre* (Richter's fiftieth) on 11 February greatly upset him. The object of his anger was the Opera's resident baritone Theodor Reichmann, and his report to Jahn described an incident which the conductor felt would bring the name of the Imperial Opera into disrepute.

Our otherwise excellent baritone is not the master of his excitable temperament on stage in proportion to what is desirable of the ensemble. Repeatedly loud insults to the prompter—whether justified or not is not my concern—have now become commonplace. In the last performance of *Walküre* the artist got so carried away that he aimed Wotan's spear at the prompter's box time and again and poked it around inside. In the third act, however, matters became worse.

During a performance it is the conductor's duty to oversee and lead the musical ensemble on stage and in the orchestra. During the whole of my time as a Kapellmeister each and every singer has thanked me for the care with which I give cues; during the *Walküre* evening I was always careful to give the singer the difficult entries, but the conductor cannot disregard the whole merely for the sake of being a look-out for an individual. So it was that at one point I had to correct the mistake of a very nervous bass clarinettist, N.B. not at a place where I had to cue Mr Reichmann with an entry after several bars' rest or after another's lines, but during a long and uninterrupted phrase. After two or three bars devoted to the errant player, my attention returned to the stage. This short distraction so upset the artist that he angrily took a few steps forward and menaced me and the orchestra with both his fists. I do not feel at all insulted by this unpleasant behaviour because I make allowances for the singer's excitement, and do not consider that a well-bred man such as Reichmann would deliberately offend. On the other hand I confess to thinking, where will this lead? Should one not expect that one evening an agitated artist will throw his wig at the Kapellmeister's head or spit at the prompter?[2]

Reichmann duly sent a humble apology to Jahn, admitting his lack of self-control whilst devoting himself with such commitment to his art. He would in future try to control his paroxysms and not get carried away. He was sorry that Richter had thought fit to report him to Jahn rather than approach him directly. The clenching of his fists was, he contended, 'a misreading. I made a gesture but it was like that of a drowning man.' He did not finish without a dig at his 'very honoured friend Herr Hofcapellmeister Richter', for he found it difficult to reconcile the conductor's concept of the honour of the Opera with his 'punishing look at the orchestral player which was so long that my attention and that of the public was drawn towards the poor sinner and his error, and meanwhile the whole scene was leaderless'.[3]

One suspects that Richter may have found the affair rather amusing and that it was at the request of the poor prompter (who must have feared for his life at his next encounter with Wotan) that he made his official complaint. His distaste of preparations for Pfeffer's *Harold* was, however, genuine, and expressed to Dvořák.

I am a worried and overworked man; this is my excuse for not having sent you a report immediately after the performance of the *Scherzo capriccioso* and the D minor Symphony. Today your kind letter has somewhat lightened the pre-dress rehearsal of that utter drivel *Harold* (by our choral conductor Pfeffer), and I gladly seize the opportunity of renewing contact with a God-gifted musician. Before I drew up my London programmes I had it in mind to ask you if you had anything new for me. Now your *Symphonic Variations* have come as a splendid addition to my London scheme, one I accept with most cordial thanks. . . . I should like very much to put this work straight away into my first programme if possible. The *Scherzo capriccioso* is already included in my schemes. How often has the Symphony in D minor been performed in London? It is very dear to me (perhaps my favourite), but I must, for my manager's sake, be careful that the same work is not given too often; although I may claim without boasting that only a *dramatic* conductor, a Wagnerian (Hans Bülow must forgive me!), can do full justice to this particular symphony. . . . Your *Scherzo capriccioso* was very much liked in Vienna, the symphony unfortunately did not, on the whole, please as much as I expected, considering it was almost perfectly played by the Philharmonic orchestra. Let us say that our public is often very odd![4]

Richter conducted the *Variations* (which had been written ten years earlier) on 16 May at the third of his four London concerts that month. Even at the first rehearsal he was 'positively carried away. It is a magnificent work . . . among the best of your compositions.'[5] Dvořák quoted Richter's report to him of the first performance in a letter to his publisher Simrock. It had been 'an enormous success; at the hundreds of concerts which I have conducted during my life no *new work* has ever had such a success as yours'.[6] Richter was also delighted with the change of management reflected in bigger houses and a better orchestra thanks to Vert. At the fourth concert on 23 May an incident occurred which reflected enormous credit on the conductor and which became a Richter legend. The programme consisted of the delayed first London performance of Bruckner's massive Seventh Symphony preceded by two extracts from *Walküre*, but it opened with Brahms's Academic Festival Overture, and it was in this work that an untypical lapse of memory took place. 'My mind was preoccupied by my departure for Düsseldorf the next day,' he confided to his diary. 'I was distracted and forgot to beat *alla breve* after the sextuplets (shortly before the Gaudeamus). There was confusion for a few bars. I could not escape taking the blame so at the end of the overture I turned around and said the following words, [Richter's English] "When we have played this overture first time it was not the orchestra's mistake but mine," whereupon we repeated it, naturally without any distractions on my part. My candour was rated very highly.'[7] The *Musical Times* ignored this incident and was indifferent to the new Bruckner symphony, concentrating more upon its length than its content, 'the audience listening with unmistakable coldness or else going away'.[8] Charles Barry (who wrote the programme notes) described Richter's 'masterful performance before a large audience' to Bruckner, and the conductor told him that the work had been 'stormily' received.[9] Neither man wanted to be the bearer of bad news.

Richter's mind had been on his début at the sixty-fourth Lower Rhine Music Festival in Düsseldorf, an annual three-day choral and orchestral event at which he conducted on the second and third days. The other conductor was Julius Tausch, whose association with the Festival had begun in 1853 as assistant to Schumann. The 1887 Festival was to be his tenth and last but for Richter it was to be the first of four (1887, 1888, 1890, and 1897). He made a triumphant début on 30 May, not surprising considering that his opening work was the Overture to *Die Meistersinger*. D'Albert then joined him to play Beethoven's Fourth Piano Concerto and the concert continued with choral works by Bach and Weber, Mendelssohn's Hebrides Overture, and the 'Eroica' Symphony. On the following day he shared the concert with Tausch and conducted orchestral and vocal items by Berlioz, Brahms, Wagner, Liszt, and Bruch (the First Violin Concerto with Robert Heckmann). His soloists were Rosa Sucher, Hermine Spies, Heinrich Gudehus, and Fritz Plank. Cosima Wagner was present with her daughters Eva, Isolde, and Daniela. Although this concert included two works by Wagner and one by Liszt, Eva had tried in vain to persuade Richter to change the programme of the following day to include 'much by Papa and Grandpapa'.[10]

The chorus numbered 650 and the orchestra 127 with its string sections led by the Heckmann quartet. The rapturous reception Richter received in England in 1877 was now repeated in Germany. Being free of the score he was able to allow his gaze to roam around the orchestra, constantly balancing the sections and highlighting the solos. Düsseldorf was as unprepared for this revelation as London had been a decade earlier. 'It was a wonderful triumph,' wrote his friend August Lesimple in the *Allgemeine Musik-Zeitung*.[11] The *Kölnische Zeitung*, whilst fulsome in its praise, also pointed out that the Rhineland had its own excellent interpreter of Beethoven in Franz Wüllner, who had recently produced an 'Eroica' of equal stature. The paper also described how Richter stopped the Scherzo of Beethoven's Symphony after a few bars because 'a sudden noise in the hall had disturbed the beginning. As a rule a rapping such as this [on the conductor's stand] would in itself produce excitement and a general murmur, but no one stirred.'[12]

Back in London, and timed to celebrate Queen Victoria's Silver Jubilee (he watched her procession from Mrs Joshua's brother's house on 21 June), Richter introduced three new British symphonies (two of which bore the same title), Parry's Second in F 'Cambridge' (6 June), Cowen's Fifth in F 'Cambridge' (13th June), and Stanford's Third in F minor 'Irish' (27 June). Parry's work had been revised since the two men had played it through in June 1886. They had another session together two weeks before its performance. 'He played the upper part and I the lower, and even at *presto* pace in the Scherzo he was hardly ever at a loss, always picking out the particular part of the score that would be prominent at the moment, and playing fiddles, clarinets, and horns with equal success. It is an astounding gift.'[13] Richter had already suggested that Parry might compose a celebratory work for the festivities but the offer was declined. 'I had a letter from Richter this morning inviting me to write a Jubilee overture. I really can't. The idea is disgusting. I'm so stupid. It's just as if all my wits were clean gone.'[14] Stanford's new work was almost a disaster. 'The society functions at the Castle very nearly imperilled the first performance (under Richter). At the last moment several of the best players in the Richter orchestra, who were also members of the Queen's band, were ordered down to Windsor, and if it had not been for the unique sight-reading powers of their deputies and for Richter's vigilant eye, the difficulties of the work might well have brought about a catastrophe; but happily no flaw was observable.'[15] All three symphonies were well received by musicians and public alike; June 1887 was a rich month for British music and Richter had played his part well. Vert had also proved a success in running

Richter's affairs though the conductor was caught up in litigation between Vert and Franke (on 1 July Richter was in court and had great problems coping with having to speak English for three hours). One of Vert's reforms was to limit the conductor's appearances in England to the one season in 1887.

From Bayreuth Cosima wrote to Richter in Vienna, asking him to return for the 1888 Festival. She had accepted advice by Levi and Mottl to give up her plan of nine *Parsifal*, three *Tristan*, and five *Meistersinger* performances; they thought three operas too many and so *Tristan* was dropped. Her invitation to Richter suggests that the remaining two operas would be divided between the three conductors, but Levi did not conduct in 1888. Richter and his family went on holiday in the Nasswald during July. He journeyed back to Vienna every weekend to fulfil his duties at the Court Chapel, and returned alone at the beginning of August when the Opera reopened for rehearsals. In his reply Richter asked Cosima to write to the Opera to obtain official leave of absence. 'As a last resort', he wrote, 'I shall turn to the emperor himself, for I *must* conduct *Die Meistersinger* in Bayreuth.'[16] Matters became more complicated when dates were discussed, for he was committed to his London season from the beginning of May until 9 July, and was wanted in Bayreuth before the end of that period. He also pointed out that his London earnings were absolutely vital to feeding and educating his six children; his Viennese posts were not in themselves sufficient for that purpose. There was an interchange of letters between the two which lasted until April 1888 before it was agreed that Richter should come to Bayreuth as soon as his London season ended. Mottl would prepare the orchestra for him in time for his arrival. By May of 1888 Cosima's involvement was becoming apparent. 'The stage director Harlacher swears that the brawl scene in the second act of *Meistersinger* needs supernumeraries. I am resisting this as I want the whole scene played by the chorus. In order not to appear obstinate, I ask you if (without my knowledge) extras were used in Munich in 1868 to fill the streets.'[17] 'As far as I remember', replied Richter, 'a few male ballet dancers danced about in the background at the end of the brawl scene in Munich; there were only a few. Too much is done in this scene on German stages. . . . In the interests of the musical entries only a few extras should be used in the background.'[18]

There was another celebration that summer of 1887 in which Richter played an important part. Mozart's *Don Giovanni* had been written a century earlier and the anniversary was celebrated by the Mozarteum in Salzburg with two performances on 20 and 22 August with Theodor Reichmann in the title-role, Heinrich Vogl as Don Ottavio, Josef Staudigl as Leporello, Marie Wilt as Donna Anna, Lilli Lehmann as Donna Elvira, and Bianca Bianchi as Zerlina. Richter conducted a predominantly Vienna Opera orchestra (sixteen violins, six violas and cellos, four basses), insisting that his principal oboist, bassoonist, and two horn-players were fetched from either that city or Munich. At a banquet he gave a speech in which he revived the idea, first mooted ten years earlier, of an annual Mozart Festival in Salzburg. What Bayreuth was doing for Wagner, Salzburg should do for its most famous son; but it took until 1920 before the Festival was launched by Max Reinhardt.

Richter's concert season in Vienna from November 1887 to April 1888 included several new works; Dvořák's *Symphonic Variations* on 4 December (the composer was Richter's house guest for two days in November), Robert Fuchs's Symphony in E flat on 18 December, d'Albert's Overture *Esther* and Saint-Saëns's *Le Rouet d'Omphale* on 8 January, octets by Svendsen and George Onslow and Mendelssohn's Overture for winds at the Nicolai concert on 2 February, Goldmark's Symphony in E flat on 26 February, a new violin concerto (played by Rosé) by Stefan Stocker, critic, teacher, and friend of Brahms and Hanslick, Mottl's orchestration of Liszt's *St Francis of Assisi's Sermon to the Birds* on 11 March, and the Good Friday music from *Parsifal* on 8 April.

The season of operas for the same period included *Lohengrin* and *Faust* with Lola Beeth, Vienna's anniversary *Don Giovanni* with only Lehmann from the Salzburg cast (Materna sang Donna Anna in Vienna) on 29 October, Massenet's *Le Cid* (surprisingly with Richter and not Jahn) on 22 November, Auber's *Le Domino noir* on 19 January, and Pauline Lucca's fiftieth *Carmen* (Richter's sixty-fifth) on 7 March. There were also four Gesellschaft concerts which included Mendelssohn's *St Paul*, Schumann's *Paradise and the Peri* (with Materna and Gustav Walter), and the first hearing of Dvořák's *Stabat mater*, plus two extra concerts for the Society of which the second was Bach's B minor Mass. There were extra events such as a concert for Pablo de Sarasate on 29 November (in which he played the Beethoven and Mendelssohn concertos as well as his own *Muiñiera*), a Mozart memorial concert on 4 January, a concert of Bruckner's works (Fourth Symphony and the Te Deum) on 22 January with a chorus drawn from both the Singverein and the Wagner Association, and a concert for the latter organization on 29 January in which Wagner's early C major Symphony was heard for the first time.

This relentless schedule led, in May 1888, to Richter's London season. It began sadly with the deaths of Hermann Franke in a lunatic asylum and of the flautist Oluf Svendsen, a close friend and fine artist, from lung cancer.

After two concerts of normal Richter fare (though at the second, on 14 May, Stanford's 'Irish' Symphony was given again, an important test for new works even to this day), he travelled to Aachen for the sixty-fifth Lower Rhine Music Festival, where with a 'brilliant chorus and splendid orchestra the whole Music Festival was wonderful'.[19] His first programme on 21 May consisted of overtures by Weber and Schumann, the closing scene from *Götterdämmerung*, and Beethoven's Ninth Symphony, the second on the following day was more enterprising with works by Berlioz, Mozart, Liszt, Wagner, Schubert, Bruch, and Brahms. When it came to programming, Richter had no qualms about mixing the New German School with the classical German Romantic tradition, even in the Rhineland, where Schumann and Brahms devotees were based. Both works by Brahms and Bruch were new to Richter; they were the Double Concerto for violin and cello by Brahms and Bruch's *Kol Nidrei*. Joseph Joachim and Robert Hausmann were the soloists, as they had been when the pieces were premièred in Cologne 1887 and Liverpool 1880 respectively.

Back in London from 24 May and now joined by Marie, Richter conducted another seven concerts that summer. One of his soloists was Henri Marteau in Bruch's First Violin Concerto on 4 June and the Introduction and Rondo capriccioso by Saint-Saëns a week later. Marteau (who had appeared with Richter just six months earlier in Vienna in the same work) was aged 14. In the first of those two London concerts Mackenzie's Overture *Twelfth Night* was given its first performance. The concert on 18 June included a performance of Siegfried's Funeral March in memory of Emperor Friedrich, who had just succumbed to throat cancer after only 100 days on the throne. One of the laryngologists consulted by the dying monarch was Sir Felix Semon, physician to the Prince of Wales and one of Richter's friends.

My experiences with Hans Richter were rather curious. I have no doubt that of the great conductors of my time—Mottl, Levi, Weingartner, Schuch, Nikisch—whose performances I attended, he towered head and shoulders above all his contemporaries. I have never observed in a conductor so wonderful an understanding of various composers, nor the power of conveying his own perception to each individual member of his orchestra. He *played* on the orchestra as a virtuoso plays on the piano or the violin. A powerful, thick-set, thoroughly Teutonic man with a reddish-blonde, full beard, piercing grey-blue eyes, quiet dignity of demeanour, subtle humour illuminating his conversation, which was usually devoted to the highest artistic ideals and illustrated by rich, ripe experience of men and matters, he greatly attracted me. The impression seemed mutual. Soon, in the German fashion, he offered me the 'thee and thou'. For many years we saw a good deal of each other in our house, in the German Athenaeum, at his concerts, and at houses of mutual friends.[20]

Richter conducted the Overture to *Tannhäuser* in a testimonial concert on 13 June for the retiring manager of St James's Hall, Ambrose Austin. Others taking part included W. G. Cusins, Zélia Trebelli, Sims Reeves, Edward Lloyd, Charles Santley, Janet Patey, Alwina Valleria, and the pianist Vladimir de Pachmann. The season ended with a customary large work by Beethoven, his *Missa solemnis*. Arthur Coleridge recalled one of the first performances of this work under Richter in 1881, illustrating the power of both the Mass and its interpreter. 'I went to the rehearsal as well as the performance; it is an event in one's life to watch the command of Richter.'[21] Engel in the *World*, whilst retaining his admiration for the conductor, was beginning to tire of Wagner and Beethoven. Richter's masterly beat and memory were again a source of wonder, though Engel remarked that at the concert on 11 June, when Richter was conducting three works new to him, it was unusual to see him with scores on his music stand. His review of Richter's first concert contained the following interesting observation which is valid today.

It is true that he indicates perhaps a little too clearly how well he knows each entry, but the fact remains that he never misses one; and let those who understand what that means look in awe at the incredible performance. Nor can a conductor be expected to do great things with his orchestra unless he accustoms them to look perpetually at his baton, and you may instantly see the power of a great conductor when, at the rehearsal, he suddenly stops conducting because he has something to say. If all the instruments do not at the same time stop, if they go on playing for four, five, six bars, then you may be sure they do not much care for their conductor, and they will never do what Richter's does. So far as this matter goes, you could hear it at the first chord when he began the *Kaisermarsch*; there was a sudden striking ensemble with a power, a spontaneity that you will never get from a band unless their eyes are riveted on the conductor. If only that conductor would leave Liszt's *Hungarian Rhapsodies* alone, which he usually takes too slow at the beginning and too fast at the end, I should have no fault whatever to find with him.[22]

From London Richter journeyed to Bayreuth for eight performances of *Meistersinger* (the third of which, on 30 July, was his fiftieth). In his first appearance since 1876 at this sixth Festival it became clear from letters and notes from Cosima to him during rehearsals that she was warily exploring ways of influencing his work at Bayreuth. Cosima was now getting into her stride running the Festival but she soon discovered that, whereas she could manipulate Felix Mottl, Richter was not so easily swayed. She was jealous of his success in handling his orchestra and singers, an art which eluded her. It was unprecedented to see a woman on stage directing a production, but she generally won her singers over and they respected her direction; in musical matters she tended at this stage to use others to convey her orders, though with Richter and Levi she found herself in a dilemma. On the one hand she needed them because they had worked directly with Wagner, on the other they were impossible to influence because they possessed that same authority.

Forgive me, my dear Richter, I forgot some things, namely: (1) Beckmesser's lute (harp). Are you agreeable to it like that? I mean that the paper is audible, and that it was better at the beginning. (2) Do the trumpeters have their music fixed to their trumpets, as I discovered from Adolf [von Gross?]; that cannot be! Can't they play it from memory? Of course no risks should be taken. Don't be angry with me, these matters are going round in my head.... P.S. Something else, I thought the little conversation between David and Sachs (Act II) at the workshop door was lost. Was that due to the delivery or perhaps *a little, a trifle* to do with the tempo? Once again, forgive me!

1. Magdalene: Act one 'Hier ist das Tuch', her re-entry was too soon.
   Act two 'Unser Junker vertan', 'vertan' was somewhat unclear. 'Vielleicht vom Sachs' a bit clearer.
2. Walther: Act two 'Näselnd und kreischend'.
3. Eva: Act two 'Gleich Lene'. 'Gleich' is addressed to Lene, on the other hand 'Zum Abendmahl' to Pogner again. 'Dem Meistergericht' more tender.[23]

There were several cast changes during the eight performances. Reichmann, Scheidemantel, and Plank sang Sachs, Sucher and Malten shared Eva, Gudehus sang Walther. The *Musical Times* had nothing but praise for this uncut staging accompanied by the 106-piece orchestra under Richter, 'whose co-operation was of the greatest service'.[24] The American Louis C. Elson was present and described how 'the curtain did not go up promptly after the prelude of the last act, and one heard Richter's bass voice shouting from the pit "Auf! auf!", a thing which does not often occur in the Bayreuth theatre.... At the end we had all of us gravitated to Angermann's.... At the particular plank which served me for a table there sat a certain fat and hearty beer connoisseur named Hans Richter... and all went merry as a marriage bell.'[25]

After Bayreuth came the Birmingham Triennial Festival during the last four days of August with two concerts daily (an oratorio in the morning, whilst the evening so-called Miscellaneous Concerts were mixed choral and orchestral events). Richter conducted *Elijah* (with Albani, Patey, Lloyd, and Santley), Dvořák's *Stabat mater*, Parry's *Judith*, Sullivan's *The Golden Legend*, Handel's *Messiah*, Bach's Magnificat, Berlioz's *Messe des morts*, and Handel's *Saul*. Parry thought 'Bach's *Magnificat* not very good. Beethoven's C minor first rate. Berlioz *Messe des Morts* the work of a huge big man, but mostly big charlatan in this case. The musical material mostly nil, though some amazing great strokes of effect. Writing for voices mostly fearfully bad.'[26] Fanny Davies played Schumann's Piano Concerto and Grieg was present to conduct two of his own works, the Overture *Autumn* and the 'Holberg' Suite. The composer conducted 'with marked success, as well as to the unconcealed amusement of an audience accustomed to Dr Richter's undemonstrative style'.[27] Parry also saw him. 'Grieg turned up to conduct his Suite. A most characteristic little object he is, with about the sweetest expression I ever saw on any man's face. And he is altogether of a piece with his music. On a tiny scale, so tiny that a big stool had to be brought for him to stand on to conduct. His conducting is very funny to look at, but is very good all the same.'[28] Richter was criticized for using Franz's version of *Messiah* and the Magnificat, and for a reading of the *Golden Legend* 'which was certainly not Sir Arthur Sullivan's'.[29] Other English composers represented were Dr Frederick Bridge, who conducted his own new cantata *Callirhöe*, and Goring Thomas in a scena and aria from his opera *Esmeralda* sung by Albani. Concern was voiced in the press at the smaller audiences at all the concerts except for the evergreen *Elijah* and *Messiah*, though the management and not Richter were held responsible for correcting this in the future.

Parry's relationship with Richter was now less formal. They had several meetings to discuss *Judith*, but Parry also sought advice on the matter of a National Opera scheme.

I paid Richter another visit after tea owing to a postcard he sent, and found him sitting in his jersey and trousers and looking rather like those comical china images of Japanese fattys, squatting all flab as to tummy etc! He was genial enough and gave me some very good points of consideration about the National Opera scheme.
   1st year: Six weeks rehearsal of a single thing and a short season, principally a run. Lose £12,000
   2nd year: Less rehearsal and more performances. Lose £8,000
   3rd year: Good long season with several works. Lose about £4,000
   4th year: Still longer season and less proportionate rehearsal, and make both ends meet.
   The object of the whole scheme must be essentially 'first rate ensemble'. Real

good performances all round. The success of the scheme as an English affair must come out of the aim to make the thing thoroughly good.[30]

Of the rehearsals for the first performance of *Judith* he wrote.

22 August: Rehearsal at St George's Hall at 10. Richter seemed in a very bad humour and evidently knew nothing about the score. There were very few mistakes in the parts but he made a great fuss about the few that were there and lay great stress on the excellence of the parts being due to Messrs. Novellos' care and the mistakes being due to me. Altogether it was a very disagreeable experience. But most of the instrumentation sounded as I meant it, which was a comfort and the people who were present seemed pleased with it. Specially Grove, elder Stainer and Squire. Found that Richter expected me to go to Birmingham with him for choral rehearsal and had to hustle back to Kensington to get my luggage and only just caught the 3 o'clock train. Found Richter at the Town Hall superintending the arrangement of the desks. . . . Richter took many of the tunes too fast and seemed not at home. He pottered, moreover, over passages between the choruses and could not get to the end, notwithstanding the readiness of the chorus to stay over their time.

25 August: Up to Birmingham early with Richter. He went in to rehearse orchestral things and I took possession of the innermost artists' room to do more correcting of band parts. In the afternoon full rehearsal of *Judith*. He still had lots of the tempi wrong and was often all abroad about it. And he seemed fagged, irritable and anxious. It appeared to me that the chorus dragged fearfully, but I found out after a time that it was owing to the fact that in the Town Hall when empty the sound travels from the Chorus to the other end of the Hall and back to the Conductor's desk, by which I was sitting, and the effect is consequently most bewildering. And it is only by watching the lips of the Chorus that one can tell whether they are singing up to time. I thought the rehearsal very bad, discouraging and at the time I very much regretted that I had surrendered the stick to Richter.

26 August: I escaped church and went for a walk with Richter. But it was a most silent walk and I got next to nothing out of him. He mostly grunted or snuffled as he walked along and answered in monosyllables. He told me he had not had a chance for even a little walk for six months, such is his incessant grind. He woke up a little talking of Goldmark whose works he does not like. Expressed his coincidence with my own opinion that Berlioz's effects are deliberately got up and not music, and was particularly contemptuous about the bass trombone and flutes dodge in the *Messe des Morts*. We walked through the town of Bromsgrove and got back just as the family were finishing midday lunch. A little later Anna Williams, Mr and Mrs Patey, [and] Santley turned up to spend the afternoon and dine. Thereupon followed a lot of pottering about the grounds looking at horses, cows, pigs etc and talking a deal of flimsy nonsense. . . . The dinner at 4.30 was a tremendous affair. Everything on the richest scale with wines of exceptional

vintages and so forth, and great profusion of hospitality coupled with too much ostentation but a good deal of kindness and heartiness to help me through. Santley was cram full of funny stories which he told with admirable élan and effect. Richter also woke up after the ladies were gone and capped some of Santley's stories, but not with so much point.

29 August: Up early to the Town Hall to meet Richter and go through all the work with him to make sure of the tempi, which he seemed at last to take better hold of, and after all the performance was excellent. The chorus was of course splendid, band ditto.... The rule that no applause is allowed after the separate numbers is rather trying, as it is difficult to tell whether people are liking it or not. But there was a good row after each half and my dear chorus shouted and waved their pocket handkerchiefs at me like mad.[31]

Herbert Thompson reviewed the Festival for the *Yorkshire Post*.

Herr Richter's reading of *Messiah* was, we are pleased to notice, devoid of the sensationalisms and tendency to exaggeration which is so often indulged in. This was notably the case with the chorus 'For unto us a child is born' in which the opening phrases were sung and played piano and not pianissimo and words 'Wonderful, Counsellor' were forte and not fortissimo, thus following the composer's intentions and not producing the exaggerated effect too frequently aimed at and, in our opinion, quite out of keeping with the character of the music.... We suspect too that Herr Richter is not the man to permit organists, any more than other players in the orchestra, to assert themselves too strongly—an awkward *contretemps* by which the effect of the concluding cadence of the *Hallelujah* Chorus was considerably damaged, was no doubt the fault of the instrument and not of the organist, so it must be laid to its charge.

... The only fault that can be found with the selection of Beethoven's fifth Symphony as the second piece in the programme was that it seemed unfair to all the other composers whose works had been performed during the week, making them appear pygmies by the side of the giant of composers. From every other point of view the choice was admirable, in the first place to have had the services of Herr Richter, the greatest of living Beethoven conductors, and the fine orchestra under his command, without including a Beethoven symphony in the scheme would have been a waste of material nothing less than wicked whilst of all the immortal nine, none is so thoroughly characteristic of the composer and at the same time so universal a favourite as that in C minor. As regards its execution we can only say we have never heard so magnificent and perfect a performance even under Herr Richter's baton. The restrained vigour of the concise yet satisfactory first movement, the sweet loveliness of the unapproachable song without words which follows the weird and mysterious scherzo leading into the triumphant and brilliant Finale with its long-delayed catastrophe were all rendered with that complete insight into the composer's meaning, which is the great secret of Herr Richter's wonderful success in his interpretation of the Bonn master's works.... We cannot refrain from noticing the wonderful pianissimo obtained at the end of the scherzo just before the remarkable transition to the Finale, the effect of which was thus much enhanced. The performance made a marked impression upon the large audience, who received it with enthusiastic applause. . . . Now that the Birmingham Festival of 1888 has come to an end, a few general remarks on the results may not be out of place. That it has been a great artistic success cannot be doubted, and whilst this may be in some measure due to the Festival Committee, among whom there are doubtless some musical people, we cannot but recognise the excellent influence of the distinguished musician who has been the conductor and the leading spirit of the Festival. This has not been noticeable merely in the interpretation of the works given, but also in their selection, and if 'nothing common or unclean' has disfigured the programmes, then we can make a shrewd guess to whom we can attribute so merciful a deliverance.

. . . Turning to the orchestra, Herr Richter's extraordinary power of producing excellent results with unpromising materials makes it but a poor compliment to say that they proved equal to the occasion. But we can fortunately go further and add that they were excellent and remarkable for spirit and accuracy as well as for richness and performance of tone.[32]

The year 1888 ended with Richter's return to Vienna, where Ernest van Dyck sang Lohengrin under him on 17 October with Lola Beeth as Elsa, Materna as Ortrud, and Reichmann as Telramund. The Belgian van Dyck had moved to Vienna a month earlier having scored a triumph at Bayreuth as Parsifal during the summer, and he was to become another jewel in Jahn's crown at the Vienna Opera, whose building was now illuminated by electric lighting. The first Philharmonic concert (11 November) included Bruch's *Kol Nidrei* played by the orchestra's principal cellist Reinhold Hummer. A week later Richter's vocal soloists were Materna, Papier, Winkelmann, and Weiglein for the first Viennese performance of *Theodora* by Handel for the Gesellschaft der Musikfreunde. He also conducted Brahms's German Requiem and Joachim's *Overture in Memory of Heinrich von Kleist* for the Society on 16 December. The pianist Bernhard Stavenhagen in Beethoven's Third Concerto (25 November) and Joachim and Hausmann in Vienna's first hearing of Brahms's Double Concerto (23 December) were his illustrious soloists at the Philharmonic, and at the Opera Richter ended the year 1888 with a *Ring* cycle (11, 14, 17, 21 December).

19

1889–1900

Vienna

THE year 1888 was the half-way point in Richter's forty-seven-year professional career from 1865 to 1912. It was also the busiest year of his life; he never again had *simultaneous* associations with so many organizations or held so many positions. In that year he had his four posts in Vienna, his London season, his return to Bayreuth, his first engagement at the Lower Rhine Music Festival, and his second Birmingham Triennial Festival. Between 1888 and his departure from Vienna in the New Year of 1900 to reside in Manchester as Hallé's successor there were no new commitments, only new works, new artists, and new composers. On 23 December 1888, at the fourth Vienna Philharmonic concert, he conducted his 2,000th performance (his first thousand had been completed on 9 January 1881, the third would take place on 17 November 1895, and the fourth on 14 May 1906). With his annual schedule firmly established by the end of the 1880s, an overview of Richter's life in the 1890s covering all aspects of his musical life now follows, with an emphasis on special occasions or events during the decade.

At the Opera the Jahn–Richter duo held sway until 1897, the year of Mahler's arrival. Jahn's great achievements in his last years were the staging of Verdi's *Otello* and *Falstaff*, and Massenet's *Manon* and *Werther* (neither or which concerned Richter, who only conducted the less successful *Le Cid*). Both were vehicles for van Dyck and Marie Renard. They were two of the new generation of singers about to replace those who had reigned supreme during the previous two decades, but they maintained the golden era with wonderful performances of their own. Amalie Materna sang her last Brünnhilde in *Götterdämmerung* under Richter on 30 December 1894, and retired in 1897. Lilli Lehmann took on the mantle of that role, whilst

Paula Mark's fame grew with her portrayal of Nedda in *Pagliacci*. The American contralto Edyth Walker made her Viennese debut as Azucena in *Il trovatore* on 11 May 1895 under Richter, and remained a member of the company well into Mahler's era. Marie Wilt and Rosa Papier completed Jahn's most successful female singers, whilst among the men were Karl Sommer, Josef Ritter, and Scaria's heir, Karl Grengg. They all sang under Richter and many were recommended by him to Cosima Wagner at Bayreuth.

Richter conducted world premières of operas which have not survived the years, such as Robert Fuchs's *Die Königsbraut* (22 April 1889), Antonio Smareglia's *Der Vasall von Szigeth* (4 October 1889) and *Cornelius Schut* (23 November 1894), Richard Heuberger's *Mirjam* (26 January 1894, although the composer conducted the first two performances as Richter was ill), and *Walther von der Vogelweide* by Albert Kauders (28 February 1896). He also conducted the first Viennese stagings of Liszt's oratorio *St Elizabeth* (Christmas Day 1889), Mendelssohn's fragment *Die Loreley* (13 September 1890), *Der Barbier von Bagdad* by Peter Cornelius (4 October 1890), Mascagni's *L'amico Fritz* (30 March 1892), Leoncavallo's *Pagliacci* (19 November 1893), Smetana's *The Kiss* (22 August 1894), and Lortzing's *Der Waffenschmied* (11 February 1896). Curiously Richter never conducted that ultra-Wagnerian opera *Hänsel und Gretel* by Humperdinck, which received its first performance in Vienna during December 1894. Works from the repertoire which were new to Richter included Gluck's *Orfeo ed Euridice* (7 September 1889), Bellini's *La sonnambula* six days later, Donizetti's *La favorita* (preceded by Mendelssohn's *Die Loreley* in a double bill on 28 November 1890), Karl Goldmark's *Die Königin von Saba*, (25 August 1891), Mozart's *Bastien und Bastienne* (in a double bill with Mascagni's *L'amico Fritz* on 4 August 1892), Flotow's *Martha* (8 September 1894), and Mascagni's *Cavalleria rusticana* (29 January 1896). There were also his very infrequent appearances elsewhere (apart from Bayreuth) as an operatic conductor, such as in Wiesbaden for a performance on 14 May 1896 of *Meistersinger* with his old colleague Franz Betz as Sachs. Then in March 1898 he gave three performances of *Lohengrin* at the Marien Theatre in St Petersburg and the first *Meistersinger* in that city. The 16-year-old Igor Stravinsky (who went to school in St Petersburg) attended these performances and, on the subject of conductors he heard at this period, recalled:

[Eduard] Nápravník had certainty and unbending rigour in the exercise of his art; complete contempt for all affectation and showy effects alike, in the presentation of

FIG. 7 *Wiener Journal*, 1 February 1890.

the work and in gesticulation; not the slightest concession to the public; and added to that, iron discipline, mastery of the first order, an infallible ear and memory, and, as a result, perfect clarity and objectivity in the rendering . . . Hans Richter, a much better-known and more celebrated conductor, whom I heard a little later when he came to St Petersburg to conduct the Wagner operas, had the same qualities. He also belonged to that rare type of conductor whose sole ambition is to penetrate the spirit and the aim of the composer, and to submerge himself in the score.[1]

There are some interesting stories of Richter as an operatic conductor in Vienna. Theobald Kretschmann was a cellist in the opera orchestra (he had also played under Richter at Bayreuth in 1876), and remembered Richter's phenomenal ear, memory, and presence of mind.

It was easy for him to identify and correct mistakes or bad tuning in the most complex music or where transposing instruments was concerned. Conducting from memory, he gave every cue to each instrument—a look was sufficient and one knew what he wanted. He was great at staying with singers who had a memory lapse by beating quickly through the bars, and at the same time one heard a few pithy comments. During the lovely but rhythmically very complex opera *Der*

Barbier von Bagdad the score became unglued when he turned a page too violently. A shower of separate pages flowed down to the floor; reacting quickly he grabbed the cello part and conducted without further ado to the end of the opera, giving all the signs and cues to the orchestra.

Wilhelm Gericke, the conductor of *Der Widerspänstigen Zähmung* by Hermann Goetz, fell ill and Richter had to do it though he did not find the time to look through the score and gauge the *tempi*. He asked the two cellists sitting to his right to correct him in case he set a *tempo* which was not usual. However he had made the wrong choice, because every *tempo* was too slow for these two men who had little enthusiasm for opera work. They constantly whispered 'faster, faster' to the conductor and as a result the opera finished about half an hour earlier than usual.[2]

Another cellist from the Philharmonic, Joseph Sulzer, recalled an incident during the 1890–1 opera season involving himself and Richter, which was reported in the *Fremdenblatt*.

In yesterday's performance of *Lohengrin* at the Court Opera, which was attended by the emperor, the widowed Crown Princess Stephanie, and the Romanian heirs to the throne, an interesting episode took place. Herr Winkelmann was singing the role of Lohengrin even though he felt very unwell. During the course of the performance his vocal indisposition became so bad that in the third act, when taking leave of Elsa, he made gestures to the conductor of the orchestra Hofopern-Kapellmeister Richter that he was unable to sing his role further. It was hard to know what to do but Herr Richter offered a solution. The well-known principal cellist of the Court Opera orchestra Prof. Josef Sulzer took over the role of Lohengrin on his instrument, whose sound bears the nearest resemblance to the male voice. The experiment worked surprisingly well. While Herr Winkelmann was limited to miming his role on stage, his vocal line, including the Grail aria, resounded from the orchestra in the vigorous singing tones of the cello, which proved to be a splendid Lohengrin in Sulzer's masterly hands.[3]

The year 1889 began with a world première in the Philharmonic concerts when Annette Essipoff (currently Madame Leschetizky) played Paderewski's Piano Concerto. 'The concerto was favourably received', wrote the composer, 'not only by the audience but by the critics as well. Hans Richter had great faith in my compositions, as he proved by giving a masterly performance of my symphony in London some years later [18 December 1909]. Richter was extremely kind, amiable, and a thoroughly good man. He was very much acclaimed wherever he appeared, but he remained all through his life a modest, accessible artist, and ready to assist a young musician in every way he could.'[4] Because of either indifferent or hostile press reaction Richter now brought hardly any English works to Vienna (unfortunately he discovered Elgar's music just as he was leaving

Austria to live in England), Mackenzie's Twelfth Night Overture being one of the last examples on 27 January 1889. Hanslick took his customary cruel pleasure at its cool reception by the public when, by playing upon the title of Shakespeare's play in its German translation *Was ihr wollt* or *What you will*, he wrote, 'had the Vienna public really been asked, "Was wollt ihr?" the reply would have been in no uncertain terms, "something else!"'[5] His scorn was not reserved exclusively for English music. When the tone poem *Phäeton* by Saint-Saëns was conducted by Richter at the last concert of the season on 7 April 1889, he was just as vitriolic. Recalling that it had been thirteen years since this work had been heard in Vienna, he hoped it would be longer before it was given again. At the same concert the blind pianist Josef Labor played Johann Rösler's Piano Concerto, which, at the time, was thought to be a newly discovered work by Beethoven.

There were, however, many more successes than failures in the Philharmonic concerts under Richter in the 1890s. Bruckner's Third Symphony got a decent performance at last (21 December 1890) followed by his First Symphony a year later (13 December 1891), the Second on 25 November 1894, and the massive Eighth, which was premièred on 18 December 1892. Grieg's music was now establishing itself in the programmes; his Piano Concerto was played by Teresa Carreño, the second of d'Albert's six wives, on 9 November 1890, and the *Peer Gynt* Suite on 4 January 1891. Dvořák's new music was loyally promoted by Richter; his Eighth Symphony was played on 4 January 1891, *Husitská* on 21 February 1892, Carnival Overture on 9 December 1894, *In Nature's Realm* on 3 February 1895, *Othello* on 1 December 1895, the 'New World' Symphony on 16 February 1896, the *Water Goblin* on 22 November 1896, the *Noonday Witch* a month later on 20 December, and the Cello Concerto on 7 March 1897. His compatriot Smetana was now enjoying the success not granted to him in life with first performances of his *Vltava* on 2 March 1890, the Overture to the *Bartered Bride* on 15 March 1891, *From Bohemia's Woods and Fields* on 29 January 1893, *Vyšehrad* on 12 November 1893, and *Šárka* on 11 November 1894. Richter's support of Tchaikovsky, whose *Romeo and Juliet* Fantasy had been hissed in Vienna in November 1876 and whose Third Symphony had, after being rehearsed, failed to achieve the confidence of both management and orchestra in the following year, at last succeeded with his Serenade for Strings on 5 April 1891; but it was two years after his death, on 3 March 1895, before his 'Pathétique' Symphony took the city by storm followed by the Fifth Symphony a year later on 1 March 1896, the First Piano Concerto at the end of the same year on 20 December, and the Third Suite on 20 February 1898. His Russian colleagues Borodin and

Rimsky-Korsakov were also introduced to Vienna by Richter, the former's Overture *Prince Igor* on 6 December 1896 and the latter's *Sheherazade* at one of his last concerts in the city, on 6 February 1898. Finally Richard Strauss began a long and successful association with Vienna when Richter conducted *Don Juan* on 10 January 1892, *Tod und Verklärung* on 15 January 1893, *Till Eulenspiegel* on 5 January 1896, and *Also sprach Zarathustra* on 21 March 1897.

There were plenty of new works by Vienna's home-grown, but second-rank, composers, who always had a regular place in the Philharmonic programmes, such as Goldmark, Fuchs, Brüll, and Volkmann, as well as newly discovered compositions by the city's famous sons from the past such as Haydn, Mozart, and Schubert. First hearings of works by lesser composers from further afield included Moszkowski's Second Suite (23 November 1890), two new violin concertos by Max Bruch (his Third on 4 December 1892 and his Second on 23 January 1898), Lalo's Piano Concerto (20 November 1892) and Rhapsody (19 February 1893), Reznicek's now famous Overture Donna Diana (15 December 1895), d'Albert's Second Piano Concerto played by his wife on 11 February 1894, and a work by a future leader of Vienna's music life, Felix Weingartner, whose Interlude from *Malawika* was played under Richter on 15 March 1896.

In the ten seasons leading to his last concert with the Philharmonic on 27 March 1898 Richter engaged many soloists who were either established artists or who stood on the threshold of great careers. There were the violinists Eugène Ysaÿe (Wieniawski 17 November 1889), Willy Burmester (Paganini, Wieniawski 17 February 1895), Henri Petri (Spohr 21 February 1897), Fritz Kreisler (Bruch 23 January 1898), and Emil Sauret (Saint-Saëns 20 February 1898), the pianists Emil Sauer (Henselt 7 December 1890), Leonard Borwick (Brahms 22 January 1891), Moritz Rosenthal (Ludvig Schytte's new concerto 16 December 1894), Josef Hofmann (Rubinstein 3 February 1895), Mark Hambourg (Chopin 3 March 1895), Fanny Davies (Schumann 1 December 1895), Ferruccio Busoni (Rubinstein 16 February 1896), Ossip Gabrilowitsch (Tchaikovsky 20 December 1896), Harold Bauer (Liszt 19 December 1897), Ernst von Dohnányi (Beethoven 9 January 1898), and Frederic Lamond (Brahms 6 February 1898), and the cellists Hugo Becker (Saint-Saëns concerto and Bruch's *Kol Nidrei* 29 January 1893, Dvořák's new concerto 7 March 1897), Jean Gérardy (Raff 14 January 1894), and Julius Klengel (Volkmann 15 December 1895). The young English violinist Henry Such had his own concert under Richter on 9 December 1895, whilst the well-established Emma Albani engaged the conductor and his orchestra for two special concerts (27 February and 10

March 1893) at the first of which she 'was recalled about twenty times during the evening, and had to sing four or five encores, thus doubling the length of my original programme'.[6] The singers Nikita (the American coloratura soprano Louisa Margaret Nicholson) and the tenor Franz Naval also had their own concert on 25 January 1896, whilst later in the year, on 28 August, the new emperor of Russia, Tsar Nicholas II, and his German wife, the Empress Alexandra, paid a state visit to Vienna and were entertained at a concert under Richter. The programme of German and Russian music was long and included vocal items sung by Edyth Walker, Paula Mark, and Theodor Reichmann.

Kreisler was a fortnight short of his twenty-third birthday when he made his orchestral debut with Richter in 1898. Having given up playing as a child prodigy to study medicine, which he also abandoned, this occasion signified his return to the concert platform. When Richter emigrated to England he provided many opportunities for Kreisler there, beginning in May 1902. At 75 Kreisler expressed his gratitude for having been 'privileged to associate with Olympians like Johannes Brahms, Anton Bruckner, Antonin Dvořák, and Hans Richter. . . . Richter was once asked why he had conducted a certain piece differently and a little faster than he had done before; and he answered "because the pulse of an artist is his only metronome—a metronome which changes according to his disposition, according to his age, and even according to the climate in which he performs".'[7] Mark Hambourg, who made his début in 1895 at 15 years of age, had been encouraged by the conductor four years earlier to move to Vienna and study with Leschetizky. Richter undertook practical help (by raising money together with Paderewski and Felix Moscheles) to move the boy and his family there, and though he did not live at the Richter home, young Mark soon took d'Albert's place under the conductor's wing.

He would often take me with him to his rehearsals and performances at the Opera, and I would sit hidden in the orchestra. Richter used to tell me that for a pianist to be a thorough musician he must know also the instruments of the orchestra intimately, and be able to detect the sonority of each different one as distinct from its fellows. What I learnt from Richter's teaching in this way has proved invaluable to me in after-life. Once however, being young and weary, I was discovered fast asleep in the second act of *Tristan*. Richter was furious with me, and swore that if he ever caught me asleep again he would debar me from setting foot in his orchestra.

Through Dr Richter I was engaged to play [Chopin's E minor Concerto] on Sunday 3rd March at the Vienna Philharmonic Symphony Concert. . . . Before the rehearsal Richter asked me to play over all the technical passages of the concerto to him. He said that generally when a pianist played a concerto with his orchestral accompaniment, he, Richter, used to compare it to washing linen in a swiftly running brook; before you knew where you were, all was carried away down stream. So he desired to know precisely what I had to play in the various developments of the passages, in order to arrange how to keep the orchestra properly with me. . . . I do not like to undertake to play a work with an orchestra unless I am guaranteed forty minutes rehearsal. . . . When Hans Richter conducted I had two rehearsals of forty minutes each as well as a private one with him so that we could go through the score together.[8]

Michael Hambourg, Mark's father, wrote to Richter to thank him for launching his son's career.

Allow me to express my grateful thanks to you for your extreme kindness and care towards my son Mark. You have done so many things for him besides giving him the opportunity of making his successful début under your baton, that I feel quite unable to thank you sufficiently, but you will, I hope, accept my heartfelt feelings as his father. Mark himself will ever remember your truly artistic sympathy to a young beginner, and if fate should make him a great artist in the future, he will certainly be kind to others as you have been to him.[9]

From time to time Richter was frustrated by his situation in Vienna. He rarely kicked against the traces and generally accepted the rules and conventions of his workplace, though there were a few exceptions. In 1869, with Wagner's powerful backing, he rebelled against the Munich Court in the *Rheingold* affair and in 1882 he resigned for a year from the Vienna Philharmonic in response to negative press reviews, but thereafter he went along with the potentially fallible system whereby orchestral votes decided the fate of new works for the Philharmonic concerts. Bruckner's complaint that Richter was susceptible to the opinions of some of the Viennese critics was justified, and at the Opera the conductor never took a stand against cuts. Nevertheless his contribution to the musical life of Vienna in his quarter century there was immeasurable. Although he was deeply respected there by public and press, it was only at rumours of his departure or once he had finally gone that it was acknowledged. Richter nearly left Vienna in 1893, when, despite his hatred of sailing, the American city of Boston almost lured him away. He already had a commitment for August at the World Exhibition in Chicago, where he was to conduct several concerts, but suddenly Boston newspapers were full of speculation that he had been offered, and had accepted, the post vacated by Nikisch as conductor to the Boston Symphony Orchestra. The Director of the orchestra, Colonel Henry Higginson, and his manager Vert (brother of Richter's London manager) were reported in the *Boston Post* as having started negotiations with Richter. Then followed almost two weeks of press silence until, on 14 April, reports dated the previous day were received from Vienna by the *Boston Daily Advertiser*, the *Boston Daily Globe*, and the *Boston Herald* that Richter had resigned his posts and had accepted Boston's offer. The *Boston Post*, whilst giving itself a pat on the back for first breaking the news two weeks earlier, was also correct in foreseeing problems which lay ahead for Richter in securing his own release from his Austrian contract; however it then wandered off into the land of make-believe.

The fact that Richter would receive a pension by remaining in his position two years more would not necessarily stand in the way of his coming to this country for he is possessed of a colossal fortune, and has already laid away $1,000,000, the interest of which will be used if need be to perpetuate the organization after his death. . . . One of the leading musicians has this to say of the future leader of the Symphony. 'In 1889 I was introduced to Hans Richter, and truly he is a real Bohemian in his mode of living. . . . When speaking of his love of music he would get rather warm in his approval of some authors, and said he used his left hand while conducting Gounod's compositions. With those he liked he said he made use of his right'.[10]

The 'leading musician' quoted in this report was probably the mezzo-soprano Lena Little from Boston who had sung Erda under Richter in concert performances of Wagner operatic extracts in London in 1890. On the same day as the *Post*'s report she was identified together with more of her anecdotes in the *Boston Journal*, ending with the *non sequitur*, 'he likes most of the German music, I believe, but he endorses very little French music. His wife is very pleasant.'[11] The description of his ambidexterity with the baton when it came to French music must not be dismissed too lightly, for an issue of the *Neue Musik-Zeitung* three years earlier quoted the Paris *Le Figaro*, which also described this habit of conducting German music (Wagner operas) with the right hand and showing a distaste for French music (*Carmen* in this case) by conducting it with the left. This reason was, however, dismissed by Richter, who would never have conducted 137 performances of an opera he disliked. According to the German paper Richter wrote a letter of explanation to *Le Figaro* in which 'he refers to the fact that conducting is a very tiring occupation. As one of the busiest orchestral conductors he had learnt through years of practice to use his baton with his left hand as well as his right. He conducted only the most well-known operas, in other words those most familiar to the orchestra, with his left hand; amongst these were not only *Carmen* but also *Rienzi* and *Lohengrin*.'[12]

The Boston papers vied with one another for biographical and anecdotal fact and fiction about Richter in the days following reports of his coming. Nikisch was quoted as describing him as the greatest conductor in the world and many lesser musicians came forward with claims of having worked with him or knowing him in Europe. Richter signed his contract with Higginson on Thursday 13 April 1893 and wrote his letter of resignation to the Directors of the Vienna Opera the following day. With effect from October 1893 he would be resident in Boston, conducting at a salary of $10,000 two concerts a week for six months of the year, leaving himself free for European engagements from May to September. American papers reported that artistic and government circles in Vienna were in turmoil, the latter issuing statements that Richter's contract had another four years to run. At the annual concert for the Philharmonic pension fund (16 April) he conducted the traditional end-of-season 'Choral' Symphony, after which he made a farewell speech and received an ovation designed to demonstrate the strength of public feeling against his resignation. According to the *Boston Post* Richter gave two reasons for accepting the American offer; first 'a feeling that it will give him a relief from the overwork now imposed upon him here, which he feels the need of, and second, a thorough and sincere admiration for and sympathy with the American people and American ideas. . . . He said with a smile that when once in America, he expected that he should become thoroughly identified with its people, manners, and ideas.'[13]

The Opera administration rejected Richter's resignation on 20 April, and on the following day he accepted their decision. On 21 April the Boston papers reported that Richter's request for a release from his contract had been refused, but that the conductor still intended to honour his acceptance of the American offer and intended to refer his case to the emperor. The next day, however, he cancelled his American contract, any acceptance of which was always dependent upon being able to secure his release from his Vienna obligations. He had no desire to forfeit his pension rights or prevent any future return to Vienna should he leave. He also cancelled his engagement in Chicago. 'Everything', as the *Musical Courier* put it, 'is in a pretty pother.'[14] The *American Art Journal*, from its office in distant New York, lagged far behind the other papers with its story. 'Boston, the American Leipzig, will see to it that the learned Doctor, his better half, and the dozen or more little Richters will be amply provided for. Richter's leavetaking . . . was a very impressive scene. . . . He explained that the work at the Opera was too much for him and that by directing concerts only, he could not provide for a large family at Vienna.'[15] The conductor himself is quoted in the *Boston Herald* under the headline 'The Richter Affair.'

> I was very hopeful of being released from my present contract, and my request was seconded by my friend the director of the Opera [Jahn]. Such a request must pass through several stages before it is finally accorded; at any one of them it may be definitely refused. The Direction [administration] of the Opera kindly supported me, but the supreme board, what we call the General Intendanz, refused to give its consent and the matter was at an end. [They wrote] . . . 'it cannot but lay stress upon the retention as long as possible to the Court Opera Theatre of his artistic services.'[16]

Among the reports was a statement that the Philharmonic, in an attempt to compensate for the financial loss incurred by Richter when he turned down the Boston offer, would increase his number of concerts per season. This did not happen for they remained at the usual eight plus the Nicolai concert. On 4 May he invited the whole orchestra to an evening meal at the restaurant Zur schönen Aussicht (best translated as Belle Vue) in Döbling. Ninety-six members of the orchestra signed up to accept their conductor's generous invitation. The Boston affair was over and, for the moment, Richter remained in Vienna.

In the 1890s Vienna experienced the deaths of its two most famous resident composers within a year of each other, Bruckner in 1896 and Brahms in 1897. Richter was no longer the 'generalissimo of deceit' in Bruckner's eyes for he had conducted the first four symphonies, the Seventh, the Eighth, and the Te Deum in the eight years 1886–94. Bruckner's name appeared in every Philharmonic season during the last eight years of the Richter regime (and in Mahler's time until 1902–3), and the Fourth and Seventh appeared fairly regularly in his programmes in Vienna and England. Bruckner never became a close friend of the conductor, who was too down-to-earth to understand such a man. Richter often irritated the composer by taunting him with plain untruths. One such occasion took place in the summer of 1887, when the first version of the Eighth Symphony was finished. Bruckner was invited to Nasswald, the Richter family's annual summer holiday retreat. They all returned together whereupon Richter, in full Nasswald costume, blew the horn calls from the Seventh Symphony on the post-coach horn. Bruckner's friend Professor Lichtenberg was told by the composer that, on a walk in the woods with Richter during the holiday, he asked the conductor which opera was his favourite, *Tristan* or *Meistersinger*. '*Dutchman*', replied Richter, 'because I don't have to conduct it.'[17] This quite deliberate provocation on Richter's part drove Bruckner into a fury, though it is doubtful if he showed it to Richter at the time.

In 1890 Richter accompanied Bruckner to an audience with the emperor at which the composer presented him with a bound score of the new Eighth Symphony. Richter recalled that Bruckner was asked by Franz Josef if he had any requests, and that the composer responded by a desire to be relieved of his duties as an organist at the Court Chapel and to receive financial support for his remaining years. The emperor could not grant such a request officially but offered to dip into his private coffers, which Bruckner graciously refused. In later years when the offer was renewed Bruckner once again refused though he did ask for, and was granted, money to travel to hear his works performed elsewhere. Richter was very moved by this incident, for he had not expected the old man to use such well-chosen and touching words when speaking to his emperor.

Richter gave two performances in Vienna of the third version of the Third (or 'Wagner') Symphony, on 21 December 1890 and a month later on 25 January 1891. This latter performance was an unmitigated triumph, exorcizing the nightmares of Bruckner's own attempt with the second version in 1877. Whereas the Philharmonic orchestra treated him so shamefully on that occasion, he told Levi after hearing the performance under Richter that they played 'as I have scarcely ever experienced in Vienna'.[18] Richter was so enthusiastic about the symphony that he promised it a first performance in England later in the year, which he duly gave on 29 June. At the end of the year he conducted another revised symphony, this time the Vienna version of the youthful First Symphony on 13 December, and it too was extremely well received. 'It is one of my most difficult and best works,' wrote Bruckner. 'Hans Richter is secretly (because of Hanslick) crazy about it.'[19]

It was no secret that Bruckner would have preferred the première of the Eighth Symphony to be given by Levi in Munich followed by Weingartner in Mannheim, but circumstances dictated otherwise and it fell to Richter in Vienna. The symphony was first heard on 18 December 1892 and was the only work on the programme. There were many of Bruckner's friends who were Richter's detractors, particularly those who resented the composer's new description of the conductor as his patron. They were quick to remember the years when Bruckner's music was not heard, quick to forget how the audience left in droves during the first hearing of the Seventh Symphony, and quick to condemn Richter for being Hanslick's lap dog. Richter was no such thing. He knew his Vienna audiences, how much and what they could take, and when they were ready to take it. Göllerich

FIG. 8 Richter conducts a Bruckner symphony, possibly the première of No. 8, 18 December 1892. Silhouette by Otto Böhler.

cynically suggests that Richter gave the Eighth Symphony simply because the emperor, to whom the work was dedicated, announced his hopes of attending the performance. When, instead, Franz Josef went off on a hunting trip, Richter consoled the disappointed composer with the words, 'that doesn't matter, the Symphony will still be performed'.[20] Richter gave the work a committed performance which did it full justice, and it was a sensational triumph. The composer was called on to the platform at the end of each of its four movements to receive an ovation from the audience, which included the Crown Princess Stephanie and Archduchess Maria Theresia. Brahms stayed in his Director's box but Hanslick left before the Finale with the adulation of the composer's supporters (intensified when his departure was spotted) ringing in his ears. Old Beckmesser now had to acknowledge his own defeat and Bruckner's triumph. When Richter left by the stage door Bruckner awaited him with a tray of forty-eight steaming hot doughnuts, a strange but typically eccentric reward for the exhausted conductor.

In November 1894, two months after Bruckner's seventieth birthday, Richter performed his Second Symphony, which had not been heard in the city since the composer himself conducted it in 1876. It was another great success, but by now Bruckner was a very sick man. His heart was failing and he was becoming increasingly housebound through his inability to climb to and from his top-floor apartment. Doctors and friends, particularly his former pupil and new secretary Anton Meissner, secured royal approval of his move to the small house at the entrance to the Belvedere Palace. Family tradition has it that Richter played a part by pleading with the royal family on behalf of the composer. Bruckner's health had recovered sufficiently by the New Year of 1896 for him to attend a performance under Richter of the Fourth Symphony on 5 January. The first performance of *Till Eulenspiegel* preceded it, a work which Bruckner found very interesting, usually a sarcastic observation on his part. He thrived on the warm reception afforded him by the audience, and despite the concern of his doctor in attendance, wanted to acknowledge applause between movements from his box and come on stage at the end. On 29 March 1896 Bruckner attended his last concert, the annual charity event for the pension fund of the Opera. The programme (which repeated Strauss's tone-poem) began with Cherubini's Overture to *Medea* and continued with Wagner's rarely heard *Das Liebesmahl der Apostel* sung by the Vienna Male-Voice Choir. The last music Bruckner ever heard was the Pilgrims' Chorus from *Tannhäuser* by his beloved Wagner. After the concert he personally congratulated Richter and was carried away on the sedan chair which had brought him to the concert hall.

Before the summer began many people came to see the dying man before they left for their holidays. Bruckner was desperately trying to complete the Finale of his Ninth Symphony, but he was too weak to work for any length of time. Richter was one of his visitors; he came to announce his intention of performing the Seventh Symphony at the opening concert of the 1896–7 season on 8 November. When the conductor saw how the unfinished state of the Ninth Symphony was distressing Bruckner, he suggested that the Te Deum could be played as its Finale. He need not look far for a precedent for a choral finale to a ninth symphony. The old man was grateful for the idea, even if he saw it only as a last resort. He survived the summer, becoming thinner and looking more like a Franciscan monk in his bed. Richter and his wife were among visitors during that period though it was becoming increasingly hard to know how to treat Bruckner, whose mind often wandered. On 10 October he felt well enough to get to his piano and do a little work on the symphony, but the next day he died quietly at three o'clock in the afternoon after three sips of tea. It was a Sunday and Richter conducted the Court Chapel choir that morning in music by Haydn and Mozart, and the latter's *Zauberflöte* (only the eleventh of his career) at the Opera that evening. Three days later, on 14 October, he attended Bruckner's funeral at the Karlskirche and conducted members of the Philharmonic orchestra in Löwe's arrangement of the Adagio from the Seventh Symphony for winds and brass. Bruckner's faithful follower Hugo Wolf was refused admission as he had no formal invitation with him. In 1903 the same music was played at his own funeral. From the Westbahnhof Bruckner's body was taken to lie under the great organ at the monastery in St Florian, where he had wished to be buried. Richter could not accompany him, for that evening he was conducting *La traviata* at the Opera, but the performance of the Seventh Symphony on 8 November was now played in memory of its creator.

Had they been on speaking terms Hugo Wolf could have had Brahms's invitation to Bruckner's funeral, for it was not used. Brahms stood at the entrance to the Karlskirche but decided not to follow the coffin into the church. He was deeply affected by the occasion and was already terminally ill. Within six months he followed Bruckner to the grave. Richter continued to perform Brahms's symphonies, concertos, and the few orchestral works penned by the composer, but nothing new was forthcoming for the orchestral concert platform after the Concerto for Violin and Cello, which was played at the Philharmonic concerts for the first time on 23 December 1888. Instead Brahms watched as works by Dvořák, Tchaikovsky, and Bruckner made the musical headlines in the 1890s. On 7 February 1897 Richter conducted his 200th concert with the Philharmonic, not part of the subscription series, but a centennial celebration of Schubert's birth at which he conducted the first complete Viennese performance of the great Mass in E flat. After the concert there was a banquet in the conductor's honour. Exactly a month later, on 7 March, Brahms attended the Philharmonic concert conducted by Richter, which began with his Fourth Symphony and ended with Haydn's Symphony No. 73 'La Chasse'. The last music Bruckner ever heard was by Wagner; for Brahms it was by Haydn. At the end the composer, yellow with jaundice and gaunt from the cancer which was killing him, came backstage where Richter had assembled his men to bid him a last farewell. Richter described the occasion to Edward Speyer.

Yes, one has to travel to Eastbourne to find the time to write! I am moved to write a few words to you and your wife about the great loss which German music has suffered. I do not know if you saw Brahms during the last few months, but it was painful to see how this strong man fell prey to such a powerful and destructive [illness]. I am left with *one* beautiful memory; the last work that he heard was his Fourth Symphony which was performed in the penultimate Philharmonic concert of the season. After the first movement the public responded with enthusiastic cheers; instead of taking my customary bow I looked up to the box in which the sick Master sat at the back. The public understood this sign and Brahms was compelled to come forward several times to the edge of the box to receive the homage of the grateful Viennese, in fact after each movement. At the end of the concert he came to the artists' room to thank me and the orchestra with moving words for the successful performance of his work. With deep emotion, I wished him a swift recovery on behalf of the orchestra and myself, and voiced the hope that he might soon present us with a new work; unfortunately a vain hope. The last sounds he heard came from the Vienna Philharmonic, at least they were the sounds of his own creation. This memory is a small consolation for us Viennese musicians who have baptized so many of his works.[21]

Richter was unable to attend Brahms's funeral on 6 April; he was asked by Stanford to send a wreath on behalf of the London Bach Choir, which he did together with one of his own. On 4 April he paid his respects publicly to the dead composer by playing Mozart's *Masonic Funeral Music* at the start of the Nicolai concert which ended the 1896-7 season. The same music had been used in memory of Josef Hellmesberger sen. to begin the Philharmonic concert on 29 October 1893. Richter's absence from Vienna on the day of Brahms's funeral was due to a rehearsal for a conducting engagement the following day in Budapest, a city with which he resumed his former association through his friend the double bass player Carl Gianicelli by conducting without fee a concert there on 14 December 1892 for the widows' and orphans' fund. In that winter of 1892-3 Richter also guest-conducted with the Berlin Philharmonic orchestra, one of several conductors called by the agent Hermann Wolff to replace the ailing Hans von Bülow, the Director of the Philharmonic. Richter's programmes were standard fare of Wagner, Liszt, and Beethoven, though the last one (6 February 1893) included Saint-Saëns's Concerto and Bruch's Ave Maria (a transcription of an aria from the choral work *Das Feuerkreuz*) played by the cellist Jean Gérardy. Von Bülow was a sick man a year before his death in Cairo, where he went in desperation to seek a cure, and his judgements were becoming more eccentric than ever. Richter, he said, was no longer the man he once knew. 'His performance of *Damnation of Faust* [London 18 June 1888] was a torture for me; no tempo correct (I heard the work performed by the composer several times, in 1852 in Weimar and 1854 in Dresden). It completely lacked the impression it made under Hallé, its first conductor, who at least was in possession of a tradition.'[22] He was nevertheless glad that Richter had enjoyed a success with his orchestra after the first concert (17 October 1892) because it left the way clear for him to absent himself further from conducting engagements. Reflecting on his deputies (Richter, Levi, Mottl, and Maszkowski) at the end of the season, he considered Levi to have enjoyed the greatest success. At the end of 1893

Richter followed Nikisch once again, this time to Leipzig, where he conducted two typically Richter concerts on consecutive days (15 and 16 December).

He returned to Budapest for two more concerts (8 April 1895 and 22 January 1896) and then directed seven more constituting the 1896–7 Philharmonic season there. Richter had to exert a basic discipline among his compatriots. 'If I agree, it must be on condition that rehearsals will be properly run. The arbitrary comings and goings, or even absences, which were the case on the last occasion, are unacceptable to me. I am used to the greatest punctuality from the Vienna Philharmonic and my English orchestra, and I expect to find the same wherever I assume our art to be practised by artists.'[23] The Philharmonic's artistic leader at the time was Nikisch, who had arrived from Boston in 1893 and left for his dual role in Leipzig and Berlin after only two years, and this unsettled period may have been responsible for the lack of professional discipline. Richter was only able to work in Budapest after telling Gianicelli to write to Jahn and ask on his behalf for official leave (Cosima Wagner was also urged to do this if she wanted Richter in Bayreuth). It is not surprising that Richter, with contractual permission for his annual trips to England and triennial visits to Birmingham already obtained, did not wish to push his luck too far with the authorities. With his time limited for rehearsals in Budapest, Richter sent instructions ahead to his rehearsal conductor Sándor Erkel for the first performance in Budapest of Tchaikovsky's 'Pathétique' Symphony.

This mainly concerns the beginning of the Allegro (violas and cellos), and the Scherzo [here Richter quotes awkward rhythms found in the second violins at letter O]. These places should be played with ease and in tempo, everything else is not too difficult, nor the 5/4 beat. I can help here. Divide the bar with red ink; it helps a great deal, particularly those with the accompaniment, but the red line must be placed exactly between the second and third beats. The division, contrary to all present custom, is such that the first half of the bar is the shorter, the second on the other hand is the longer [i.e. a two plus three division of the five beats]. . . . The red dividing lines only as far as the fourth line [of the part]; *very exactly* otherwise they will be confusing; naturally the bars rest should not be divided. That's a good afternoon's work.[24]

A postscript to the letter (including further demands for 'punctual attendance at rehearsal!') shows diagrams of Richter's beating pattern of the 5/4 rhythm, a long down-beat, up again for the second, a shorter down-beat for the third, out to the right for the fourth, and up to the original starting-point for the fifth. He laid stress on the need for clarity in giving a long down-beat, much longer than the third which lay in the same plane, in order to help players who were counting bars rest. At his second concert of this 1896–7 series the soloist in Paganini's Fourth Concerto was the 23-year-old Hungarian Carl Flesch, who wrote, 'In Budapest I played under Hans Richter for the first and only time in my life. . . . [He] belonged to the noble class of intellectually primitive, eminently natural musicians of genius whom the Austro-Hungarian monarchy had always produced in astonishing numbers. He too had been born in the West-Hungarian musicians' corner, i.e. at Raab (Györ)—a supremely competent conductor of the old, solid school.'[25] In support of his emphasis on the number of musicians born in that area of Hungary the author listed Haydn, Liszt, Nikisch, Richter, Dohnányi, the composer Mosonyi, the Wagnerian singer Katharina Klafsky, and Flesch himself. The violinist, who was a student in Vienna during Richter's heyday, had another memory of the man. 'I was to play to Hans Richter at the Vienna Opera House and had to wait in the antechamber. A gruesomely cacophonic piano duet reached my ears from the direction of the directorial chambers, and when Richter eventually received me he told me that together with Dvořák he had just run through the latter's symphonic poem *The Wood Dove*.'[26]

Most of Richter's letters to Gianicelli concerned programme planning, travel arrangements, and rehearsal schedules. Occasionally there were some musical points such as in Brahms's Haydn Variations. 'N.B. The winds must practise the parts thoroughly alone at home; in the last variation the piccolo player must produce the most sensitive *pianissimo*.'[27] By the time his second concert of the series was over, Richter was well pleased with the orchestra's performance. They had obviously taken his strictures to heart and improved their rehearsal discipline. He also made a point of playing as many Hungarian works as he could, trying for one in each concert, and using native soloists where possible. A month after his last concert of the season in Budapest, Richter conducted a three-day festival in Stuttgart (15–17 May 1897), which he described to Edward Speyer as 'a great joy. Schubert's great E flat Mass, Brahms's Second [Symphony], and Beethoven's Ninth were performed. The chorus was superb and the orchestra excellent.'[28]

The American author Mark Twain was a welcome visitor in December 1897 and attended Richter's performances at the Opera during that month (*Traviata*, *Walküre*, *Carmen*, *Meistersinger*, *Fidelio*, *Götterdämmerung*, and *Tannhäuser*). Twain made a point of hearing Richter in Vienna or London, and on this occasion presented Marie with a copy of *More Tramps Abroad* inscribed as follows. 'To Frau Dr Richter with the cordial greetings and sincere regards of the author. It is begged that notice may be taken of the following statistics, to wit: To be good is noble; but to show others how to be good is noble and no trouble. Truly yours, Mark Twain Vienna, December 1897.'

Richter returned to Budapest for three concerts in the 1897–8 season; at the first (on 17 November) his soloist was the 20-year-old Hungarian composer/pianist Ernst von Dohnányi in Beethoven's Fourth Concerto. A consequence of the pianist's huge success was that Richter took him to Vienna (where he repeated Beethoven's concerto at the Philharmonic concert on 9 January 1898) and to England, where he made thirty-two appearances in two months. Jenö Hubay (Hans Koessler's Violin Concerto on 15 December 1897) and Edouard Risler (Liszt's Second Piano Concerto on 9 February 1898) were the soloists in the remaining two Budapest concerts of that season. Two months later Richter made his debut in Paris in two of the Concerts-Colonne at the Théâtre du Chatelet on 3 and 8 April 1898. The second ended with a triumphant performance of Beethoven's Ninth Symphony, which cheered him up after his Vienna concert two weeks earlier on 27 March, for, unknown to anyone else at the time, that had been his final appearance on the concert platform with his beloved Philharmonic. It was his 243rd Philharmonic concert and ended with Beethoven's 'Eroica', the symphony which had concluded his first concert with them twenty-three years earlier on 7 November 1875.

20

1897–1900

Richter and Mahler

WHEN Richter returned to Vienna from his summer break in 1897, it was to a very different musical landscape. With Jahn's departure on health grounds in October he lost a true friend, and after the deaths of Brahms and Bruckner within six months of each other, the unsettled years of 1898 and 1899 began. Everything about him was changing fast, and he appeared undecided whether to stay or to go. Having decided to stay for reasons concerning his pension until he had served twenty-five years in post in 1900, Richter then showed signs of wanting to go by touring as much as he could. Gustav Mahler had arrived at the Opera and put down his marker with a revelatory performance of *Lohengrin* on 11 May 1897. By then Jahn was ill, overworked, and artistically drained. Of his three conductors Fuchs also directed the Conservatoire and Richter was frequently absent. Another man was needed, so Mahler was appointed Kapellmeister. It then became only a matter of months before Jahn withdrew, leaving the way for Mahler to be appointed Director of the Opera. Richter's behaviour over the next two and a half years is fascinating. Having seen Mahler leapfrog over him to the position of Director (a post Richter twice refused, never wanted, and still did not covet despite outrage from some quarters of the press that he had been ignored), he now found a man appointed who was younger, brilliant, and temperamentally as different from him as chalk from cheese. Jahn was a personal friend of a similar age (eight years older than Richter, who was 54 when the 37-year-old Mahler took over in Vienna). Richter saw out his next season as Director of the Philharmonic concerts, but he was uneasy as claques in the audiences formed in favour of Mahler.

On 10 May 1898, at the annual meeting of the eighty-five players to select the conductor for the forthcoming season, there was a unanimous vote in favour of Richter. The conductor took the chair at committee meetings, but at the next one, on 22 September, he announced his resignation as soon as proceedings began. A letter dated the next day states that he did so for health reasons; he was, he wrote, suffering from muscular pains in his right arm and feared a recurrence, through stress and overwork, of facial shingles. Richter proposed Mahler or Löwe as his successor, and at an extraordinary meeting two days later (24 September) the committee elected Mahler. Judging by Richter's schedule, which was as full as ever, his reasons for resigning were not genuine. The fact was that he had no spirit for a fight and felt his own position undermined at the Philharmonic by Mahler's power base at the Opera. Although Mahler's ambition to take over at the Philharmonic was barely concealed, Richter also sensed his own need for a change after nearly a quarter of a century with the same orchestra. On 28 September another meeting was called at which Mahler took the chair for the first time (though he soon absented himself from such meetings once established in his post, attending only four in his nine-year tenure) and proposed his programmes for the forthcoming 1898–9 season. The Richter era was at an end. On 4 August the freedom of the city of Vienna was conferred upon Richter, but despite this new honour and added prestige he turned his back upon his beloved Philharmonic with a fair amount of bitterness. The manner of his resignation and the interest it aroused in the press was not the way he would have orchestrated his departure.

The matter did not rest there but resurfaced a year later when Mahler's re-election took place. At a heated committee meeting on 30 May 1899 the pro-Mahler and pro-Richter factions met head on. Tempers were roused and the meeting threatened to break up in disorder. Calm was restored by the second violinist Franz Heinrich, who tried to explain the reasons for the bitter arguments which were taking place. 'The majority of the Philharmonic have, so to speak, grown out of their infancy in the orchestra at the hands of Hans Richter. Two-thirds of them owe their musical upbringing in it to Hans Richter. He is both a father and an example to us, and the love towards a father is always stronger than that shown to another beloved relative, in this case Director Mahler.'[1] Heinrich's pacifying words had their effect and a vote taken to adjourn the meeting to let tempers cool was passed. Strangely Richter's supporters had not asked him before that turbulent meeting if he would take his place once again at the head of the Philharmonic, but only approached the conductor afterwards. Their hopes were dashed when his reply came from Bayreuth on

1 August gently refusing their offer, this time saying that he would be absent from Vienna for two periods of six weeks each in October/November and February/March, and this made such a plan impossible. Three weeks later Mahler was reappointed, but not unanimously (Mottl and Hellmesberger were also proposed, which, together with three abstentions, deprived him of 25 per cent of the votes).

Richter's wanderings throughout Europe continued during the years leading to his departure from Vienna at the end of January 1900. He conducted two concerts at the Concertgebouw in Amsterdam on 3 and 6 November 1898 and returned to Budapest on 21 December, where his soloist in Leonore's aria from Beethoven's *Fidelio* and the closing scene from *Götterdämmerung* was the rising star (and Mahler's former lover in Hamburg) Anna von Mildenburg, followed by another on 11 January 1899 in which Dohnányi gave the first performance of his E minor Piano Concerto. The young composer had played it through to Richter in London on 15 October, when the conductor found it a 'splendid work'.[2] Later in January he travelled to Russia, conducting in Moscow (28 January) and St Petersburg (4 February). In Moscow he performed Tchaikovsky's 'Pathétique' Symphony and the dead composer's brother Modest was in the audience. 'I shall never forget the impression produced by the incomparable performance,' he told a friend.[3] After conducting his final *Ring* cycle in Vienna (14–20 February 1899), Richter returned first to Budapest (22 February) then to St Petersburg for Beethoven's Ninth Symphony on 4 March, and then took the Berlin Philharmonic to Hamburg six days later. Pablo de Sarasate played Saint-Saëns's Introduction and Rondo and his own transcription of *Carmen* in Budapest on 18 March under Richter, who conducted Beethoven's Ninth in the last concert of the season nine days later. On his way to London for the 1899 summer season he stopped off in Brussels to conduct a *concert populaire*, and his last concert in Europe before leaving Vienna took place in Hamburg on 12 January 1900 with the Berlin Philharmonic. By the end of 1899, embroiled in negotiations with the Hallé orchestra in England and the Vienna Opera, he wrote to Gianicelli pessimistically, 'I cannot come to Budapest any more this year, and probably not later on either. If I should sign my new contract, I shall not get any more leave; if I do not sign, then I shall be in England and unable to come from there either. All good things must come to an end.'[4]

Mahler's accession to the Directorship of the Philharmonic concerts has been described, but now it is time to turn to Richter's departure from the Vienna Opera. Mahler, unhappy as first Kapellmeister in Hamburg with his deteriorating relationship with the theatre director Pollini, began making overtures to Vienna during the summer of 1896. He forged strong links and friendships with prominent members of the city's musical life, among them Rosa Papier, whose lover Eduard Wlassack was a highly placed official in the Opera Directorate. Another ally, though Mahler did not know it at the time, was Hanslick, if only because the critic was strongly opposed to Felix Mottl's appointment to any post in his native city. Mahler met Jahn in Dresden in February 1897, and the old man promised no more than to bear his name in mind if he needed an assistant. That depended upon the outcome of a cataract operation he would have to undergo later in the year. The die was cast, however, for Jahn's departure was a certainty and Hanslick's support of Mahler crucial. Events moved swiftly in the spring of 1897; Jahn resignd as early as January, sensing that his dismissal was imminent as part of sweeping changes to be implemented by the Lord Chamberlain, Prince Liechtenstein, but his resignation was not acknowledged. By the beginning of April it was made public and Hanslick wrote some tactical attacks on him to smooth the path for Mahler's nomination. No one really considered this a serious possibility if only because he was a Jew, but by the end of the week his appointment as a staff Kapellmeister was official. He hoped to circumvent the anti-Semitism rife in the Austrian capital by his conversion to the Roman Catholic faith on 23 February 1897. He arrived in Vienna on 27 April.

Two weeks before he moved to Vienna he wrote from Hamburg to Richter. He had, he was sorry to say, been unable to meet him during his last visit to the city. His appointment, he insisted, was as unexpected to him as it had been to others, but he wished above all else to establish good relations with his senior colleague.

Since my earliest youth you have been the model that I have tried to emulate through all the trials and tribulations of my life in the theatre. With what rapture did I listen to and watch you in performances at the Opera and at the Musikverein! Later when I took up the baton, which in the hands of most of us is nothing but a simple cane, but in yours is a magic wand, I was reminded of your great deeds. If something was unclear to me, I asked myself the question, 'How would Hans Richter do this?' My admiration for you has remained undiminished to this day and is something I shall be proud to voice to the last. Now that I shall have the honour of working beside you in the same place and of emulating you under your gaze, so to speak, I feel compelled to write to you today to express all that has lain within me for so long.

I hope you will not consider this an intrusion on my part, but I could not bring myself to come to Vienna, and visit you as I would any other. I *had* to say this beforehand. When I meet with you soon, you will know my feelings, hopefully for ever, and, once and for all, how I shall regard my relationship with you.

I shall earn no higher praise than that which, if I can obtain it, I receive from you. And if I do not obtain it, then I ask you for your masterly teaching. I place my entire person at your disposal. It will now be my pleasure to relieve you of any task which is either unworthy of you or in any way tiresome to you. I ask for your trust, which it will be my honoured task to *earn*.[5]

Richter wrote a terse reply, limiting himself to an assurance that, as a colleague, Mahler would not find him 'ill disposed towards you. On the contrary, when I have convinced myself that your activities will profit the Imperial institute and promote the cause of our Art, you will find me well disposed and obliging.'[6] Though this is rather pompous in tone, Richter might have suspected Mahler's excessive obsequiousness. He knew the younger man's ill-disguised views on traditional attitudes and conventions in performance, and he may also have been aware of Mahler's epithet for him as 'honest Hans'. Soon Mahler would bring discipline to the house, which had been allowed to become lax under Jahn, Hellmesberger, and Fuchs, if not under Richter. Mahler was effectively acting Director due to Jahn's increasing ill health (he died three years later in 1900) and was appointed Director on 8 October 1897. From now on singers would not be permitted to depart from the musical text. There would be no more liberties taken with the composer's original intentions, and artists no longer capable, who were living on their past successes, would find that their contracts would not be renewed. As well as instilling a musical discipline among the singers and orchestra, he made a serious attempt, not entirely successful in the long run, to abolish the claque system, where members of the public were bribed with money or free tickets by singers for their ostentatious support in the performances. His successful innovations included the staging of uncut operas, even the lengthy Wagner works. *Götterdämmerung*, for example, was played complete with the Norn and Waltraute scenes for the first time in Vienna on 25 September 1898. What surprised everyone was the acceptance by the public of this change; the people stayed to the end. Whereas *Tristan* had attracted only half-full houses, and *Meistersinger*'s audiences had left in droves after the famous quintet, the public now stayed, fully attentive to Mahler's inspired conducting. Neither were they allowed to arrive late. Those who did had to wait until the end of the first act or, in the case of such one-act operas as *Rheingold*, miss the performance altogether.

The success of Mahler's changes irritated Richter, who resented the ease with which they were accomplished. In Vienna he had always acquiesced to bad traditions such as cuts, audience behaviour, and the star system. In Bayreuth, and to a large extent in London, he did not have to put up with them. He put himself in a weak position through his frequent absences, and the press at the time and writers since have been quick to point that out. In the 1898–9 opera season, for example, Mahler conducted 111 performances, Fuchs 57, and Richter 39, laying himself open to charges of losing interest in the Opera. What they failed to understand was that he was not absent through laziness, but hard at work conducting. As well as harbouring a degree of resentment at the adulation Mahler was receiving from the public and at the fickleness of some parts of the press who were quick to forget his own past triumphs in Vienna, Richter also took great exception to Mahler's tampering with the orchestration of *Fliegende Holländer* and *Tannhäuser*. Two years after his departure from Vienna he visited the city on his way to Manchester from Baracs in Hungary, where his daughter Richardis lived with her husband. During his short stay in the Austrian capital he gave an interview for the *Neues Wiener Tageblatt* in order to correct misinterpretations of his relationship with Mahler leading up to his departure.

The former Hofkapellmeister, or as he amusingly now signs himself, 'Hans Richter out of service', looks splendid and has a freshness about him which many a youth would envy. . . . 'You know', began Richter, 'that I am no friend of interviews, but if I depart on this occasion from my normal principles, I do it because I still take the liveliest interest in the musical life of this city, in which I have spent the greatest part of my life. Also I consider it my duty not to keep silent about certain incidents which have occurred in Vienna, in case my silence is interpreted as approval; and finally because I am of the opinion that matters might perhaps be improved through my intervention. Naturally this primarily concerns the Director of the Opera, Gustav Mahler. To avoid any misunderstanding I consider it necessary to make a statement of loyalty, that I have always had the best of personal relations with Mahler, and that nothing apart from the concern for the further development of Vienna's musical life guides my actions. On the other hand there is my fear that it could be understood that what Mahler undertakes, he does as part of an inheritance from me. This I must unequivocally countermand. What I as an artist cannot forgive Herr Mahler are his worsenings through improvement, to use an expression of Schopenhauer.

As an example I shall use Mahler's retouching of *Fliegende Holländer*. You will know that this work was composed by the young Wagner. When he heard the opera he saw at once that it had been too loudly scored. He set to work at once to re-orchestrate the score, removed all unnecessary trombones, muted wherever possible, and reduced the power of the sound except where it was definitely necessary to bring up his big guns. Henceforth this version was performed in

Dresden. Later, when Wagner was living in exile in Zürich, he often wrote to Uhlig about the 'noisy instrumentation' of *Holländer*, and there is something about it in his writings. It is known that Wagner organized some theatre performances in Zürich. As an exile he was denied access to the corrected Dresden score, so he reworked it from memory and as a result made even more small alterations to the scoring. That version is performed nowadays in Bayreuth, Esser accepted it for Vienna, and I inherited it from him; but now Mahler has to do something else. He reinstates the original score. When I heard about it, I told Mahler all I have related here, but he would not give way and replied, 'What I am doing would be all right with Wagner if he were alive today.' That was the first rebuff I received from Mahler, but another example ... is even more stupid. ... What Mahler did in *Tannhäuser* is simply absurd, I mean the chorus of younger pilgrims at the end of the third act. It is absurd because it proves that Herr Mahler is unclear about the characteristics of the scoring of *Tannhäuser*, and has not given it any thought. Throughout the opera the religious, serious, and, let us say, moral elements are constantly represented by the wind instruments. If Wolfram or one of the other 'virtuous' singers gets to his feet, his song will be accompanied by the low-registered violas or cellos together with the woodwinds as well as the main accompaniment on the harp, of course. If Tannhäuser sings, violins are used in a sensuous fashion. The deliberate contrast between these two groups is apparent in the Prelude to the second act, the chaste rejoicing of Elisabeth (oboe) interrupted by the scornful laughter of Venus (violins). ... Now Mahler has cut the woodwinds in the chorus of the younger pilgrims, and they appear to the sound of the violins characteristic of Venus. That sets a bad example for the youth of today to emulate. Mahler's influence would be of much greater value were he to shed his arrogance in improving the great masters, among them Wagner, whose skill in orchestration was acknowledged by even the most vicious journalistic hack.

... But I must add that, in spite of everything, I am of the opinion that Mahler is the right man for the job of Director at the Court Opera. It is simply laughable to underestimate his great abilities; I have always recognized them, in particular his awareness of scenic matters. ... I must honestly acknowledge that, with the exception of Felix Mottl, I know of no-one who is more suited to his post than Gustav Mahler. His temperament is much to blame, for example there is the affair with the Philharmonic. Those people are first-class musicians through and through and should not be treated in a bureaucratic fashion. A conductor must inspire his musicians and transmit his ego to the orchestra. Artists with whom one performs the works of Mozart, Beethoven, and Wagner are the conductor's respected collaborators, who must be treated as such. One could, together with Brünnhilde, ask Mahler, 'Dich selbst liessest Du sinken?' [Act III scene iii of *Die Walküre*, to Wotan: 'Would you so demean yourself?'] precisely because he does not transmit his ego to the orchestra.

Mahler is said to have expressed the view that 'for me there is no such thing as tradition'. That is quite a wrong standpoint. All those who have worked here,

Nicolai, Esser, Dessoff, and I, have inherited one from the other what was worth upholding, and have then built further upon it. I am considered to have been strict with lazy, careless, and untalented people, but only with these. I regarded all others as being my artistic collaborators and treated them as such. I want an orchestra to respect me, not fear me. He who has power must show how to make use of it otherwise he is considered a tyrant before whom one trembles and whom one eventually wishes to be rid of. Under those circumstances there is no room for any artistic collaboration.[7]

On 19 January 1900 Richter had his one and only encounter with Nellie Melba when she sang Violetta in his twenty-seventh *La traviata*, and performed the Mad Scene from Donizetti's *Lucia di Lammermoor* as an encore (both works were the choice of Emperor Franz Josef). The occasion was for charity and in the presence of the emperor on his first outing to the Opera since the murder of his wife. Melba's success earned her the title of Kammersängerin from the emperor. At the performance she was permitted to sing in Italian, but her schedule precluded any rehearsal beforehand.

When I came back from Budapest I found Richter, the conductor, waiting for me at my hotel with a broad smile on his face. 'You look cheerful', I said; 'I'm not. How would you like to sing without a rehearsal?' 'Don't be afraid', [he replied], 'I'm not afraid of you. And I hope you're not afraid of me. In any case I do know this—that you will sing what the composer wrote. You won't sing Melba-Verdi. You will sing Verdi'. . . . Richter asked me a few things about certain passages in the Mad Scene, whether I made a pause here or an *accelerando* there, and then departed saying that 'It will be all right'. I cannot sum up the performance better than saying it was 'all right' in every sense of the word.[8]

Mahler, it must be said, had done everything he could to keep Richter at the Opera. He was short of staff conductors and was fast becoming overworked (the sick Fuchs died in October 1899 and his replacement, Ferdinand Löwe, was not living up to the promise he had formerly shown). Richter was never denied his choice of repertoire and his requests for leave of absence, for whatever reason, were always met. Mahler promoted Richter's financial interests with unprecedented success by having his salary of 6,000 gulden a year doubled, but to no avail. By the beginning of the new century Richter had had enough and virtually begged Mahler in a letter from Manchester to cancel a newly signed contract and release him. Richter addressed a similarly worded plea to the Directorate of the Opera. He described his years in the theatre (from 1859 as a player and from 1868 as a conductor) as years of war not years of joy. 'My whole life in the theatre has been a battle against lack of appreciation, unrefined attitudes,

FIG. 9 Hans Richter conducts Wagner's *Götterdämmerung* at the Court Opera in Vienna.

and incompetence. It has been too rarely interspersed with the joys of true artistic attainment, distinguished and high artistic ideals, and well-intentioned, selfless assistance.'[9] He then outlined his complaints against the press which had harassed him over the years and justified his cry, 'I am tired of the theatre!' Now, he said, his grounds were ill health, his nerves could no longer stand the strain, and he appealed for his release as a sign of recognition of his past services. He turned to Mahler 'in the name of humanity', and asked him to intercede on his behalf with the Directorate so that he could 'go in peace and not force me to carry on with duties which I am not strong enough to carry out and from which I no longer derive any pleasure'. His letter ended with a musical quotation from the end of the first scene of Act I of *Siegfried*, in which the young hero anticipates his happiness at being free of his guardian Mime, 'Wie ich froh bin, dass ich frei ward.'[10]

Richter conducted at the Opera for the last time on 26 January 1900. The opera was Marschner's *Hans Heiling*, the thirtieth time he had conducted it. Six years earlier Richter had written underneath a handwritten quotation of the opening bars of the *Meistersinger* Overture, 'I began my conducting activities in Vienna in this way on 1 May 1875. I shall fill in the music stave below with the notes which I will have conducted to end my Kapellmeister duties. Hans Richter (Imperial Court conductor in service).'[11] Underneath the music stave (which is empty) are the words 'in retirement'. He either forgot to fill it in or did not consider Marschner worthy of being a musical bookend to Wagner.

By the time Richter's interview concerning Mahler appeared, he was well established in Manchester in his new appointment as the Hallé orchestra's conductor, and was writing of his new-found happiness to his Viennese colleagues and friends. Nevertheless this pleasure was overshadowed by the manner in which he left Vienna. His wife Marie told the piano manufacturer and family friend Ludwig Bösendorfer that 'my poor [husband] has experienced nothing but ingratitude in Vienna, and looks back on the long time he spent there with melancholy and bitterness'.[12] A decade after his departure he told his children, 'Leave me in peace as far as Vienna is concerned. I shall never conduct there again, and that's that!'[13] Mahler brought in Franz Schalk as Richter's successor and, before long, Bruno Walter. The new generation of conductors quickly established itself in the city, although Richter's name was not forgotten, either then or now, and there were many who seriously regretted his departure. The Philharmonic gave a banquet in honour of the Richters' silver wedding anniversary on 27 January 1900. The Wagnerian menu consisted of Haus Krafftsuppe, *Holländer*—Schell (haddock), *Grane*—Lungenbraten (loin roast), Gebratene Schwäne von *Monsalvat* mit Blumen und Früchten aus *Klingsors* Zaubergarten (Roast swan from Monsalvat with flowers and fruit from Klingsor's magic garden), and, after ice-cream, Süssigkeiten (sweets) from the *Venusberg*.

Richter's duties at the Court Chapel were tinged with sadness in his last years as Hofkapellmeister. On 10 September 1898 the Empress Elizabeth was assassinated by an Italian anarchist on a landing stage beside Lake Geneva, and it fell to Richter to conduct the music at her burial service in the Kapuziner church a week later, followed by a memorial performance of Mozart's Requiem on 23 September. Over the years he had conducted the funeral services for many archdukes and archduchesses, counts and countesses, though not for Crown Prince Rudolf, whose suicide at Mayerling had shocked the world in January 1889. One of his final duties was for his erstwhile colleague Johann Nepomuk Fuchs, whose funeral took place on 8 October 1899, and for whom, as a deceased musician, the almost obligatory *Masonic Funeral Music* by Mozart was played. Richter conducted his last church service as Imperial Hofkapellmeister in the Court Chapel on 14 January 1900, at which all the music was by Michael Haydn.

Richter was accused at the time of deserting Vienna for Manchester in pursuit of Mammon. Even Hanslick joined those in the press who accused him of seeking a much higher financial reward by his guest appearances abroad. Such accusations hurt Richter deeply, as he told the more sympathetic Joseph Scheu of the *Arbeiter Zeitung*. His underlining of the word Director somewhat diminishes Richter's praise for Mahler and is symptomatic of his complaints which led to the article two years later, and quoted earlier, in the *Neues Wiener Tageblatt*.

I expected most of the Vienna critics to view the whole affair in the most unfavourable light. As if to order and by common consent they have presented me as if I were obsessed by money, even though six months ago I declared publicly in the *Fremdenblatt* that material considerations had not made me seek retirement. Now I have to put up with the fact that these gentlemen have portrayed me to the Viennese public in an unfavourable light. I was therefore all the more pleased to see from your article that you showed fairness and understanding for my case. Everyone must concede that I am tired of the theatre after forty-one years—thirty-two of them as a conductor. You have pointed out the other purely artistic reasons which have made this weariness intolerable; I want to leave Vienna peacefully. If, however, these envious people do not stop spreading wicked lies about me in order to support their favourite, I could lose my patience and say things when I would prefer to keep silent out of consideration for Mahler, who in his capacity as *Director* has achieved much of excellence.[14]

By the end of 1900 he was able to tell Ludwig Bösendorfer.

I am very happy in my new 'circumstances'. An excellent, well-appointed orchestra with whom I can perform the most difficult works and a wonderful well-balanced choir of over 360 well-trained voices are at my disposal. My orchestra loves me and I am moved when I see that the splendid artists do their best to please me both at rehearsals and performances. Rehearsals are my, and I may say their, main pleasure. They are happy when I explain a new or less-known work to them, but they derive their greatest pleasure when I throw new light on well-known works such as Beethoven's symphonies. Ah, so very different from that to which I had been accustomed hitherto. I cannot understand at all today how I endured looking at the apathetic faces of certain gentlemen who sat in front of me for twenty-five years (they were not wind-players). Thank God that is all over!

The Manchester public is equally well disposed towards me and enthusiastic, and above all they are not prejudiced against new works. They, and those of the neighbouring towns, are very musical, the credit for which must go to the late Hallé, who for nearly forty years educated the north of England in music. The critics here are respectable. I do not say that because they praise me, but because it is true, and [Moritz] Rosenthal, Busoni, and others can verify that.

I do not live in smoky Manchester itself but in charming surroundings [The

Firs, Bowdon, Cheshire] with wonderful parks and no air-polluting chimneys, about the same distance [from Manchester] as Mödling lies from Vienna, and with numerous and frequent connections. I seldom hear of matters in Vienna; I asked my friends to be silent on the subject. If you wish to cheer me up with a note, please write only about yourself and what concerns you; anyway something from and about a Viennese, that kind and highly gifted creature, which, today, is extinct in that metropolis.[15]

21

1889–1890

England

RICHTER'S activities away from Vienna during the 1890s were, at the beginning of the decade, largely confined to the familiar areas of London and Bayreuth, together with the Festivals at Birmingham and on the Lower Rhine. As the new century approached, and with it his growing unease at working with his new Director Mahler, Richter spread his wings and made guest appearances elsewhere in Europe. His schedule for 1889 in London began with the regular spring series of Richter Concerts, which programmed much of the familiar repertoire with which these concerts were now associated. It was becoming his practice to repeat his programmes in London and Vienna, though his championing of works by Austrian and British composers was left to performances in their respective capitals. After negative critical reaction in Vienna to the music of Cowen and Stanford, and in London to new works by Fuchs and Grädener, it was Brahms and Bruckner who provided the novelties for London. Unfortunately it was a one-way traffic, for Richter could take nothing of similar musical worth back to Vienna; his association with Elgar began too late to benefit the Austrian capital.

Richter's London season in 1889 began in May with five concerts of regular fare. What followed in June was, however, something new. A friend for many years was the painter and amateur composer Professor Hubert Herkomer who lived at Bushey in Hertfordshire. There, in the garden of his house Dyreham, he built himself a small theatre with the help of local workmen and students from his school of painting. In December 1888, after successfully staging his fragment *The Sorceress* that summer, he wrote to the conductor with a proposal, suggested to him at table by his 12-year-old daughter.

I am going to ask a tremendous favour of you, which you must refuse without hesitation. I ask if you will conduct the orchestra for my twelve performances next year. In return for this boon I would ask to be allowed to paint a half-length portrait of Mrs Richter for you.

The artistic experiment I am making with my theatre is, I firmly believe, of some real use in the world. I am trying to perfect scenic effects on the stage, from another point of view to the ordinary stage traditions. I call these 'pictorial music-plays'. First the picture, then the music to put you in the right mood to look at the picture, then the play or story, which is to be the excuse for this display. I only aim at making the music fit on to the scene and action. The play I have on hand is in three acts, each of forty minutes duration. The story is of my invention and the words are written by Joseph Bennett, the musical critic. The time is 14th century in England. The songs and choruses are only introduced where they would naturally fall in, the rest is pantomime, illustrated by orchestral music. This never ceases but there is no dialogue. . . . I am most anxious to have as perfect an orchestra as possible, and this is difficult to obtain where only a few play. I have written it for two flutes, one oboe, two clarinets, one bassoon, two horns, three trombones (to be put far back in orchestra), harp, and complete strings of course, six firsts and six second violins. Joseph Ludwig will lead (thirty performers). Everything depends on the conductor. I want you my dear friend. Thus I know the utmost would be made of the music. Armbruster is a good and clever fellow, but has no power of inspiring his players. Last time they played anyhow, but this time, as I have written every note, I cannot bear the idea of indifferent players; and with a few exceptions they will not care until they have a conductor they respect and honour. Perhaps I can answer for your heart, but not for your time.

Performances are in June 1889. One requirement would be to see one or two piano rehearsals, to conduct two complete orchestral rehearsals (each taking a whole day) and twelve performances. These take place in the afternoon, at about 2.15. You could be back in London after each performance by 6.18. They are private performances. All the guests are invited, and payment will only be for one charity, but these invitations include the most interesting people in art, science, literature, and music. Last year we had about 1,000 in eight performances. That the performers are not professionals of course you know. But last year's training greatly helped them. They are enthusiastic and do all in their power to make the best of everything I tell them. The finest feeling prevails amongst my dear students. They honour you and have the highest admiration for you. For Armbruster they never had any respect. I count the days for your answer so I need say no more.[1]

Herkomer's pictorial music-play was called *An Idyl*, a non-committal title, as he described it in his autobiography. 'There was just story enough to give reason to the changes of pictorial effects, and music enough to attune the mind of the spectator to the pictures. It was based upon the fantasy of a painter who used as pigment living colours, and a magic canvas, his mind all aflame with the excitement inseparable from a new experiment.'[2] From Vienna Richter agreed to conduct and spirits rose at Bushey, for he threw himself whole-heartedly and with full commitment into the project. In the end there were nine performances, and he moved to Bushey in June, commuting to London for a concert during the run. At this time an incident occurred which fuelled another story about Richter's idiosyncratic English. When he and Marie travelled to Bushey from London he ordered his railway tickets, 'one for me to come back and one for my wife not to come back'. The composer and conductor Joseph Barnby (at the time Precentor of Eton College) attended a performance and told Richter that 'I should certainly have given myself the pleasure of speaking to you at Bushey but that I saw you were engaged in conversation with friends and in a language in which alas I could not participate. I had intended waiting for the opportunity of shaking hands but time and trains wait for no man. Anyway pray be assured it will always be one of my greatest pleasures to shake hands with one for whose great gifts and kindliness of heart I have so high an appreciation.'[3]

The choristers for *An Idyl* were art students from Herkomer's academy, and the painter brought in singing teachers for their training. His own music, amateurish and intuitive as it was, had to be carefully corrected in harmony, chording, and instrumentation, and he willingly brought in a musical acquaintance for that purpose. At early rehearsals (string quintet, harp, and piano) he was asked by his players to conduct ('I can't do that yet, I forget all about time and want to listen. Of course I know best how it ought to go and can help them that way').[4] At the performances themselves the composer sang the role of the village blacksmith, father of the heroine, but naturally it was the painting of the set, a medieval village, which caught the admiration of the critics. He and his wife made the costumes themselves. Herkomer even had 'three electric time-beaters behind the scenes that work from a knob that you touch with your left foot'[5] for Richter to communicate the tempo to his back-stage assistant conductor in co-ordinating the off-stage music. The final three performances were for charity, the seventh and eighth for the Bushey Village Nurse Fund, and the last for the Beethoven house fund in Bonn. Novello published a limited edition (676 copies) of the illustrated vocal score, and the work was 'affectionately and gratefully dedicated to Dr Hans Richter'.

Herkomer persisted in writing music. His next play, entitled *Found*, was scheduled for 1892 and Richter gave every intention of continuing his involvement with activities at Bushey. Letters from the artist to the conductor between November 1890 and the summer of 1891 refer in detail to piano and orchestral scores, but then Richter advised Herkomer that the music was not good enough and that he should stop composing.

I still think day and night of your good and sound advice. I only wish I had stopped when I so often felt that I was not strong enough yet for such a subject. But failure has been such a rare thing in my life, that I feel this disappointment with an intensity that I little dreamt of. I shall have to endure many painful things from the outside world who will be only too glad to laugh at me now for having attempted such a thing. . . . I cannot come to the concerts. Music hurts me now.[6]

Instead of composing, Herkomer concentrated his efforts on building himself a bizarre-looking house in Bushey, which he named Lululaund, though his love for music was soon revived. He even composed some pieces for zither ('I have played this instrument for a great number of years, but never so seriously as of late' he wrote).[7] He remained close to Richter (his sons were called Siegfried and Hans, his daughter Gwenyddydd was named after the heroine of his aborted music-play) and his letters continue until the conductor's return to Germany in 1911. There are some interesting observations from Herkomer on London's operatic life in June 1900 which he sent to Richter in Vienna.

Last night I heard the *Meistersinger* at Covent Garden in Italian, and I must tell you how splendidly it was done as far as singing and acting went. The orchestra was not good enough. The singing throughout was by far the best I have ever heard. The beautiful phrasing of these Italian trained singers showed me again how much Wagner needs the finest trained singers and not merely declamatory shouters. Here is this man de Reszke singing four times this week—twice in the *Meistersinger* and none the worse for it. Madame Albani sang Eva very sweetly and just right. The Hans Sachs by Lassalle was excellent, after Gura by far the best, the Beckmesser of Isnardon capital. Pogner good and the mens chorus excellent, the men overpowered the women in the last chorus. The 'Streit chorus' was sung more than acted, and proved a good idea because they could not act and sing correctly. The whole thing was placed on the stage much better than at Bayreuth, especially the last act, which was excellently managed here and so deplorably (to my mind) at Bayreuth. I was only sighing for your baton all the time. You would have changed the orchestral 'timbre' very soon, and kept the tempi more as they were intended. It is to your having paved the way by giving some of the striking things from the *Meistersinger* that has *ripened* the public's taste for it.[8]

Last night I witnessed Arthur Sullivan's new opera *Ivanhoe* of which so much has been said and written, and in D'Oyly Carte's new theatre too, which added additional interest to the public. Henry Irving, Mackenzie the musician, and others seem to hold the idea that operatic art is an unnatural thing and can never be made true in its relation to drama. Well *this* one certainly *did* feel last night. Anything more undramatic and unpoetic I cannot imagine. For passion there was noise, for sentiment there was affectation. Not one interesting musical phrase worth remembering, except where in places it was more like Wagner. . . . This opera has been written very quickly and is music that after all sounds as if it were made so much per yard. Even the lighter parts—that were quite suggestive of the Savoyard style—have failed to my mind, because he wanted grand opera. All the declamatory portions are empty phrasings. There is some skilful writing in places, but as all attempt at symphonic work in the orchestra was abandoned on principle, the result can only be a tiresome series of merely varied accompaniments. . . . Ballads strung together do not make a lyrical drama. . . . The gauntlet has been thrown down in direct challenge to the Wagnerian school, or even the theories worked out in Verdi's *Otello*, in fact against all foreign influence.[9]

Herkomer died in 1914 and two of his last surviving letters reveal the depth of respect he had for his friend.

The Brahms [First Symphony] was a revelation in many ways, and I think the most wonderful revelation was you yourself. I do not believe that any conductor who ever lived so completely carried out a self-effacement as you did. It was a curious feeling; I heard only music, I saw no conductor and saw no players. That, I think, is the highest compliment I can pay you.[10]

In tonight's *Globe* I see a notice that you are retiring from your labours. . . . You are the father of all modern conducting; you have set an example of interpretation of great works which had as its basis *respect* for the masters. Never for a moment did you strive for a sensational effect, for novelty for novelty's sake. A perfect musician combined with a magnetic personality is rarely found. You held your orchestra in the palm of your hand. Nobody could tell you anything about music that you did not know, try as they would; hence they felt the master hand over them. These qualities form the laurels that will be placed on your dear brow as long as the memory of music lasts. I love you Hans, as you know, and one of the proudest moments of my life was when you conducted my little effort.[11]

Back in London a Wagner night, presented in conjunction with the Wagner Society on 24 June 1889, at which the soloists were Edward Lloyd and Max Heinrich, was sold out. Richter's next concert on 1 July included the first performance of Parry's unpublished Fourth Symphony in E minor. Preparations for the première were recalled by the composer.

15 June: At three to Richter, who was to go through the Symphony again. Found him engaged with a lady pianist from Vienna who wanted to play at one of the concerts, and as he found it almost impossible now to get her into the programmes, she simply sat there mum as if she expected he would cave in if she

stuck to him long enough. But after about half an hour of it his patience won the day and she departed, but the old chap was left in no fit state to attack the Symphony then so I had to go away again and make another engagement.

17 June: Off to Richter. Programme too long. An enormous long list from *Walküre* on top of Dvořák's long Variations. Bored the audience and I felt tired myself. Moreover the performances were not so perfect as they might have been. Saw Richter for a minute or two.

26 June: In afternoon to Richter where we went through the new Symphony. Barry was there and I think rather put him off and he didn't read the score so well as usual.

29 June: To St James' Hall early. Found Richter hard at work on Brünnhilde's great scene in *Götterdämmerung* which must have taken longer than he expected as it was not done till 10.30. Fillunger sang it very musically, but missed a few points here and there and was at a loss. Richter absolutely at home in every detail of it without book. My Symphony didn't get along well and I felt very dissatisfied with it. Slow movement doesn't come off as I hoped, scherzo the only effective movement. Last movement quite ragged.

1 July: At Richter concert in the evening. The Symphony he asked me to write for him at the Birmingham Festival time. Parts of it came off pretty well, first part of first movement, development of slow movement and I think all the scherzo. Middle of first movement and development of slow did not please me, nor last movement either. It was much better received than I expected and after scherzo I had to go up and make a bow or two.[12]

The London season ended with a performance of Berlioz's *Damnation of Faust* on 8 July, and it was reviewed by George Bernard Shaw in the *Star* under his pseudonym of Corno di Bassetto. Shaw had been writing music criticism for a year and took over from Louis Engel at the *World* in 1890. The tone of his attitude towards Richter was set in this review. 'I never unreservedly took my hat off to Richter until I saw him conduct Mozart's great symphony in E flat [No. 39]. Now, having heard him conduct Berlioz's *Faust*, I repeat the salutation. . . . When the scene on the banks of the Elbe began—more slowly than any but a great conductor would have dared to take it—then I knew that I might dream the scene without fear of awakening a disenchanted man. As to the dance of the will o' the wisps in the third part, Richter's interpretation of that most supernatural minuet was a masterpiece of conducting.'[13]

Richter's contribution to the 1889 Bayreuth Festival was five performances of *Meistersinger* (though after the third he had to return to Vienna for a Court Chapel service and *Lohengrin* on the Sunday, and Ignaz Brüll's *Das goldene Kreuz* the following day). Felix Weingartner, whose early career at Bayreuth was cut short after he criticized Cosima's interference in artistic matters, dubbed Richter the custodian and born conductor of *Meistersinger* who 'by power of his authority turned away every uncalled-for interference if it were ventured upon' after the 1888 performances.[14] A year later he felt that the performances lacked drive, which he put down to Richter's late arrival at Bayreuth from London, leaving him little time for rehearsal with his cast. The *Musical Times*, on the other hand, in a comparison with the 'tamely correct' Covent Garden performances in Italian under Luigi Mancinelli and praised by Herkomer, considered it 'a genuine treat, all the piquancy and delicate points being brought out with masterly skill under Dr Richter'.[15]

In May 1890 Richter conducted at the sixty-seventh Lower Rhine Music Festival on his way to London for the summer season. Hermine Spies sang Brahms's Alto Rhapsody and Bernhard Stavenhagen played Beethoven's Third Piano Concerto. Rosé came with Richter from Vienna to play Goldmark's Violin Concerto and Carl Perron sang extracts as Sachs and Wotan. Dvořák's Eighth Symphony was the only new work in that London season on 7 July (the composer had conducted the first performance in the capital in April) but there were several English musicians as soloists in his seven concerts including the 22-year-old pianist Leonard Borwick (a pupil of Clara Schumann) and the singers Lena Little, Anna Williams, Pauline Cramer, Andrew Black, and Edward Lloyd. Others included Marie Fillunger, George Henschel and his wife Lillian, and Max Heinrich. With the conclusion of the 1890 season Parry wrote regretfully:

14 July: Dined with Balfours and after to the last Richter concert. A very good one with a remarkable performance of Ninth Symphony. I went down to wish him goodbye in the interval and he was very affectionate and gave me a warm hug. I'm sorry they are over for not only is the music of the best but one always sees the best of one's friends at the concerts.[16]

22

1891–1895

England

In his 1891 London season Richter engaged Paderewski to perform his own Piano Concerto in A minor, after its première had been demanded by Annette Essipoff in Vienna two years earlier. The concert took place on 22 June (a year after Paderewski had played its first performance in London under Henschel). Apart from the interest of hearing Beethoven's three Leonore Overtures in the order of their composition (i.e. 2, 3, 1) on 8 June, the first London performance of Bruckner's Third Symphony on 29 June, and the première of Stanford's choral ballad *The Battle of the Baltic* on 20 July, there was nothing new in Richter's repertoire in 1891. This was noted by Shaw in the *World*. Engel had had his blind spots, Wagner being a particular target for this musical dilettante who, in his chatty columns, often appeared more interested in society gossip than in any serious writing. He also repeated himself with his descriptions of Richter's near infallible memory or his status as a Napoleon of the orchestra. Shaw's weakness was his bigoted dislike of Brahms (the folly of which he was ready to acknowledge and recant forty years later), but, though an admirer of Richter, he was critical of programme planning motivated by commercial interests, or of any lack of orchestral preparation as a consequence of the conductor's busy schedule. 'Richter', he wrote in June 1891, 'has no right to stuff a programme with the most hackneyed items in his repertory in order to save the trouble of rehearsing. . . . Nothing can be artistically meaner than to trade on the ignorance of those who think that the name of Richter is a guarantee for unimprovable perfection. As a matter of fact, the orchestra is by no means what it ought to be; and it has been getting worse instead of better for some years past.'

Fortunately Richter was no composer, or else Shaw would probably have had him in his sights. The socialist critic waged many a campaign against the knights of the music world, Parry, Stanford, Mackenzie, Cowen, and Sullivan, often because of their academic backgrounds at music colleges or universities. When Elgar arrived, devoid of knighthood and not formally trained in music, Shaw pronounced the revival (since Purcell) of English music. Although he had few kind words for Richter's advocacy of Brahms, even the conductor's renditions of Wagner's operatic extracts were not immune from criticism when it came to shoddy performances. Shaw could be as defensive of Wagner as Richter himself.

The wild gallop of the Valkyries was upon us with a heathenish riot. And I can unreservedly assure Richter that a more villainous performance of it never was heard before in St James's Hall. To offer us such an orgy of scraping, screeching, banging, and barking as a tone-picture of the daughters of Wotan was outrage to Wagner.... When it comes to depending on the reputation of the band and the conductor to dispense with careful preparation, and to snatch popular victories with exciting pieces like the *Walkürenritt* by dint of what I can only describe as instrumental ruffianism, then it is time for every critic... to warn Richter that unless he promptly takes steps to bring the standard of quality of execution in his orchestra up to that set by the Crystal Palace orchestra [or] the Manchester band [in performances of Berlioz in the winter 1890–1], he will lose his old pre-eminence in the estimation of all those who really know the difference between thorough work and scamped work in performing orchestral music of the highest class.[1]

[Of the Ride of the Valkyries] everyone knows how Richter charges headlong through the whole piece from beginning to end, aiming solely at a *succès de fou hullaballou*, with the result that the tone, strained to the utmost from the first, cannot be reinforced at the climax, which gets marked by a mere increase of noise, and that the middle wind parts lose their individuality, the wood and horns jumbling together into an odd, dry sound which strikes the ear like a compound of bugle and bass clarionet.[2]

The Birmingham Festival of 1891 was a mixed success for Richter. Albani withdrew through illness and her replacement, Margaret Macintyre, proved inadequate in the traditional opening *Elijah* on 6 October (in which the *Musical Times* also questioned some of Richter's tempos). That evening Mackenzie conducted his own *Veni, Creator Spiritus* whilst Richter accompanied Joachim in the Beethoven Violin Concerto ('Richter was in his true element and the fine orchestra played up to him with enthusiasm.'[3]) and conducted Brahms's Third Symphony, Sterndale Bennett's *Naiades* Overture, and a new duet by Arthur Goring Thomas. The following day Joachim played the violin solo obbligato in Bach's *St Matthew Passion*, a

1891–1895: England 289

'monumental performance.... Richter and his people "went for it" in downright earnest, and had their reward in consciousness of a good thing well done.... The performance of the Passion should be marked with a red letter in the history of the Festival.'[4] Shaw was by no means in agreement with his colleague from the *Musical Times* (though he was in a black mood because he was seated at the back of the hall and next to a steward reading the daily newspaper).

In the opening chorus the plaintive, poignant melody in triple time got trampled to pieces by the stolid trudging of the choir from beat to beat.... Richter, whom we have so often seen beating twelve-eight time for his orchestra with a dozen sensitive beats in every bar, made no attempt to cope with the British chorister, and simply marked one, two, three, four like a drill-sergeant. The rest of the performance did nothing to show any special sympathy with Bach's religious music on Richter's part.[5]

For Shaw the highlight of the festival was Stockley's conducting of *Messiah* (ignored by the *Musical Times*), but as well as having to put up with listening to Mackenzie's new work, his mood was not improved either by Parry conducting his *Blest Pair of Sirens* or Stanford his new *Eden*, which he dismissed as 'brilliant balderdash'. The various items at the Miscellaneous Concert (8 October) included Joachim playing his own 'Hungarian' Concerto, but the Ride of the Valkyries, which concluded the concert, elicited from Shaw the opinion that 'our gifted Hans must be slightly mad considering the outrageous position he gave it in the programme'. Another composer left to his own devices by Richter was Dvořák, who conducted his new Requiem, 'which bored Birmingham so desperately that it was unanimously voted a work of extraordinary depth and impressiveness', said Shaw. Richter followed Dvořák to the podium ('and the performers steadied themselves, as they could not help doing', reported the *Musical Times*) with the Prelude to *Parsifal* and Beethoven's Seventh Symphony. He conducted the final concert, consisting of Berlioz's *Faust*, on 9 October. Thompson made several reports for the *Yorkshire Post*, including a description of a rehearsal before the festival opened.

Tuesday 6 October [Rehearsal] Joachim [had] just arrived from Berlin in time to take part in the rehearsal, which it may be said was so satisfactory to give prospect of a remarkably fine performance, the refined and artistic singing of the chorus being the subject of remark on all hands. After an interval, the shortness of which was not relished by some members of the band and chorus, work was resumed with an Offertorium on a Tantum Ergo which are the latest Schubert finds and have only last year been published, after which Dr Dvořák took his seat at the conductor's desk and spent most of the afternoon in rehearsing his *Requiem*. Dr Richter, sitting close at hand, took a lively and practical interest in the proceedings and was materially helpful in correcting the occasional inaccuracies and imperfections, sometimes the result of mistakes in the copies, but more frequently rising from the inherent difficulty of the music, which makes great demands upon the singers, especially in the matter of intonation. After this the chorus was dismissed, but there was found time to take the band through Beethoven's Violin Concerto, Dr Joachim being of course the soloist.

Once more the Committee have secured the services of Dr Hans Richter, the greatest of living conductors, and though much may doubtless be said in favour of the appointment of an English musician to so important a post, still the object to be attained is a performance as near perfection as possible, and the Birmingham authorities would show wisdom in sticking to Dr Richter until they can find an Englishman who is his equal.

7 October: *Elijah* The orchestra played with splendid spirit and power under the conductorship of Dr Richter, of whom we can only say that if, as it has been said, he is not in sympathy with Mendelssohn's music, we greatly prefer his indifference to the enthusiasm of most other conductors.

8 October: *St Matthew Passion* We think there is a point at which an improvement might have been made in the arrangement of the orchestra for this morning's performance, and that is the balance between strings and woodwind. In the orchestra of today the latter are outnumbered by the former to such an extent that the music of Bach's time, when the numbers were more equally divided, cannot be properly rendered by it, the important often elaborate flute parts, for instance, being completely lost when played by a couple of performers pitted against thirty or forty violins. Another point: the extent to which the organ and cembalo, now replaced by the pianoforte, should be employed is of too controversial a measure to be entered upon now. We must confess that the edge of our enjoyment of the superb unaccompanied singing of the chorus in the chorales was somewhat blunted by the fact that Bach certainly intended the voices to be supported by the band and organ and not to be sung in the manner of a modern sentimental part-song.

10 October The chorus had been made the most of by the superb conducting of Dr Richter, who has shown himself not only to be an ideal orchestral conductor, but now that he has acquired a closer acquaintance with the English language, a no less masterly conductor of the choral works which form the staple of an English Festival, able both to teach his choir at rehearsal and lead them to victory at time of performance. It has been pleasant to see too that Dr Richter is thoroughly appreciated by, and is a prime favourite of all who work under him, a factor of which there has been abundant evidence at both the full rehearsals and the concerts themselves.[6]

Among Richter's correspondents at this time was the actress Ellen Terry who wrote, 'I am leaving Birmingham as you enter it—alas!—for I should rejoice and be glad at heart to witness your triumphs. I want you to be kind enough to do me a favour, and I feel sure you will if you can. [Pier Adolfo] Tirindelli, a young violinist from Venice (Ysaÿe knows and admires him greatly) is coming to London next month. The favour I ask of you is to hear him and praise or condemn him, remembering that your introduction would be of supreme value to him in case arrangements could be made for him to play in London next season.'[7]

If anything 1892 was less enterprising than ever regarding Richter's London programmes, a performance of Dvořák's tone-poem *Husitská* on 13 June and repeated 'by general desire' a fortnight later being the only novelty. Ironically, while Richter opened his season with two Wagner programmes, Mahler was in London conducting Covent Garden's first *Ring* in a German season for Augustus Harris. 'He knows the score thoroughly', wrote Shaw, 'and sets the tempi with excellent judgement.'[8] There was just one cycle and a single performance of *Tristan*. For this Harris had resorted to hiring the company of the Hamburg Opera lock, stock, and barrel (including its Kapellmeister Mahler), and at the same time dropped the word 'Italian' from the Royal Italian Opera, Covent Garden, for good. A letter from Arthur Goring Thomas in October 1891 (after Richter had conducted his duet at the Birmingham Festival) suggests that an approach may have been intended to get him to conduct for Harris's project. 'Mr Higgins,' he wrote, 'who is on the Committee of the Royal Italian Opera, Covent Garden, is most anxious to see you before you leave England, as he has some important proposals respecting the opera season of 1892 to lay before you.'[9] Higgins later became chairman of the syndicate which took over the running of Covent Garden on Harris's death, and began a close association with Richter a decade later.

Ellen Terry was one of many who sent new musicians to Richter. At this time Ethel Smyth had first met him in Vienna in the spring of 1892 at Hermann Levi's recommendation. She was anxious to have her new Mass performed, but unfortunately for her 'Richter was conducting no choral concerts in London. "But perhaps elsewhere in England it might be arranged?", he added. I, however, had heard a good deal in Germany about the great man's "peasant astuteness", and knew that once in England he would realise that I was not in the swim, and then good-bye to his enthusiasm.'[10] Frederic Lamond wrote in March 1891, offering Brahms's Second Piano Concerto for a London or Vienna engagement, but met with no success at this time. On the other hand the American soprano Lillian Nordica, resident in London by 1892, made her début at a Richter Concert on 4 July that season singing the closing scene from *Götterdämmerung*. He was so delighted with her success that he arranged for her to sing to Cosima later that month in Bayreuth, and as a result she sang Elsa in *Lohengrin* in the 1894 Festival. Nordica repeated her concert triumph as Brünnhilde in Richter's 1893 season in London, and became, along with David Bispham and Eugène Oudin, one of the few American singers with whom he worked. Bispham appeared on 8 October 1894 when he sang Sachs's Monologue and Wotan's Farewell 'with the best of all Wagnerian interpreters, Hans Richter, with whom it was my privilege to work many a time afterwards, deriving the greatest benefit from association with him. Taking it all in all, and looking back upon a long line of orchestral conductors, I consider him, to be the chiefest of them all. It is much to be regretted that he never came to America, for Richter said he would come if Joachim came, and Joachim said he would attempt the journey if Richter did, but as a matter of fact neither of them wanted to cross the ocean even to visit the New World.'[11]

Oudin, actually a French-Canadian but resident in America, sang the role of Eugene Onegin in the first English production of Tchaikovsky's opera in 1892. In his first appearance for Richter he sang an aria from Marschner's *Hans Heiling* and Wotan's encounter with Erda in the first scene from the last act of *Siegfried* (with Agnes Janson) on 26 June 1893. He worked with him again a year later at the 1894 Birmingham Triennial Festival when he sang Dr Marianus in the third part of Schumann's *Faust* on 5 October. He was only 36 years of age when, just over a fortnight later on 20 October, he 'was struck down by apoplexy in the artists' [Richter's] room at Queen's Hall after the Richter concert'.[12] He died on 4 November.

The 1894 Birmingham Festival began on 2 October with *Elijah* and ended three days later with Beethoven's Ninth. Apart from these regular ingredients it also included Cherubini's Mass in D minor and Wagner's edition of Palestrina's *Stabat mater*. Thompson's reports for the *Yorkshire Post* were as follows.

Elijah: The orchestra of course did its work admirably, nothing else could be expected of so able a body of players in a work so familiar and conducted by so masterly a hand as Dr Richter's. In only one thing was it disappointing, that is in string tone which was not so good either in weight or quality as might be expected from a body of over eighty players whose names are, for the most part, those of well-known London musicians. Of Dr Richter's conducting little need be said. He is universally admitted to be all-round the most eminent of living conductors, and he devoted to Mendelssohn's oratorio the same care and watchfullness that he gives to everything. He is generally supposed to be less in sympathy with Mendelssohn than with some of the other composers, and this may perhaps be the case for even a conductor may have his likes and dislikes, but even if this be so, it must be allowed that he does not permit favouritism to interfere with the conscientious performance of his duties. In his reading of *Elijah* there are, here and there, deviations from the tempi that use has consecrated in this country, but being hidebound by tradition is surely not an unqualified blessing whether the tradition be that of Birmingham or Bayreuth.

3 October: Nowhere is the advance more marked than at Birmingham where since the advent of Dr Richter the miscellaneous programmes have been models worthy of the imitations they have provoked. This evening the second part of the programme began with one of the noblest of modern symphonies, Brahms' second Symphony in D . . . it is impossible to notice all these performances in detail, but a word must be given to Dr Richter's superb and sympathetic reading of the symphony, a work that may be ranked among the strongest and most individual symphonies since Beethoven. . . . This evening's performance brought out its manifold beauties in the most favourable light, and deserves unequalled praise. It was a triumph for both composer and conductor.

5 October [morning concert]: Cherubini's Mass in D minor. . . . Save that in the matter of intonation the chorus fell to pieces in the Kyrie, the performance was a very fine one, Dr Richter's reading of the work being admirable in bringing out the dignity and melodic beauty of the music.

Mozart Symphony No. 39: That Dr Richter does not share the popular opinion that Mozart's music is out of date was shown by his highly sympathetic reading of the symphony. It was particularly satisfactory to find that he avoided the detestable tradition of scrambling through the first allegro, an instance of the 'naïve allegro' of which Wagner writes in his essay on conducting.

[Evening concert] Beethoven Symphony No. 9: Schumann's work was preceded by the *Tannhäuser* overture and followed by Beethoven's *Choral* Symphony, both works calculated to display Dr Richter's powers to the greatest advantage. The symphony, with its final Hymn to Joy, was a most appropriate ending for a great Festival, especially when the conductor happens to be *facile princeps* in his rendering of Beethoven's music. But it was hardly anticipated that the chorus, whose task in the *Choral* Symphony is, as is well known, of supreme difficulty, would come out of the ordeal so brilliantly considering that during the week, we have too often had occasion to remark, they have shown a want of staying power so necessary in their work. But they remembered their conductor's instructions at rehearsal when he told them that enthusiasm, rather than mere voice, was necessary to attack the *Choral* Symphony with any hope of success, and they simulated enthusiasm with the best results. Save for one passage where the tenors once again proved fallible, though very pardonably so, the chorus sang admirably and finished their labours with what was on the whole a most successful effort.[13]

The *Musical Times* regretted that Wagner had meddled with Palestrina's *Stabat mater* by, among other things, adding a quartet of soloists to the double choir. There were two new works performed over the three days. First was Stanford's orchestration of Arthur Goring Thomas's *The Swan and the Skylark*. Thomas, a manic depressive, had shocked the musical world two years earlier when he committed suicide at only 41 by throwing himself in the path of a train at West Hampstead station. His music was 'pretty, graceful, sentimental, and engaging'[14] and the chorus was joined by Albani, Brema, Lloyd, and Brereton as soloists. Richter appropriately chose Sullivan's *In memoriam* to follow and the concert concluded with Mendelssohn's *Hymn of Praise*. Chorus master Stockley conducted *Messiah* on the second morning of the Festival (after the 1885 and 1888 Festivals at which Franz's edition had aroused a storm of protest, it was abandoned in favour of Costa's version). That evening the other new work, Henschel's *Stabat mater*, was conducted by its composer as part of a vocal and orchestral programme directed by Richter. The solo singers that year raised already high standards to greater heights. As well as Albani, Lloyd, and Henschel, other names were appearing which would come to dominate English concert platforms for several years to come, such as Andrew Black, Anna Williams, and Marie Brema. Brema went on to sing Fricka as the only English singer in the revival of the *Ring* at Bayreuth in 1896 under Richter.

An interesting occurrence in the spring of 1894 had been Felix Mottl's visit to London to conduct two concerts (mostly Wagner) under Schultz-Curtius's management. The Karlsruhe-based conductor made a great impression, particularly upon Shaw.

I have one other strong reason for desiring to see Mottl established here as a conductor. His greatest rival, Richter, is so far above the heads of the public that he has no external stimulus to do his best in London. Only a very few people can perceive the difference between his best and his second best; but the difference between his second best and Mottl's best would be felt at once by a considerable body of amateurs. Now I do not suggest that Richter ever consciously does less than his best; but I am materialist enough on these matters to believe that even the best man does more work under pressure than in a vacuum.[15]

Richter concluded his visit to England in October 1894 with eleven concerts in several towns ranging from Brighton to Edinburgh, and Liverpool to Leeds. In his conducting books covering the thirty years from September 1865 to March 1895 Richter listed sixty-six towns and cities in which he had conducted. His 1895 spring concerts in London included the first performance of Stanford's Second Piano Concerto with Leonard Borwick, Dvořák's Overture *In Nature's Realm*, and Smetana's symphonic poem *Šárka* (the third part of *Má vlast*). Moritz Rosenthal's first appearance in England took place when he played Liszt's First Piano Concerto under Richter on 10 June. With Hermann Levi and Arthur Nikisch now also part of the London musical season, together with Mottl, the *Musical Times* seized upon the chance to compare the familiar Richter with his colleagues.

Highly as connoisseurs have for long appreciated the magnificent combination of qualities united in the artistic personality of Hans Richter, they have of late found it necessary to place still higher value on his splendid services. Comparisons may be odious, but criticism after all forms its judgements by their aid. English concert-goers had begun to believe that Hans Richter was but one of quite a number of great conductors to be found abroad; recent years have clearly demonstrated the fallacy of this idea. Three of the greatest have been heard in London within the last three months and their merits have been freely acknowledged in these pages; but the highest compliment we have been able to pay these eminent artists was that, in certain qualities of their work, those which distinguish the conducting of Richter were approached.[16]

These were heady days, for Nikisch had his series of concerts in the same month at the Queen's Hall. 'His baton is employed rather to indicate the effects of accent, phrasing, and expression rather than to beat time', ran the review in the *Musical Times*, ending with a tart comment about Nikisch's principal oboist brought specially from Budapest for the concerts, 'until we heard this gentleman we had no idea that so disagreeable a tone could be produced from the instrument'.[17] Levi had his Wagner concerts during April 1895, and his chorus master was the young Henry Wood. Though Mottl was the greatest influence upon Wood, he acknowledged that although 'it is true that I had attended Richter's concerts as a very young man, I failed to appreciate his great qualities. I fear his performances left me cold and unmoved. Yet strangely enough when I heard him in Vienna, Berlin, Munich, and Bayreuth, he thrilled me with his masterly grip.'[18] This year of 1895 was a watershed for music-making in London, for that August the manager of the Queen's Hall, Robert Newman, engaged Wood to conduct the first of the so-called Promenade concerts. He did so when the greatest conductors of the day were in the capital (Siegfried Wagner, though not in their class, also had a concert there in June), but this only spurred Newman on in his belief that the time was ripe for a British conductor to make his mark, and before long the man he chose proved to be the right one.

23
1895–1900

England

Richter's summer season of 1895 was a short one and he returned for another tour of England and Scotland in the autumn. His itinerary from 19 October began with a concert at Brighton's Dome followed by London, Nottingham, Edinburgh, Glasgow, Manchester, Liverpool, London, Oxford, Birmingham, Sheffield, Bradford, and ended in London on 4 November. Once out of London, all the concerts were on consecutive days. The programmes included little new apart from Goldmark's Overture *Sakuntala*, the staple fare being Wagner extracts, Tchaikovsky's 'Pathétique' Symphony, and the Eighth Symphonies by Beethoven and Schubert. By a curious coincidence Richter arrived by train in Manchester for his concert that night on the very day that Sir Charles Hallé suddenly died. His diary records 'a really boring journey, snow. Shortly before Manchester I learned of Hallé's death. I began the programme with the *Funeral March* [Siegfried's from *Götterdämmerung*] by *the man whom the dead man mocked*. It was nice that, whilst listening to the March, the public stood to honour their teacher. A splendid reception of the Tchaikovsky!! Loud cheering! Discussion about the future!!!!'[1] A member of the audience was Edith Hall, who lived in Bowdon, where Richter would settle. She kept a diary in which she made occasional references to her concert-going, her most common complaint being the inability of the concert planners to satisfy her insatiable thirst for music and longer programmes. She first heard Richter in the autumn of 1892, when she wrote:

The hall was not very full, most of the stalls were taken but for a concert like this it ought to have been crammed. Hallé, Lady Hallé and Olga Neruda were there

1895–1900: England 297

and Mr Dawson. It was a splendid programme, only much too short—it was over before 9.30. . . . The *Walküren Ritt* is magnificent, so stirring it carries you away. The audience tried hard for an encore, but Richter would not give it. . . . But the whole concert has been glorious, oh if only I could hear it again! It has been so splendid, but anyway it is a great deal to have heard him this time, and with such a programme.[2]

She also attended Richter's concert on the day of Hallé's death.

At the concert this evening . . . Mr Edghill told us that Sir Charles Hallé died this morning. It is very sudden and very sad. . . . Richter played the *Trauermarsch* first for respect to Hallé's memory. It was simply beautiful. A noble and fitting march for a hero.
  . . . After the third movement [Tchaikovsky 'Pathétique'] Richter got tremendously cheered, but he disclaimed it all and gave it to his band. They really did play marvellously . . . the way the strings swept up and down was really splendid. The only drawback to the concert was it seemed so short . . . if only one could hear that music oftener.[3]

Despite wishing not to be too hasty in discussing the question of a successor to Hallé so soon after his death, Herbert Thompson, writing in the *Yorkshire Post*, clearly nailed his colours to the Richter mast from the outset.

The greatest tribute to the importance of Sir Charles Hallé's dual position as head of the music school he founded in Manchester and as conductor of the famous band is that it is impossible to discuss his death without thinking of the difficulty that will attend the appointment of a successor to him in either capacity. It's too early, and would hardly be in good taste to suggest possible names, but it may be pointed out that nothing less than a musician of absolute distinction will suffice to sustain the reputation of the Manchester band. In the matter of beauty of tone, really a comparatively minor point but one to which English musicians are apt to attach too much importance, the Hallé orchestra is, it must be confessed, not quite the equal of the best London orchestras, but in energy and fire it is unsurpassed in this country, or was so until the advent of Richter, the greatest of living orchestral conductors. To maintain this well-deserved reputation it will be necessary to secure a musician who has made orchestral conducting his speciality, and will not have to gain the necessary experience at the expense of the Manchester band.[4]

Gustav Behrens, James Forsyth, and Henry Simon were the three guarantors of the future of the Hallé concerts upon Hallé's death. It is probable that Richter met Behrens in the green room after the concert on that fateful day. The discussions, deals, correspondece, etc. which took place over the next four years before Richter was in place at Manchester have been documented in great detail by Michael Kennedy in *The Hallé Tradition* (Manchester, 1960). In short, Richter was initially in demand by both Liverpool and Manchester, but soon Liverpool got cold feet at the delays and prevarications emanating from Vienna, and went ahead and appointed Frederic Cowen to the Philharmonic Society. At this point (February 1896) Richter suddenly told Behrens that for a figure of £3,000 he would leave Vienna at once as this would compensate for his loss of pension rights. This sum was quite out of the question and so they joined Liverpool and appointed Cowen for two seasons as their joint conductor. Matters rested there with Cowen at the helm, though he was a tactless man with a short temper so that trouble was always simmering just below the surface. For the moment, Richter was out of the picture and would remain so until the summer of 1898.

Richter returned to London for his season in 1896 which began with a concert on 18 May and included more Goldmark, this time the Overture and Entr'acte from his opera *The Cricket on the Hearth*. Two other concerts followed (after a quick trip to Brussels for the city's first performance of Tchaikovsky's 'Pathétique' on 22 May) and these included Strauss's *Till Eulenspiegel*, Dvořák's 'New World' Symphony and *Othello* Overture, and Tchaikovsky's *Romeo and Juliet* Fantasy (not heard in England since its first performance twenty years earlier at the Crystal Palace), all new to his own concerts. There was little new during the autumn tour of the country apart from the first English performance of Dvořák's *The Golden Spinning Wheel* at London's Queen's Hall on 26 October. This work was not well received by the press despite its masterly orchestration. It offered not 'a single stirring moment, delight changed to indifference and indifference to irritation, and the audience all but "declined" the work'.[5] It had been Richter's intention to perform Dvořák's other two new symphonic poems, *The Noonday Witch* and *The Water Goblin*, during his English tour that autumn, but these were not ready in time from his publisher Simrock. The Scherzo capriccioso was played in their stead and he conducted the two new works in Vienna immediately upon his return (22 November and 20 December).

His 1897 season opened on 24 May with a familiar programme but included Strauss's tone-poem *Don Juan* for its première in England five years after Richter had conducted it for the first time in Vienna. His second concert a week later began with the three 1891 Dvořák concert overtures which belong together, *In Nature's Realm*, *Carnival*, and *Othello*, as a group entitled *Nature, Life, and Love*; nowadays they are rarely programmed together according to the composer's intentions. Ossip Gabrilowitsch then made a brilliant English début with Tchaikovsky's First Piano Concerto, after which Richter premièred Cowen's new Sixth Symphony, the 'Idyllic' (Parry, who was there, thought it 'very pallid and pointless').[6] Richter confided to his diary that Strauss's new work had an 'average success'.[7] Gabrilowitsch, on the other hand, 'had a superb and deserved success. Dvořák's works were liked but Cowen's symphony failed completely, *and it is good*. I will perform it in Vienna. The English just do not want their own people; for eighteen years they have defeated all my attempts to promote English composers.'[8] That judgement was a little harsh, for it was the commercial considerations of his English managers over the years that were more responsible for such a situation than the British public or press. The latter, on the contrary, were becoming weary of even excellent and definitive performances of Beethoven, Wagner, and now a handful of Tchaikovsky's works. Even his habit of leaving the orchestra to play the second movement of the Sixth Symphony in 5/4 time alone was becoming commonplace. 'We have our own opinion of this "no conductor" joke,' grumbled the *Musical Times*.[9]

Richter's pattern of a short London season in May/June followed by an autumn tour of about a dozen concerts in many towns and cities but beginning and ending in the capital was, by 1897, well established. Brahms's First and Fourth Symphonies served as a memorial to the recently deceased composer and Edith Hall recalled the Fourth at the Free Trade Hall in Manchester on 20 October.

[Richter] With his complete London orchestra of 90 performers. Brahms fourth Symphony in memoriam. The heat was fearful, so it made my head rather bad. The Symphony was not as interesting as what he usually plays. The rest I enjoyed, and a good deal of the Brahms, but the first movement I did not care for. *Leonore* he played splendidly.[10]

The year 1897 brought another Birmingham Triennial Festival, and the four-day event included three new English works, Stanford's Requiem, Arthur Somervell's *Ode to the Sea*, and Edward German's symphonic poem *Hamlet*. Joseph Bennett, writing in the *Musical Times*, praised Stanford's work, thought that Somervell's might have benefited from being conducted by Richter rather than by its composer, and then made a curious error regarding German's work, 'the composer conducting in this case also. . . . Mr German secured a performance with which little fault could be found and received the usual compliments.'[11] This was not the case. Richter, who described German as 'a nice and talented man'[12] when they met on 26 September, conducted *Hamlet*. Not only is it listed in his conducting book (Stanford's and Somervell's pieces, which they themselves conducted, are not), but other reviews and letters from German prove it. On the day of the performance (5 October) he asked Richter to 'take all the broad parts a little slower. This, I believe, is your own wish, so it shall be mine.'[13] The following day he wrote to 'offer you my sincere gratitude for all the trouble and interest you have taken in my *Hamlet* music'.[14] Richter's true opinion of *Hamlet* was that it was a bad work. 'It is fatal', he wrote, 'when nuances only occur to a composer when he is standing before the orchestra.'[15]

The *Musical Standard* was present at some of the rehearsals for the Festival held at the Queen's Hall in London and produced a lengthy and vivid account of the proceedings.

The talented young composer [Somervell] was extremely nervous and, judging by the score, he did not at all realize his intentions. Madame Albani . . . smiled her encouragements and looked as if she would have liked to conduct. Dr Richter wandered about the hall and also looked as if he would like to conduct the work. Indeed it is a pity the composer did not relinquish his baton to the master of the orchestra. . . . Mr Edward German was wise to leave his symphonic poem in the hands of Dr Richter, satisfying himself with giving general directions.

Under the baton of a conductor such as Richter a work grows before one's very eyes. There is no rest for the orchestra until it has given him what he wants. A passage is played monotonously, mechanically; Richter stops the orchestra with a gesture that plainly means 'this won't do at all'. He satirically imitates the tame way in which the passage has been played, and then in his strange guttural yet effective way, he sings it as it should go. . . . Though it is doubtful if Richter can sing in the ordinary sense of the word, he certainly can express vocally his idea of how a melody should be played. He is always singing. When the band is playing fortissimo he roars out the melody; in a word, acts the music so that his men can be under no misconception of what he wants. And when he is not singing, he is shouting directions as to expression, or, at an important point, the note the violins have to play. When everything is going well, his face wears a happy smile, but he is none the less 'on the pounce'. The brass is unsatisfactory. Down goes the baton on the desk and he shrugs his shoulders. The orchestra stops dead, except for a straggling note here and there. 'Do not be so short-winded; hold on to ze note longer. So'—and the conductor sings the passage as it should go. Then we start again, and when the same point is reached Richter pounces on the trombones and by the expression of his lips imitates what he wants them to play. Then the violins get too loud. 'Piano, piano', he roars. It's no good, they have not done it. 'This is to be played softly, like a piece of beautiful chamber music', he explains when the orchestra is silent. In Beethoven's *Leonora* No. 3 he was particularly keen on strength of emphasis. Over and over again the orchestra was stopped because the emphasis was too tame. 'pom-pom-pom-POM; pom-pom-pom-POM—not pom-pom-pom-*pom*, as you play it'.[16]

Plate 1 (*a*) Hans Richter's mother Josefine.

Plate 1 (*b*) The boy Hans with his father Anton Richter.

Plate 1 (*c*) Hans as an Imperial Chapel chorister.

Plate 1 (*d*) Hans Richter, second from right, as a teenager, with his uncle (Anton's brother) and cousins.

Plate 2 (*a*) Hans Richter in 1868 after his appointment as Music Director in Munich following Hans von Bülow's departure.

Plate 2 (*b*) Hans Richter with his fiancée Marie von Szitányi, Budapest 1874.

Plate 3 (*a*) Richter (centre) with members of the 'Nibelungen Chancellery' at Bayreuth in 1872 (left to right, Hermann Zumpe, Demetrius Lalas, architectural assistant Karl Runckwitz, and Anton Seidl).

Plate 3 (*b*) Hans Richter's children (left to right, Hans, Mathilde, Ludovika, Marie, Richardis, and Edgar) about 1884.

Plate 4 (*a*) Hans Richter with the score of Wagner's *Die Meistersinger*, London 16 June 1898. The photograph is dedicated to Pedro Tillett, nephew of the conductor's agent Narciso Vertigliano.

Plate 4 (*b*) Richter outside Birmingham Town Hall during the Triennial Music Festival in October 1909, photographed by his future son-in-law Sydney Loeb. The musical quotation is the bassoon part taken from the introduction to the Prisoners' Chorus from Beethoven's *Fidelio*.

Plate 5 (*a*) Richter at The Firs, Bowdon in 1908.

Plate 5 (*b*) Richter and his daughter Mathilde on board ship crossing the Channel.

Plate 6 (*a*) Richter with George Bernard Shaw and his wife Charlotte at Bayreuth in 1908.

Plate 6 (*b*) Richter with (left to right) Eva, Isolde, and Siegfried Wagner, Daniela and Blandine von Bülow, Bayreuth 1890.

Plate 7 (*a*) Richter holidaying at Baracs, Hungary.

Plate 7 (*b*) With Marie and his granddaughter Eleonore in the garden of Zur Tabulatur in Bayreuth 1914.

Plate 8 (*a*) The familiar sight of Richter with shopping bag during his retirement in Bayreuth. The musical quotation from *Die Meistersinger* was the family whistle.

Plate 8 (*b*) Richter in his Bayreuth study on his seventieth birthday, 4 April 1913.

Richter was very pleased with his orchestra, which he described as 'splendid, a real joy for me to work with these sensitive people'.[17] Besides Albani, Richter's singers included Marie Brema, Ada Crossley, Anna Williams, Ben Davies, Edward Lloyd, Andrew Black, Plunket Greene, and David Bispham. Cowen and Parry were also there to conduct their own works, the former his *scena Endymion*, and the latter his *Job*. 'Went very well,' Parry noted. 'Richter very sympathetic after. Praised the scoring very warmly. Said it was "wonderful" and we had an affectionate hug and kissed both cheeks cordially.'[18] On 22 September Richter travelled to Birmingham for his first chorus rehearsal. It was his first meeting with the new chorus master Dr Charles Swinnerton Heap, who was judged by the conductor to be 'a splendid musician' after the rehearsal of Schubert's E flat Mass.[19] Once again a member of the press was at the rehearsal, this time a local reporter who recorded Richter's opening words to the chorus as 'I thank you for your signs of sympathy [possibly a mistranslation by either Richter or the reporter of the German word *sympathisch* meaning "pleasant" or "nice"]. To me sympathy is music. It is always a great pleasure to work with the Birmingham chorus.' According to the reporter, Richter found little to criticise in Heap's preparation of the chorus, his most frequent comment being 'Bravo' or 'Excellent'.[20] At this Festival Richter conducted *Messiah* (Mozart-Costa version, 'whilst conducting it I had to maintain respect for old Handel'[21]) as well as *Elijah*, Berlioz's *Damnation of Faust*, Bach's Cantata No. 34 'Oh Light Everlasting' (Robert Franz's edition of 'O ewiges Feuer'), Fuller-Maitland's edition of Purcell's *King Arthur* ('harpsichords Mr and Mrs Dolmetsch' according to his conducting book entry for 6 October), Schubert's E flat Mass, an arrangement by the conductor himself of Beethoven's 'Abendlied' (a song written in 1820 to Goeble's words 'Wenn die Sonne nieder sinket') sung by Marie Brema, and a host of typically Richter orchestral fare. In the second movement of the 'Pathétique' he is again reported to have sat down whilst the orchestra got on with playing it alone. 'This has been described as clap-trap', wrote a reviewer, 'though we think it is simply due to a desire to let the orchestra speak for itself for a time, and the results have justified the experiment.'[22]

Herbert Thompson wrote on the Festival for the *Yorkshire Post*:

If the Birmingham Festival of 1897 is distinguished for nothing else, it will be favourably remembered in history as the first English Festival at which our insular high pitch was given up. The organ in the Town Hall has been lowered to the diapason normal [A = 435], which is now being very generally adopted as the standard or classical pitch. It will be remembered that Dr Richter advised the adoption of the still lower Viennese pitch, which, being just half a tone below our high pitch, will obviously involve less practical difficulty.

Ever since the wise step was taken of securing Dr Richter as Sir Michael Costa's successor, there have not been wanting the praisers of past times who deplored the fact that the new conductor was not in sympathy with English traditions concerning the performance of the two oratorios which, to the musical middle class, represent serious music. Even now we find the critical representative of a local paper sternly warning Dr Richter that, though he may give us his readings of such small fry as Wagner, Brahms, Dvořák or Tchaikovsky, he is to leave the *Messiah* and *Elijah* 'to the heart of the English people'. It is true that the heart resembles a conductor, in that his office is to beat, otherwise one might be at a loss to understand how the heart, even of the English people, could do anything to maintain a traditional reading. The common or garden metronome would be more in point one would think but without contesting this it must be pointed out that Dr Richter has hardly had a chance of impressing his individuality upon *Elijah*, considering that he has probably never conducted a full rehearsal of the work at Birmingham. For our own part we should not mind risking tradition for the sake of hearing what the greatest conductor of his time can do with the work. It must certainly be allowed that today there is nothing sensational in the performance save that it was an exceptionally finished one, and that Dr Richter's thoughtful care was shown or given to every passage capable of it. Nothing was thrown over, and the way in which each climax was kept well in hand, so that its full force was felt, is deserving of the highest praise. We cannot understand how the heart of the English people could have done much better.

Two great improvements connected with the performance and with the Festival in general have to be noticed. Two things have been lowered, the pitch of the organ and the headgear of the lady choralists; as regard to the latter point it only now remains for the lady principals to doff their hats at morning as well as evening performances for a sensible reform to be made thorough. We can hardly hope for it to extend to the ladies in the audience, though there are times when one would rather see the conductor or vocalist than even the most profuse of artificial flower gardens. As regards the altered pitch to which reference is made elsewhere, it can hardly be said to have seriously detracted from the brilliance of the performance, while it must have eased the chorus and conveyed more accurately the composer's intentions.

The all-important question of the chorus comes next. It would be time to judge of its musicianship, its attack and the like, when it has been heard in less familiar music. In respect of its tone quality, however, it may at once be said that it is worthy of the highest Birmingham traditions. It has extreme refinement and finish, and a certain vocal quality that seems peculiar to Birmingham, at least one never meets it elsewhere, and though forces and weight of tone are not its leading characteristics, as they are at Leeds, it has sufficient. Witness the splendid effect of such passages as 'Hear and answer', 'He shall die' and the like. The sopranos and basses are excellent, the tenors are up to the high standard of 1891 and far above that of 1894. The altos are the weakest section, even allowing for the comparative ineffectiveness of the alto voice. Their lead in 'Though thousands languish' was wanting in bite. Their numbers are Sopranos 107, Altos 80 (including 9 male altos), Tenors 80, Basses 88, a total of 355 voices. Against these the band numbers 124 players, Messrs Burnett and Schiever being the joint leaders.

After the unique performance of Beethoven's C minor symphony at the Birmingham Festival of 1888, it was almost risky to tempt fortune by giving it again, for even the lapse of nine years has not blotted out that superb performance from the minds of those who heard it. Still tonight's was a most noble and inspired rendering, bringing out all the force and dignity of the work. It would have been practically flawless had it not been for the undue prominence of the bassoon in the latter part of the scherzo.[23]

After Richter's subscription concert in Bradford on 22 October as part of his annual national tour, Thompson wrote:

Not only is Richter the greatest of Beethoven conductors, but the Symphony [No. 7] in A is especially suited to his powers and we recollect many supremely fine readings of it under his baton, notably one at a Birmingham Music Festival. This time the tempi seemed to drag in both the first movement and the scherzo to an extraordinary extent. One hardly likes to criticise Richter's tempi in Beethoven's music, which he has made his own, but we compare him with himself and according to this high standard it must be confessed that the symphony lost something of the verve and rhythmical vigour that should characterise it. The slow movement was indeed perfectly played but even it lost a little by want of the necessary contrast with its surroundings. As usual at a Richter concert Wagner was much in evidence; the Meistersinger Overture opened the concert . . . 'Wahn, Wahn' and the gorgeously coloured final scene of the *Valkyrie*, both of which were finely sung by Mr Andrew Black. In the latter, by the way, Dr Richter showed that even without the veil of a concealed orchestra, the soloist need not be drowned by the orchestra, for Mr Black's distinct enunciation enabled him to be followed without difficulty. Dvořák's *Carnival* overture was played with great brilliance but lost some of its effect owing to a piano-organist in the neighbouring street, who by some mischance escaped with his life.[24]

Richter's 1898 concerts contained new works and soloists. *Sheherazade* by Rimsky-Korsakov was included at the first on 23 May, whilst the second opened with Robert Fuchs's Overture *Des Meeres und der Liebe Wellen*. Ferruccio Busoni made his first appearance with Richter on 4 June with Liszt's arrangement of Schubert's Wanderer Fantasy and his own of Liszt's Spanish Rhapsody. The concert also included Tchaikovsky's *Nutcracker* Suite for the first time. Richter found accompanying the piano 'difficult.

Busoni splendid. The *Nutcracker* is somewhat embarrassing, especially the final Waltz, perhaps the Polonaise from the Suite would make a better ending. Excellent concert, the *Nutcracker* the greatest success—oh you public!—it was however excellently played. I spoke with Saint-Saëns before the concert. We have known each other now since Munich 1868.... Bennett criticizes—with justification—the performance of the *Nutcracker* suite. Nice as it is, it is not for me. Away with such trifles, there's not enough time to spend rehearsing the refinements.'[25]

Svendsen's *Carnival in Paris* was played at the third concert on 13 June and Cosima Wagner attended the last concert a week later which ended traditionally with Beethoven's Ninth. At its final rehearsal Richter is reported to have said to the chorus, 'When you come to the hall on Monday night will you bring a little joy, enthusiasm with you. The widow of the greatest composer of music for the stage will be present and she has not heard this Symphony since the laying of the foundation stone of the theatre at Bayreuth in eighteen thousand, seventy-two.'[26] Richter's diary makes no mention of her reaction to the Ninth. She could not, however, 'stand the Brahms [*Song of Destiny*]. We played *Les Préludes* in her honour, which pleased her this time. She found some accents in the extract from *Tristan* too piercing; in fact she sat too close and seldom hears an orchestra. On the whole she found the performance "clear" throughout, and spoke well of it to others.'[27] London was in the grip of three *Ring* cycles and *Tristan* with glorious performances by Nordica, Ternina, Brema, Schumann-Heink, the de Reske brothers, van Rooy, and van Dyck. Hermann Zumpe (an original member of the Nibelungen Chancellery but one who never raised the baton at Bayreuth) and Mottl were conducting. Despite the presence of so many of the world's finest singers, Cosima was 'little pleased with Bayreuth in London'.[28]

Joseph Bennett's review of these June concerts took Richter to task.

We fear Fuchs' Overture... will not prove an addition to the list of great masterpieces amongst overtures.... Dr Richter had a surprise the reverse of pleasing in store for his audience in his performance of Tchaikovsky's *Casse Noisette* Suite, which lost a vast deal of its *esprit* and daintiness through the leisurely tempi adopted for most of the movements.... The *Trépac*, on the other hand, he took at a breakneck speed, which... completely obscured the melody. Needless to say he was himself again in the *Parsifal* prelude (a wonderful performance) and Dvořák's E minor symphony.

The review goes on to cover Richter's reading of Tchaikovsky's 'Pathétique' Symphony in his next concert. Bennett liked it no better than Lamoureux's and, though greatly preferring Wood's interpretation, he exhorted his readers not to forget the best ever in his opinion, that of Stanford at the Royal College of Music on the occasion of the third and fourth performances in England—despite the limited standard of his student orchestra. His review continued:

May we ask why Dr Richter gave the last four notes for the bassoon in the *Adagio mosso* . . . to another instrument? The bassoon *can* be played quite softly on the 'loud' bassoon, for we have often heard it so played. And why should one of his woodwind players be allowed to play the identical wrong note in the triplet passage of the *moderato mosso* which he played wrong *last year*? In the delicious 5/4 movement Dr Richter once more put down his baton, and at its conclusion went through his customary display of dumb show to convey his appreciation of his men's cleverness. The Finale was played with the intensest heart-moving expression. This, at any rate, was a great performance.

Today it is common practice to replace the bassoon's four notes leading down to low D and marked *pppppp* with the bass clarinet (and presumably this was also Richter's method of securing a quiet dynamic). Bennett's knowledge of the problems encountered by trying to make reeds speak whilst playing low notes extremely quietly was obviously limited by a purist's rather than pragmatist's approach to the score. It would have been interesting to know where and how often he had heard a bassoon produce a satisfactory tone and dynamic at that place. He was, however, better pleased with Richter's *Tannhäuser* Overture and Mozart's 'Haffner' Symphony. 'The conductor revelled in it and so did we.'

Dr Richter's last concert . . . was an emphatic declaration of his greatness. He was in his grandest mood, and secured performances of such magnificence that the voice of criticism is silenced, and hero worship and eulogy pure and simple may have their way. Whether the presence of Madame Cosima Wagner incited that venerable lady's friend to a supreme effort we know not, though it seems only natural. . . . The *Tristan* prelude was a matchless performance. Dr Richter's crescendo leading to the *fff* was a veritable simoon of burning passion, his stupendous climax in the finale the very acme of overwhelming pathos and delirious ecstasy. Were these the hackneyed *Vorspiel und Liebestod*? They seemed fresher, fairer, and greater than ever. . . . Here was a programme of great masterpieces and such a one calls forth all those superb qualities by virtue of which he still remains the greatest living conductor.[29]

The autumn tour contained no new music but Ernst von Dohnányi made an impressive début in Beethoven's Fourth Piano Concerto on 24 October at the Queen's Hall. Excitement lay elsewhere with a revival of Richter's interest in the Hallé post during the summer of 1898. Frederic Cowen had realized for some time that he was just keeping the seat warm for Richter and, with a degree of justification, he strongly resented his treatment at the hands of the Hallé's management. The row spilled over into the public domain because, despite requests for confidentiality, Cowen broke the story to the press in September 1898. He knew that if Richter came to Manchester he might lose not only his joint post with Liverpool but also many engagements at other northern towns. It was certainly difficult for the management to maintain any moral high ground. On 19 July the Hallé Concerts Society was established after the three-year period in which Hallé's executors had enjoyed financial benefit from the proceeds of the concerts. Cowen had undeniably done a good job in improving orchestral standards, but he had always been on a yearly contract and his position was thus weakened. He decided to appeal instead directly to the public and press. Despite describing Richter yet again as 'the greatest living conductor', the *Musical Times* took the view that 'we do not want our orchestral conductorships to fall entirely into the hands of foreigners'.[30] Cowen at least salvaged his Liverpool base, for on 26 September they decided to have nothing further to do with the offer to Richter and appointed Cowen as conductor to their Philharmonic Society. Cowen persisted in his attempts to stay with the Hallé by writing to all the executors and even appealing to Richter himself by painting as black a picture as he could of what awaited him in Manchester if he accepted the offer (small choir and orchestra, endless touring, concerts on one rehearsal, and the retention of Hallé's name, which would always overshadow any incumbent of the post). At a meeting in London on 12 October with his old friend Adolph Brodsky (who led the orchestra for Hallé's last concert but relinquished the post in 1896 to concentrate on his tasks as Principal of the Royal Manchester College of Music) and Behrens, Richter was briefed on the legal position. In his diary he simply wrote of Cowen 'a liar'.[31] On 18 October the Hallé executive committee officially informed Cowen that his services would no longer be required after the spring of 1899.

Herbert Thompson, having already declared his support for Richter within days of Hallé's death, now weighed in with his latest opinion.

Who shall be conductor of the Hallé orchestra? This is the burning question of the moment. It is one that agitates many minds for it is of far more than local importance. Apart from the fact that the orchestra supplies the whole of the north of England with its orchestral music, it is practically the only first class permanent symphony band in the English provinces, and it is needless to insist once again on the supreme importance of cultivating local orchestras which has been so repeatedly urged in these columns. On the broad question of the advisability of enlisting Dr Richter in the ranks of English musicians there can be no possible doubt whatever. He has been a familiar figure in this country for full twenty years. As an orchestral conductor he is, taking all things into consideration, without an equal or at least a superior all the world over, and though his sympathies are notoriously not so strongly on the side of choral music, he certainly does not deserve the epithet that has been applied to him of a bad or even an indifferent conductor of oratorio. He has not, it is true, the English traditions at his finger ends, but if this be a shortcoming it is one that can be removed by further experience. At the same time it is an objection that has weighed against him in many minds with regard to the Birmingham Festivals, where choral music is of course the *raison d'être* of the event.

The whole question is unfortunately mixed up with personal considerations that cannot be ignored. No matter what the precise understanding between Mr Cowen and the managers may have been, the broad fact remains that he has conducted these concerts for two and a half seasons, has greatly increased the efficiency of the band which had become distinctly 'slack' in Sir Charles Hallé's old age, has given all kinds of music with all-round efficiency and has generally more than maintained the reputation of the orchestra. If he is to be regarded merely as a sort of warming pan to keep a place ready till Dr Richter is at liberty to step into it, he is much to be commiserated. Of course national prejudices are also being imported into the matter, but the plain duty of the Manchester people is to procure the best man, providing they can do so without acting unfairly. That Dr Richter has no desire to act unfairly to a brother musician would be the firm belief of all who know him, and the terms of his letter of January 26, 1896, show that he has always considered himself bound to let the Manchester Committee know if at any time he could free himself from his Viennese engagements. Mr Cowen too is an honourable man, more truly than was Brutus, and the present difficulty would seem to have arisen from an absence of a definite understanding at the time of his first engagement that it was not even in the most general sense to be regarded as a permanent one.

At the meeting of Guarantors it was decided by a majority of three to one and afterwards by a resolution that was carried *nem. con.* that Dr Richter should be invited to accept the conductorship. The merits of the question cannot however be decided quite so easily, and we must confess to shrinking from deciding definitely with either party. Mr Cowen is an excellent musician, an able conductor and 'greatly to his credit' an Englishman. His supersession, just after he has done so well for the Hallé orchestra, is an undeniable hardship. On the other hand the opportunity of securing the residents of the north of England for six months out of the twelve a conductor of world-wide reputation seems, as one of the speakers put it, 'like a dream' too good to be true. The most satisfactory solution of the difficulty is to provide room for both, but is the time ripe for this?[32]

By the time Richter got to Manchester for his concert on 21 October 1898 the controversy was boiling up to fever pitch, as Edith Hall confided to her diary.

Cowen got a very fair reception. There has been a great excitement lately as Richter has offered himself as conductor of the Hallé concerts. Of course we are dying to have him, but he hasn't signed yet and till he does we don't feel safe. Cowen is awfully annoyed at being given up.[33]

Cowen got a very good reception. The people clapped and cheered for quite a long time. However we intend to do better for Richter tomorrow.[34]

The Richter concert and it has been a splendid one. He got a splendid reception. The whole of the unreserved gallery rose and cheered, and a good many people in the other part of the gallery, and after almost everything there was cheering. . . . I wanted to give a cheer for Richter at the end, but we had to go as we had arranged to catch the ten o'clock train. It really has been a magnificent concert. I do hope we shall get him for our concerts. It will be too disappointing if he does not come after all.[35]

Richter had his own memories of that October night when he wrote, 'Manchester—unequalled! the greatest cheering at my reception and throughout the performance; the same at the end. I was near to crying through joy. I *must* come here.'[36] Go there he did, but during the months of 1899 there must have been moments when Behrens and his colleagues wondered if it was all worth while. Richter prevaricated on more than one occasion in the many letters and telegrams which flew back and forth. Matters were not helped by anonymous letters, some of them threatening, from Cowen's supporters in Manchester nor by his problems with Mahler in Vienna. On 11 May 1899, after lunch with Mrs Joshua, he met Behrens, Brodsky, and Alfred Rodewald (Behrens's opposite number in Liverpool), after which he entered in his diary 'wait for 14 March 1900', the date his full pension rights and official retirement from Vienna became effective. In fact the outcome was different and there was an overlap of his posts. His first concert with the Hallé took place on 19 October 1899, and on the date of his official retirement five months later he was conducting a concert at the Theatre Royal in Dublin.

Meanwhile his visit to England in the summer of 1899 was to prove significant for English music for years ahead. Five of his six programmes included works new to his concerts. Glinka's *Jota aragonesa*, also known as his First Spanish Overture or Capriccio brillante, was in the first on 15 May, and two weeks later he ended his second with Glazunov's Sixth Symphony. On 5 June he performed the Overture to Siegfried Wagner's opera *Bärenhäuter* and an Entr'acte and Air de Ballet from Tchaikovsky's opera *Voyevode*. The same composer's *Hamlet* Overture was sandwiched among Wagner and Beethoven on 12 June but a week later, on 19 June, Elgar's Enigma Variations were heard for the first time in a programme which also included Svendsen's legend for orchestra *Zorahayde* and Rimsky-Korsakov's orchestral suite from his opera *Snegourotschka*. His last concert on 26 June contained no new compositions. That same summer F. G. Edwards wrote a long biographical article on Richter for the July issue of the *Musical Times*.

As in the case of Kennedy and the Hallé, the relationship between Richter and Elgar has been well documented, not least by Kennedy himself but also by Jerrold Northrop Moore in his many books on the composer. Elgar had first heard Richter conduct in October 1881 and attended several other concerts over the years including the English premières of Brahms's Third, Bruckner's Seventh, and Parry's Fourth Symphonies, and Dvořák's *Symphonic Variations*. In February 1899 Elgar made contact with Richter's manager Vert, who promised to send the score of the Enigma Variations to the conductor in Vienna. A similar approach was also made by Elgar via A. J. Jaeger (of his publishers Novello), who asked Parry to recommend Elgar to Richter. The conductor first studied the score upon his return to Vienna from a concert in St Petersburg on 4 March. His reaction was instantly positive and he resolved to 'promote the work of an English artiste'.[37] There is no mention in his diary of an evening visit in pouring rain from Parry urging him to conduct the Variations, which Plunket Greene recounts in his biography of Stanford. Northrop Moore in *Letters of a Lifetime* (Oxford, 1990) quotes Dora Penny's confirmation that both Jaeger and Parry made the visit but Richter was abroad.

Elgar and Richter probably met initially at the first rehearsal of the work on 3 June 1899, though Richter mentioned the composer only two days before the concert when he wrote, 'Elgar a nice man.'[38] After the première he was glad that 'Elgar had a handsome and deserved success'.[39] Parry thought the new work 'first rate, quite brilliantly clever and genuine orchestral music'.[40] A solution to the Enigma puzzle has been proposed by Joseph Cooper. Bearing in mind Elgar's love of Mozart among several other arguments, he suggests the Andante from Mozart's 'Prague' Symphony, which appeared at the end of the programme at which Richter premièred the Enigma Variations. The similarity is striking, but it can only be a matter of speculation whether Richter programmed the two works deliberately or not; certainly there is no evidence to support the idea that Elgar asked Richter to do so. Before long the relationship between composer and conductor became a friendship which lasted until Richter's death, but in the spring of 1899 it would have been inconceivable for a comparatively inexperienced composer based in the provinces not only to ask an internationally renowned conductor to perform his new work but also to offer further programming suggestions. Alternatively Richter may have guessed Elgar's Enigma (he was certainly a fine enough musician to do so) and teased him by juxtaposing the two works. If he was party to the secret he never told another soul, not even members of his close family.

At the after-concert supper Alexander Mackenzie, seated opposite Richter, 'heard the enthusiastic terms of admiration, shared by us all, which the conductor addressed to the composer'.[41] Bennett, sceptical of the new works by the Russian composers featured in the first concerts, changed his tone when it came to a discussion of Elgar's composition, which was given 'a splendid performance under the greatest living conductor. Here is an English musician who has something to say and knows how to say it in his own individual and beautiful way.'[42] Elgar told Herbert Thompson that 'the work has had a glorious reception and Richter is going to play it "everywhere"'.[43] After the première Jaeger urged Elgar to reshape the work's conclusion and he agreed. 'All the critics, including Richter, said my work deserved a "big" ending so I've put it,' he told him.[44]

Richter played a huge role in promoting Elgar's fame when he gave the première of the Enigma Variations, and throughout the following decade his regard for the composer grew inestimably. Though they saw little of each other in the last few years of Richter's life when first distance and then war separated them, the conductor often told family and friends that he had devoted his musical life to the service of two great composers; one was Wagner, the other was Elgar.

24

1890–1899

Bayreuth

THERE were six Festivals at Bayreuth during the 1890s and Richter conducted at four of them, *Meistersinger* in 1892 and 1899, and the *Ring* in 1896 and 1897. Four performances of *Meistersinger* were scheduled for 1892 but he was barely able to rehearse as he could not free himself of his London commitments, which lasted until 4 July, nor from his operatic duties in Vienna, which took him back to the capital for the first eight days of August. There were also other complications. Over the winter 1891–2 Marie fell seriously ill. She made a very slow recovery and remained unwell for most of the first half of 1892. Richter wrote from London to Cosima in June 1892 asking for understanding.

According to the rehearsal schedule I have my first orchestral call on 13 July; allow me to hold piano rehearsals with the singers (mainly the Masters and the apprentices) not before the 11th or 12th. I must first go to Vienna to fetch my wife, who fell seriously ill on 24 October 1891, and take her to Carlsbad. Also my youngest daughter, Mathilde, went down with scarlet fever during my time in London, but though she is feeling better, I would like to convince myself in person of the situation; only then would I have peace of mind to work in Bayreuth. As my mother is now too old to run the house successfully, and as my poor wife only made her first attempts at walking a few days ago, the household had to be entrusted to my two oldest daughters, Richardis and Ludovika. . . . You have no idea what sort of winter I have spent, continually worried for the life of the mother of my children and coping at the same time with an amount of work such as I have never had before.[1]

Cosima was very sympathetic, though she was also having to cope with Richard Strauss's illness (a recurrence of pleurisy and bronchitis), which forced him to withdraw his services as a conductor. Instead she called upon

Karl Muck to prepare *Meistersinger* for Richter, with a promise of two performances. After hearing him rehearse the orchestra she changed her mind but implied to Muck that it was Richter who had raised objections to his engagement as a conductor. Muck suspected the truth, that any objections lay with Cosima, who was not prepared to risk someone who was a relatively unknown quantity to her, despite his protests that he was having to cope with a strange orchestra and singers as well as accustoming himself to the unique Bayreuth pit. His appeal was ignored, however, and he had to wait until 1901 before he made his début with *Parsifal*, after which he inherited Levi's mantle with that opera until his retirement in 1930. Meanwhile Julius Kniese was made responsible for preparing the cast as well as the chorus for the absent Richter. In mid-July Richter attempted to withdraw completely from the 1892 Festival because, when he got to Vienna, he found the health and mental state of his wife and children worse than he had suspected. Cosima begged him to conduct at least the third and fourth of the four performances to which he agreed on the understanding that she would ask Jahn in Vienna for his release. Between the two performances, which took place on 18 and 25 August with Carl Perron as Hans Sachs, he returned to Vienna to conduct *Lohengrin*. From their correspondence it is clear that Richter was in no way responsible for the withdrawal of Cosima's invitation to Muck; she played safe and gave the first two performances to Mottl. On his way home Richter wrote to Cosima from Nuremberg:

Once again they were wonderful days. *Parsifal* and *Tannhäuser*. In *Parsifal* it was a real pleasure for me to have seen the constant and faithful Levi at work in the second act; we won't find a better one than him anymore. And there's our dear, wonderful Mottl! . . . And now to Siegfried. Dear Frau Meisterin, make the right decision and send him to me in Vienna. He must learn his craft. The Master himself said as much in the *Meistersinger*: 'die Meisterregeln lernt bei Zeiten' ['take time to learn the master rules']. . . . Send Siegfried to me in Vienna. When he was born, I wrote to Tribschen that I would be as true to him as 'Kurwenal'; I want to keep that promise. Even if he becomes no 'Tristan', he has a good deal of healthy, tough Mastersinger blood in him.[2]

Cosima was touched by Richter's offer and genuinely pleased with his performance of *Meistersinger*, despite du Moulin Eckhart's suggestions that she was basically anti-Richter and only asked him to revive the *Ring* in 1896 because she saw no way around him. There was always distance between them, from their first meeting at Tribschen in October 1866. A quarter of a century later, however, Richter was one of a decreasing number who had been musically close to her late husband, and this link became more precious to her over the next twenty years. She may not have approved of everything he did, she resented his commitments to other musical centres (Vienna and London) in order to earn a living, and she tried to embroil him (without success) in the power-play and internal politics of Bayreuth, but sentiments such as those expressed in her reply to his offer to tutor Siegfried as a conductor were genuinely expressed and heartfelt.

See, dear Hans, despite all life's distractions we meet up at *Meistersinger* performances, and find that our bonds are as strong as ever before, Tribschen lives on similarly in us both; you possess the spirit which enables you to produce *our Meistersinger* under the most abnormal circumstances. . . . You remind me of Siegfried's birth and your dear promise. I have never forgotten that you were with us at that time, and that you are the godfather of both my Tribschen children. As you came to our house and made your mark, so shall Siegfried come to yours and learn what he can from only you. Thus the bond which binds us, and which can never be broken, will be shown to all the world, and together we shall show what faith means.[3]

By this time Siegfried had abandoned all thoughts of architecture as a career and was becoming more involved with composition, supporting his mother in running the Bayreuth Festival, and developing his abilities as a conductor after assisting with Bayreuth's first *Tannhäuser* in 1891. He studied score-reading with Humperdinck and came under Richter's tutelage in the New Year of 1893 when he visited the conductor in Vienna. Siegfried, who held the baton in his left hand, made his London début in November 1894 with works by Wagner and Liszt in the first of Mottl's series of concerts. In February 1895 Richter urged Cosima to send Siegfried to Vienna to make his début there rather than accept an invitation from the Berlin Philharmonic, 'which I know from my own experience; they are fine, industrious and one can achieve a good deal with them, but Siegfried must conduct the Vienna Philharmonic at his début'.[4] Richter went on to suggest that the annual Nicolai concert for the widows' and orphans' fund would be ideal.

Meanwhile Siegfried had another London concert, this time on 5 June 1895. It included his own composition *Sehnsucht* (Longing), which received a negative critical response. Richter met his pupil at Pagani's restaurant (his favourite London haunt) for lunch on the day before the concert. Having attended the morning rehearsal that day, he recorded in his diary that he was 'still very immature'. Of the concert itself he wrote that 'he conducted the second and third movements of [Beethoven's] Eighth

Symphony quite well, a few stupidities at the end of the first and fourth movements. His [own] composition... was rubbish'[5] (he told Siegfried that it was too much like Liszt). Two days after the concert they had a heart-to-heart conversation and Richter told his pupil that though he had talent as a conductor, he should 'avoid à la Bülow stupidities'.[6] The *Musical Times* described his performance of the Overture to *Der Freischütz* as 'the worst... it has ever been our misfortune to hear'.[7] The timing of Siegfried's London concert was unwise, for it proved a tough task for him to be conducting there at the same time as concerts given by his Bayreuth models, Richter, Levi, and Mottl. Cosima believed whole-heartedly in her son from the start. 'To me', she had told Richter in the New Year of 1894, 'he exemplifies words you may remember from Tribschen: one does not *become* a conductor. Having become a musician, one *is* a conductor either immediately or never. I would like it so much if Siegfried could appear in Vienna under your patronage. Do you know of such an opportunity?'[8] For that Siegfried had to wait a further two years before his mentor considered him ready to lead the Philharmonic orchestra at the Nicolai concert on 19 January 1896. His programme included the same Beethoven symphony which Richter had criticized, together with works by his father and grandfather.

Siegfried, in his short memoirs, had some kind words for his teacher and friend as well as some apocryphal tales. Some of the following comments are hard to believe—those about Brahms's scoring could have been motivated by Wagner–Brahms politics, but his knowledge of Richter's attitude to money was based only on what he knew of him at Bayreuth, not elsewhere, which was very different.

In our orchestral master Hans Richter I found a true patron. From my earliest youth he was the dearest of playmates and cared for the children of Wahnfried to the end of his days in unspoiled and trusted friendship. When Richter was announced, ten eyes shone with joy. I learned many good tips from him in conducting technique. When he saw me conduct for the first time in London he told me, 'I am pleased that you have the score in your head and not your head in the score.' Here are some typical remarks which I gladly recount. When rehearsing an ultra-modern composition on one occasion, he said, 'Now we must take care that we play the correct wrong notes.' The orchestra belonged to him above all else, he could only stand singers if they were musical. When one of the most unmusical was making mistakes on stage, he was heard at his podium saying, 'When the curtain goes up all my pleasure is over.' His concert activities in London were dearer to him than conducting operas in Vienna. In order to get as much leave as possible for England he undertook everything there was to conduct in Vienna so that his colleague Fuchs did the same too.

At the beginning of *Tristan* he called out to the cellos, 'Gentlemen, forget that you are married men!' He once got very angry with an unmusical singer at Bayreuth. When I apologized on her behalf, saying that she was a splendid actress, he retorted, 'What? I couldn't care less about her acting!' He observed of my *Bärenhäuter* score, 'Do you know, when I look through a new score I always glance at the bassoon part. If it is always the same as the cello part, then it's nothing. Yours is good.' Brahms's scoring he found unbearable, 'the sun never shines with him'. My mother was always urging him to conduct *Parsifal*, but he never took the bait and always replied, 'I am saving two things until the end, the National Gallery in London and *Parsifal*.' He said of my father's theoretical writings, 'What's in them is wonderful, but, for me, another score would have been preferable.'

Of all my father's works it was *Meistersinger* which lay closest to his heart and he liked conducting it the most. That is understandable, for he was predestined for it. In money matters he was governed by a sweet modesty and nobility. Not for the world would he ever have had his costs reimbursed. 'That *would* be fine,' he said. 'I have the Master to thank for everything, and for that I should be paid!' After he had reared his family on the money he earned in England (which he could never have done on his miserable salary in Vienna), he moved to Bayreuth to end his days, and shunned the outside world. A last great pleasure for him, shortly before his death, was the announcement of a new grandchild of the Master, but unfortunately he did not live to see the birth of my son Wieland.

With Hans Richter a type of German musician has disappeared and appears to be almost extinct, the orchestral master who originates from the orchestra itself. He was no conductor in the way of Bülow. He always remained a musician. His way of conducting was splendid, powerful yet flexible; and he had shining blue eyes and a gold-blond full beard which suited him so well.[9]

The operas at the 1894 Bayreuth Festival were *Parsifal*, *Lohengrin*, and *Tannhäuser*, conducted by Levi, Mottl, and Strauss, though Richter was engaged to share some performances of *Parsifal* with Levi. A letter written in the New Year of 1894 reveals Richter's joy at the prospect of conducting the opera, hopefully the first performances so that he would have a rest before beginning the new opera season in Vienna during August. The same letter brought news of the engagement of Richardis, the Richters' oldest daughter and godchild to the Wagners, to Marie's younger brother Béla von Szitányi. This marriage to her uncle was achieved only by some fiscal arrangement with the Church, but having been settled it took place in Vienna's Karlskirche with members of the Philharmonic orchestra providing the music. The couple then moved to Béla's property at Baracs in Hungary, which Richter and his wife visited annually until the end of his life. In the months after writing to Cosima, Richter enjoyed poor health and reluctantly withdrew from *Parsifal* on 3 June 1894. His doctor had

FIG. 10 Richter at Bayreuth in 1889. The notice in his hat reads: 'Please do not ask me for dress rehearsal tickets as I do not have any.'

forbidden him to undertake any work involving long periods of rehearsal or stress and forced him to reduce his London concerts from nine to four, with no new works on the programmes. As far as Bayreuth was concerned, he could only look forward to the revival after twenty years of the *Ring*, scheduled for 1896.

Richter's participation in the preparations for the *Ring* was nothing compared to his input in 1876. Whereas he had scoured Europe for players and singers two decades earlier, now he was able either to write recommendations or to direct others to hear artists he thought suited to conditions at Bayreuth. A year before the *Ring* revival he wrote to Adolf von Gross (effectively General Manager at Bayreuth) with a list of names and addresses of players (many of them of German origin from his London orchestra and elsewhere) who might be engaged in Bayreuth in 1896.

Violin I: Johannes Nalbandian. First class performer who does not play in any orchestras but would like to take part at Bayreuth. An Armenian who speaks some German. V. V. Akeroyd. Excellent [Liverpool].

Violin II: Edward Lardner, plays first violin for us, is reliable in the [Wagner] works, has a good instrument and could be engaged as a first violinist. Meyer van Praag, also plays first for me, talented man with a good instrument. L. Bauer, excellent, dependable second violinist, graduate of Vienna Conservatoire. Frank Stewart, very good experienced second violinist, who also often plays first, would be more than up to a second desk position in Bayreuth.

Viola: Henry Lewis. First rate with an excellent Italian viola. Knows it all. If the first desk is not fixed I recommend [him] F. A. Baker, a quite superb orchestral player, splendid technique, very good, *large* instrument; perhaps together with Lewis at same desk. Franz Hackenberger and W. Laubach, both good orchestral players; if there's a shortage of viola players, they are both recommended.

Violoncello: Carl Fuchs, a chamber music and orchestral player, a fine musician, beautiful Italian instrument, is another recommended for a front desk position. H. Trust, excellent orchestral player, *good* technique, good instrument, very reliable. Hans Bronsil, a very good musician with a lovely instrument. Walter Hatton, excellent cellist with a very good instrument. Leo Taussig, very talented principal and rank-and-file player, excellent technique, good instrument. Here are some names of other cellists, all excellent people rarely found among cellists in German orchestras, Mr Bucknall, Alfred Gallrein, Adolf Schmid.

Double basses: C. Winterbottom and C. F. Maney. These two are *quite superb* with excellent huge instruments; it would be difficult to find better.

Harp: Otto Mosshammer, graduate of Vienna Conservatoire with a good tone and good technique.

Flute: A. P. Vivian, first class artist but he must receive at least the same fee as Mühlfeld [Brahms's favourite clarinettist], because he would have to get a lower pitched instrument. Vivian knows the works exactly and really plays incomparably well. Siegfried must remember him.

Oboe: Désiré Lalande, a wonderful young artist, graduate of the Paris Conservatoire. A full, poetic tone, faultless technique; plays cor anglais just as well, I almost prefer it to his oboe playing, but excellent overall. Should also be engaged at Mühlfeld's rate.

Horn: Adolf Borsdorf, my excellent first horn in London. I would like him on third or fifth horn, as we have the incomparable Wipperich from Vienna in mind for playing first. Borsdorf is safe, has a good tone, and is very familiar with the works.

Bassoon: Edwin F. James, very fine, beautiful sound, good technique, recommended as second or third bassoon. Mr Conrad, excellent player, especially good on contra-bassoon, has an excellent instrument. Recommended as contra doubling third bassoon.

Contra Bass Trombone: Albert E. Matt, plays the new, improved contrabass trombone excellently; I heard him before my departure from London and was surprised by the wonderful instrument as well as its player who has been entrusted with all the [Wagner] works for years in my orchestra.[10]

Richter concluded by urging these young, idealistically minded players on Gross, in particular the cellists who were apparently so much better as players and possessed better instruments than their German counterparts. They were all agreeable to the terms of free board and travel and a subsistence of 500 marks. Of those he recommended only the violinist Akeroyd, cellist Fuchs, harpist Mosshammer, and oboist Lalande were invited to Bayreuth that year. There were no other operas staged in 1896, the five cycles of the *Ring* from 19 July to 19 August (each opera on consecutive days) being divided between Richter, Mottl, and Siegfried Wagner. Richter offered suggestions regarding singers and was asked detailed questions by Cosima. By the spring of 1896 the question of the sound of the anvils in *Rheingold* was being addressed. Whilst Franz Fischer (who was responsible for organizing and conducting them offstage in 1876) was being consulted, Richter's opinion and memories of the first performances were also valued. Experiments with nine short sections of railway line, six genuine anvils, and three damped gongs were not proving a success. Did he really want eight harps? What about the fanfares to call the audience before each act of the operas, what were they in 1876 and would he be responsible for them again now? A shock, though not unexpected as it turned out, awaited Cosima when, as late as *June* 1896, Richter asked to be allowed to conduct only the first two cycles. Although she reluctantly agreed to his request, he actually conducted the first and last cycles, Siegfried directed the second and fourth, and Mottl was at the helm for the third.

Shaw visited Bayreuth and attended Richter's opening cycle. Having damned Cosima's production and accused the singers of over-stylized acting, he turned his attention to the music and had some very telling criticisms to make about the conductor.

The performance was, on the whole, an excellent one. Its weakest point was Perron's Wotan, a futile impersonation. The orchestra, although it was too good, as orchestras go, to be complained of, was very far from being up to the superlative standard of perfect preparedness, smoothness, and accuracy of execution expected at Bayreuth. We in London have taught Richter to depend too much on his reputation, and on his power of pulling a performance through on the inspiration of the moment. The result of our instruction is now apparent. The effect of the *Das Rheingold* score was not the Bayreuth effect but the London effect: that is it sounded like a clever reading of the band parts at sight by very smart players, instead of an utterance by a corps of devotees, saturated with the spirit of the work, and in complete possession of its details. The strings were poor; the effects were not always well calculated—for instance the theme of the magic helmet was hardly heard at first, and in the prelude the great booming pedal note—the mighty ground tone of the Rhine—was surreptitiously helped out, certainly with excellent effect, by the organ.

The difference [in *Walküre*] was very noticeable in the orchestra. Richter was in his best form, interpreting the score convincingly, and getting some fine work from the band. In the scene of the apparition of the Valkyrie in the second act, the effect of the wind instruments was quite magically beautiful. The deep impression made ... was due very largely to the force with which Richter, through his handling of the orchestra, imposed Wagner's conception on the audience.[11]

Shaw dealt firmly and harshly with most of the singers, among whom Heinrich Vogl as Loge was the only survivor in his original part from the 1876 première and 'played it again with a vocal charm which surpassed the most sanguine expectations'. Marie Brema, the only English singer amongst the cast, was a 'very fine Fricka'. Carl Perron sang Wotan rather indifferently but was at his best as the Wanderer in *Siegfried*, Lilli Lehmann, a Rhinemaiden and Helmwiege in 1876, excelled as Brünnhilde although Richter and the prompter had to guide her through *Walküre* because she was unwell. Shaw considered her voice uneven, brilliant at the top but weak in its middle register. Her sister Marie was another survivor amongst the singers from 1876 (when she also sang a Rhinemaiden and Ortlinde) who now sang the third Norn. Wilhelm Grüning as Siegfried was a disappointment but Alois Burgstaller, who took over the role for *Götterdämmerung*, proved, among the male singers, to be the find of the season. Ernestine Schumann-Heink was cast in three roles, Waltraute, Erda, and the first Norn, and made a deep impression in all three. Rosa Sucher, in the twilight of her career, sang Sieglinde. The *Musical Times* considered that 'to hear Herr Richter conduct the trilogy is to receive a pleasure that can never be surpassed or forgotten'.[12]

Lilli Lehmann had little sympathy for Cosima the despotic stage director, nor for her son's talents as a conductor. 'Hans Richter, the embodiment of unselfishness,' she wrote of the latter, 'did for the young son of his old master what no one else indeed would have done. He had already held forty-six orchestral rehearsals that Siegfried had diligently attended, so as to learn, by listening to Richter, how it should be done. Then Richter conducted the first stage rehearsal, Siegfried the second, and Mottl the final one.... The Festival was concluded with the fifth cycle in which I sang again [Ellen Gulbranson had made her début meanwhile], and our dear Hans Richter conducted magnificently.'[13] One of Richter's recommended musicians, the cellist Carl Fuchs, remembered that 'the excellent orchestra steered [Siegfried Wagner] safely past many a rock. He held the baton in his left hand and his powers were but moderate.'[14]

At the opening of the fifth cycle on 16 August, sitting in seat No. 1006 with Siegfried Wagner just behind him, was a young man of 20 named Sydney Loeb. An impressionable youth devoted to Wagner's works, Loeb was taken by the scenery, the magnificent sound of the orchestra, and the singing of Vogl, Brema, and Schumann-Heink. He remained for the whole cycle, marvelling at the performances by Sucher and Lehmann (the latter had in his opinion 'a more powerful voice than Albani's, but Albani's attack is better'[15]), collecting autographs, and spotting artists such as David Bispham, Jean de Reske, Cosima, Siegfried, and Richter on the terrace during the intervals ('I was disappointed', he thought, that de Reske looked 'a bad-tempered, commonplace little man, with very well developed shoulders. I expected a good-tempered little man, with fair moustache, not dark').[16] Loeb got nowhere near Richter during his visit to Bayreuth, but he had met him in London three months earlier after a Richter Concert. 'The preludes [*Meistersinger* and *Parsifal*] and everything, *he conducted marvellously*, but although the orchestra is hardly recognisable as the Philharmonic (to which part of its members belong) yet it has not the unity and shade of Lamoureux's, although for the preludes I much prefer Richter's reading. He conducts with his right hand and does not get excited, and only occasionally uses his left. . . . After the concert went behind, *shook hands with Richter* and *asked him to write his name*, to which *he answered "Ja, ja!"* and *did so*. As I took out my pen my hand shook very much.'[17] Sixteen years later (in March 1912) Sydney Loeb, born whilst Richter was busy preparing the first *Ring* in 1876, became the conductor's son-in-law when he married his youngest daughter Mathilde.

With his return to Bayreuth in 1896 Richter renewed an association which would bring him to every Festival until 1908 inclusive and, apart from the year 1899 when he conducted *Meistersinger*, he only conducted the *Ring*. In 1897 he took charge of just one cycle (Siegfried Wagner then took over) but it was with an impressive cast. Gulbranson replaced Lehmann, van Rooy replaced Perron, but Schumann-Heink, Brema, Sucher, Grüning (Siegfried), Friedrichs (Alberich), and Breuer (Mime) repeated their successes of the year before. After singing Loge in *Rheingold*, Vogl went on to triumph as Siegmund in *Walküre*. Once again Richter was fleeting in his attendance, arriving at Bayreuth from Vienna where he was involved in Court Chapel appointments. The strenuous winter schedules, his increasing unease with Mahler's activities at the Opera, and the growing press campaign against him in the capital seemed to take their toll by each spring

FIG. 11 Hans Richter's autograph at the 1896 Bayreuth Festival, the year in which the *Ring* was staged for the first time since its première in 1876. Above his signature are the opening bars of *Das Rheingold*, below the closing bars of *Götterdämmerung*.

and early summer, so that he had to replenish his energies with a June holiday in Lower Austria every year. 'Thank God there's no piano in the hills,' he told Cosima, 'so I can enjoy an undisturbed rest, though I am not entirely without music; the most wonderful woodbirds delight us with their magical song, trilling away indefatigably until late in the evening.... Here in the woods I am more aware than when I am at my conductor's desk of how wonderful the motion and yet such holy stillness of the *Waldweben* [Forest Murmurs] in *Siegfried* is portrayed; a miracle!'[18]

Their correspondence in the spring of 1898 contains discussion of several male singers, the de Reske brothers, the Danish tenor Erik Schmedes (who began his operatic association with Vienna in that year), the bass-baritones Leopold Demuth, Anton Sistermans, and Anton van Rooy; apart from the de Reskes, the rest had roles in 1899. Both Cosima and Richter were shocked by the sudden death on 28 March 1898 of Anton Seidl, not yet 50 years of age. It was Richter who had recommended his younger Hungarian colleague to Wagner, who thought highly of Seidl's sense of theatre. The composer in turn recommended him to Angelo Neumann and his touring Wagner company, and Seidl took the composer's works to England and

America as well as throughout Germany and elsewhere. Despite being a founder member of the Nibelung Chancellery, Seidl only conducted once at Bayreuth, in *Parsifal* in 1897, the year before his death. 'After you', Cosima told Richter, 'he was the first and last friend of the children in our house, and even though we communicated little, he was faithful to us. His interpretation of *Parsifal* was solemn and fervent. . . . even as I write of him I weep and must break off.'[19]

For 1899 Cosima gave Richter the choice of two cycles of the *Ring*, seven performances of *Parsifal*, or five of *Meistersinger*, though as late as May or June 1899 Richter was still down to conduct the first *Ring* cycle, asking for all the *Meistersinger* performances, and expressing pleasure that Franz Fischer was being recalled to conduct *Parsifal* in place of Seidl, because he had shared the work with Levi at its first staging in 1882. Once again it is remarkable how, just two months before the opening of the Festival, no firm plans had been made regarding the allocation of the operas to the various conductors (eventually Richter conducted *Meistersinger*, Fischer *Parsifal*, and Siegfried the *Ring*). Demuth and van Rooy shared the role of Sachs, Fritz Friedrichs repeated his 1888–9 triumphs as Beckmesser, Johanna Gadski and Ernestine Schumann-Heink were Eva and Magdalena respectively, Sistermans sang Pogner, and Ernst Kraus sang Walther. *Meistersinger* was the highlight of the Festival, with neither Siegfried Wagner nor Franz Fischer able to compete with Richter's compelling authority and vast experience as a conductor. It 'fully sustained the traditions of Bayreuth; the perfection of the ensemble and the wonderful clearness, with which even the most complicated choruses were given, being quite remarkable'.[20]

25

1894–1899

Richter's Diary

BEFORE entering the final years of Richter's life, which, from the turn of the century, were devoted primarily to his activities in England and at Bayreuth, a glance through the surviving diary extracts, which his son Edgar later selected for his sister Mathilde in England, makes fascinating reading. Unfortunately Edgar stopped at June 1900 presumably because Mathilde knew all about the final decade in Manchester because she lived, together with her sister Ludovika, with her parents in Bowdon. Her diaries contain little more than a perfunctory account of her father's music-making in the north of England and elsewhere. As always, the faithful Mrs Joshua would have a basket of flowers or fruit awaiting Richter upon his arrival in London when he entered No. 11 Bentinck Street. Soon the conductor was doing the social rounds, meeting his friends and entertaining his soloists or orchestral players at either the Café Royal, the German Athenaeum Club, or his favourite restaurant, Pagani's. He heard singers and players recommended to him either by his manager Vert or by other musicians and teachers. He went to rehearsals of other musicians to pay his respects (on 7 June 1894 'went to the Philharmonic rehearsal to speak with Saint-Saëns and heard his symphony [No. 3]—organ and piano duet—"much ado about nothing"'). He had endless meetings with administrators from the Birmingham Festival or (after Hallé's death) from the Hallé orchestra.

On 11 June 1894 he recorded his nervousness before a concert because he was conducting Schubert's 'Great' C major Symphony from memory, but it went well and the next day he could relax with a visit to his old friend Hubert Herkomer at his new house Lululaund, nearly completed at Bushey. At the end of the day he wrote (and despite his earlier advice that

Herkomer should leave composition alone), 'spoke into the phonograph: "Dear friend! What you have painted and what you have built are quite wonderful, but if one has such beautiful musical ideas, one must not be lazy; when you have finished another score, I will deem it an honour to conduct it."' Unfortunately there is no trace of this short speech (made in German) having survived the years. It would have been the only audible record of the great man, as he never conducted a note of music for the growing gramophone industry.

He often visited painters, such as his friend Burne-Jones, or the theatre. Of Eleonore Duse in Dumas's *La Dame aux camélias* at Daly's Theatre on 14 June 1894, he wrote: 'A superb artist. Sarah Bernhardt was applauded as she made her entrance and sat next to us in the box. Stage managed?' There are entries showing the care with which he managed his finances (he banked £150 on 6 June 1894) or his domestic tastes ('after lunch went to Mappin and Webb's to buy the fish knives'—this after a full morning rehearsal on 18 June and with a concert that evening at which the young Josef Hofmann played Rubinstein's D minor Piano Concerto). At a party given by Alma Tadema the next evening, he played billiards before accompanying Eugène Oudin, William Shakespeare, and George Henschel in songs for the guests. In the summer of 1894 he was visited by the conductor Walter Damrosch from America, who came with a request that he should conduct a season of opera in New York the following New Year, but once again Richter refused to visit the United States, even though he was still discussing it with Vert later that autumn. His final days in London in June 1894 were spent visiting his programme note compiler Charles Ainslie Barry for lunch at Sydenham Hill, accompanying the Joshua family to Richmond (where he watched his first game of golf), and, after supper at the Café Royal, visiting the Empire Theatre for a programme of 'decorated parrots, dogs, a mandolin player, and *tableaux vivants*'. Among these June entries, four months after von Bülow's death, is the curious comment, 'Bülow said that I conduct better from an unmutilated score.' Henschel played through his *Stabat mater* for Richter ('very clever mediocrity', he wrote in English about the work to be performed at that year's Birmingham Festival) after which the conductor paid a visit to his old friend Wilhelmj (leader of Bayreuth's 1876 orchestra and the 1877 London Wagner Festival) before leaving for Europe.

Later that year he returned for the Birmingham Festival and a country tour. Apparently obsessed with fish cutlery, he purchased another service, this time in Birmingham after a lunch at the Clef Club. He was naturally fêted by the dignitaries of the city such as the Lord Mayor and Lord

Dartmouth. After a brief few days in London (7 October: 'Today tiredness really overcame me. It really was *a lot* of work to undertake') he set off on tour on 9 October 1894.

Orchestral train from St Pancras at 10.30. Delayed in Wakefield (the place where the novel was set) and in Mirfield. Lovely hall in Huddersfield where we arrived at last after a delay of more than an hour. At first the playing was very average, only the *Ride* and No. 5 went well. St George's Hotel good.

10 October: the reviews were excellent. They didn't hear what was bad. Others come here with fewer strings, we had very many. The sound covered any weaknesses. Arrived 4.30 in Sheffield. Very good concert in the evening, they played splendidly. Great, enthusiastic reception. Victoria Hotel.

11 October: Left early at 9. Once away from the smoky area we had wonderful weather and a lovely journey via Carlisle, arriving at 4 in Edinburgh. Windsor Hotel. The concert less well attended, but the applause was nevertheless warm.

12 October: In the morning by car around the city, castle, and Holyrood with Deichmann and Schiever [principal first and second violinists of Richter's orchestra]. Travelled to Glasgow after lunch. Splendid concert. St Andrew's Hall was full and the most enthusiastic to date. Very cultured public. Windsor Hotel.

13 October: Early 6.40 departure for Liverpool. Excellent concert at the Philharmonic. Large numbers in the evening, the people very nice. Billiards. Talked with Rodewald about a Liverpool Music Festival. Received a white stick, made from the spine of a shark, from Fletcher.

14 October: Rested very well after the strains of the previous week. In the morning heard [Emil] Hedmondt—who sang a performance of David in Bayreuth in '88 and wants to join a German opera house. Travelled to London at 11.20 with Deichmann and Schiever. Fletcher was so nice and gave us a luncheon hamper. Arrived in London at 5. Large gathering at the Makowers in the evening.

15 October: 9.30 rehearsal—quite tiring, a whole programme with difficult works to rehearse in three hours. [Wagner's *Fliegende Holländer* Overture, Schubert's Unfinished Symphony, Weber-Berlioz *Invitation to the Dance*, Smetana's *Bartered Bride* Overture, Grieg's *Peer Gynt* Suite No. 1, and Beethoven's Fourth Symphony.] Afterwards heard Miss Pool at Vert's, alto in the ninth Symphony. A superb concert in the evening. The musicians achieved excellent results, hardly anyone else would get such results. Well done artistes!

16 October: 10 o'clock to Manchester, my wife stayed behind with the Joshuas [at 57 Cadogan Square], in the afternoon visited Hallé's music school [later Royal Manchester, now Royal Northern, College of Music]. In the evening it appeared that the Free Trade Hall was empty, but it soon filled up with a clap-happy public. Met my old friend from [Vienna] Conservatoire days Risegari. Queens Hotel.

17 October: Midday to Newcastle, 4 o'clock arrival, splendid Station Hotel.

Annual Police Concert at the Olympia Hall, wonderful public, but a less good acoustic and the dreadful noise of an electric machine which intruded.

18 October: Walked alone over both bridges. Arrived in Leeds at one o'clock, received a tobacco pouch from a Mr Hudson at the station. Queens Hotel. 3½ hour rehearsal of [Mackenzie's] *Britannia* Overture. Difficulties with the stands. In the evening a splendid concert, very good performance, good sound, enthusiastic public. Afterwards the old music dealer Archibald Ramsden told us funny stories.

19 October: 10.45 departure. 3.20 arrival in London. Four-hour rehearsal of the *Forging Song* [from *Siegfried*] and the ninth with soloists. Evening party at the Studenmunds.

20 October: 9.30 rehearsal [this was Richter's first experience of the Queen's Hall, which had opened in November 1893]. The fine oboist [W. M.] Malsch came despite Manns and gave up the Crystal Palace. 3 o'clock concert in the Queen's Hall. Apparently it sounded very good from the auditorium. It was a wonderful performance. Afterwards invited by Mackenzie to the Garrick Club. Beautiful paintings. Mackenzie promised to intercede on Malsch's behalf. Oudin had a stroke.

21 October: A restful day. In the afternoon to Bennett, who especially praised the orchestra's performance at this concert. In the evening to the Café Royal.

22 October: Withdrew £300 from my bank. Vert paid me £600. Oudin not yet recovered. Lunch with Mrs Joshua, Agnew, Duckworth, Semons, Frantzen. A good concert in Brighton, boring journey home, Makower accompanied us.

23 October: 10 o'clock from Charing Cross. Frantzen, Vert, Westrop, and Makower were at the station. Bad Channel crossing.

Such was the hectic pace of Richter's touring in Britain. He expected no easy life when he left Vienna six years later, and was fully prepared for the tiring schedule and busy social life which invariably accompanied such tours. He rarely had a good crossing and must have dreaded that part of his trips to Britain. On 16 May 1895, when his next visit began, he wrote, 'In the evening 8.30, miserable crossing. Oh was I sick! The ship was called the *Prince Albert*, certainly the worst crossing I have ever had.' Mrs Joshua's flowers awaited him as usual, though this time he stayed at 31 Upper Gloucester Place and his landlady was a Mrs Potter. Once again an approach was made from Higginson, this time in person from Boston because he was not happy with Emil Paur, the man hired to succeed Nikisch in 1893 when Richter turned the offer down. At his first concert of the season (20 May 1895) he met the Duke and Duchess of Coburg '(formerly Edinburgh). Both of them, particular the duchess, said how much they looked forward to hearing good music again. In the interval I smoked a cigarette with the duke. Brahms's Haydn Variations were too long for him.' The next day Richter paid a visit to the Alhambra where 'a most delightful chap sang the *Last Rose of Summer* from the piano with a dressed-up black pig, two roosters, a donkey, and a poodle'. The 'Child of Tribschen' did not seek Cosima's approval before celebrating his Master's birthday (22 May) at such a doubtful place of entertainment. Neither would she have favoured his attendance at *Il trovatore* the next night to hear Francesco Tamagno sing. The Italian tenor had sent his card the day before, and Richter wrote, 'a huge voice, temperamental, but likes singing flat. [Margaret] Macintyre good, otherwise the performance was really miserable. A single-minded young Italian conductor [Armando Seppilli] who would not give way and allowed the orchestra to play *forte* throughout.'

In the summer of 1895 Richter paid a visit to the Guildhall School of Music at the invitation of its Principal, Sir Joseph Barnby. There he heard 'a little female pianist, a little violinist, a soprano, an alto, a tenor (good!) and a baritone'. Richter's friends Wilhelmj and Wilhelm Ganz were professors of violin and singing respectively at the school. He then spent a restful few days at Felixstowe, where Marie Joshua had a country home (though he stayed at the Bath Hotel, a 'really nice and, thank God, still rural hotel'). The agenda included brisk walks before breakfast, games of golf, time spent writing letters to Cosima and to Nikisch, promenades along the beach, watching a regatta in which the Prince of Wales took part, and many long talks about 'the English, their weaknesses, and the Jews' (despite his hostess's origins).

Mrs Joshua took Richter to the Houses of Parliament on 11 June 1895 at the invitation of the Speaker and his wife. On the following day he recorded that he heard the Sutro sisters (American piano duettists and pupils of Heinrich Barth) playing Schumann and Bach on two pianos. His only comment was that they were plain. The day was enlivened, however, by his enrolment as member No. 182 of the 'You-Be Quiet Club', whose founder Archibald Ramsden (a piano merchant in Bond Street) had regaled Richter with funny stories in Leeds the previous autumn. For five shillings 'you have the right at anytime that you are passing up Bond Street with a friend to enter Mr Ramsden's premises and demand two whiskies with soda'.[1] Its members included Mackenzie, Elgar, Vert, Bennett, and the singers Edward Lloyd and Ben Davies among the musical fraternity. In short it was a social club 'of downright good fellows', with two bottles, a cudgel, and a box of cigars on its heraldic shield. Small wonder that Richter noted, after an evening with Ramsden at a club dinner on 18 October 1895, that 'he has fine paintings, but smells strongly of whisky'.

His enrolment on 12 June that year was followed by a visit in the evening to the Indian Exhibition where he heard a concert of Burmese music and dance, including 'a splendid finger-drummer on tuned drums, graceful and comic dances, and strange wrestling to agitated music'. In between 'oh, these parties!' he went to hear Nikisch for the first time in the afternoon of 15 June 1895 after a morning rehearsal and lunch at the Café Royal. 'A lot was excellent, the C minor [Beethoven's Fifth] Symphony not broad enough, the last movement sloppy, nevertheless a great conducting talent. A *very* nice evening at the Garrick Club with the Beale brothers, Macrory, Holzmann (secretary to the Prince of Wales), Chitty, Schiever, Deichmann. The English really are always the nicest.' Poor Richter nearly embarrassed himself on the return crossing of the Channel that summer when he found himself in the company of the Duke of Coburg and the Grand Duke of Hessen-Darmstadt, both of whom had been at his last concert in London the previous evening. He was compelled to withdraw midway through the luncheon to which he had been invited for fear of disgracing himself by his inevitable seasickness. Later that year on his return trip to Vienna he overheard two Englishmen talking over lunch in the dining car. One was travelling to Vienna 'to hear the splendid orchestra. He mentioned my name and added, "No doubt he is the first [the best]." At least someone is looking forward to it.' This dread of leaving England and returning to Vienna becomes more apparent a day later (7 November 1895), when he wrote after a piano rehearsal of *Das Rothkäppchen*, 'I'm not in Vienna yet, my heart and soul is still in dear England.'

The early summer of 1896 took him to England for a short season before Bayreuth's *Ring*. In between concerts he visited the horse show in Regent's Park, a circus at Kensington Olympia, and travelled to the lighthouse at Beachy Head for the day. In London, after a morning at the National Gallery with Marie Joshua, he paid an afternoon visit to the Royal College of Music (5 June 1896). 'Heard a symphony by a nigger—Coleridge. Afterwards I conducted the *Tristan* excerpt.' Parry also recorded the visit:

Richter came in the afternoon to hear the orchestra play [Coleridge] Taylor's Symphony, and was persuaded to take the College band through the Vorspiel and Finale of *Tristan*, which they played to perfection. He seemed very pleased with them, especially so with the harpist.[2]

When he left his 1896 lodgings, 28 Montagu Place off Russell Square, for Bayreuth, Richter placed in the care of his landlady, Mrs Henke, six white shirts, twenty-seven collars, and his hat in anticipation of his return there later in the autumn. He was beginning to note that the Scottish venues were unevenly filled, the cheaper seats sold out and spaces appearing in the higher priced parts of the hall. It was true of Edinburgh in 1895 and Glasgow in 1896. Once again the noise of electric machinery disturbed the quieter passages of the music (it happened in Newcastle in 1894 and again, particularly in the Prelude to *Parsifal* in Glasgow's St Andrew's Hall two years later). In Oxford's Sheldonian (28 October 1896) he had to conduct in his doctoral robes ('Pathétique' Symphony, the Ride of the Valkyrie, and *Till Eulenspiegel* were on the programme), whilst the next day he was in Birmingham trying out the 'wonderful' organ in the Town Hall before the afternoon rehearsal. '[Charles] Perkins played some dreadful piece by Widor,' he wrote. On his way back to Vienna he gave his customary rendition of Beethoven's Ninth Symphony, this time at the Queen's Hall. Sir George Grove, who attended the rehearsal on 2 November, commented that he had never heard a better performance of the Adagio. Until 1884 Grove had been Director at the Royal College of Music and was always grateful for Richter's encouragement of young talent.

I have heard from my *Kinder* [children] of your charming behaviour towards them, and I can't help writing a line to say how heartily and deeply I thank you for it. Ach Gott! if all great musicians would be as condescending and kind to students it would be well indeed![3]

From Lower Sydenham in south London Grove kept in touch with Richter after his retirement (he died in 1900), sometimes simply reporting musical news, sometimes asking him to help new musicians, often signing with two notes G.G. under *fermatas* in the tenor clef. 'We have had plenty of music by strange orchestras this year', he wrote in 1897. 'Lamoureux, Mottl and the others, but *none* to my thinking so good as Hans Richter and his band. Joachim's quartet has made a very great success, and most deservedly. Goodbye my dear fellow. P.S. What a calamity is Brahms' death!'[4] Rather than attend evening concerts at his age (approaching 80), Grove preferred to be present at Richter's morning rehearsals. He attended a London rehearsal of Tchaikovsky's 'Pathétique' scheduled for the Birmingham Triennial Festival a week later.

I could not stop today, but I must write to say what an extraordinary pleasure and benefit you gave to me by the rehearsal of the Tchaikovsky. I was quite unprepared for it. I have heard the Symphony four or five times from other conductors but it has always left me cold; *now* I have suddenly woken to find a masterpiece which I can feel towards as I do towards those of Beethoven. I would have given *anything* to be able to hear it once again. Allow me to give you most hearty thanks for conferring upon me so great an event as I have this day received through your goodness, and believe me to be always your sincerely grateful, faithful, obliged servant G. Grove. Please forgive me for expressing myself so imperfectly.[5]

Another who was impressed by Richter's rendition of the Tchaikovsky symphony was the writer Arnold Bennett, who received the conductor's permission to observe the proceedings from within the orchestra.

Tuesday, 24 October 1899: Richter Concert. I sat in the orchestra, between the kettle-drums and the side-drum. You can't be too close to an orchestra. The sound is quite different, more voluptuous, more significant, when you are in the middle of it. Everything takes on a new aspect. The orchestra becomes a set of individuals delicately interrelated, instead of one huge machine.

Richter has all the air of a great man. He seems to exist in an inner world of his own, from which, however, he can recall himself instantly at will. He shows perfect confidence in his orchestra, and guides them by little intimate signs, hints, suggestions. When pleased he shows it in a gay half-childlike manner; smiling, nodding and a curious short wave of the forearm from the elbow. Having started his men, he allowed them to go through the second movement of Tchaikovsky's *Pathétique* Symphony without conducting at all (I understand this is his custom with this movement). They played it superbly. At the end he clapped delightedly, and then turned to the audience with a large gesture of the arms to indicate that really he had had nothing to do with that affair.[6]

The Birmingham Festival of 1897 provided Richter with another chance (on 23 September) to inspect the organ in the Town Hall. It had been newly pitched and tuned and the city organist, Charles Perkins, played him 'a brilliant composition [the sonata] by Julius Reubke. He died very young and was a pupil of Liszt, a splendid work with quite wonderful harmonies, remarkable!' He felt, however, that London was beginning to suffer from a surfeit of music. 'The Queens Hall was not quite full,' he wrote on 18 October upon his return from Birmingham. 'There's too much music-making. Still, the public must find that at least with me the orchestral playing is first class.' In Sheffield on 21 October he had a visit backstage from Ellen Terry ('she is still an interesting woman') and Henry Irving's son. A few days later (27 October) he was visited at Bristol's Colston Hall by Michael Balling, who trained as a viola player and played at Bayreuth before becoming an assistant conductor there in 1896 and 1897. Richter wrote of him on this occasion, 'Balling, the splendid orchestral man was there, a future Kapellmeister for Vienna.' In fact he succeeded Richter, not in Vienna but as conductor of the Hallé when Richter retired and later at Bayreuth. From Bristol (after an inspection of the Clifton suspension bridge) Richter travelled to Nottingham where, on 28 October, there was 'a thick fog and the cab-driver had to lead the horse. Greatest admiration for the people of Nottingham, for despite the fog the hall was full of the most enthusiastic public.' The concert the next afternoon was in Oxford's new Town Hall and 'it sounded magnificent', but the best acoustic appears to have been at Brighton's Dome two days later (30 October). 'The sound is simply splendid, the *piano* full, the *forte* powerful, no noise.'

'These English are just capital people,' he exclaimed in his diary after a read-through of Rimsky-Korsakov's *Sheherazade* on 21 May 1898. The English orchestral player was justly famed for his prowess as a sight-reader and Richter marvelled that it was 'superbly read straight off'. The first mention of 'my neuralgia pains' from so much conducting is made on this date. The concert (on 23 May) was back at St James's Hall. 'I wanted to do the *Funeral March* [from *Götterdämmerung*] in memory of Gladstone, but Vert was against it, the extra instruments were not available. Lazy man!' His daughters Mitzi, Ludovika, and Mathilde were with their parents on this trip, and received daily bicycle-riding lessons when the family took a few days off and stayed at Eastbourne. Here Richter took his favourite walk to Beachy Head on several occasions, weather and rheumatic headaches permitting. On 27 May he learned from Vienna that 'Buchta, the Czech, is dead. Thirty years ago and more I played viola on the same desk with him in the Philharmonic concerts when I was not required as a horn-player.'

Back in London he received yet another offer from America, this time with a remuneration of a quarter of a million marks. 'Oh God!' was the only comment in his diary for that day, 6 June 1898. It came just months after his last Philharmonic concert in Vienna, and the behind-the-scenes manœuvring there and at the Opera, together with the uncertainties of the Hallé contract, must have made this a very attractive and tempting prospect. As it was he stood by his decision to move to England, a country which, after twenty years, had taken him as much to heart as he had taken it to his. With his appearance on the concert platform at the Free Trade Hall in Manchester on 19 October 1899 and his trade mark opening work, the Overture to *Meistersinger*, Richter began his final dozen years as a conductor. His discovery of the music of Elgar could not have been better timed.

26

1899–1900

Hallé Orchestra

ONCE established with the Hallé, Richter found himself conducting in new towns and cities in the north of England. Middlesbrough was one, on 3 November 1899 (an all-Wagner programme and Beethoven's Fifth Symphony). On the way back to London the train stopped in York where 'one of the stationmasters, who had been at the concert, said "it was grand"'.[1] On this tour Richter worked with Dohnányi and Busoni as solo pianists, the singers Louis Fröhlich, Ella Russell, Marie Brema, Kennerley Rumford (Clara Butt's husband), and Elison van Hoose (who sang Wagner in Hull where an 'aural' rehearsal had to be held in the hotel because there was no piano). Another was Blanche Marchesi, who sang Beethoven's 'Ah! perfido' and Liszt's *Die Loreley* at the conductor's official début in Manchester on 19 October when 'a very large audience assembled to greet Dr Richter and to give assurance of goodwill undimmed by recollection of any past misgivings and preferences'.[2] Richter's rendition of Brahms's First Symphony in London on 30 October was greeted with enthusiasm in the press. 'Never has he conducted it with happier results. There was a dignity about his reading that caused the music to stand more than ever aloof from all things mundane and common. No nervous excitement was there, no frantic hurrying up to sensational climaxes, no explosion of musical powder magazines, so to speak; but order and strength, beauty and nobility reigned supreme and braced us up after the day's labour instead of merely exciting and irritating us as so much modern music does. It was a great performance of a great work.'[3]

Another venue was Lord Fitzwilliam's castle, Wentworth Woodhouse, in Rotherham, where he conducted on 17 November. The concert only began at 9.30 in the evening and fog prevented many of the audience from arriving. After spending the night in the castle as the guest of Lord Fitzwilliam's daughter-in-law Lady Alice ('a fine lady'[4]), Richter returned to Manchester to rehearse with the Hallé and Sarasate and then off to the Palace of Varieties for an afternoon's entertainment (the music hall was now a favourite haunt). On 24 November he gave Newcastle its first taste of Beethoven's 'Choral' Symphony. He was well pleased, the soloists (Agnes Nicholls, Muriel Foster, Joseph Reed, and Daniel Price) and the Newcastle and Gateshead Choral Union were excellent and the public response overwhelming. The programme was huge, beginning with the symphony and then continuing with Tchaikovsky's *Hamlet* Overture, Wagner's Good Friday Music from *Parsifal* and Siegfried's Funeral March, Brahms's *Song of Destiny*, and the Academic Festival Overture. Of the symphony Herbert Thompson wrote:

Now this evening's performance was a peculiarly happy one in this, that it soon inspired a confidence that was in no respect misplaced. First of all there was a pilot whom to know is to trust. He knows the Beethoven symphonies like the proverbial old song. He is in perfect sympathy with the music and he conducts with a minimum of gesticulation that conveys a maximum of expressiveness. With the Hallé band, which is now becoming in close touch with a chief of whom its members are obviously proud, the success of the orchestral part of the symphony was assured. The singers, both soloists and chorus, were excellent. The chorus, a large one numbering some 350 voices, achieved a success quite unequivocal. Their fresh tone was never unduly forced, they had been well-drilled in their task and responded to the conductor's beat with an energy that never flagged. We well remember hearing Dr Richter tell a chorus that they could not sing the *Choral* Symphony with the voice, they must sing it with enthusiasm, and enthusiasm these Novocastrians had in abundance. Helped by the merciful low pitch, now happily adopted, the sopranos sang the highest passages without screaming and sustained the fearful high A without sinking.

 . . . To come to a few points in detail, the first movement was given with all the majesty which its outlines demand and the sublime pedal point in the coda which marks the culmination of the movement had a superb effect. The scherzo gained by not being taken at the breathless scramble effected by many conductors; the subject had a chance of telling and it was possible to make the proper contrast between it and the trio. Further some of Wagner's suggestions had been adopted, very greatly to the gain of the spirit of the music, if in contradiction to its precise letter. In the slow movement the melody was insisted upon with Richter's usual sympathetic care and the perilous horn passages came off satisfactorily, while the recitatives introducing the Finale were right in feeling, though they would have been better for a little more rehearsal.[5]

A short visit to Ireland for three concerts followed. On 26 November he travelled through 'beautiful Wales via Chester to Holyhead, on the *Connaught* to Kingstown—without seasickness! Stayed that evening in the Shelbourne Hotel and saw a bit by tram.' The next day he arrived in Belfast 'but found no rehearsal because boxes containing the music had been left behind [in Dublin]. They arrived shortly before the concert on an extra locomotive. It was a good concert, though not quite full.'[6] The two Dublin concerts followed, on 28 November at the Rotunda and the following afternoon at the Theatre Royal, where Richter was entertained during the interval by Ireland's Lord Lieutenant in his box. His final concert of the tour was back in Manchester and was dominated by two choral works by Parry, *Job* and *Blest Pair of Sirens*. No narrator had been engaged for *Job* and Plunket Greene saved the performance by adding it to his singing role. Richter thought 'Parry's compositions very heavy. The people breathed a sigh of relief at the *Meistersinger* [vocal extracts]. I got a laurel wreath from the orchestra. Left at midnight, the station was full of reservists and crying people'[7] (soldiers being seen off by relatives as they left for South Africa and the Boer War, which had begun on 11 October).

The next day Richter sailed for Europe and his last appearances in Vienna. His absence from England at this time was to have great significance for the future of a 20-year-old youth in the nearby Lancashire town of St Helens who was destined to become one of England's greatest conductors. Thomas Beecham's father Joseph had been elected the town's mayor and decided to hire the Hallé orchestra for a celebratory concert on 6 December 1899. When it was disclosed two days before the concert that Richter would not be conducting, Thomas proposed that he should take over. After some hesitation his father agreed, then stubbornly insisted when the orchestra's leader, Risegari, refused to play under such a novice. Thomas, armed with the threat that his father would send for a London orchestra if the Hallé refused him as conductor, was sent to negotiate with the orchestra's manager Forsyth, who agreed to his terms, and Risegari made way for a deputy leader. The concert went ahead with Richter's original programme, a typical fare of the *Meistersinger* and *Lohengrin* (Act III) Overtures, the third movement of Tchaikovsky's 'Pathétique', Beethoven's Fifth Symphony, arias by Delibes, Gounod, and Verdi sung by Lillian Blauvelt, and Berlioz's Hungarian March, all of which Beecham conducted from memory (and without baton). The concert was a success, though the orchestra could have managed that programme with anyone at its head. Beecham's genius became more apparent within a few years in London. His attitude to Richter was hostile and he was quick to condemn him because he was an obstacle, and a foreign one at that, to Beecham's chosen career. On the other hand he greatly admired Nikisch (and modelled himself physically upon him at this time). Beecham said of Richter, 'a few things he interpreted admirably, a great many more indifferently, and the rest worse than any other conductor of eminence I have ever known'.[8] Beecham's tastes were very different, with his love for French and Italian music and his championing of Richard Strauss and Delius. Beecham was at home in *Tristan*, not in *Meistersinger*, which gives insight into the temperaments of the two men.

Herbert Thompson made some interesting observations at this time on Engligh conductors and their art, prompted by Sullivan and Cowen.

Sir Arthur's rule is an excellent one if you offer me two men of equal merit, I take the one who is born and probably educated in England. Unfortunately the rule is sometimes applied by eager patriots when our own countryman is not the superior or even the equal of the foreigner. . . . It was this consideration that prompted the Manchester orchestra to accept with avidity Richter as their conductor, for to anyone who possesses the materials for forming a dispassionate conclusion on the matter it is simply absurd to maintain that we have any native conductor of anything like equal all-round ability. He has made conducting the business of his life, which is more than can be said of any British conductor with one possible exception, Mr H. J. Wood and his experience is both longer and wider than that of any of his contemporaries, while he enjoys in addition the sufficient enormous advantage of not being a composer.[9]

The paper on conducting, which Mr Cowen read at Scarborough yesterday, proved, as we expected it would, a valuable contribution to the subject. . . . Mr Cowen does not seem to have referred to the enormous advantage to the beginner in conducting which exists in Germany, in the many subordinate posts which the opera houses that exist in all large German towns afford him. Here it is the chief organist who is commonly the centre of musical life in our smaller towns, there it is the court or town Kapellmeister. The result is that our organists, thoroughly efficient though they may be, and often are in their own branch, are fishes out of water when, in their capacity as conductors of the local choral societies, they come into contact with an orchestra. This was of less importance in the old days when the orchestra was used chiefly to support the voices, but now when the band in every modern cantata has its own independent part to play, and that often one of great difficulty, it's quite another matter.[10]

He also described the French conductor Charles Lamoureux, who died aged 69 in December 1899, 'as one of the foreign musicians who, during the past twenty years, had taught us what are the possibilities of an orchestral conductor. If Richter has shown us dignified and sympathetic readings enforced by a personality of exceptional power and persuasiveness, Lamoureux has given us a not less necessary lesson in the advantages of perfect organisation and incessant drill, as conducting to a hitherto unheard of degree of polish in every detail.'[11]

Immediately after his final performance at the Vienna Opera (*Hans Heiling* on 26 January 1900) Richter returned to England. He had concerts throughout the north from the end of January 1900 until the middle of March, beginning with the young William Henley (a pupil of Wilhelmj and 'despite a splendid technique, too immature for Beethoven's violin concerto'[12]) as soloist in a concert at Leeds which also included the first performance there of Tchaikovsky's Fourth Symphony on 31 January. In Manchester he used the Grand Hotel as a base from where he took the Hallé from Birmingham to Dublin and from Liverpool to Huddersfield. It would appear from Richter's diary that at the beginning of February Behrens was 'pestering' the conductor to sign a long-term contract. The entry continues, 'but it is understandable that I can do nothing until . . .'.[13] Presumably these dots signify the final break with Vienna made on 14 March 1900 (twenty-five years after his arrival) when he was able to claim his full pension rights. Behrens and his fellow directors had to wait until June before they could rightfully claim Richter as the Hallé's conductor. Meanwhile his spring tour began in earnest with rehearsals and performances almost daily. He wrote of his annoyance at absenteeism and of the dreadful deputy system which led Henry Wood to read the riot act to his London players a few years later. 'This will have to be put in order,' Richter wrote ominously.[14]

Elgar, now hard at work on *The Dream of Gerontius* for the 1900 Triennial Festival, joined him in Birmingham on 7 February for the Enigma Variations (now with their new ending) 'and had to take two bows at the enthusiastic applause.'[15] He was also present in Manchester the next day where the odd programme displeased Richter, particularly Esther Palliser's solo arias. One was by Liza Lehmann and of the other he grumbled, 'Palliser sings rubbish; an immature, talentless scoring by Fuller-Maitland of an aria by Scarlatti. And a critic writes that! . . . Elgar was greeted warmly. He is a dear and excellent man, his wife very fine. In the hotel afterwards we had a really nice chat.'[16] He was further vexed when Charles Santley (now 66 years old) joined him in Liverpool for a concert on 13 February in which he sang a Purcell aria and one of his own called 'Christmas Comes'. The concert was 'good and well attended despite the snow. Only Brossa spoilt the [Enigma] variations, the ass cannot look at the conductor. Santley composes Handel-type arias, oh! As a singer he's already impossible, but he still got enormous applause. . . . These Liza Lehmann and Santley compositions must be dropped, too stupid!'[17] Brossa was the orchestra's principal flute but his name only appears in the 1899–1900 season, because he did not survive the purge of players which Richter insisted upon when he signed his five-year contract after his first season. The Hallé orchestra was playing Schubert's 'Great' C major Symphony in these concerts, the Manchester performance on 15 February giving its conductor 'a quite particular pleasure. I have *never* heard it played with such fire.'

Elijah (22 February) in Manchester required little rehearsal with the orchestra and chorus, for 'the good old English know the thing from memory'.[18] Along with a purge of the orchestral players Richter also intended to overhaul the programmes, Bach being in his mind to stretch the choir under their excellent chorus master R. H. Wilson, who held the post for thirty-six years from the conductorships of Hallé to Harty. When he came to do Bach's Mass in B minor in November 1901 the review began by stating that 'under Dr Richter's irresistible generalship [it was] the most arduous task ever yet undertaken by the Hallé Choir'.[19] The writer of this review was Arthur Johnstone, music critic of the *Manchester Guardian* from January 1896 until his early death from appendicitis at the age of 43 in December 1904, the first of a line of eminent critics which continued with Ernest Newman, Samuel Langford, and Neville Cardus. Johnstone was an admirer of Richter and came down firmly in his favour during the difficult period of Cowen's interregnum. On the occasion of Richter's Manchester concert on 20 October 1897 he wrote a lengthy article on the conductor.

To Richter's influence and example, far more than to anything else that could be named, is due that prodigious improvement in the standard of orchestral performance all over the world, which is the most notable feature in the history of music during the past thirty years. Principally owing to Richter's matchless combination of artistic enthusiasm, practical mastery, and genial good sense, we now hear things that musical prophets and wise men, such as Beethoven, desired to hear and had not heard.

. . . Such was the impression made by Richter upon all who were concerned, either actively, or merely as spectators and listeners, in the inaugural Festival of 1876 at Bayreuth that they recognised him as a new phenomenon in the world of art. The period of modern orchestral conducting may be said to date from that occasion. It was then brought home to everyone that conducting was a great art worthy of independent cultivation. The public began to take an interest in the style of different conductors, and to show some sensitiveness as regards interpretations of the great masters.

. . . Of late Richter has conceived a certain dislike to the theatres, where he finds his work beset with small worries. He is coming to regard the concert hall more and more as his special sphere of activity. Upon Richter's art as a conductor a good-sized book might be written. Here I can attempt no more than to enumerate a few of his qualities. Practical knowledge of the technique belonging to all the more important instruments; mastery of musical theory in all its branches; an unerring rhythmical sense; judgement and insight with regard to every possible musical style, enabling him always to find the right *tempo* for any movement or section of a movement; mastery of the principles discovered by Wagner respecting orchestral dynamics, such as the necessity of equably sustained tone without crescendo or diminuendo as a basis to start upon the conditions determining proper balance of strings and wind, the nature of a round-toned *piano* delivery (to be studied from first-rate singers), the manner of producing long crescendos and diminuendos, also of producing a true *piano* and a true *forte* (Wagner having pointed out that old-fashioned orchestras never played anything but *mezzo-forte*); mastery of Wagner's system of phrasing, his far-reaching investigations with regard to *cantabile* passages, his treatment of *fermate*, his distinction between the naïf *allegro* and the poetic *allegro*; mastery and practical realisation of all Wagner's other ideas concerning musical interpretation or public performances, a subject in which Wagner took a far more deep, expert, and fruitful interest than any other of the great composers.

Finally Richter is distinguished from most other conductors by his personal behaviour at the conductor's desk. He is free from antics; every movement has significance and every attitude has dignity.[20]

Although Johnstone allowed his imagination a free rein here (it is unlikely that Richter ever analysed his own phrasing, dynamics, and such matters as naïve and poetic allegros) the conductor had certainly studied and assimilated Wagner's *Ueber das Dirigieren* (On Conducting) from the start of his career. Meanwhile concerts towards the end of February and the beginning of March 1900 were preceded by the National Anthem in celebration of various victories in the Boer War (Cronje's surrender at Paardeberg was announced at Huddersfield on 27 February and the Relief of Ladysmith celebrated on 1 March when Hugo Becker played d'Albert's Cello Concerto and Tchaikovsky's Rococo Variations). Another famous German artist playing with Richter at this time was the violinist Willy Hess, who played a formidable array of solos at the Gentlemen's Concert on 5 March (Bruch's First Concerto, Saint-Saëns's Introduction and Rondo capriccioso, and five movements of Mozart's Haffner Serenade).

As the day of his irreversible break with Vienna neared, he still had a few remaining doubts in his mind. As if to convince himself how far he had gone down the road of no return, he confided to his diary, 'but what can I still do in Vienna? Is there still a place there for me? Think about the talentless, pointless, and heartless fools at the Opera, and of how the Court

Chapel has gone to seed. Well?', he asked himself.[21] Then he had consultations with various key orchestral players (Paersch, Fuchs, and Hoffmann, principal horn, cello, and double bass respectively) about a shake-up amongst the personnel and his spirits seemed to lift. He was rootless at this point, staying at Manchester's Grand Hotel and without the domestic comforts of his wife and family, nor with people close to him with whom he could confide or discuss his worries. His spring season involved another three concerts in Ireland (12 March in Belfast, 13 and 14 March in Dublin), and once again the trip to the Emerald Isle was not without incident when first Paersch's horn went astray and then further disaster followed.

After the concert a dinner which would have bad consequences for me. A wonderful sea crossing, calm and with a full moon, but the damned players would not let me sleep. When we arrived at 4 o'clock in the morning I was already a bit feverish. At dinner the night before I had been poisoned by anchovies on toast. When I awoke I was in a bad way, could hardly stand up. The rehearsal [for the concert that night in Manchester which ended the tour with Beethovens' Ninth Symphony] was cancelled and I lay down feeling very ill. Above all Cowen's *Ode to the Passions* had to be taken off the programme. At 6 o'clock I still felt miserable, but this final concert had to go ahead. I pulled myself together and got through the concert with great energy. The public could feel content with what they got in exchange for Cowen's watery soup, Schubert's *Unfinished* and Wagner's *Good Friday* Music. Weakness and fear almost overcame me, but steadily I brought the concert to an end. Great cheering, enthusiastic applause.[22]

From Manchester Richter travelled via London to Berlin where he conducted the Philharmonic orchestra in a concert for its own pension fund (Wagner and Beethoven's Ninth on 26 March). Two days later he had a concert in Budapest, a week later one in Hamburg, and then back to Budapest for the *St Matthew Passion*. Beginning 19 April and ending in Hanover on 14 May he undertook an exhausting tour with the Berlin Philharmonic Orchestra (organized by the Berlin impresario Hermann Wolff) to eastern Germany, Austria, Italy, France, Switzerland, and concluding with more German cities. There were twenty-five concerts in twenty-six days, with one free day in Zürich. The programmes were mainly limited to the usual orchestral extracts from Wagner's operas, Beethoven's 'Eroica' Symphony (eighteen times), Tchaikovsky's 'Pathétique' four times, Strauss's *Don Juan* five times, and the rest overtures or other pieces by Liszt, Dvořák, Smetana, Berlioz, Beethoven, and Weber. There were no soloists on any of the programmes. After a single *concert populaire* in Brussels on 24 May, Richter returned to London for three concerts in as many weeks. At the last one (18 June) he introduced the young Czech violinist Jan Kubelik to the British public in Wilhelmj's edition of Paganini's Violin Concerto in D. 'He is an amazing violinist,' he wrote. 'Whether he is also a musician, time will tell.'[23]

Richter was delighted to be back in London. 'Had a very fine rehearsal,' he wrote. 'It is a pleasure to see with how much love these splendid musicians play under my direction, a wonderful sound. My London players really are my favourites.'[24] A few days later he travelled north to Manchester on a house-hunting trip. In the company of Alfred Rodewald, James Forsyth, and Henry Ettling he travelled to Bowdon, about eight miles from Manchester, where he 'looked at the house "Firs" in Firs Street and made a note to take it. Hopefully it's the very thing.'[25] That evening he spent at Behrens's house in the company of Ettling (a timpanist nicknamed Klingsor—the magician from Wagner's *Parsifal*—because of his interest in magic), Brodsky, and principal cellist Carl Fuchs. The next morning, 2 June 1900, in the presence of Behrens, Brodsky, both Forsyths, and a lawyer he signed a five-year contract with the Hallé from October 1900 to March 1905 for a minimum of forty concerts per year at a fee of fifty guineas per concert, with permission to conduct the Triennial Festival in Birmingham and 'three to four London concerts if they do not interfere with our Manchester concerts'.[26] Behrens's dream of securing Richter bore fruit five years after Hallé's death. A few days were spent relaxing by playing Mozart's Viola Trio and a four-hand piano version of Tchaikovsky's Fourth Symphony, and travelling with Rodewald to Colwyn Bay and Llandudno, where smoking one of his companion's strong cigars gave the conductor a restless night.

On 5 June he returned to Liverpool to stay with Rodewald, who, with Schiever, witnessed his signature to the contract renting 'The Firs' in Bowdon. There was a complication when a surveyor declared that the house required drainage repairs to the cost of £600 despite the production of a certificate by the department of sanitation that all was in order. This, together with raging toothache, did not put Richter in the best of moods. Dr Davenport of Wimpole Street tackled the errant molar for five guineas, the housing crisis was resolved, and Richter conducted his last concert of the season in London before returning to Vienna to collect his family for his summer holiday, first to his country chalet at Weibegg in Lower Austria, then to his daughter Richardis and her husband at Baracs in Hungary. There was no festival at Bayreuth that summer, so for once Richter had a genuine rest from music for three months. A pity, therefore, that the score of Elgar's *Gerontius* was not available for him to prepare thoroughly in time for the Birmingham Festival in October, his next engagement.

When Richter returned to England at the end of the summer he brought with him his wife and two of his daughters, Ludovika and Mathilde. The other two daughters were already married, Richardis in Hungary and Mitzi to Richard Richter (no family connection this time), an actor who often came to England with a German company. Of the Richters' two sons, Hans was studying shipbuilding at the polytechnic in Charlottenburg, whilst young Edgar remained at boarding school at Krems on the Danube. There are no more glimpses of Richter from the diary extracts made later by Edgar for his sister Mathilde about their father's life and activities in England. These went only as far as his return to Austria in the summer of 1900. On or about 15 September the family moved to Bowdon and Mathilde remained with them until her parents returned to Germany in 1911. She remained in England and married Sydney Loeb in 1912, whilst her sister Ludovika never married and returned to live in Bayreuth with her parents. In response to later requests by Mathilde to her brother (then in America) he made two more transcripts, one for February–March 1912 when Richter and Marie attended her wedding, the other (in a letter of 28 October 1952) in response to a request for his diary entries from 28 September until 6 October 1900, covering the period of the Birmingham Triennial Festival.

Hans Richter has often been blamed for the failure of the first performance of Elgar's *Gerontius* on 3 October 1900, and undoubtedly it was not a happy occasion. It is easy to criticise today when, like Wagner's *Ring*, *Gerontius* is part of the repertoire. The case must, however, be put in his defence that the full score of this totally new work only reached him on 23 September, the night before the one and only rehearsal with the orchestra at the Queen's Hall. It was personally delivered by Elgar at the Great Central Hotel in Marylebone Road at 6 p.m., as requested by Richter in a letter from Bowdon. He wished 'to look through seriously your score of *Gerontius*'.[27] Despite the short time available to the conductor to learn it, E. A. Baugham was able to write in the *Morning Leader* (critics as well as such eminent musicians as Parry and Manns were allowed to attend the session) that 'Dr Richter had a fine grasp of the score and was indefatigable in his endeavours to get light and shade from his band, but four hours in the morning, with a couple in the afternoon for the soloists [Marie Brema, Edward Lloyd, and Plunket Greene] with orchestra, are not sufficient.'[28] This day of rehearsal, with an orchestra consisting of 122 hand-picked players from Manchester and London, could hardly be regarded as more than a read-through, with a lot of the time spent on clarifying notes and other queries in the parts. Nevertheless the press praised Richter's rehearsal technique and his grasp of the new and complex work.

Arthur Johnstone considered that 'so far as Dr Richter was concerned the rehearsal was a veritable *tour de force*, for the gifted conductor had had no opportunity of studying the full score until the previous evening, yet so complete was his mastery of every page that not a single slip was allowed to pass uncorrected'.[29] Another wrote,

> at ten o'clock Dr Richter called the band to attention. Mr Edward Elgar took his seat beside the conductor, meeting with a cordial reception and proceedings began.... The whole of the morning was devoted to reading through the band parts alone, and a wonderful piece of sight-reading it was. It was, so to speak, the final revision of the instrumental parts. Mistakes were found here and there but there were surprisingly few. The score is most intricate and the music extremely difficult. Apart from the errors in the copies, stoppages were few and those mainly in regard to the matter of tempi.... Dr Richter was unremitting in intention and seemed to know every detail of the score.[30]

One gentleman anticipated events rather too accurately for comfort.

> Dr Richter is a fine conductor in his mastery over a score and over his men, and what he managed to get out of his forces on Monday last was quite wonderful. It did seem to us that all he could do was to pay attention to the difficult points and on these he concentrated himself. There was room for much more light and shade, better graduated *crescendos* and more delicate *pianissimos*. The spirit of the thing was there no doubt but more could have been got out of the work. How all this is to be obtained from a single rehearsal at Birmingham we do not understand, but rest in the placid hope that it will be all right 'on the night'.[31]

Another was more carried away.

> There is a crowd of demons in the Dream, and as they are contrapuntal devils, they gave a good deal of trouble. But Dr Richter was equal to all emergencies. His quickness of ear is wonderful and the manner in which he pounces down on an erring sharp or flat amidst a Babel of sound is marvellous. Each note in one octave above middle C gives hundreds of vibrations per second, and each sound generates other sounds. In many passages of the Dream there are from 20 to 30 notes vibrating simultaneously and a wrong semitone would mean but a difference of from 20 to 30 vibrations per second amidst thousands. To find the proverbial needle in a bottle of hay is consequently child's play to what a conductor's ear accomplishes momentarily. I should like to see a biograph of Dr Richter's auditory nerves when he is on the warpath with an orchestra of 120 braves.[32]

Everything augured well according to those present at the rehearsals, but then the chorus was not yet part of the proceedings. On 11 June that year they had suddenly lost their chorus master when Charles Swinnerton Heap died. His replacement was the former Festival chorus master William Cole

Stockley, then 70 years old and totally out of touch with Elgar's music. The chorus only began rehearsing *Gerontius* upon their return from summer holidays, giving them just six weeks, and it was written in a choral idiom with which they were entirely unfamiliar. Richter travelled to Birmingham on 28 September and rehearsed other Festival music that evening. On the afternoon of the following day the only combined rehearsal for the full company took place, and this was virtually a public dress rehearsal. That Saturday was an exhausting one for Richter; in the morning Bach's *St Matthew Passion* was rehearsed, *Gerontius* took all afternoon, and the evening was devoted to Coleridge-Taylor's trilogy *Scenes from the Song of Hiawatha*, which had only received its first complete performance in March at London's Royal Albert Hall. It was during the rehearsal of Part II of *Gerontius* that Elgar came to the podium and addressed the chorus on their performance of the Demons' music. The press wrote about this.

The rendering of this part did not satisfy the composer, and he addressed the executants on the subject from the front of the orchestra. We venture to think that this was a mistake and showed want of tact more especially as the baton was in Dr Richter's hands. A repetition apparently brought out all that was desired.[33]

This report upset Richter so much that G. H. Johnstone, chairman of the Festival Committee, wrote on his behalf to the *Birmingham Post*.

We have received the following letter in reference of a comment made in this column yesterday on Mr Elgar's indiscretion in lecturing the chorus on the conclusion of Saturday's rehearsal of his work. Mr Johnstone tells us that Mr Elgar addressed the choir at the request of Dr Richter, but there was no evidence or intimation of the fact at the time. And in any case, of course, Dr Richter is not responsible for Mr Elgar's criticisms, which have been the subject, we understand, of much discussion among the members of the choir:—

To the Editor of the *Daily Post*: Sir, Your musical critic in this morning's issue assumes that Mr Elgar made a mistake in addressing the choir while the baton was in the hands of Dr Richter, and charges him with want of tact. Dr Richter desires me to say that it was at his request that Mr Elgar addressed the choir, so as to give them his view of how the work should be rendered. I have no doubt you will be willing to correct the wrong impression conveyed by your notice.[34]

Elgar wrote a comment alongside this last cutting: 'Atrocious and insulting paper! No apology at all.'[35] Despite the bad omens and the desperate need for rehearsal there simply was no time. Richter had to rehearse *Hiawatha* that same evening. Herbert Thompson was there. 'Soon after 7.30 Mr Coleridge-Taylor's *Hiawatha* trilogy was taken in hand, and it lasted till after ten, in spite of the fact that the composer, whether from satisfaction with the results or because he was too diffident to put forward his views, did not take the initiative in suggesting to Dr Richter any improvements. The mere enumeration of the day's doings will indicate how closely it was occupied and it is safe to say that only the extraordinary business-like methods of Dr Richter kept the interest of band and chorus so well as they were in fact kept. Those who perform under him have the satisfaction of knowing that their time will not be wasted.'[36] That the choir was not up to the mark was even more apparent the next day when, in addition to rehearsing solo parts of a Byrd Mass and those in *Elijah* and *Messiah*, Richter had to go into basic detail with them in Bach's *St Matthew Passion*. Again Thompson attended and showed how the choir were coping with works other than *Gerontius*.

Considering that only the choruses were taken, it occupied a long time, but this is not surprising seeing that Dr Richter was not infrequently compelled to hark back and teach the chorus their notes in the chorales. It may be that the singers, like some other Festival choruses I know, are inclined to think so well of themselves as to regard these 'hymn tunes' as beneath their powers, but the results showed that this is a mistaken opinion. It is of course too early to pass judgement upon the chorus, and I do not wish even to seem to do so, but it may be well to bear in mind that Birmingham has been somewhat handicapped in the matter of training for its chorus with the present Festival. It is proverbially inadvisable to swap horses in the middle of a stream, but this is what had to be done, the untimely death of that very able musician Dr Swinnerton Heap occurring in the midst of the work of preparation, though Mr Stockley very considerately came to the rescue and resumed his old post as chorus master. It is obvious that such a break of continuity is trying. . . . Dr Richter is the most indefatigable of men, and though his work is by far the most responsible and continuously exacting of any performer at the Festival, he allowed only half an hour's interval for lunch, and punctually at its close resumed work with the German Requiem of Brahms.[37]

The Festival began on 2 October with the customary morning *Elijah* and an evening Miscellaneous Concert of a choral work (Parry's twelve-part Psalm *De profundis*), orchestral pieces (Schumann's *Genoveva* Overture, Mozart's 'Jupiter' Symphony, Tchaikovsky's *Romeo and Juliet*, and Wagner's *Tannhäuser* Overture), and some vocal solos by Plunket Greene and Clara Butt (who sang four of Elgar's *Sea Pictures* under the composer's baton). *Gerontius* was performed the next morning (followed by parts of Handel's *Israel in Egypt* and Schubert's Unfinished Symphony) and its calamitous results were well documented in the press. Blame was laid correctly at the feet of the chorus, whose entries were often hesitant and whose intonation, particularly in the unaccompanied passages, was woefully inadequate.

Thompson 'inclined to reserve judgement till, let us say, the Sheffield Chorus undertake the work, which would be worthy of their powers. . . . The band was excellent and Dr Richter, who is not to be blamed for the shortcomings of the chorus, conducted admirably, getting as much out of his forces as they were capable of giving, and not endangering the performance by attempting too much.'[38] Richter wrote very little in his diary, but what he did conveyed Elgar's distraught mood. 'The chorus in *Gerontius* very bad, the poor composer was in great despair; but the solo performances and the orchestra were excellent.'[39] He was more upset that *Hiawatha*, which concluded an exhausting day, was a greater success. 'In the evening conducted that rubbish *Hiawatha*. It was well received, Oh!'[40] After the concert Coleridge-Taylor recalled that at a supper party Richter played the Overture to *Tannhäuser* on a pianola. 'The juxtaposition of the greatest orchestral exponent of Wagnerian music and the mechanical piano-player struck him as being irresistibly funny.'[41]

It is wrong to criticize Richter as a bad choral conductor. As Henry Wood said, '*Gerontius* ought to have created a far greater impression when it was originally performed . . . than it did. I know how anxious Richter was over it. I can see him pacing up and down his bedroom with the score on his mantelpiece, but I shall never believe he was to blame for its failure. I do blame him, however, for not postponing its performance. . . . The choral idiom was so new, so strange, and so excessively difficult for a chorus brought up on the *Elijah* style of writing that at least another six months of choral preparation would not have been too much.'[42] Coleridge-Taylor was unhappy with Richter as a choral conductor who obviously had little sympathy with *Hiawatha*, particularly after the events that had taken place that morning. Richter's choral experience went back to 1868, when he had sixty-six rehearsals in Munich for the première of *Meistersinger*. His choral repertoire (detailed in the Appendix) scarcely leads one to the conclusion that he was either a bad or an unsympathetic choral conductor (that 1900 Festival alone contained nine choral works, including the difficult *Spectre's Bride* by Dvořák on the day after *Gerontius*), and he went on to conduct Elgar's work nine times in the years ahead. As Arthur Johnstone wrote after a performance in Manchester on 12 March 1903, 'It was doubtless the most carefully prepared of the performances that have been given thus far in this country. Dr Richter was, for various reasons, peculiarly anxious that it should go well.'[43] Richter never forgot the débâcle of Birmingham in 1900, and wrote on Elgar's manuscript of *Gerontius*, 'Let drop the Chorus, let drop everybody, but let *not* droop the wings of your original genius.'[44] (The *Musical Times* quoted this inscription incorrectly in its November

1900 issue as 'let *not* drop the wings of your original genius'.) Fortunately the genius of *Gerontius* survived the inadequacies of its first performance, and Richter did more than enough during the coming decade to make amends to its composer.

27

1900–1902

England

In the autumn of 1900 Richter was installed at Bowdon with his wife and two daughters, and opened his first season as the Hallé's regular conductor. After a quick trip to Hamburg to conduct the Berlin Philharmonic concert (12 October) he began with a Gentlemen's Concert at the Free Trade Hall in Manchester on 15 October, in which Muriel Foster sang three of Elgar's *Sea Pictures* and Stanford's arrangement of Purcell's *Mad Bess*. The Gentlemen's Concerts Society had been founded in 1770 and continued until 1920. When Hallé, a refugee from revolution in Paris in 1848, arrived in England he was persuaded by Hermann Leo, a Manchester business man, to come north and revive the city's musical fortunes. He took over the Gentlemen's Concerts in May 1850, and then established his own series. Richter's season of twenty Thursday concerts (the traditional concert night since 1861) began three days later on 18 October. His soloist was the soprano Blanche Marchesi who sang *La Cloche* by Saint-Saëns and Leonore's aria 'Abscheulicher' from *Fidelio*. A year earlier she had also sung Beethoven's aria at the first concert of the season (19 October 1899, the fourth anniversary of Hallé's death).

[Richter] turned to me in the green room and said, 'Woman, why on earth are you not singing in opera? You would make a great Wagner singer'. And on my reply, 'Do you think my voice would be sufficiently powerful to represent Wagner's dramatic heroines?' he answered, 'It is just such singers as you that Wagner desired and wished for. He wanted classic style and perfect vocal method, and it is a great mistake to think that all the people who did sing his works, ignorant of methods, were to his liking'. . . . When we parted that night at the Manchester Free Trade Hall I went home a new-born, perfectly happy creature, seeing a bright future before me. To be brief, after the blessing of Richter I made my début at Prague in the *Walküre* as Brünnhilde.

... To my astonishment at the rehearsal, when Richter had beaten the first bars and I came in at once with all my power with the famous words 'Abscheulicher, wo eilst du hin'? the orchestra did not come in on the bar. Richter lifted his baton and stopped dead. Throwing a deeply astonished look around him, enveloping all the players, he said, 'Tchentlemen, ton't you know the *Fidelio* air?' Chorus answered 'No' on which he turned round to me and said 'Unglaublich' [unbelievable]. ... It was after this incident that Richter wrote to Frau Cosima Wagner to arrange for me to sing Wagner at Bayreuth.[1]

Marchesi was not always happy with Richter, in particular she resented his exclusion of herself and other highly paid vocal soloists from Hallé programmes once he had established himself in Manchester, but she never questioned his musicianship, integrity, and sensitivity when he accompanied singers. She continued, 'After the rehearsal Richter said to me, "I am astonished, Madame Marchesi, that you sing these *appogiaturas* in *Fidelio*."' She explained that her mother's aunt, Baroness Dorothea von Erdtmann, had been a close friend of Beethoven, and that her mother culled 'from my aunt's lips all the information and directions as to how the master wanted his music to be performed'.[2]

Richter's autumn season of London concerts took place on three consecutive Mondays beginning on 22 October at St James's Hall under Vert's management, with a special concert at the Crystal Palace on Saturday afternoon 17 November. All the programmes were without soloists and contained the usual Richter items (Glazunov's Sixth Symphony on 29 October being the most novel). The Manchester concerts, on the other hand, still billed solo performers, often singers, and between Thursdays the orchestra would travel to such towns as Burnley, Leeds, Liverpool, Newcastle, and Bradford, as well as undertaking a regular trip to Ireland, where two concerts would be given in Dublin. Herbert Thompson reported on several of these out-of-town events and was pleased to note the changes that Richter was implementing.

[The 'New World' Symphony] ... which was given for the first time in Newcastle was delightfully and sympathetically interpreted by Dr Richter, while in it the band showed the advance in refinement and finish of detail that has characterised it since he took the reins. ... The finish, which is now so noticeable in the playing of the Hallé orchestra, was perceptible in nearly everything, notably in the Beethoven overture of which a dignified and expressive performance was given. The power of the individual players was, however, most felt in the Elgar Variations. These Dr Richter has taken pains to polish in every detail and the result is a strikingly brilliant piece of virtuosity; the superb tone of the violoncellos in the 13th variation was so conspicuous as to deserve very special mention. A good deal of new blood has been infused into the orchestra since last season, not before it was wanted, and it wanted little imagination to find traces of youthful energy in the Smetana overture, in which some of the strings seemed like young greyhounds impatient of the leash. The result was that though this was not the most finished of the performance, it had abundant spirit, and a similar criticism applies to the 'Tannhaueser' piece, though this, as we have suggested, is so emphatically theatre music that it cannot produce its full effect in the concert room.[3]

Ysaÿe and the *Eroica*, these were the features of last night's concert at Bradford. . . . Both reached the highest possible level of interpretation. Beethoven's great work can hardly have been often played with equal perfection, even under as great a Beethoven interpreter as Dr Richter, and the Hallé Band showed very strikingly the advance it has made in finish and refinement. There was plenty to be done in this direction when Sir Charles Hallé died and much was accomplished by Mr Cowen during his tenure of the conductorship. Dr Richter has gone still further and has now effected some changes in the personnel of the band, the fruits of which are gradually making themselves felt as the newcomers are finding their feet. Certainly no more finished playing could be desired than was heard in the *Eroica*; the Funeral March was notable, not only for the dignity and deep feeling with which Richter always invests it, but for refinement of phrasing especially noticeable in the strings, though a word must be given to the very expressive playing of the first oboe. Perfection and delicacy was again a remarkable feature in the Scherzo, while throughout there was the closest possible attention to detail and the balance between the different sections of the orchestra was maintained with quite exceptional care and success; altogether it was a very notable performance.[4]

It is rather significant how Dr Richter's personality has dominated the concerts of the Hallé orchestra since he became its conductor. The organisation retains its original title, but its concerts are looked upon as Richter concerts, and, as in the case of that which was given at Hull last evening under the management of a local firm, are actually advertised as such. Certainly the programme justified the title for it was typical of those which have made Richter's concerts famous in London for the past twenty years, and which the London public demand so exclusively that any departure from their well-defined lines is impossible. A Beethoven symphony, a liberal allowance of orchestral pieces from Wagner's works with perhaps some safe classic thrown in, this describes the average Richter concert. It applies with equal force to yesterday's programmes. . . . The crowning effort of the concert was the seventh symphony of Beethoven, which brought it to a close. Here Richter's sympathy with all that is broad, virile and noble in music had the fullest scope. His peculiar deliberate forcefulness is just what is most required in Beethoven's music for which slippery brilliance is by no means the ideal treatment. The abrupt contrasts that are so marked a feature of this particular symphony, and its vigorous rhythms, all received due expression, and, as a matter of detail, it must be added that the doubling of the wood wind effected by Dr. Richter has an especially good effect in Beethoven's music, restoring the historical balance which the greatly increased number of strings in modern orchestras has tended to destroy. The playing of the wind, by the way, was quite inaccurate but the strings were excellent and could not be surpassed for rush and vigour. There was a fairly large audience, the people of 'musical Hull' having responded to the implied challenge of the announcement of the concert, and the enthusiasm was great and seemed genuine, albeit people who have paid half a guinea to hear a concert are naturally disposed to believe they are enjoying themselves. In this case it was their own fault if their enthusiasm was in any degree feigned.[5]

On 12 December 1900 the tenor Edward Lloyd gave a farewell concert in London's Royal Albert Hall to an audience of 7,000. Richter was one of several conductors who took part, accompanying the tenor (still at the height of his powers) in the Prize Song from *Meistersinger* and 'Lend me your aid' from Gounod's *Die Königin von Saba*. Ben Davies, the veteran Charles Santley, and Emma Albani also took part, and Elgar conducted some of his *Sea Pictures* with Clara Butt as soloist. At the end of the concert there were spontaneous renditions of 'For he's a jolly good fellow' and 'Auld Lang Syne' from platform and audience. No mention is made of whether Richter sang along. A less happy occasion took place in the New Year of 1901 upon the death of Queen Victoria on 22 January. Verdi's death took place five days later but the occasion was not marked and his name did not appear on a Hallé programme in Richter's time. In memory of the Queen, the 'Marche funèbre' from Beethoven's 'Eroica' Symphony and the aria 'I know that my Redeemer liveth' from Handel's *Messiah* began the programme in Manchester on 24 January. A week later, at a memorial concert, Richter performed two movements from Brahms's Requiem followed by Beethoven's *Missa solemnis*. 'God save the king' and the slow movement of the 'Eroica' began the programme at a two-day Festival to celebrate the opening of the new Victoria Hall in Halifax on 8 February 1901.

Highlights of this first season were the first performances of Strauss's *Don Juan* (instantly hailed a masterpiece by Arthur Johnstone), Rimsky-Korsakov's *Sheherazade*, and the first *St John Passion* at a Hallé concert. Richter's soloists included Lady Hallé (for whom the orchestra stood in respect as she entered to play Beethoven's Violin Concerto), Ysaÿe, and the pianists Fanny Davies and Busoni (the latter in Liszt's Second Piano Concerto) on 28 February. Richter met Busoni in a London street in the summer of 1899 and took him off to one of his favourite haunts, Gambrinus. Busoni reported Richter's words to his wife (in his original German Busoni used Richter's Viennese dialect).

'Unfortunately I was not in Vienna when you played there, but I have heard that Mahler gave you a lesson in the rehearsal. That is the limit! He doesn't like soloists because he has no routine and cannot conduct at sight. But a conductor must be able to do that just as well as a pianist, mustn't he?'[6]

Included in the programme of Bach's *St John Passion* was a letter from Richter to the public extracted from a recent issue of the *Manchester Guardian*, which, although its purpose was to launch a pension fund for members of the orchestra, also gave notice of his musical aspirations for the city.

There is a fine motto, *Musica lux in tenebris* [music brings light into darkness]. I am not sure where I read or heard that motto, but it is here that I have learned to appreciate its full significance. I shall scarcely be accused of ingratitude or hostility to Manchester if I venture to say that we are not exactly spoilt by sunshine. But as a makeweight for that, the genius of the city has given to the inhabitants a certain warm sensibility to the eloquence of tone. Living in a climate rather unfavourable to the delight of the eye, they seem to be all the more keenly alive to the delight of the ear. Fortunately we have in our midst the resources necessary for the satisfaction of that musical sense—an excellent choir with a most able choir-master, and an orchestra of the highest ability and devotion to duty. There is zealous work at our rehearsals and no loss of time. So much is certain, the results being recognized on all hands. The maintenance and improvement of this orchestra is the object that I now have most at heart, and the first condition of success in that object is stability. There must be a nucleus of experienced musicians about which such newcomers as may from time to time have to be admitted will be grouped till they can combine with the rest on an equal footing.

To bring about that stability we require an old-age pension fund. Orchestral players are not, as a rule, in a position to make a fortune or to lay by any considerable savings for the support of their old age. They are therefore easily enticed away from one appointment by the offer of another with slightly better remuneration. I require complete devotion to the matter in hand, and I recognize with pride and pleasure that our orchestra, almost without exception, give the very best of their power and ability for the sake of the works performed and for the honour of the Hallé Concerts. It is my firm intention to promote the formation of a fund from which, in case it prospers as I hope it may, members of the orchestra disabled by old age or illness would draw a pension. Such a fund would greatly strengthen the bond which unites the musicians of the Hallé Orchestra, and would make it easier both to obtain and keep talent of the highest quality.

For this good and charitable purpose it is my intention once a year to give a concert, the entire proceeds of which will be devoted to the fund in question. I venture to hope too that the amount of the fund may be increased by free-will offerings, and that it may thus in no long time begin to serve its purpose.

I propose that the names of charitable contributors should be printed in the programme books as 'Promoters of the Pension Fund'. In the firm belief that my appeal to the friends of music in Manchester will not be in vain, I sign myself yours etc. Hans Richter.[7]

The first concert for the pension fund concluded the season on 21 March 1901, and consisted of three Wagner orchestral extracts and Beethoven's 'Choral' Symphony. A three-day Festival in Blackpool (15–17 May) was followed by three Richter Concerts in London as a short summer season, after which he travelled to Bayreuth to conduct one cycle of the *Ring*, twenty-five years after its first performance. Once again Cosima had expected him to conduct *Parsifal*. Letters between the two of them during the summer of 1900 discussing his request to conduct the jubilee *Ring* (Cosima resisted any attempt to celebrate any such anniversary) and *Parsifal* were very genial. Cosima was delighted to accept his services (Siegfried, she said ruefully, would much rather operate the stage lights than conduct) along with those of Mottl and a new conductor to Bayreuth, Karl Muck. At one point she offered Richter two performances and the rest to Muck, but it was not to be. Though Richter was a frequent visitor to London's National Gallery, the other treat he was saving for himself, conducting Wagner's last opera, still eluded him, though apparently by his own choice. Muck therefore began his association with *Parsifal* and, like Levi, made it his own for many years to come. Richter only conducted the first *Ring* cycle, after which the reluctant Siegfried took over. Richter still sent singers to Bayreuth for audition, such as Lillian Blauvelt, who went in the autumn of 1901. She made an enormous impression on Cosima and her daughter Eva, not only because of her musical ability but also due to her striking personality and intelligence. In the American soprano they also thought that they had found a wife for Siegfried, but it was not to be.

The 1901–2 season began with a Gentlemen's Concert in Manchester, a bizarre mixture of Mozart, Grieg, Wagner, Delibes, Coleridge-Taylor, and Haydn. After concerts in Liverpool and Middlesbrough, Richter started the Manchester Thursday night series with a purely orchestral programme of Weber, Nicodé (*Symphonic Variations*), Wagner, Dvořák, and Beethoven. Elgar's new *Cockaigne* then made a first appearance in Manchester at the second concert on 24 October, after which Richter began touring the north and making trips to London for his autumn season in the capital, all during his regular Manchester series. When he conducted *Cockaigne* in Newcastle on the day after its first Manchester performance, Thompson thought it was 'exactly suited to him and his orchestra, and its intense vitality, picturesque colour and flashes of humour were perfectly reproduced by them'.[8] The pianist Harold Bauer played Brahms's First Concerto (14 November), and the orchestra extended its touring to the south-west with a visit to Bristol for a purely orchestral programme four days later. When he conducted Liszt's *St Elizabeth* in Newcastle on 27 November Richter was so impressed by his bass soloist, Francis Harford, that he immediately wrote to Cosima and urged her to audition him for Bayreuth. At his Manchester concert on 12 December (and at Liverpool on 12 January 1902) an orchestral suite, named the 'Heroic', was given its first performance, the composer being the young Cyril Scott. In his autobiography, a bad-tempered tome with an appropriate title *Bone of Contention*, Scott at the age of 90 recalled his meeting with Richter at Bowdon.

> This great conductor... was one of those fat, thick-necked, grunting, square-headed, gauche, professional types who do nothing to set one at one's ease, but give the impression that the whole business of meeting one is a dreadful bore.... Having been shown into his music room, I waited some time for him to appear. It was a showery afternoon, and when he finally did appear in a loose jaeger suit, of which he had not even buttoned the trousers, he grunted something about having been out and got wet. I felt anything but comfortable inside myself, and I fervently wished that he could have been a little more gracious as he opened the piano and indicated that I should proceed. Nevertheless, by the end of my performance, his grunts had given way to, for me, very gratifying expressions of enthusiasm, the upshot being that he promised to produce the work both in Manchester and Liverpool—a promise which he faithfully kept.
>
> ... My own view about Richter's opinion of the suite ['a fine and original work'] is simply that it appealed to him because it was outside the rut of those academic compositions which, especially in England, were being produced at that period. Satisfactory as regards form and orchestration, it had just that amount of unusualness, and no more, which he at his age and at that time demanded from a new work. Even so, with all due respect for the opinion of so eminent a man, it soon came to displease me because it had ceased to be representative of my musical aims and ideas.[9]

Scott may have held this rather ungrateful opinion in his dotage but it was quite a different Scott who wrote to Richter after the Manchester performance of his suite.

> I feel I must write to you and thank you for all you have done for me. The performance on Thursday was perfectly splendid and I can never express my gratitude to you enough in spite of all odds to take me up and give me a hearing. The criticisms have amused me very much. It is almost with pleasure that I see they are anything but enthusiastic for in the whole history of art no good work has

ever received a really good newspaper criticism [Scott ignored the excellent notices for Elgar's most recent works]. Of all the conductors I ever saw in England, you are the only one that had the courage to take me up and I thank you very very much, for with you as my guide I go fearlessly through all the hostile forces.[10]

Scott had genuine grounds for gratitude towards Richter, who did much to help young artists in need of assistance. The young London-based pianist Donald Francis Tovey, destined to become one of the country's most eminent musicologists and writers, also contacted him in 1902. 'I venture to send you the enclosed letter of introduction [not extant] from Dr Joachim together with some of my concert programmes. I need not try to express how grateful I should be for the honour of meeting you and of taking any opportunities of music-making that your kindness and discretion may allow you to give me.'[11]

Richter took four years to reward the young man, who played the second performance of his own Piano Concerto, Op. 18, at a London Symphony Orchestra concert on 17 December 1906. Tovey often repeated to soloists who played with him at Reid Concerts in Edinburgh Richter's persistent question to him on that occasion, 'Have you a vish? *Come on*, I am not touchy!'[12] The singer Frederic Austin was also grateful to Richter, who had no hesitation in recommending artists to his colleagues in other parts of the country (Austin was proposed by him to the Leeds Philharmonic Society in May 1902). Someone who appears to have been unsuccessful with Richter was Rutland Boughton, who wrote in the autumn of 1902, 'Please pardon me for bringing to your remembrance the fact that you have a score of mine which did not displease you when we ran it through earlier in the year. May I hope for the pleasure of hearing it under your baton in London, do you think? Such pleasure would be all the greater to me inasmuch as I have had very serious troubles and disappointments these last few months. Of course I would gladly wait, wait, wait, if ultimately your previous opinion of a possible performance could be brought to fruition.'[13]

There were also letters such as this one from Leicester, where he could do nothing to help.

I am writing to ask if you could possibly arrange for my husband, Herr Willibald Richter, to play with your orchestra in Manchester or any other town you are playing at with them. Unfortunately my husband bears the same name as you and it has been a great drawback to him because people ask at the ticket offices 'Is it *the* Richter? When they are told 'No, it is the young pianist', they go away, no doubt thinking it is an imposter trying to trade on your name. We had a concert tour and to our great regret found that the name 'Richter' was for us a great loss for the reason I named above, and so lost a *great deal of money* on the tour. I think therefore it would be a great help if he could play under you and in time would live down the prejudice which no doubt exists. . . . P.S. Our son, who promises some day to become a good pianist and who is already a very fair performer for one so young, may have to suffer under the same delusion.[14]

Huddersfield had its first taste of Glazunov's Sixth Symphony ('exceedingly well played under the conductorship of Dr Richter, who made the most of its rhythmic force and vivid colouring') on 25 February 1902, but more surprising was the first performance in the same concert there of Schumann's Piano Concerto written over half a century earlier, 'Mrs Edward Haley undertook the solo part and acquitted herself creditably, though her bravura passages were not always as distinct as they might be and her modifications of tempo did not invariably seem justified. The worst defect, however, was that the intense poetry of the work was somehow left out and Mrs Haley seemed far more at home in Rubinstein's Valse Caprice. It should be added that neither was the band heard to advantage in the concerto, appearing hardly to comprehend the soloist's reading of the music with as much sympathy as could be desired.'[15] There were other novelties in this season. Dohnányi's performance of Beethoven's 'Emperor' Concerto was preceded by his D minor Symphony on 30 January, played from manuscript parts. Strauss's *Till Eulenspiegel* was so popular at its first performance in Manchester on 13 February that it had to be repeated the following week in a concert which also included Manchester's first 'Polish' Symphony (No. 3) by Tchaikovsky, and Bruch's Second Violin Concerto with Ysaÿe. Wilhelm Backhaus made his début, substituting for Siloti, in Beethoven's Fourth Piano Concerto (on 27 February). In the same programme Richter gave Manchester Cowen's *Phantasy of Life and Love* and Elgar's new Pomp and Circumstance March No. 1. Brodsky played the last new work of the season, Christian Sinding's Violin Concerto, on 13 March 1902.

At the end of the season Richter went on a short European tour. Two concerts were given in Copenhagen (a Beethoven programme with Wilhelm Stenhammar as solo pianist), two in Stockholm (much Wagner), and two in Turin (more variety here with Strauss, Dvořák, and Mozart added to Beethoven and Wagner). His short London season of three concerts followed, noteworthy for the début in England of Fritz Kreisler (12 May) in Beethoven's Violin Concerto and Bruch's G minor Concerto (2 June). The critic of the *Musical Standard* described Joachim, Sarasate, Ysaÿe, and Kreisler as 'the four artists whose memory violin lovers of the future will cherish. In some ways perhaps, Kreisler is the most remarkable of the four.'[16] The other violinist in Richter's London season (in Joachim's 'Hungarian' Concerto at the second concert on 26 May) was the 18-year-old Bohemian Jaroslav Kociân, a pupil of Dvořák, Ondricek, and Kubelik.

On 4 June 1902 Richter went to Paris for three performances of *Götterdämmerung* at the Paris Opéra beginning five days later. He was in correspondence with Cosima at the time (yet another detailed set of arrangements for him to conduct *Parsifal* as Muck was unwell) and gave a mixed report on standards in the French capital. Though delighted with his Alberich and Hagen, he was far less happy with other aspects, such as the operators who could not cue lighting changes at the exact moment in the music, and a lack of musical assistance such as he received at Bayreuth. From Paris Richter continued to Bayreuth, where he arrived on 15 June and prepared one cycle of the *Ring*. Herbert Thompson attended.

So far as my present experience goes the performances have more than maintained their high standard, which three years ago seemed in some danger of being lowered. It has been said that Madame Wagner does not regard the conductor as a factor of prime importance in these performances, and certainly the fact that his name never appears on the programme seems to bear out this assumption. It's true that he is of less importance at these Festivals undertaken by artistic people for the advancement of Art than he is at an English Festival conducted by a commercial Committee in the interests of some local charity. . . . Now this time the principal work given has been the four-days music drama of the *Ring of the Nibelungens*, used to be heard twice during the Festival, the conductor of the first performance on July 25th–28th was Hans Richter, who should know Wagner's intentions as well as any living person, since he was brought up under his influence from sometime in the 60s and was chosen to conduct the first performance of this very work in 1876. Perhaps too, it may be whispered, he is too old and experienced a conductor of Wagner's music to allow any interference, however, well meant, with his peculiar function. Certainly there was a gripping force in the playing of the orchestra that I have never known exceeded. The richness and energy of the music made itself felt without being unduly boisterous.[17]

The 1902–3 Hallé season was full of new pieces, beginning with Strauss's *Tod und Verklärung* at the opening concert on 16 October. The Gentlemen's Concert four days later had some interesting works, Bennett's Overture *Naiades*, Elgar's *Dream Children* (new to the city), Handel's Overture *Parthenope*, and Agnes Nicholls singing 'Caro nome' (Gilda's aria from Verdi's *Rigoletto*) with Richter at the piano. Tchaikovsky's *Francesca da Rimini* was also new (13 November), whilst a week later the Manchester pianist Frank Merrick made his début in Litolff's Piano Concerto (of which only the Scherzo is played today). In the same concert Richter conducted the city's first *Faust* Symphony by Liszt (uncut and with the choral ending). In advance of this he wrote to Cosima asking for her blessing in a matter of a slight re-orchestration on his part. 'For the solo in the first movement', he wrote, 'I use the much more sonorous and noble bass clarinet instead of the bassoon. Siegfried agreed with me, and I hope I am not contradicting the Master's intentions.'[18] Cosima sanctioned this change. Johnstone considered that the work 'fell flat',[19] but Richter, when sending the review to Cosima, did not agree. 'This is not quite right. The public was quiet after the first movement (in Vienna and anywhere else there would have been hissing and laughter); after the *Gretchen* movement there was warm applause, but at the end of the powerful choral section at the end of the *Mephisto* movement there was loud applause. I took this to mean that they wanted to make amends for their silence after the first movement.'[20]

Bayreuth was in turmoil at this time because *Parsifal* was being planned for performance in America contrary to Wagner's thirty-year ban from his death on any performance elsewhere. If Siegfried had not unwittingly given the publishers Schott permission to produce a miniature score of the opera, the crisis would not have arisen. Cosima drew up a petition to keep the opera sacrosanct for Bayreuth, and asked for signatures from her closest collaborators. Richter's was sought after those of Humperdinck, Glasenapp, Mottl, Klindworth, and Fischer were received. Richter had in fact already given his support in a letter just a few days *before* she wrote to him. 'The German genius will still have enough power to protect Bayreuth, the only place which worships him with a pure heart.'[21] This incident came hard on the heels of a remark by Emperor Wilhelm II that Wagner's music was noisy. This offended Richter so much that he wrote a letter to the *Allgemeine Musikzeitung* urging Kapellmeisters to observe Wagner's highly detailed dynamic markings, particularly *pianos* and *pianissimos*. He shunned an invitation to Berlin to attend an unveiling of a statue of Wagner because he was engaged elsewhere (at the 1903 Birmingham Triennial Festival), but he told Cosima that he would never have attended had he been free.

A choral concert on 27 November 1902 was devoted to the first performances in Manchester of Stanford's Te Deum and Brahms's *Triumphlied* (the latter a work which Richter is said to have described to the choir as 'a great big tough ham sandwich'). On 21 December Brodsky and Richter celebrated the twenty-first anniversary of their première of Tchaikovsky's Violin Concerto. It was also appropriate that Manchester University had conferred honorary doctorates of music on the two friends earlier in the spring of 1902 for services rendered to music. There was another celebration before the year ended, but it was inaccurate. The so-called centennial performance of Berlioz's *Damnation of Faust* on 11 December 1902 was a year premature as the composer was born in 1803.

Richter gave three concerts in London in the autumn of 1902, and they were the last of the Richter Concerts begun twenty-three years earlier in 1879. Their popularity and the universal regret at their demise is best summed up by a letter he received from a total stranger.

> I see in the newspaper a rumour that this is your last series of concerts in London. I can only hope the rumour is untrue; if however, it is true may I beg and implore you to reconsider your decision. Since 1885 I have attended your concerts regularly and no other conductor (and I believe I have heard nearly every celebrated conductor in Europe and America) has ever given me the perfect enjoyment and satisfaction that you have. They are all far behind you, and very many things I do not care to hear interpreted by anyone but yourself. In saying this I am expressing the feelings of very many people besides myself. In their name and my own I ask you with all my heart not to give up your London concerts. I only wish you would still give us nine concerts in the summer as you used. Apologizing for troubling you with this letter, I sign myself one of your humblest and warmest admirers.[22]

Richter concluded the year 1902 with a trip to Cologne to conduct the famed Gürzenich orchestra in two concerts. Its conductor of eighteen years, Franz Wüllner, had died in September at the age of 70 and the orchestra needed some guest conductors until Fritz Steinbach was free to take up the post early in 1903. Weingartner, Strauss, d'Albert, and Richter were asked to appear. Richter's first concert was Haydn's *Creation* on 16 December, but two weeks later on 30 December he scored a triumph when his programme included Ysaÿe as soloist in Vieuxtemps's Fourth Violin Concerto, Elgar's Enigma Variations, and Berlioz's *Symphonie fantastique*. 'He is the sort we love in Cologne,' it was reported.[23] Once home he wrote to Elgar with quaint spelling that 'the Variations had *no* bad success at Cologne; after the first reading the orchestra were delighted and in the performance the audience enjoied them very hearthily'.[24]

28
1903–1904

England

IN 1903 Richter celebrated his sixtieth birthday, and although he may have given up his Richter Concerts in the previous autumn, the volume of work he undertook each year showed little sign of abating. There was no Bayreuth Festival in 1903, but instead he conducted for the first time at the Royal Opera House, Covent Garden, and only a year later his lack of Richter Concerts had been more than made up for by his commitments to the newly formed London Symphony Orchestra. His 1902–3 season with the Hallé from October to March consisted of exactly forty concerts with the New Year more or less the half-way point. The second half of the season was also sprinkled with works which were new to the city; some were even new to the conductor. There was Glazunov's Carnival Overture and a Violin Concerto in E flat (K. 268), then attributed to Mozart but now considered spurious, with Lady Hallé as soloist. 'Dr Richter conducted like a giant refreshed with Christmas holidays.'[1] A choral and vocal concert followed on 15 January with Sullivan's *Golden Legend* and Elgar's new *Coronation Ode* on the programme.

Richter and Elgar were now close friends, and the composer presented his champion with one of three full scores of *Gerontius* he received from Novello's (the others went to Sheffield Public Library and Granville Bantock). In it Elgar wrote, 'To my dear friend and musical Godfather Hans Richter from his affectionate Edward Elgar, Oct. 1902.' They were now on Christian name terms, and Elgar's letters reveal the customary, natural warmth that he showed towards his closest and most trusted friends. Richter visited Elgar at Malvern on 12 January 1903 to go through the *Coronation Ode* and to have an early look at the *The Apostles*, down for Birmingham that year. Alice Elgar wrote in her diary that day, 'After lunch he first came into the drawing room and looked round with evident satisfaction, and with a gesture "Ach, wie gemütlich!" ["Ah, how cosy!"] E. and he went through the *Ode* and some of the *Apostles*. He was much impressed and said to me, "Ach, grossartig, eine so heilige Stimme, aber er ist ein ganz famoser Mann, und es ist so wunderbar (or something like that), ein so sehr ehrenvoller Mensch."'

On 22 January 1903 Stenhammar played his own Piano Concerto in B flat minor and Stanford's Irish Rhapsody concluded the first half of the concert. Another new solo work was Bruch's Serenade with Willy Hess on 29 January, followed by Mackenzie's Overture *Cricket on the Hearth*. The concert on 26 February was also notable for new works. Glazunov's tuneful Seventh Symphony began the programme, after which Alexander Siloti gave Manchester its first hearing of Rachmaninov's Second Piano Concerto, and Carl Fuchs the Cello Serenade by Volkmann. The last concert of the season was the first Manchester performance of *Gerontius* with which Richter took such care, particularly with the able R. H. Wilson as his chorus master. Scheduled for 5 March, it had to be postponed for a week when the tenor John Coates fell ill. The event was a sell-out and a triumph. Elgar, who was to have attended on the original date, could not do so a week later because he was rehearsing the same work with the North Staffordshire Choral Society in Stoke-on-Trent, but sent a card of congratulations. 'My heart was with you and I send you my thanks. I hear the performance was the *best*—again, true thanks, your Edward.'[2]

Towards the end of January 1903 Richter gave the people of Newcastle the chance to decide the programme themselves from a wide selection offered to them. The results proved interesting.

The principle according to which he who pays for the piper may call the tune has been effectively put into operation at Newcastle, where the programme of the orchestral concert organised by the Choral Union for 11 February has been settled by the votes of the ticket holders. The result is interesting though it expresses local taste only to a limited extent, since the area of choice was restricted to a specified number of works suggested by Dr Richter with symphonies by Beethoven (Nos. 3, 4, 7 & 8), Brahms (1 & 2), Glazunov, Liszt, Mozart, Schumann, Stanford, and Tchaikovsky. The *Eroica* was far ahead receiving 137 votes, while Tchaikovsky (not with the *Pathétique* however, but the less familiar work in E minor) made a bad second with 81, Brahms' second Symphony came next and attracted many more votes (47 as against 27) than the more sombre first. For the fourth place Schumann in D minor tied with Stanford's *Irish* Symphony, the only work of British origin in the whole list. Of overtures by Beethoven (3), Berlioz, Brahms, Cherubini, Dvořák (2) Glazunov, Mendelssohn, Mozart, Tchaikovsky, Wagner (4) and Weber (3),

Parsifal comes out first with 107 votes, a rather noteworthy choice since it is by no means a composition of superficial attractiveness, nor is it indeed at all suited for a concert performance. Mendelssohn's *Midsummer Night's Dream* Overture won the next place to which it was well entitled, not merely on account of its intrinsic merits as one of his original and fanciful creations, but because Dr Richter gives such a masterly interpretation of this music. A point deserving note in this section was the extraordinary disproportionate share in the comparative popularity of two of Weber's Overtures, *Freischütz* receiving 37 votes while the maturer and more masterly *Euryanthe* had only 3 to its credit. The only work in the list without a solitary vote to cheer it was Cherubini's *Medea* Overture. The third class was of miscellaneous pieces of which, as in the case of the overtures, two had to be chosen. Here Wagner's *Walkürenritt* may not have inappropriately be said to have romped in with a grand total of 240 votes, by far the highest number recorded in the plebiscite. Next, we are pleased to note, came the symphonic poem by Richard Strauss *Tod und Verklärung*. English audiences are so inclined to adopt as their motto *stare super vias antiquas* that one recognises with satisfaction any sign of readiness to hear what is new. Strauss, however, only just managed to scrape in, a Liszt rhapsody being but just four votes behind and one of Dvořák's coming next. Tchaikovsky, Brahms and other works by Wagner and Dvořák complete the list in the order of their popularity with the subscribers. The general result is, it must be admitted, an exceedingly good programme and as long as a choice of an audience is carefully confined to good music so that Philistine propensities are excluded, there seems to be no reason why the principle should not be more frequently adopted. Of course there's nothing new in the idea, which used to be adopted by Mr August Manns for his Benefit Concerts at the Crystal Palace more than twenty years ago.[3]

The concert itself received the following notice.

That the Newcastle public had, by their votes, selected such a programme as this is some evidence of their advanced taste. It was certainly an excellent one, almost too full of good things, or at least of strenuous ones. The effort of listening to such a continuous succession of strongly emotional, highly coloured pieces was exhausting and one felt the desirability of a few moments of repose, but for those who could thrive on such highly spiced meats it was indeed a lordly banquet. It was also peculiarly suited to the genius of the Hallé orchestra and their conductor Dr Richter. Though his worst enemies would not style him Mendelssohnian in his tendencies, he shows that he appreciates Mendelssohn by the very sympathetic rendering of the *Midsummer Night's Dream* Overture. The humour and delicate mystery, the like of which he realised perfectly. Tchaikovsky [*Francesca da Rimini*] suffered from an unaccountable misunderstanding, some of the band being apparently provided with wrong copies which made it necessary to hark back, but after this *da capo* a fine performance of the vigorous, if almost too unrestrained music was given. Dr Richter took great pains with Strauss, whose powerful and impressive work was finely given. That the conductor was at home with Wagner goes without saying, and though *Parsifal* could not be as effective as at Bayreuth, the Valkyrie's arrangement was perhaps almost more so. But Richter's power is never more apparent than in Beethoven's music, and the nobility and serene power of the *Eroica* was splendidly demonstrated. Its great merit was that it was an evenly sustained effort, not a spasmodic one, with a few brilliant points to atone for many waste spaces. It was indeed a worthy performance of a great work and coming last it was the more grateful after the storm and stress that had characterised much of the earlier part of the programme.[4]

The principal cellist of the orchestra, Carl Fuchs, began noting what he called *Richterisms* about this time. In later years he would often bring some of them out. There were traditional stories about Richter, often apocryphal to begin with and then embellished with the passage of years, but there were also many that he noted from his front desk at the conductor's right hand (as Sigmund Bachrich and Joseph Sulzer had done from their viola and cello chairs in the Vienna Philharmonic).

At a rehearsal of *Lohengrin* [orchestral extracts] on 19 February 1903 he said 'Once they found this no music' (meaning there was once a time when this was not considered music). . . . In the slow movement of Beethoven's fifth Symphony (25 November 1903) there is a violin passage in the slow movement which ends rather abruptly. All the violinists automatically put their fiddles down for a little rest, but often a considerable noise is caused by the instruments touching watch-chains. This is particularly noticeable in this place as a silent bar follows. Dr Richter to the violinists, 'In Beethoven's time no musician had a golden watch-chain. Must the violin like a little child rest a little at your stomach?'[5]

The Grand Opera Syndicate, with its energetic secretary Neil Forsyth and enterprising managing director Harry Higgins, engaged Richter to conduct the German season of three cycles of Wagner's *Ring* at the end of April 1903. Though the scenery impressed the critics, that was the one element that displeased Richter, as he told Mrs Joshua. 'Hopefully it will go well musically and vocally tonight [the opening *Rheingold* of the first cycle]; the scenery is lamentable but there was nothing to be done. Inability and unwillingness cannot be overcome. If the music had not been so far advanced and its progress so excellent, I would have cancelled after the first stage rehearsal; but as it was so late I could not let those down who had made it their job and done everything to ensure an excellent performance.'[6] A year earlier he had bemoaned the lack of competent musical assistance in Paris, but now he acquired the invaluable and long-time services of Percy Pitt, with whom he would form a close friendship. He persisted with his Bayreuth habit of not appearing before the curtain at the end of the evening, which did not go unnoticed in the press. 'Dr Richter surpassed himself and the orchestral playing was truly magnificent,' said the *Musical Times*. 'The eminent conductor modestly declined to appear before the curtain at the close of the work, but throughout the three cycles the audience cordially applauded his successive appearances at the conductor's desk.'[7] Parry attended the final *Götterdämmerung* and thought it a 'superb performance. Richter indicated his right to be the foremost of Wagner conductors and Ternina was very fine indeed, though she seemed to tire by the last act.'[8]

Herbert Thompson travelled down from Leeds to report on the cycle.

Die Walküre . . . the orchestra, though not so large as at the model performances at Bayreuth and Munich, was exceedingly good in quality, the strings being especially fine in colour and force and only the brass showing any imperfections. It is composed of the ordinary Covent Garden orchestra reinforced chiefly by members of the Hallé orchestra to a total of ninety-eight. Hans Richter, who has grown up with the Ring, has made the orchestra a part of himself and no one makes it quite so big in its lines and imposing in its effect as he, and he knows how to keep it down so that the voices may tell as and when they should.

Siegfried . . . Fortunately one could always find a refuge in the orchestra, whose superb playing under Dr Richter was quite the finest feature of the performance, and for vigour and colour could hardly be surpassed at any opera house. The glow and forcefulness of the splendid music in the last act was quite memorable, and it was not surprising that Richter was greeted with marked cordiality each time he made his appearance in the conductor's seat.

Götterdämmerung . . . Of the interest excited by the performance there could be no doubt, and the applause, happily suppressed while it was going on, burst forth at the end in repeated calls for the actors but most of all for Hans Richter, who was very properly felt to be the hero of the occasion. The orchestra was indeed the one feature of the production which gave complete satisfaction, and it afforded incidentally a most interesting illustration of the comparative value of opened and concealed orchestras. The strings certainly gained in richness and grip, the brass suffered slightly from having no obstruction and such things as the Abschied in the *Walküre*, the impetuous opening of the last act of *Siegfried*, and the Funeral March in *Götterdämmerung* I've never heard more effectively and impressively in any theatre; and then there was Richter's reading, which realises better than that of any other conductor I know, the elemental forceful bigness of the work. It conveys to the full the majestic gravity of the music, but has no need to resort to what I may call the Bayreuth device of slowing down the tempi till one loses touch with rhythmic movement. Dr Richter has, I believe, often been invited to conduct the Ring in this country, but he has consistently declined until he could be allowed as many rehearsals as he deemed necessary. These he has had on the present occasion and the result is generally admitted to have been the finest reading and the most convincing orchestral performance the work has yet received in this country.[9]

The renowned critic Hermann Klein wrote in the *Sunday Times* of

the rare excellence of the series of performances now being presented under the aegis of the old Bayreuth giant. And why not? There is only one great, one incomparable Wagnerian conductor in the world, and his name is Hans Richter. Give him the right material, with the time to mould it in, and he will bring you forth a model that shall reproduce and interpret the ideas of the master more closely, more accurately, more amazingly than that of any other surviving disciple of the original school. . . . The orchestra is not only quite first-rate, but it knows the work in hand more thoroughly, perhaps, than any other English band that ever performed it. What a contrast to the uninitiated players whom Richter had to drill in the Wagner scores when he first conducted at Covent Garden nineteen years ago! [It was in fact at the Theatre Royal, Drury Lane.][10]

There was unanimous praise for Richter's commanding presence, for the detail he brought out in the playing of his orchestra, and the superb balance he achieved between stage and pit. Voices were never drowned and he inspired his singers, not all of whom were first class, to give of their best. The finest accolade appeared in the press at the conclusion of the cycle.

The man of the moment, musically speaking, is Dr Hans Richter, who has been engaged to conduct the three complete performances of Wagner's *Ring des Nibelungen* at Covent Garden. Dr Richter has for so many years played a prominent part in London music that it may seem superfluous to praise him at this time of day. But among those who wish to win an easy reputation by posing as the pioneers of musical thought there has lately appeared an inclination to speak of Dr Richter as if he belonged to a past age, and to a school of conducting that has had its day. It would not be worth while even to refer to such ebullitions of ignorant folly save to point out how triumphantly they have been dispelled by the magnificent work done by Dr Richter at Covent Garden during the last week. No more triumphant vindication of his claim to rank as the chief of living conductors could be desired. It is only to be hoped that the Covent Garden authorities, having for once put the right man into the right place, will never again consent to make shift with an inferior substitute. . . . Londoners have learnt to look upon Richter almost as one of their own fellow-countrymen, while in others of our cities, such as Manchester and Birmingham, he is now as well known as in London. What England owes to Richter it is difficult to estimate. His influence upon the development of our musical taste has been prodigious, and always in the direction of what is grand and noble. He has never had a better opportunity of showing us his mettle than in the present series of performances at Covent Garden. To see him at the head of his orchestra, controlling his vast forces with a look or a gesture, recalls the words in which Wagner summed up his genius: 'Richter's accomplishments as a practical musician are simply enormous. To him it is given to interpret my intentions to an orchestra with the most absolute distinctness and sympathy. His power over an orchestra is unlimited, and he is almost more at home in my scores than I am myself'.[11]

Percy Pitt was a great find. Between Richter's annual German opera seasons at Covent Garden (seven consecutive years from 1903 until 1910) Pitt, in addition to his duties as musical coach, adviser, stage conductor, and keyboard player at the opera-house, travelled everywhere on Richter's behalf, looking for players and singers. He fulfilled the function of the Richter, who, thirty years earlier, scoured the length and breadth of Germany on Wagner's behalf. The two men remained in close touch for the rest of Richter's life, with the older man doing much to help the career of his younger colleague. Besides Ternina the cast of the *Ring* were familiar faces to Richter such as van Rooy, Olive Fremstad, and van Dyck, but the find of the season was the mezzo-soprano Louise Kirkby Lunn.

Richter enjoyed an unusual rest that summer of 1903. The last performance of *Götterdämmerung* was on 16 May and he next raised his

FIG. 12 Hans Richter conducting at the Vienna Court Opera.

baton before the public at the opening of the Birmingham Festival in *Elijah* almost five months later on 13 October. That evening there was a Miscellaneous Concert of Stanford's *The Voyage of Maeldune* (conducted by the composer), and, in the second part, a mixture of orchestral and vocal works (with Kirkby Lunn, whom Richter took up immediately after the *Ring*, and Ffrangcon-Davies). Herbert Thompson had written an interesting preview of the Festival for the *Yorkshire Post*, in which he also appraised Richter's current worth on the English musical scene.

One of the most wholesome reforms that have taken place of late years in our Musical Festivals is the increased time spent over the work of preparation. No doubt the character of modern music is something to do with it. . . . One result of this is an increase in the demands made upon the performers and hence a greater need for thorough rehearsal. There is a second reason and this is the vastly higher standard of orchestral playing during the past twenty-five years. This period is not an arbitrary one; its commencement may be readily fixed by reference to the advent of Hans Richter, who came here imbued with Wagner's ideas and effected a prompt revolution in orchestral playing. It is worthy of note that Wagner himself had not only preached but practised the same methods in England more than twenty years before, but with so little effect that when Sterndale Bennett took up the Philharmonic baton as his successor, he was careful to tell the band that he would play Beethoven not *à la* Wagner, and he was as good as his word. But the times were not ripe in 1855 as they were in 1877, and the young Richter had not irritated the pedagogues as had his master, so he came and conducted and conquered. Incidentally too he raised the standard of performance so that we are no longer satisfied with lifeless playing, even though distinguished by the precision for which we must not forget to thank Costa, the cold-blooded martinet and absolute Philistine as he was from an artistic point of view. . . . The programme of the present Festival contains a more than fair proportion of lengthy choral works but the strain upon the chorus is not so great as might be imagined for two of them are thoroughly hackneyed, *Messiah* and *Elijah*, and another, *The Golden Legend*, is familiar. In Bach's great *Mass*, Beethoven's *Choral* Symphony, Elgar's new work [*The Apostles*] however, there remains ample work for the chorus to do. This year a change in the chorusmastership was necessitated by the death of Dr Swinnerton Heap, but to fill his place the choice of the Committee fell upon Mr R. H. Wilson, a choice which had much to recommend it, for Mr Wilson has proved his efficiency and fulfils the important condition of having the all-complete confidence of his chief, with whom he has been associated at Manchester since Dr Richter became conductor of the Hallé orchestra. More of the Manchester element was introduced by the engagement almost *en bloc* of the Hallé orchestra and this is naturally something of a blow to the *amour propre* of Birmingham. Probably the Manchester strings are not quite as rich in tone as the Londoners, but on the other hand they have an *esprit de corps* which only a permanent orchestra can possess, and it is to be hoped that the Birmingham people will see that the best way of meeting the difficulty and repelling this northern incursion is to develop the orchestral resources of their own town until they can supply at least the nucleus of a Festival band.[12]

The morning of the second day was the focal point of the Festival that year, for another new choral work by Elgar was on the programme. At the beginning of the year it was quite apparent from Richter's visit to Elgar at Malvern that he went, not only to discuss the forthcoming performance of the *Coronation Ode*, but also to peruse what was already composed of *The Apostles*. At that point Richter was to be its conductor. Delays and distractions throughout the year meant that by the time the chorus (now under the able R. H. Wilson from Manchester) began its rehearsals Elgar was still hard at work orchestrating his new composition, leaving Richter with very little time once again to study the new score. No one, neither Richter, the Festival Committee, nor Elgar himself, wanted a repetition of events three years earlier, and so it was decided with full unanimity among the protagonists that Elgar should conduct the new work himself. From his holiday home in Lower Austria Richter allayed Elgar's fears that he might be offended.

It is quite a natural feeling that the public like to *see* the man, whose work they have enthusiastically enjoyed, therefore I recommend Mr Johnstone and other members of the Committee to invite you to conduct your work; this was long before the Committee meeting. I am happy that I shall have the great pleasure to hear your work as a listener. In the course of the coming season I shall conduct the *Apostles* in Manchester, then it will be a great advantage for me, that I have heard them under the creator's guidance. I am sure you will not be angry when I promise you that I shall do my utmost to try to conduct the *Apostles* still better than the composer himself.[13]

After the triumphant première on the morning of the second day, Richter was among those congratulating Elgar in the green room. Chorus master Wilson recalled that Richter described *The Apostles* as '"the greatest work since Beethoven's Mass in D" [*Missa solemnis*]. And when I said – knowing his intimate connections with Wagner and his works — "excepting Wagner?", he replied almost savagely, "I except no man,"'[14] and left the room to complete the morning's concert by conducting Brahms's Fourth Symphony. A few months later Richter lunched with David Ffrangcon-Davies. The singer told Elgar that the conductor 'evinced deep emotion while he discussed your work. "The greatest musician since Purcell in England. Genius not talent. Individuality and loyalty to the truth as he sees it. Always himself and always English, glorified by the Universe in his compositions."[15] Nothing in the rest of the remaining programmes of the 1903 Birmingham Festival (Sullivan's *Golden Legend*, *Messiah*, Bach's Mass in B minor, the first performance in England of Bruckner's Te Deum, or one of Richter's famed renditions of the 'Choral' Symphony, 'one of the finest ever given in this country', said the *Musical Times*[16]) quite matched the excitement of that morning. Richter's *Elijah* and *Messiah* were now frequently attacked as coldly and unsympathetically interpreted, whereas the piece which had an exclamation mark beside it in his conducting books was Sullivan's *Golden Legend*. His *Messiah* was criticized because of his adherence to Franz's edition, which he was now having to drop in favour of Costa (Prout's edition was not yet in use, having appeared just a year earlier). Of this performance Thompson noted that 'the most individual feature of his reading is perhaps its obvious effort to realise the dignity of the music, and it was rather amusing to find how he broadened the introductory passage of the airs, which the vocalists, when they began, promptly hurried up to suit their convenience'.[17]

The programming of the 1903/4 season at the Hallé was the apogee of Richter's decade in Manchester. A purely orchestral concert began the season on 22 October, Goldmark's Merlin Overture being the only novelty. *Šárka*, from Smetana's *Má vlast*, Borodin's *Prince Igor* and Berlioz's *Beatrice and Benedict* Overtures, Bruch's *Ave Maria* for cello and orchestra (Carl Fuchs), and Saint-Saëns's Fifth Piano Concerto (Busoni) were all heard for the first time at subsequent concerts until the end of the year. His Gentlemen's Concert (30 November) included Goetz's Overture *Der Widerspänstigen Zähmung* and the first performance of Mozart's Symphony No. 33 in B flat, K. 319. The concert in Liverpool on 5 December was dedicated to the memory of his (and Elgar's) close friend Alfred Rodewald; Muriel Foster sang Brahms's Alto Rhapsody and the Angel's Farewell from *Gerontius*. Mozart's *Masonic Funeral Music*, his old Viennese way of saying farewell to friend and colleagues, began the programme which also included the 'Eroica' Symphony, the Prelude from *Parsifal*, and Strauss's *Tod und Verklärung*. Rodewald's death affected Richter deeply. He described him to Arthur Johnstone as 'a master in the art of living in the best sense of the word. I made his acquaintance some sixteen years age, and the friendship between us became closer in the course of a musical festival in Aachen [sixty-fifth Lower Rhine Music Festival], where as a volunteer he handled his double-bass effectively enough in the orchestra. But it was in Bayreuth that we became thoroughly intimate. He was one of the most zealous and most intelligent among the frequenters of Bayreuth. I shall never forget the tall, fine fellow with open-hearted, kindly and loyal nature that he was.'[18]

Another double-bass playing friend of Richter's, the Hungarian Carl Gianicelli, was still doing his best to entice him to guest-conduct in Budapest but Richter's schedule precluded it (he managed a quick trip to Brussels for a single concert on 18 December, his last engagement in 1903). Richter still held fast to his Hungarian roots, and if he could not go back to his homeland at this time then he at least did his best for any of his compatriots who came to England, such as Ernst von Dohnányi. Gianicelli also introduced Richter to another Hungarian pianist and composer when he was holidaying in Hungary during the summer of 1903. This was the 24-year-old Béla Bartók, who had just graduated from the Budapest Academy of Music. Richter was shown the piano score of a symphonic poem entitled *Kossuth*, a work with a Hungarian nationalistic flavour influenced by *Ein Heldenleben* of Bartók's current hero Richard Strauss. Richter was impressed, but he insisted that it be well scored before accepting it for performance with the Hallé in the forthcoming season. The year 1904 began with *Elijah*, but the Overture to Siegfried Wagner's opera *Der Bärenhäuter* opened the next concert on 14 January. Coincidentally another vast tone-poem by Strauss, *Also sprach Zarathustra* (an 'absurd farrago' Sir George Grove had called it), received its Manchester première (only the third performance in England) on 28 January. Richter then took the Hallé down to London and, on 2 February, repeated it before an audience which included the new Queen Alexandra ('she stayed until the end', he noted proudly in his conducting book). Then it was back north for concerts on consecutive days with Berlioz, Wagner, Strauss, and Brahms in Liverpool, followed by an operatic evening in Manchester consisting of the *scena* 'Abscheulicher' from *Fidelio* sandwiched between the second acts of *Fliegende Holländer* and, for the first time, Cornelius's *Barbier von Bagdad*. On 11 February Manchester was given its first taste of Bruckner's Seventh Symphony, which 'was condemned before being heard, as sometimes happens here', the *Musical Times* cynically remarked.[19]

Bartók arrived at Bowdon on the next day, 12 February, his *Kossuth* having received its first performance a month earlier at a concert in Budapest. He was fully accepted into the Richter household, as d'Albert had been twenty years earlier, mothered by Marie and befriended by Mathilde. The head of the house, meanwhile, had taken his orchestra off to London again and, on 16 February, gave yet another young artist the chance to show his mettle before the public; this time it was the 22-year-old Austrian Artur Schnabel in Brahms's Second Piano Concerto. As a boy Schnabel had often attended Richter's Philharmonic concerts in Vienna and he went on to become a pupil of Leschetizky. At this London début Queen Alexandra was present once again 'and sat at the end of the balcony [in the

Queen's Hall], just behind him, and as the audience applauded after each movement, he had to make a complete *volte-face* each time and make a special low bow to her. At the end, after the sixth recall, he could not help feeling a little foolish. As to the performance, he said he played with much animation, that the orchestra produced beautiful tone throughout and that the cello solo [played by Carl Fuchs] in the slow movement was "wonderful".'[20]

Bartók featured as piano soloist in the first hearing of Liszt's Spanish Rhapsody (orchestrated by Busoni) and, at Richter's insistence, played Volkmann's *Variations on a Theme by Handel* as a piano solo. The concert began with Schubert's Unfinished Symphony and ended with Dvořák's 'Czech' Suite; everything else in between (including Bartók's *Kossuth*) was, according to the critic of the *Musical Times*, not to the taste of the Manchester public, whose reception of Bartók was at best 'tolerably cordial'.[21] He emerged in the press as a better pianist than composer, many taking offence at the blatant anti-Austrian sentiment of his music in which that country's national anthem is roughly treated. Richter had singled this part out for praise with 'Bravo and grossartig [splendid]!'[22] when Bartók played the work through to him in 1903. After the concert, however, it appears that Richter cooled in his attitude to *Kossuth*. There were signs of his old sensitivity to press criticism, stemming from his years in Vienna. Any attack on the choice of programme or the works themselves (and this included Volkmann's slight piano solo which he had selected) was considered an attack upon him. Bartók complained bitterly to his mother, accusing Richter of not introducing enough new music into his programmes. This was not true as almost every Hallé concert at this time included first performances in the city if not in the country.

Bartók stayed with Richter long enough to witness just such a première, this time Manchester's first *The Apostles* on 25 February. It was 'remarkably fine', said the *Musical Times*. The concert on 10 March also included some new music, Busoni's Violin Concerto (with Brodsky) and Tchaikovsky's 'Manfred' Symphony. From there Richter went to London for the Elgar Festival where, at Covent Garden, he conducted on three consecutive nights (14–16 March) *Gerontius*, the first performance in London of *The Apostles*, and a Miscellaneous Concert in which Elgar conducted his new Overture *In the South* and the first two Pomp and Circumstance Marches. History repeated itself; Richter was to have conducted *In the South*, but with his crippling schedule in March and with the score and parts delayed at the publishers, he insisted (as he had done with the Birmingham *The Apostles*) that the composer was the best man for the job. Instead Richter

1903–1904: England

conducted *Froissart*, Enigma Variations, selections from *Caractacus*, and *Sea Pictures* with Clara Butt. The whole Hallé orchestra (100 players) and chorus (275 singers) came to London for the event but the venue of the Royal Opera House was ill chosen. The full effect of the choral singing was dampened because the chorus was stuck right at the back of the vast stage with their sound disappearing upwards into a blue canopy beneath which were festoons of roses with electric lights in their midst. The effect of the Demons' Chorus in *Gerontius* suggested to the critic of the *Referee* 'a carnival celebration heard through a telephone'.[23] The orchestra was not in the pit but on stage in front of the chorus, and the plush furnishings of the auditorium further deadened the acoustrics. Nevertheless the event was hailed as 'a triumph for English music'. Richter conducted *Gerontius* 'with sympathetic intuitiveness and consummate mastery' and *The Apostles* with 'rare authority and dignity'. The king and queen attended the first two concerts, the queen attended the third alone. After it 'Her Majesty the Queen sent for [Dr Elgar] and Dr Richter to express to them her pleasure at

FIG. 13 Richter conducting the *Dream of Gerontius* at the Elgar Festival, Covent Garden 1904.

A SHOW OF HANS.
[RICHTER interprets ELGAR's *Dream*.]

the success of the Festival.'[24] It was a unique occasion for English music; not even Sullivan had earned himself a three-day Festival in death let alone in life. Elgar told his father that 'Richter did it all splendidly',[25] and the *Manchester Courier* considered that 'the Hallé chorus has indeed given our London societies a much-needed lesson in choral singing'.[26]

Richter's Manchester season ended with Beethoven's 'Choral' Symphony ('magnificently done by the orchestra and choir conducted by Dr Richter with sacerdotal impressiveness'[27]) and a Wagner night in aid of the pension fund. After a single concert in Huddersfield he said farewell to the Hallé until the autumn and, in April, exchanged Manchester for two orchestral concerts in Turin and a Beethoven/Wagner concert in Antwerp. Then it was back to London for another Covent Garden season, this time a mixture of Mozart (*Don Giovanni* and *Figaro*) and Wagner (*Lohengrin*, *Tristan*, *Tannhäuser*, and *Meistersinger*) from 2 May until 18 June, twenty-six performances in six weeks as well as rehearsals for the first concert of the newly created London Symphony Orchestra on the afternoon of 9 June followed that *same* evening by *Tristan*.

Richter's notices for his Mozart performances were mixed, though, predictably, he was praised to the skies for his revelatory Wagner. It seems he may not have been at ease in Mozart, having none of Mahler's quicksilver conducting; but, heavy-handed though he may have been, he cannot be denied a place among the worshippers at the altar of Mozart. An apocryphal but illuminating story makes the point. He was asked for his nomination for the greatest composer and his choice was, not unexpectedly, Beethoven. When, however, the supplementary question 'What about Mozart?' was asked, he is said to have replied that he did not realise that Mozart was under consideration, assuming that the original question referred only to all other composers. Despite the announcement of *Don Giovanni* as uncut, the final sextet was not performed, causing considerable discussion in the press, particularly as Wagner's operas were performed in their entirety. The season was notable for the débuts of several singers who were at the threshold of great careers. Richter conducted Emmy Destinn as Donna Anna (*Don Giovanni*) and Elsa (*Lohengrin*), Selma Kurz (in a single performance as Elisabeth in *Tannhäuser*), and Robert Radford as the Commendatore in *Don Giovanni*. He had first heard Destinn as Senta at Bayreuth in 1901 and 1902 and immediately recognized an enormous talent. He told Carl Gianicelli in Budapest, 'take her; a wonderful, well-trained voice, nice youthful appearance which will seem frankly beautiful when she sings, a large voice, sings everything; but ensure that she sings good music for the young girl sings absolutely everything. If you have room for her, she will do you credit.'[28]

Richter's contribution to the German season of opera at Covent Garden in 1904, generally acknowledged as a high-water mark in the opera-house's history, is best summed up by a Sunday newspaper review which appeared on 19 June.

It was, we regret to say, Richter's last night of conducting at the Covent Garden opera this season, and we understand he has already left London for Bayreuth. The work that Richter has done during the time he has been with us this year has been of enormous value and of the highest quality; his thoroughness, his complete command of the orchestra in its every phase, his powerful personality, and the strength with which he knows how to exercise that personality over his forces, are all matters which go to make him the superb conductor we now know him to be. He leaves England with the congratulations of every musician who has been able to follow all the details of the work which he has this year done for us all at Covent Garden. There are so few magisterial conductors now alive (and in that character Dr Richter has only two or three rivals) that the pleasure which one derives from his large knowledge of the orchestra, both in its most intimate places, as we have said and in the broad massing of it, that one wonders how that in recent days there has grown up a small school that is by way of belittling the position of a conductor; they appeal to the past, and point out with accuracy that the art of conducting as we know it now was scarcely known in the very old days of music. The fact still remains that the personal magnetism of any conductor worth his salt has an enormous influence upon the band under his direction. Very few . . . can change the individuality of its playing by pure personal power. We congratulate Dr Hans Richter over the accomplishment of the Covent Garden orchestra when it is he who is actually directing there. Let all honours go to Richter by reason of his wonderful work.

All the reviews of *Tristan* agreed on the finely restrained but phenomenally powerful rendition of the uncut work (it took nearly five hours). They spoke of a rich, sonorous, and perfect balance between stage and pit, and of the intense eloquence and depth of meaning which he elicited from his players. The massive Richterian climaxes had all the required force and the passion and poetry as well as the exquisite subtlety of the score were brought out superbly, particularly in the final act. The performance of *Lohengrin* (with Destinn, van Rooy, and Kirkby Lunn) was described by the *Daily Telegraph* as the best London had ever known. His reading was described as dignified yet immensely vigorous, perfectly catching its poetic atmosphere. The *Daily Express* got to discussing Richter's contribution only after highlighting the activities of the royal party which attended.

Lohengrin with the usual cuts restored made rather a lengthy performance last night at the Opera, and at the early hour of seven, at which it began, the house would, but for the enthusiasts in the amphitheatre, have looked rather empty. The first act was, however, not far advanced before the Queen and Princess Victoria entered the royal box, and by half past eight the house was well filled. The King arrived about nine, and, after sitting for a little time with Her Majesty, went down to the omnibus box, which was occupied by the Prince of Wales. The chief merit of an excellent performance was undoubtedly the splendid playing of the orchestra under Dr Richter.[29]

The first night of the season (2 May) was overshadowed for Richter by the sudden death the day before of his friend Dvořák. The last exchange of letters between the two men had been two years earlier in 1902, when, on impulse, Richter wrote to the composer after performing the Sixth Symphony (dedicated to him by Dvořák) in Manchester on 23 January.

After the great delight which your D major Symphony gave us all yesterday, I feel I must write you a few lines. I do not know whether you know that Manchester and the towns to which I travel with my orchestra are very much taken with your compositions; your name is among those that appear in my programmes most frequently. I am writing this to you because I think it gives you pleasure, not to earn your praise. It is my *duty* to devote all my talent to the propagation and support of good and beautiful works, and you make the fulfilment of that duty easy and delightful. So no word of thanks or I shall not send you any more programmes. . . . I am proud that you should have honoured me with the dedication of this magnificent Symphony.[30]

Dvořák's reply concluded with the words, 'I know you despise compliments, but you will forgive me if I say once again; accept my best thanks for all that you have done for me and for my works.'[31] Richter remained loyal to his friend and his music for the rest of his life. During the afternoon of 9 June the new London Symphony Orchestra gave its first concert. The unusual time of day was chosen to accommodate Richter, its first principal conductor, and those of its players who were currently playing in the German opera season. The programme was purely orchestral and bore every resemblance to the Richter Concert programmes which had dominated London's musical life each summer from 1879 to 1902; Wagner, Bach, Mozart, Liszt, and Beethoven. The only newcomer in this programme was Elgar and his Enigma Variations. The orchestra had been formed as a result of an edict issued by Henry Wood to his players in February 1904 that 'after this season no deputies will be allowed at any of the concerts and rehearsals given by the Queen's Hall orchestra' (its season ran for eleven weeks from the beginning of August). No outside engagements were to be permitted without express permission from the Wood/

Newman management. The effect of this ruling was catastrophic. Many of the players earned as much in a week playing at provincial festivals such as Birmingham or Leeds as they did in a month playing for Wood in his Promenade season. A two-guinea fee in London for a principal player (one guinea for rank and file) was doubled in the provinces, whilst the suburbs of the metropolis and many amateur orchestral societies in need of stiffening offered one-off engagements; in short the freelance player, then as now, needed the freedom to pursue his work anywhere and anytime he could find it. What with the occasional concerts given by Wood's orchestra outside his summer season, a rank-and-file player was expected to earn about £120 per year, a principal player about £250. Even Richter's Hallé players were only in secure employment for about six months of the year; for the rest they had to depend on summer engagements at seaside resorts and north country spa towns.

The immediate result of Wood and Newman's action was the resignation of forty-six players led by a quartet of brass players, three (Adolf Borsdorf, Thomas Busby, and Henri van der Meerschen) from the horn section, together with the principal trumpet John Solomon. A formal meeting of the dissidents had taken place on 19 May in the middle of the opera season, illustrating once again how quickly events unfolded thereafter. Richter, having been approached, agreed to an association with them, providing the orchestra was increased to 100 players. This target was achieved and the first rehearsal, at which Richter exhorted the players to strive for artistic discipline in their quest for success, took place and culminated in a brilliant début for the orchestra. Thereafter Richter, along with his younger Hungarian colleague Arthur Nikisch, maintained a long association with the orchestra, but it also developed very close ties with Elgar from its earliest days and invited guest conductors from home (Cowen, Mackenzie, Stanford, Ronald, and Pitt) and abroad (Steinbach, Safonov, Colonne, Weingartner, Fiedler, and Schuch) to work with it from the outset.

One of the LSO's players, the sub-principal violin Wynn Reeves, had this to say about playing under Richter at Covent Garden.

He had magnificent poise and mastery of every detail. He never glanced at the score yet gave every important cue without fuss. He had a gadget on the side of the desk containing a moist sponge into which he meticulously dipped his thumb before turning each page. During the ten seasons I played under him I remember him making only one mistake. In Act III of *Tristan*, Isolde's first words on her entry are 'Ich bin's', preceded by two silent beats. Richter omitted to indicate the two silent beats, a thing we were quite unprepared for. The result was utter chaos for some bars.... If this had happened with any other conductor the orchestra would have jumped to the conclusion that he had omitted to indicate the two beats and would automatically have skipped them, but we had got into the habit of depending absolutely on Richter's infallibility.... He was one of the greatest musicians I have ever played under, with a wonderful insight into whatever he was interpreting.... He never descended into the slightest hint of effect for effect's sake.[32]

29

1904–1906

England

After the 1904 Covent Garden season Richter travelled to Bayreuth to conduct one cycle of the *Ring*. Correspondence between him and Cosima in 1903 had been largely devoted to the American performances of *Parsifal*, although Richter also reported very favourably on his new, annual association with the Royal Opera House. His letters revealed that he had been given twenty-five orchestral rehearsals for the *Ring* and gave a detailed assessment of the principal singers should they be considered for Bayreuth. His letters describing the opera season of 1904 are similar, Destinn and Ternina being singled out for the highest praise. Richter did not get much pleasure out of conducting the Italian singers in the two Mozart operas.

Only the Donna Anna of Destinn was splendid; *she* has learnt something at Bayreuth. . . . Ternina as Isolde and Elisabeth outshone everybody else. . . . Van Rooy has retained a lot from Bayreuth in his Sachs, but he barked his Telramund and Kurwenal. As Tristan and Tannhäuser Burrian was, vocally and musically, a model of security, but he is too short and his face has a comic look about it. A pity! Otherwise he is a splendid and dear fellow who involves himself enthusiastically.[1]

Richter shared the 1904 Bayreuth *Ring* with Cosima's son-in-law (Isolde's husband) Franz Beidler and Muck shared *Parsifal* with another newcomer to the podium, Michael Balling. Balling, a viola-player at Bayreuth since 1886, assisted Richter over the years and took over the Hallé after his mentor's retirement in 1911 until the First World War intervened. Richter approved highly of Balling, not just because of his innate musicianship but also because, like Richter himself, he had emerged as a conductor after years in the orchestra. The last time Richter played the viola in public (rather than chamber music at home) was on 25 February 1884, when he played in Liszt's *Hungarian Coronation Mass* in Pressburg Cathedral under the composer. He still maintained a great fondness for the instrument (Berlioz's *Harold in Italy* was often in his programmes), although Archie Camden recalled his outburst, 'Violas, what are they? Horn-players who have lost their teeth.'[2] The young Lionel Tertis recalled Richter:

I remember in my early days playing the Benjamin Dale *Suite* for viola and full orchestra, the scoring of which is as follows: in addition to some sixty string players there are two flutes, two oboes, cor anglais, two clarinets, bass clarinet, bassoon, four horns, two trumpets, three trombones, tuba, timpani, every known species of percussion, celesta, and two harps—that's all! Hans Richter was conducting it and I recall that he complained to me after the rehearsal of the huge orchestra employed, saying 'I suppose the next addition they will find necessary for accompanying the poor dear viola will be a battery of exploding air-balloons!'[3]

Tertis actually performed the suite with Nikisch (who came to the rehearsal totally unprepared and took little care with it) for the first time in May 1911 after Richter's retirement from the concert platform. Not surprisingly, no record of the work exists in Richter's conducting books; nevertheless, despite being wrongly attributed to him, it is a story worth preserving and its sentiment might well have been Richter's. A suite which did come his way was Bartók's Op. 2, which the composer showed to him in the summer of 1904 at Bayreuth, largely through the mediation of Mathilde Richter. Although Richter thought it a successful work, he also added the caveat, 'you mustn't expect it to be generally liked though'.[4] Bartók the composer was shunned from now on, although as a pianist he was to make one more appearance in November 1905, again largely thanks to Mathilde.

The 1904 Bayreuth Festival was overshadowed for the Richter family by the severe illness of their daughter Mitzi. On 23 August she gave birth to a daughter Marie, and all seemed well until three days later, when she developed a high fever and had to undergo an operation. She was suffering from blood poisoning due to the carelessness of the midwife. It was two weeks before she recovered enough to receive visitors and before Richter could take up Cosima's offer to care for her. She remained in Bayreuth for several months in a flat rented by her father. Meanwhile he had returned to Manchester, his nerves shattered by the worries of the past month. Richter told Elgar all about it in rather quainter English than usual (the midwife described as 'mischievous').

We shall be home again [from Hungary] on the 10th of October. I shall bring a Hungarian instrument—Tárogató—which I should like to be heard by you, probably you will find it good enough for the 'Schofar'. Although it is a wooden instrument, it has a strong and telling sound; the Hungarian shepherds play it. At all events it is an interesting instrument. It was played at Covent Garden in the third act of *Tristan*—'die lustige Weise'—it made a very striking effect and could not be overpowered by the greatest *ff* of the orchestra.

My dear friend, I hope you did not notice that I was silent after you have been knighted, but was it possible for me to congratulate all the 'Sirs' who are honoured by your companionship? As a sign of high estimation of our noble King I am happy about the honour he has conferred upon you, but the highest title is: Edward Elgar! Thousand thanks for the Schreibmaschine [typewritten] letter, God bless the inventor![5]

Elgar had mentioned the Schofar or ram's horn which introduces Psalm 92 (the Morning Psalm) with a flourish in *The Apostles* at the end of August 1903. On that occasion Elgar asked Richter if he might have Walter Morrow, the principal trumpeter for his London concerts, to play at the Birmingham première. There was no further mention of the Tárogató in any performance of the work.

There were fewer first performances in the 1904–5 season by the veteran conductor, as reviewers were now beginning to call him; nevertheless there were some novel ideas. The second half of the second concert of the season (27 October) consisted of a concert performance of *Fidelio* with Agnes Nicholls in the title-role. Before the next concert Richter took the orchestra to two new towns, Cardiff and Cheltenham. Kreisler joined him for the next Manchester performance (3 November) in Joachim's Variations, preceded by the first hearing there of Elgar's *In the South*. The next première was Henselt's Piano Concerto played by Busoni on 24 November, this concert also including Cowen's Indian Rhapsody for the first time. On 1 December Bach provided the surprising source of new works, his Fourth 'Brandenburg' Concerto (with Brodsky and the orchestra's two flautists Needham and Redfern) and the motet for double chorus 'Singet dem Herrn'. A few days later (5 December) his soloist in Tchaikovsky's First Piano Concerto in Sheffield was Mark Hambourg, whose career he had launched nearly ten years earlier. Hambourg never forgot his debt to Richter, 'who was always so helpful and sympathetic to fellow-artists, so generous and unselfish in his professional outlook'.[6]

Plunket Greene sang Stanford's *Songs of the Sea* on 8 December and Edward Isaacs played (Beethoven's First Piano Concerto) with the Hallé for the first time. Isaacs was a local man like Arthur Catterall, the soloist at the next concert, a Tchaikovsky evening anticipating a Sunday night at the

Royal Albert Hall eighty years later. Whereas Isaacs promoted much chamber music in the city in future years, Catterall (a Brodsky pupil and currently sub-principal first violin of the orchestra) was destined to lead the Hallé orchestra from 1913 to 1925. On this occasion Risegari was ill and Brodsky stood in for him as leader, so Catterall was given a baptism of fire when playing Tchaikovsky's concerto, for he stood a few feet from the two men who first gave it to the world.

The year 1905 continued with works new to the Hallé. There was Smetana's *From Bohemia's Woods and Fields* (12 January), Hugo Wolf's Italian Serenade (16 January at a Gentlemen's Concert), Goldmark's Overture *Penthesilea*—was it a coincidence that Richter played Goldmark's version of this formerly contentious work only days after playing a piece by Wolf?—Glinka's Fantasia *Souvenir d'une nuit d'été à Madrid*, and Strauss's *Burleske* with Backhaus (19 January), Percy Pitt's Oriental Rhapsody (23 February), and Sibelius's Second Symphony on 2 March, its British première.

Richter had been influential on Sibelius's behalf years before in 1890 during the young man's student days in Vienna by arranging for him to study with his old friend Robert Fuchs. '[Richter] received me in a very friendly way', recalled Sibelius, 'and listened attentively to what I had to say, in spite of my having called on him during a rehearsal of *Tristan*. What I most wanted at that stage of my development was guidance in instrumentation. Richter was a convinced Wagnerian, but not a fanatic, seeing that he recommended a teacher for me from the opposite camp.'[7] The *Musical Times* could muster little enthusiasm for the work, saying that it was 'played without creating any pronounced impression'.[8]

Richter fell ill immediately after this concert and was unable to complete the season (the remaining two concerts were shared between Wilson and Brodsky). According to a letter to Percy Pitt from Marie, her husband was suffering from erysipelas, a bacterial infection of the skin, which, before the days of penicillin, could be very serious. By 8 March (the date of her card to Pitt) the doctor assured the family that his patient was out of danger. By 21 March he was well enough to write a letter of copious thanks to Mrs Joshua, who had bombarded him with all kinds of good things to eat. He also told her that the doctor had allowed him to start smoking again (he still had Christmas cigars left to which he now looked forward). On 16 April he set off for London and another Covent Garden season of purely Wagner operas. Between 1 May and 14 June he conducted two cycles of the *Ring*, *Lohengrin* (4), *Tristan* (2), *Tannhäuser* (3), and *Meistersinger* (3), twenty Wagner operas in six weeks. He received wonderful notices. 'He approached [*Meistersinger*] with the vigour and enthusiasm of youth. His responsibility sat lightly upon him.'[9] 'There were moments [in *Tristan*] when the human voice seemed lost in the passionate exaltation of the music, as though Dr Richter held that the emotion of the lovers had soared to heights beyond the reach of vocal expression.'[10] 'The orchestra was excellent and the fine quality of the brass in the refined effects entrusted to them was most noteworthy. Dr Richter, who was received with enthusiasm by an audience who crowded every corner of the huge auditorium, conducted with a power of command which gave pleasing evidence of his complete recovery. . . . [In *Walküre*] the orchestra was quite superb and it was worthy of note how, under Dr Richter's direction, the voices were never overwhelmed but stood out clearly.'[11] He showed his customary modesty, striving to transfer the manners and etiquette of Bayreuth to other opera-houses. 'One of the most successful *Ring* cycles yet heard in London came to a triumphant end, and the audience was not too fatigued by the long performance of *Götterdämmerung* to render to Dr Richter the tribute he deserved so well and was quite reluctant to accept.'[12]

Ernest Newman took over as music critic of the *Manchester Guardian* when the 1905–6 Hallé season began on 19 October. Another of Bach's 'Brandenburg' Concertos (No. 3) was heard at the concerts for the first time on this occasion, and the following week Richter made a rare excursion into French music with César Franck's *Symphonic Variations* with Raoul Pugno as soloist. The concert ended with Schumann's Second Symphony. On the train home to Bowdon Richter frequently travelled with Daisy Jordan (the sister-in-law of Carl Fuchs and a teacher of violin and viola). Miss Jordan recalled that Richter considered Schumann's Fourth Symphony his best. 'He scored it quite beautifully and finely, but then he performed it in Düsseldorf where he had his post. The violas and cellos there were so heartily bad that he stuffed the work with horns, bassoons etc. In the whole symphony there is only one bar without wind instruments, and that is an upbeat consisting of one note.'[13]

Bruckner's Fourth Symphony (2 November), César Franck's Symphony (16 November), Bach's *Wedding Cantata*, a Mozart divertimento, an anthem by Purcell, and a ballet suite by Lully (23 November) followed. There were no new works at the other November concert (16th) when Bartók made his final appearance with Richter playing Liszt's *Totentanz* and Bach's Chromatic Fantasia and Fugue. Richter was no longer interested in him as a composer, but it is known amongst the family that Bartók entertained hopes of marrying Mathilde Richter. She had no feelings for him beyond friendship, so her father breathed a sigh of relief when the young man left Bowdon on the day following his concert. Two years later the subject of his suite (which Bartók had shown him at Bayreuth in 1904) came up in correspondence with Gianicelli. Richter, on his way back to England from summer holidays at Baracs, intended to stop in Budapest long enough to hear Hungarian works proposed by Gianicelli for inclusion in his orchestral programmes in Manchester or London. 'Come now dear Professor', chastised Richter, 'what are you thinking of? No question of a play-through!!! If Bartók's *Suite* is "mad"—and assuredly it is when I consider that he could memorize *Heldenleben*—then I cannot perform it.'[14] Of the composers he did hear, Mézaros, Bürger, and Kerner, there is no sign in the 1907–8 season (he played safe with Liszt), and Bartók stayed away from England for the next sixteen years.

Richter's Hallé concert on 7 December was notable for the first performance of Elgar's new Introduction and Allegro for strings (with the Brodsky Quartet taking the solo parts) and for the appearance of Percy Grainger as solo pianist in Tchaikovsky's First Concerto. Richter was so unimpressed by the audience's moderate response to Elgar's work that he promptly repeated it, much to their astonishment. Grainger, recommended to Richter by Cyril Scott, first appeared in Manchester earlier in the year (27 February) at a Gentlemen's Concert (not part of the subscription series) playing Tchaikovsky's Second Concerto in G major. Grainger told his friend Herman Sandby, 'Had huge success. Old Richter shouting himself hoarse with bravos. Reengaged.'[15] In 1909 he wrote, 'And I've played with Richter, mother! And, oh that's a privilege. Do you know that Richter is so absolutely an artist, so utterly unworldly, and without the banal instincts of a showman.... His constant striving for perfection in connection with every bar of music he touches is a fount of inspiration for us younger men, and a challenge. But Richter is unique. He has been wonderfully kind to me—and helpful—as only Richter can be.'[16]

Egon Petri, who came to live in Bowdon and teach piano at the Royal Manchester College of Music in 1905, made his début in the 'Emperor' Concerto on 14 December, and in the New Year of 1906 Richter took the orchestra to Yorkshire and Scotland for four concerts. The first notable event of the year was on 8 February, when he conducted the first performance of Strauss's *Sinfonia domestica*. Newman, who had given Richter a very bad notice (without directly naming him) for a Berlioz concert a week earlier, told Herbert Thompson:

I find a lot of people didn't think the *Domestica* so very bad after all. One reason for this is that Richter's performance does not make the ugly parts so ugly as they should be. In this sense his reading of it is not at all like Strauss', nor does it indeed follow the plain directions in the score, as to *ff* etc. You will find, I think, that I have always admired the beautiful portions of it, but there is so much actual wastage in it, there are so many things in it that come to nothing. The means employed are so dreadfully out of proportion to the end attained that it can't be called a first-rate work. Great works are harmoniously balanced throughout. I am afraid there is something wrong with the man who now can only get us to say that the beautiful parts of his symphony outweigh the ugly parts and the stupid parts, and the ineffective parts of the music, and the miscalculated parts. Isn't that really damning it with faint praise, isn't there something wrong when you can only commend him for doing something or other and apologise for his having done something else?[17]

Richter then went to London for a concert with the London Symphony Orchestra at which he gave the first performance of York Bowen's *Symphonic Fantasia* on 15 February. The young Adrian Boult, who, as a 6-year-old, had attended Richter's concert in Liverpool on the day of Hallé's death in 1895, was spending his schooldays at Westminster in London and was present at many LSO concerts including this one. He kept a diary in which he wrote, 'Fine LSO concert with Richter. Elgar's *In the South* received a magnificent performance, what a fine thing it is. *Tod und Verklärung* well played, but ensemble sometimes a little ragged between wind and strings. Bach's fourth Concerto played very deliberately and not so fine as Steinbach would have been. A dreadful new thing by York Bowen, half an hour of nonsense. A glorious performance of Beethoven No. 7.'[18] Boult admired Steinbach's Brahms and Nikisch was his god, but he had a high regard for Richter. In 1909 he wrote, 'Wonderful though [Nikisch] is, it seems to me that he is too exciting to be a good permanent conductor for an orchestra. The LSO have to mind their technical p's and q's when playing under Richter and people.'

When Bowen's piece was repeated with the Hallé on 8 March, Thompson 'thanked Providence that he is not original. . . . The best influences are his immediate forerunners Strauss, Tchaikovsky, and most of all Wagner in his *Tristan* vein. . . . It was exceedingly well played under the direction of Dr Richter, who took great pains with it and it seemed to give the audience great pleasure.'[19] At the same concert Manchester heard Wagner's *Das Liebesmahl der Apostel* for the first time, a very different composition from Elgar's *The Apostles*. The 16-year-old Mischa Elman was among Richter's soloists for the final concerts of the season. He played Glazunov's Violin Concerto on 1 March (another first performance for the Hallé), and the concerts ended with the 'Choral' Symphony. In February 1906 Granville

Humphreys attended a Richter rehearsal with the Hallé and described the event for the *Musical Herald*.

Entering the Manchester Free Trade Hall any Thursday during the concert season at about half-past-one o'clock, the one hundred performers composing the magnificent Hallé orchestra will be found grouped on the spacious platform, engaged in the necessary though cacophonous proceeding of 'tuning up'. But when the famous conductor takes his stand on the insignificant box which does duty for a rostrum, the babel of sounds ceases, and an attentive and expectant hush succeeds. . . . Dr Richter is great in every way. His appearance is commanding. His leonine head indicates great power. His profound knowledge ensures respect for every word he utters, and effectually represses any tendency to 'talk back', even in the most irrepressible and self-opinionated performer. Then he is master of the technique of conducting, a detail some much less eminent conductors seem to think unworthy of serious attention. . . . Here are no unnecessary movements, no *Sousaisms*, every gesture is significant and full of expression. . . . The right uses of wrist and forearm, and shoulder movements are continually exemplified. The left hand and arm are properly retained to indicate the more important entries and grades of expression. The beat is firm and clear. There is no sawing of the air with the baton, and the doctor uses his feet to stand on simply, not to raise dust. Energy is conserved both in respect of gesture and language. Dr Richter's English, if not literary, is pungently expressive. 'Make a mark', he cries, when some point of expression has been overlooked by a section of the orchestra, and immediately pencils are busy on the copies. 'Nothing' or 'soundless' conveys his idea of a *pianissimo*. 'You *must* look at me', he shouts when the second violins are inclined to be 'too previous' in taking up a lead, at the same time looking at them with a glance well calculated to quell any rebellious tendencies. 'No solos' at once suppresses a too pushful instrument. . . . The Doctor frequently sings passages from the score to illustrate his idea of the expression required. His voice is of great volume, extensive range, and very flexible. His sense of absolute pitch is unerring. . . . The soloists . . . occasionally have rather a rough time of it, the Doctor being often better acquainted with their selections than they themselves appear to be. . . . Dr Richter is kind and patient with young artists who have sincere aims, and he has already helped many a local singer to fame.[20]

The Hallé orchestra had a new leader from the beginning of the 1905 season when Rawdon Briggs took over from the ailing Risegari. Years later, when Mathilde and Edgar were looking for a biographer for their father, they gathered personal reminiscences from people such as Briggs.

I think it must have been in 1890 that I first met him, at the German Club in London, and the first impression of a genial, simple but overpoweringly strong personality was not altered by years of close contact in travel and in musical activity. No other conductor that I ever worked with knew so exactly when to

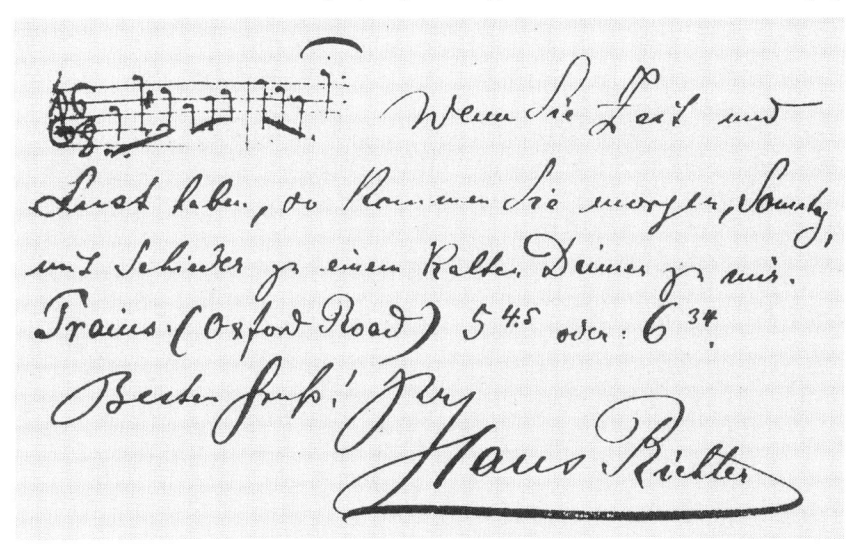

FIG. 14 A postcard from Richter to the Hallé Orchestra's principal horn Franz Paersch dated 27 February 1904 inviting him and the principal bassoon Otto Schieder to a cold supper at Bowdon. The musical quote is Siegfried's horn call.

compel and when to allow liberty. I remember in my early days as 'leader' trying to hold my first fiddles from undue hurry. To my dismay the 'Governor' (as we used to call him) thundered at me, 'One single one cannot do this'. The natural human weakness of wanting to *appear* powerful was not in him. He was the dominating power all the time and that fact led to difficulties with solo players of an inflexible type, who could only play their concerto in just one rigid manner, and to whom Dr Richter could not easily give way, knowing as he almost always did, how the piece ought to go. I remember a performance of the Schumann concerto in which the 'tuttis' were at one tempo and the solo parts at another. It was very hard for him to give way. I remember also a disastrous occasion with [William] Dayas in the E flat Liszt. Neither could accommodate so chaos ensued. All the tuttis were at one speed and the solo parts at another. An amusing conflict of will took place once when Adolf Brodsky was 'leader'. An incidental solo in one of the *Brandenburg Concertos* was the occasion of disagreement. Brodsky's idea of it was both freer and faster and he rather heatedly demanded freedom. In the artists' room after all had left Dr Richter stormed up and down saying, 'certainly he shall have freedom, but not the freedom he wants but that which I shall allow him!' Perhaps we are all like that, only that few are strong enough to hold our own. Let me add that at the performance all went well. The two great men met half way.

Of course in the nineties and even early in this century, the exaggerated effects and startling readings of recent and present times had scarcely begun. Virtuoso conductors, under whom we poor orchestral players suffered many things, were unknown in Manchester and Liverpool. In my memory of early years of Richter in Manchester rehearsals were mostly short and never did they (then) seem long; they were so interesting. How he would make the relative value of the parts clear! 'There is piano of the melodic, piano of the accompaniment [Briggs underlines "acc" twice as if Richter stressed the initial syllable], and piano of the ornament' he would explain when some misguided woodwind performer followed the letter of the law and was inaudible through conscientiousness. Then the marvellous hearing and accurate pitch! One of my colleagues used to say 'compared to him the others are tone-deaf'. In a simple passage this same player once put in a few octaves for fun. Like a flash Richter spun round. 'Octavs', he said, 'who did play octavs?' It was not safe to play tricks! I think Richter could hear differentials, for on occasion he would accuse the basses of playing a note below their lowest.

We loved our 'old man' best in Wagner; he made it all so easy. The rhythm sailed along like a ship with fair wind and favourable tide. Only the vocalists (some of them, not the opera-singers) found his inexorable rhythm too spacious for their limited lungs. It was my misfortune that I never played in opera under him. I think it was only the inadequately trained who used to fear him. Dr Richter really did have the score in his head, and not his head in the score. That was what made him so easy to follow. It may be confessed that, in later years, he sometimes studied his scores *at* rehearsal, reinforcing his memory by pointing out to us various facts of structure and of instrumentation. I shall not easily forget the long rehearsals for the *Dream of Gerontius* into which work he threw himself heart and soul. He seemed then to forget that horn players and woodwind had lips that tired, and all of us brain and fingers of men and not of demi-gods. However a word of protest always sufficed to bring him to recognise our mortal frailty and weariness.

I always felt up against a great man when in Dr Richter's company, though he seemed to me to be more like a big brother than the terrifying energy incarnate that alarmed some of my colleagues. He had much sound, practical advice for the young musician. Above all he wanted them to learn accurate rhythm and to that end he said all should take the tympani for a time. Then he never wanted slavish or wooden adherence to marks of expression, but a real alive feeling and consequent interpretation of great music. Above all he urged the study of Haydn and Mozart, because the very simplicity of this older school makes demands of feeling which to many players (not artists but mechanics) seem quite unattainable. Beethoven, he would say, was so tremendous and so vital that even the most matter-of-fact bandsman had to respond. I remember how he used to say that Mozart and Schubert wrote with their heart's blood. Of course the sapient critics objected that he put more into the music than was allowable. On the other hand the power of sheer simplicity made more of the *Pathétique* symphony of Tchaikovsky than the effects sought by ambitious conductors who seemed to desire notoriety at all costs. My own greatest delight was to play Bach's *Brandenburg Concertos* under him. No one else ever made the important parts stand out so clearly or perceived so clearly the human heart beneath the learning.

Our strange English habit of conservatism used to annoy him greatly. Why must one play the *Messiah* at every Festival and every Christmas (four or five times for the band remember) and always the *Pathétique* or the *New World* and the *Peer Gynt* Suite when visiting some other towns? I have mentioned Richter's memory for his orchestral and opera work, but he seemed to know all the Beethoven and Mozart quartets, concertos, and arias equally perfectly.

He was a bad sailor, but by going through the *Rheingold* mentally (without visible score of course) he used to beguile the fifty minute crossing between Dover and Calais and forget to feel unwell until landing. He was a great man, and had he not been a musician he would have come to the top in some other capacity, of that I am certain. If 'thoughts are things', as I verily believe, our Manchester Free Trade Hall must yet vibrate in some mysterious way to the masterful music which carried some of us to heights of wonder and adoration which seem to us now to have been finer than any level of existence since then experienced through orchestral music.

The outstanding feature in my memory of Dr Richter is that of his clarity and sanity. The economy of gesture and the irresistible force of his will. He made Brahms (I think) too austere (I played his fourth Symphony under Richter's baton and it was quite different). I have not mentioned Brahms because I never felt he really sympathised with the softness and sentimental (in the best sense) side of that *Wienerseele*. The rhythmic rigidity was quite different from Brahms' own floating, ecstatic interpretation. I fairly worshipped your father at one time, and now in memory wish to recall no other time.[21]

The summer of 1906 followed the pattern of recent years, namely a season of German operas at Covent Garden (*Tristan* (4), *Tannhäuser* (3), *Meistersinger* (2), *Fliegende Holländer* (3), two cycles of the *Ring* (with an extra *Walküre*), and a solitary performance of *Barbier von Bagdad* by Peter Cornelius. A single cycle of the *Ring* at the end of July in Bayreuth preceded his holiday in Lower Austria and Hungary. At Covent Garden Richter was able to promote yet another young English artist and reward his faithful assistant at the same time. By getting Augustus Harris to give Percy Pitt a single performance of Poldini's one-act opera *Der Vagabond und die Princessin*, Richter, who conducted Cornelius's work in the second half, ensured the first appearance by an English conductor during the so-called Grand Season (on 11 May).

Dr Richter, quite simply and informally, asked the band to give a very talented and enthusiastic English musician the same support and attention they had ever extended towards him. They would readily own, he added, that the conductor's chair was not the most comfortable seat in the opera house, but he was sure that Mr Pitt would fill it with the greatest distinction. Dr Richter then shook Mr Pitt very warmly by the hand and the orchestra applauded vigorously.[22]

Richter continued to advise Pitt as his career progressed, supporting his stand against taking a performance of *Carmen* at short notice and without rehearsal. 'Never deputize before you possess the complete confidence of the personnel and a solid well-established reputation. Fischer, with all his gifts ... conducted *Tristan* quite admirably when Levi was on leave, but no one really liked it except the musicians who were lazy about rehearsals.'[23] Richter was merciless upon orchestral players who crossed him and it was Pitt who was charged with enforcing his master's will. 'I forgot something in my last letter,' he told Pitt from Switzerland. 'The blonde cellist who also displeased Messager by not always being attentive. Show such people no mercy! How many splendid, hard-working musicians there are just waiting for work! Such *just* severity taken against slackness will give a salutary shock to those inclined to laziness.... Regarding the engagement of new members, ... we must look for the best and get as many Englishmen as possible; until now they have always proved themselves the best and most disciplined.'[24] He also used his presence at Bayreuth to Covent Garden's advantage. From the 1906 Festival he wrote to Pitt suggesting names for London's 1907 season, several of whom were taken up by the opera-house and went on to make regular appearances over many years thanks to Richter. His list gives a good indication of his casting ability.

Brünnhilde—Gulbranson; Sieglinde, Isolde, Elisabeth perhaps and Elsa—Leffler-Burckhard; Siegmund, Walter, and Lohengrin—Peter Cornelius; Loge, Tristan, Tannhäuser (possibly young Siegfried)—Carl Burrian; Siegfried—Ernst Kraus; Mime, David, Cadi, Shepherd etc.—Hans Bechstein. P.S. There is a wonderful first Rhinemaiden here, Miss [Frieda] Hempel from Schwerin, I've heard nothing like her since Lilli Lehmann back in 1876, extremely pleasant and talented.[25]

Because Richter did not trust his own ability with the English language, when it came to negotiating adequate rehearsal time and obtaining the best players and singers, he preferred Pitt to tackle Higgins on his behalf. The performance of *Barbier von Bagdad*, for example, was not to be underestimated. 'This opera should be the main novelty of the season, and it will be if it is performed as it deserves to be and as we can do it; *but for that there must be rehearsals*. Have them in the second week of the season. All the money and time would be squandered if I were not given the time I need to bring about a successful performance. I only demand this in the interests of the work and of the theatre, which in fact are one and the same.'[26]

The 1906 season at Covent Garden got off to a shaky start with Carl Burrian delayed in Germany. Several of the singers were unwell, and even some of their replacements also succumbed. A notable débutante was

Anna Bahr-Mildenburg as Isolde and Elisabeth, and there were excellent performances from the ever-reliable Kirkby Lunn, Destinn, and van Rooy. Richter received the usual accolades. 'Every instrumentalist has been specially selected by Dr Richter, and right well all of them played, especially in the second act [of *Tristan*]. . . . As for the veteran conductor who directed the performance, it is only necessary to say that he was in his usual form, conscientiously careful throughout.'[27] It was in *Götterdämmerung* that Richter always excelled in the *Ring*.

The enthusiasm of the audience may be said to have been roused almost exclusively by Dr Richter and the playing of the band. As on many former occasions, the great conductor gets every atom of expressiveness and meaning out of the music in *Götterdämmerung*, and every bar makes its full effect. . . . On Tuesday *Tristan und Isolde* was repeated, with Frau Reinl as the heroine; she did it very well, but there was not much inducement to do anything but listen to the orchestra. It is worth recording that Herr Burger sang the parts of Tristan and Siegfried on two consecutive evenings.[28]

The reviewer here forgot that, taxing though it may be to sing two such roles on consecutive evenings, it was just as tiring for a man of 63 to conduct *Tristan*, *Rheingold*, and *Walküre* or *Siegfried*, *Tristan*, and *Götterdämmerung* on *three* successive nights, which he did at the beginning of May. *Walküre* on 14 May 1906 was Hans Richter's 4,000th performance recorded in his conducting books.

30
1906–1908

England

RICHTER told Mrs Joshua that, between production rehearsals and his performances of the *Ring* cycle at Bayreuth in 1906, he went to Lenk in Switzerland to attempt a sulphur cure for 'this devilish rheumatic-neuralgia which plagues me in the winter with an unwelcome visit'. He then travelled to Zermatt to enjoy the air at high altitude before returning to Bayreuth. Richardis and Mitzi then joined the Richters at Weibegg. In thanking her for some pictures of religious themes which she had sent him, Richter revealed that he had put them in a 'Christian corner of the Weibegg chalet where such things which I have received from friends are placed'. They were there 'more for decoration than for one's edification, for it must be said I am with *Wotan*. Frau Wagner does not like to hear that, but it would be wrong to think one thing and say another; I am more grateful to Wotan than to any other god, for there's none that sings so well as he! God forgive me my sins.'[1]

Richter's 1906–7 season began with the forty-second Birmingham Triennial Festival, opening on 2 October with the traditional *Elijah*. Herbert Thompson was allowed to attend a rehearsal the day before of orchestral items included amongst the programmes and was impressed by 'his quiet mastery of the situation'[2] and by the detailed way in which he rehearsed even the most familiar Wagner overtures. As to *Elijah*, 'he put real warmth and force into [it] and gave us a fine interpretation, neither too polite and cold on the one hand nor yet tricked out with a sensationalism quite false to the ideals and character of Mendelssohn. Dr Richter certainly had both chorus and orchestra in the hollow of his hand today.' Later that evening Thompson noted that in Elgar's interpretation of his own *The Apostles* 'his tempi were nearly always quicker than he formerly employed, much quicker than he adopted at the first performance three years ago, and it's rather curious that the time he took last evening approximates closely to Richter's'.³ The next morning Elgar picked up his baton again to conduct the first performance of *The Kingdom* (with Richter conducting a Bach motet and Brahms's First Symphony after the interval).

Later that evening he conducted a mixed programme beginning with another first performance, this time Josef Holbrooke's *The Bells* (to a text by Edgar Allan Poe). Mischa Elman then repeated his virtuosic rendition of Beethoven's Violin Concerto, Agnes Nicholls sang an aria from Mozart's opera *Die Entführung aus dem Serail*, Percy Pitt conducted his own *Sinfonietta* (a Festival commission obtained for him by Richter), and Richter concluded the concert with Strauss's *Don Juan* and Berlioz's Overture *Carnaval romain*. Festivities continued next morning with *Messiah*, followed that evening by another Miscellaneous Concert. In it Granville Bantock conducted his own *Omar Khayyám* in the first half and Richter contributed orchestral and vocal pieces in the second. The last day of the Festival (5 October) began with the *Missa solemnis* in the first half of the morning concert. Thompson thought that 'Richter is at his best in Beethoven's music and he always makes one feel its bigness, its warmth and its majesty. I have heard performances which may have been more brilliant but none that got at the heart of the music better or were more truly impressive.'⁴ In the second half Elman played Tchaikovsky's concerto, Muriel Foster sang the first English performance of a solo cantata by the eighteenth-century German composer Christian Ritter, and Wagner's *Parsifal* Overture was played, a miscellaneous programme in every sense of the word. Stanford's *The Revenge* and Mendelssohn's *Lobgesang* were included in the concert which ended the Festival that evening. Thompson wrote a perceptive account of Richter, which took a strong stand against criticisms which were inevitably being levelled against him after eight Festivals.

One is sometimes asked wherein lies the supremacy of Richter as a conductor, and I think that anyone who followed carefully the rehearsals at Birmingham on Monday last would not have found it difficult to supply a satisfactory answer. To begin with he has the whole technique of the orchestra at his fingers' ends. Then he is a genuine artist, emotional but having his emotions well under control. He has his limitations, as every man with an artistic temperament must, but they are wider than those of most musicians, and as he is not a composer, they are not so obviously hedged in by the strong personal prejudices which are an essential characteristic of the creative genius. As regards his interpretations he is a master of effect, but he always makes effects servient to expression and never cultivates it for his own sake. Nor does he allow any emphasis on details to obscure the main lines of the composition. He realises the value of rhythm and emphasis, but never permits it to become spasmodic. Although his readings of music with which he is in sympathy (and this covers nearly the whole range of nineteenth century music) are strongly vitalised and have blood in their veins, they are always dignified and impressive, and if he does allow himself licence, he uses his freedom with a reticence of a strong masculine nature and he knows how to join his 'flats'. He does not present us with a series of purple patches clumsily cobbled together, but he works them into a consistent, logical, closely-wrought whole. Then, as a disciple of Wagner, he remembers to nurse the melody, which is the vital principle of all music from Bach to Wagner, and he knows the value of sustained tone, and that to make a crescendo or diminuendo in every phrase is simply a vulgar substitute for emotion. In a word, Richter occupies a middle place between the sensation-mongers on the one hand and the dull pedants on the other. There is of course no shadow of truth in the reports spread in some foreign papers and recently referred to in the *Yorkshire Post* to the effect that Richter is no longer in touch with Bayreuth, and that at the recent Festival he was not allowed to conduct more than a portion of the *Ring*. He conducted an entire cycle as at several former Festivals, and his close friendship with the Wagner family, begun in the 1860s at Wagner's Swiss home, remains uninterrupted. Frau Wagner's autocratic methods have naturally made her enemies, even among those who admire Wagner's genius, and consequently any rumours of dissension in the Wahnfried camp find ready circulation where the wishes fathered the thought.[5]

The occasion of August Manns's death on 1 March 1907 gave rise to a lengthy article in which Thompson put several leading conductors of the day and their activities in England into historical context.

Manns was not a great conductor, at least according to the standards of today but he was a triton among minnows compared with the average conductor of thirty years back. Costa was of course supreme in Handelian oratorio and Rossinian opera but his range was severely limited and his inability to appreciate or interpret the greatest masters like Bach or Beethoven, to say nothing of Wagner, was colossal. He once said to Sir George Grove 'Beethoven has no melody'. The others were simply time-beaters; Cusins at the Philharmonic was an amiable and cultivated musician whose demeanour reminded one of the bedside manner of a popular physician, then he was said to have an assortment of three tempi, fast, slow and medium. His interpretations of symphonic music were therefore decorous but dull. In the traditions of the Society he made many incursions into fresh ground impossible. He was indeed credited with a sneaking interest in Wagner, though of course his responsible position made it necessary for him to dissemble it, which he did most effectually when he conducted that composer's music. . . . Occasionally a foreigner would come over and conduct one or two concerts as Lamoureux did rather later than the time of which I am writing, and I well recollect what a revelation was afforded when von Bülow in 1878 volunteered to conduct a concert on behalf of the Norwood Academy for the Blind and gave a wonderful interpretation of the C minor Symphony. It was in the following year that the Richter Concerts were established, and our eyes were fully opened as to the possibilities of orchestral music, but up to then and till they had been in the field long enough to make a definite impression on orchestral performances, the Crystal Palace concerts held the field. Nowhere else could one hear the great classics of orchestral music so sympathetically interpreted and there can be no doubt that these concerts, which on Saturday afternoons used to draw all the most earnest lovers of music in London, did more to train audiences than any other institution. Not only were there the performances which, under Manns, had a vitality one missed elsewhere, but there were the enthusiastic analyses with which Sir George Grove awakened an intelligent interest in the music. . . . Manns had a great advantage in the fact that the nucleus of his band—all the wind and a portion of the strings—was the permanent orchestra which played daily in the Palace. Consequently he had opportunities of 'dress rehearsals' denied to most conductors, and could in an informal way run through any specially important work that was about to be performed at the Saturday concerts, and it was no doubt this daily co-operation that enabled the band to keep in touch with a conductor whose method—if he had one—was of the most erratic kind. His use of the baton was indeed difficult to follow and his beat was eccentric and spasmodic; a down beat would as often occur in the middle of a bar as at the beginning, and when he demanded any special energy, his gyrations were inexplicable to the uninitiated. Indeed it's not too much to say that one had to avoid looking at him if one wished to follow without distraction the course of the music.[6]

The 1906–7 Hallé season included some fascinating programmes. Among works new to the Hallé concerts were the *Variations and Double Fugue on a Merry Theme* by the Director of the Berlin Singakademie, Georg Schumann (6 December), Humperdinck's Moorish Rhapsody (13 December), Act II of *Tristan* (31 January 1907), York Bowen's Piano Concerto with the composer as soloist and Strauss's *Ein Heldenleben* (7 February), Elgar's *The Kingdom* (14 February), and Bruckner's Third Symphony (28 February). In London he gave the first performance of Tovey's Piano Concerto with the London Symphony Orchestra and the composer as soloist on 17 December 1906. Three days later Manchester heard his interpretation of *Messiah* for the first time, with its customary repetition next day. Richter used Mozart's edition and Samuel Langford, Newman's successor at the *Manchester Guardian*, wrote that 'Dr Richter conducted, or rather was conducted, for often after giving one tempo he accepted another pretty quickly. Dr Richter is not wont to give quite so much deference to the judgement of others. And yet his presence and reputation did wonders in improving the performance compared with other years. Those who expected he would throw fresh light on the work expected the impossible; the light has been too long on it, and it is faded a little. But Dr Richter did great service by checking much of the nonsense (called tradition) that has accumulated in the performance of it.'[7]

Officially listed as fourth bassoonist amongst the Hallé players at the beginning of the 1906–7 season was the young Archie Camden, destined for a double career as soloist and orchestral player. Richter had not been happy with the standard of bassoon playing in the orchestra for some time. In 1903 he succeeded in persuading Otto Schieder to follow him from Vienna and join the orchestra. He had written a series of letters to Schieder both reassuring him and putting him fully into the picture of what awaited him if he uprooted and came to Manchester. 'You must sign up for three years at least; you are guaranteed £200 for the six winter months, but that's only for Manchester, other engagements [concerts in other cities] will run to at least £60, which of course cannot be guaranteed though they are virtually sure. I can guarantee you an engagement for the London opera season for as long as I am there of course, but that should be for a few years yet because the gentlemen of the Syndicate know that the public trusts me.'[8] With Schieder firmly established in Manchester both as player and teacher (he was still there when Richter left in 1911), Richter decided to fund a bassoon scholarship at the Royal Manchester College of Music and to create an English school of bassoon playing to avoid having to send abroad for the particular sound he liked. Camden won the award through his innate musicianship rather than any prowess on the bassoon, for all he could do was squawk his way up (but not down) an F major scale, and that without the necessary B flat. 'I also noticed that a screen placed across a corner of the room was rocking dangerously,' he wrote. Behind that screen and trying hard not to laugh was Hans Richter, but it was to be a few years yet before Camden was to know the identity of his benefactor.

When I had been at college two years I received a letter signed Hans Richter. It said, 'Please come to my house on Friday next at 11 o'clock and bring your bassoon'.... At exactly two minutes to eleven I opened the gate and walked down the drive. From the house came the sound of a horn playing the *Siegfried Call*; it was Hans Richter, as I saw through the window. When it was finished I knocked on the door, and it was opened by the great man himself, carrying the horn in one hand. 'So you have come', he said.... 'Come inside and play for me'. I unpacked my bassoon and played the Finale of the Weber Concerto. Richter nodded and said, 'Do you know what this is?' I recognised passages from the Beethoven symphonies. Then he said 'Do you know when to play your solos?' He played a few bars from several works before the entrance of the bassoon. Fortunately I happened to come in correctly each time. . . . 'I am well pleased, You will come next Wednesday to the Memorial Hall in Albert Square to play the fourth bassoon part in the Hallé orchestra in the *Sinfonia Domestica* by Richard Strauss'. He closed the piano and I was dismissed.

Camden happened briefly to touch a wrong note during the rehearsal of Strauss's gigantic work, but Richter did not appear to notice. Before the next rehearsal, however, Camden, who was delighted to be told that his appointment as fourth bassoon had been confirmed, suddenly felt a hand on his shoulder as he was about to enter the building. It was Richter 'towering above me. He looked at me and with a smile he said, "E sharp would be nicer". He had noticed my slip, but did not humiliate the young, inexperienced player in front of the rest of the orchestra. This was kindness indeed, but the greater kindness was to let me know that in the professional orchestra you do not get away with any mistakes.' Camden tells many a tale of Richter in rehearsal and performance. There was the principal flautist who, when Richter asked him to 'dance like a fairy' in his solo in Mendelssohn's *Midsummer Night's Dream* Overture, did so excessively in the concert. It transpired he was trying to avoid a wasp. Richter apologized to the audience once again when he gave a premature gesture to his band after a solo section in Tchaikovsky's Piano Concerto, and the less experienced players followed him. The Leeds audience, who were not used to demi-god conductors admitting to any human fallibility, applauded him vociferously. As a child Camden had been taken to a Hallé concert at which Richter had accompanied Sarasate in the Mendelssohn Violin Concerto. After the concert the child asked his father why the violinist played in one tempo and the orchestra in another. This was dismissed as nonsense but years later Arthur Catterall confirmed that he had been quite right. 'Richter began the first *tutti* in a rather steady, comfortable time. Then Sarasate's entry came. He went off at an alarming rate, twice as fast. There was chaos for a moment, then Richter put down his baton. "Mr Sarasate, this is the Mendelssohn", and he beat what he considered the right tempo. Sarasate [replied], "I am the soloist, you must follow me."' Neither man would give way, so when Sarasate played they accompanied him at his tempo but in the orchestral *ritornelli* they slowed to Richter's speed.

One day we were rehearsing with a lady singer and, obviously irritated with her frequent hold-ups on notes she wished to show off, he said 'Why you make pausa on high note?' She replied, 'I always make one there'. Richter answered sharply, 'If Beethoven wish a pausa he would have wrote. He did not write, so we do not make'.

Camden discovered Richter's penchant for a glass of ale at the post-concert refreshments in Bradford. ' "Fourth bassoon, what are you drinking?" I answered "Lemonade, Dr Richter". His voice came booming across the room. "You cannot play the bassoon on lemonade. You should have beer." ' On the other hand Richter did not like spirits, and after checking a whisky-smelling brass-player's part for wrong notes, he returned to his podium complaining loudly, 'Whisky. It is terrible to be in the neighbourhood.'[9]

Whereas Camden's stories were primarily of the conductor's dealings with wind- and brass-players Carl Fuchs had some more of his Richter stories concerning the strings.

In Dvořák's *Carnival* overture there is a passage [for cello] written in treble clef. Dvořák sometimes used the old and sometimes the new treble clef, the former being played an octave lower than the modern notation. The new clef was meant in this case, the passage had to be played in the higher octave, which on the cello is effected by the left-hand moving further down the finger board, so Dr Richter promptly shouted, 'Higher down please'. . . . Richter hated harmonics on account of their weak and hollow sound, and would not allow any, especially when they followed loud ordinary notes. 'Imagine a man with a row of beautiful teeth but one missing in the middle. Dat's ze harmonic'. Perspiration and resin often cause the thin violin strings to stick to the finger of the player so that when the finger is lifted quickly a *pizzicato*-like noise results. Dr Richter's very keen ear detected this, even when it happened at the back of the platform. Then he used to shout 'Please use Pear's soap'. I remember him saying, 'It's a strange thing about concert halls. Nobody ever thinks of cleaning them, but as soon as I begin to rehearse with my orchestra, all the charwomen of the district flock into them and make a terrible noise'. [This charwoman story appears elsewhere when Richter is reported to have addressed the poor unfortunate with the words, 'Wife, do not care!' What he meant to say, translated from the German 'Frau, kehren Sie nicht!', was 'Woman, do not sweep!'][10]

On 28 January 1907 the young American violinist Albert Spalding made his début in London with the London Symphony Orchestra under Richter. Saint-Saëns had recommended the 18-year-old boy, suggesting that he play his own Third Concerto. 'Richter', he recalled, 'proved formidable-looking. He growled rather than spoke his "Good morning" in response to my timid salutation. "He has a head," I thought, "somewhat like Socrates—though probably he hasn't a Xantippe for a wife, or he'd have gentler manners!" His reddish-brown hair and beard, shot with grey, surrounded a face that seemed to have been shoved into slovenly shape by blunt instruments. It had the effect of unfinished masonry. It did not invite assurance.'[11] Spalding was a success and Richter invited him to Manchester to perform the same work shortly afterwards on 7 March. At this second concert the northern press compared him unfavourably to the work's dedicatee Sarasate and to Lady Hallé, who had played the composer's Introduction and Rondo capriccioso a few weeks earlier.

Richter found me the following morning half hiding myself in a dark corner of the hotel lobby, miserably reading and re-reading the uncomforting news. It was a gloomy morning. 'So', he exclaimed, 'I find you wasting your time. The papers they are bad—so you are unhappy. Another time they are good—so you are happy. That is no way for an artist. I thought you were an artist. Have you forgotten that? Perhaps you think newspaper print more important than *my* opinion?' I found his irritation more comforting than a tonic and managed to say so. He went on in a kindlier tone. 'Don't read notices! Let your manager do that, it is part of his business. It is not part of the artist's business'. I murmured that I realised I had played less well than in London. 'Yes,' Richter agreed, 'it was better in London. It is good that you know it, but it was not bad last night. A musician is not a machine, *Gott sei Dank*. If one were always sure of the same performance, it would lose life, lose interest. These fools', pointing disdainfully to the mass of strewn papers, 'see only the things on the surface. They are superficial. Don't read them'. 'But', I dared to point out, 'you yourself have read them this morning'. He gave me a quick sidelong glance and I wondered if an explosion was coming. Then, with as near an approach to a smile as the old round head could manage, 'Yes, *Esel* [ass] that I am, I did read them this morning. They spoiled my breakfast. Na, let us have a cup of coffee and forget it. It will be bad English coffee, but perhaps it will be hot'. The coffee they brought us looked like treacle, thick and syrupy and tasted worse. Richter philosophised on the unmusical nature of those countries that had no proper respect for the coffee bean. Germany, Austria, Holland, the Scandinavian countries, even Russia—all lands of excellent coffee—music flourishes. England, France, Italy, Spain, Belgium—miserable coffee—music languishes. I longed to point out some notable exceptions, but forbore.

We journeyed back to London together. . . . 'Don't play the Brahms Concerto too soon', he advised. 'Above all, when you do play it, don't be too impressed by its reputation of having been written *gegen die Geige* [unviolinistic]. It is written *for*, not against, the violin, and one day—perhaps I shall live to see it—the Brahms Concerto will be as popular as the Mendelssohn.

. . . I went backstage [after *Meistersinger* in May 1907] to see and thank Richter for the beautiful evening he had given me. 'Na', he said, and he was in a very bad humour, 'it was not a good performance. The Overture, yes! The Quintet, yes! The rest of it was bad, very bad. Yet', he added with a touch of sardonic humour, 'look at the reviews in the morning. They will be very fine. That is because I am old and have a beard'. He stroked my chin. 'Grow a beard! You will see what an easy way it is to eminence'. He was quite right about the press notices. They went into raptures over what they regarded as flawless perfection.[12]

After a single concert on 8 April 1907 for the widows' and orphans' fund in Budapest (Liszt's *Christus*), Richter had his annual German opera season at Covent Garden consisting of two *Ring* cycles, *Lohengrin*, *Tannhäuser*, *Meistersinger*, and *Fliegende Holländer*. His singers included Gulbranson, Bechstein, Cornelius, Hempel, Destinn, van Rooy, Lunn, Whitehill, Kraus, Agnes Nicholls, and Edna Thornton. There was no change in the excellent notices Richter continued to receive for his Wagner performances. Adrian Boult attended the second *Ring* cycle and noted the timings as follows. *Rheingold*: 140', *Walküre*: 65', 80', 65', *Siegfried*: 85', 75', 70', *Götterdämmerung*: 110', 65', 75'. Richter's *Meistersinger* was 75', 120', 110', and *Tristan* (a year later) 75', 65', and 70'. In *Meistersinger* Richter was 'gorgeous'; of the *Ring* he wrote, '*Rheingold* magnificent performance . . . it was all round excellent and the orchestra under Richter splendid. *Valkyrie*: How magnificent it all is. Of course the orchestra was superb and the Walküren (75% English) all sang beautifully. Stockhausen ill for Gunther so Zadwill doubled, doing both Alberich and Gunther very finely. . . . We suceeded in getting Richter before the curtain after the last act.'[13]

From his long holiday in the mountains of Lower Austria Richter wrote to Mrs Joshua that his nerves were calmer, and that he was not short of visitors.

At the moment Deichmann is here (together with his Stradivarius). The eighty year-old is climbing around quite valiantly. In the evening we play Mozart and Beethoven Sonatas, my little Broadwood cottage pianino is adequate for that. Later Pitt, Makower and Frantzen are coming. At the beginning of September we are off to Hungary. Until now Wotan has been quite kind to us, as far as the weather is concerned. Only once did he allow Donner to give forth a proper rumble. At the beginning of October we shall be back in England.[14]

Donner may have rumbled only once in Lower Austria, but in Manchester he was frequently audible at this time and during the forthcoming 1907–8 season. It was the Hallé's fiftieth jubilee season and the committee made a pretty poor show of celebrating it. Richter must take his share of the blame for some highly unimaginative programming, and it was unfortunate that he was unable to conduct the celebration concert itself on 30 January 1908. In his place, and for three other concerts in Manchester, he provided Franz Beidler as substitute, an uninspiring choice and an inadequate deputy

(Bantock also guest-conducted his *Omar Khayyám*). As to programming, it was decided that Richter would conduct a cycle of all nine Beethoven symphonies during the season as Hallé himself had done in 1870. The deaths in 1907 of Grieg and Joachim were marked with works by the two composers. Grieg's *Old Norwegian Romance*, Bruckner's Te Deum, and Bantock's work were the only first performances of the season. New soloists included the violinist Marie Hall, the tenor John McCormack, and the cellist Pablo Casals. McCormack and Casals both appeared with Richter in Bradford, though the latter did get to Manchester even if he had to make do with Beidler. Richter was enormously impressed by the cellist, and told Carl Gianicelli, 'At all costs engage the cellist Casals for your next season: unique! He played Dvořák's Concerto and, as a solo piece, an unaccompanied Bach sonata. Unbelievable, *all* others are children by comparison. Also a nice chap.'[15] For his part Casals's memory wrongly placed the concert in Manchester rather than Bradford, where it occurred on 17 January 1907.

[Richter] became one of my great friends. It was during one of my first tours. I was playing in Manchester under his direction, and a performance of the Schumann concerto with him was one of those that most satisfied me. He was a great conductor. I played with him every year and he always said before the concert, 'Remember, we must have a long talk after the concert.' He used to take me by the arm and we would walk to a tavern (always the same one), where we talked till the small hours of the morning. The owner of this place used to lock the door at the appointed time and leave us in peace, waiting patiently until we went home. He obviously had a great respect for Richter! And we discussed all sorts of musical subjects. . . . The fact that we had so many ideas in common used to excite me in a delightful way: I felt intoxicated. In spite of his age he always spoke with the enthusiasm of a young man, which always pleased me so much. It was when he spoke about Wagner that I was most thrilled, for I knew how intimate the two had been. Whenever he quoted any observations Wagner had made to him during rehearsals, I collected them as a treasure. I gazed at him, taking in every word. . . . After his death I bought a large part of his scores.[16]

Although there were some fine works on the season's programmes (Act III of *Tannhäuser*, all Beethoven's symphonies, and four Strauss tone-poems) Richter was often accused in Manchester of neglecting English music (apart from Elgar whom many thought the country's natural successor to Brahms) and of totally ignoring French works (apart from Berlioz, Saint-Saëns, and César Franck). He may have been guilty of treating Manchester disdainfully in the New Year of 1908 just when he should have been a major part of the jubilee celebrations, but what he was doing instead made a major contribution to the history of opera in England. For some time, together with Percy Pitt, he had been planning a cycle of Wagner's *Ring* to be sung in Frederick Jameson's new English translation at Covent Garden under the auspices of the Syndicate and its chairman Harry Higgins. Wagner himself had always approved of the idea of singing his works in the vernacular in order to reach a wider audience. There were only three non-English-speaking singers in the cast, Borghyld Bryhn (Brünnhilde in *Götterdämmerung*), Hans Bechstein (Mime), and Peter Cornelius (Siegfried); the rest came from Britain, America, and Canada. Charles Hedmondt directed and sang Loge, and the chorus master was Emil Kreuz, formerly a viola-player under Richter. There were no dinner intervals during the performances, neither were any German-language performances of the *Ring* scheduled for the summer season. Richter outlined his plans in a communiqué to the press published on 5 December 1907, which became the testament to his operatic work in England.

An artistic success with the *Ring* would open a wide prospect: *the foundation of permanent opera*. By that I mean performances in English. When, on my last visit to Bayreuth, I spoke to Frau Wagner of our English *Ring*, she greeted the plan with great enthusiasm, and added that a profound influence on the public was only possible through the *national language*. But there must be no narrowness; the great masterpieces of the classical and romantic schools, the excellent existing works by English composers must be conscientiously performed, and thereby native talent encouraged to produce new excellence. If this goal is reached, or at least the way prepared for it, I shall have achieved one of my highest aims and embodied my gratitude for the hospitality which has been unstintedly and unceasingly bestowed on me in this country for thirty years.

He showered Pitt with orders, instructions, and recommendations. From Bradford he wrote in the autumn of 1907,

Today a Miss Edith Evans sang the *Freischütz* [Agathe's] aria (in E major) and Elsa's Dream. She showed so much talent and feeling that I am convinced she could sing and act Sieglinde. She has such enthusiasm in her really big voice. I think it's worth an attempt. She should study some things with you, Kreuz, or Waddington straight away so that in two weeks she can audition for us on stage. She should learn two extracts [from *Walküre*]: in act one 'Der Männer Sippe sass hier im Saal' until the end of this monologue, and also the vision scene in the second act 'Weiter, weiter' until the end 'Siegmund' where she collapses. Miss Evans will look good, [she has] expressive eyes. To begin with at rehearsals she was not very promising, but she was uplifted by the orchestra, became fiery and sang with verve and temperament. If it is too late for Sieglinde, she could be used very well as

Freia or Gutrune, especially Gutrune. Let her start straightaway on learning the above scenes, if she were ready I'd come earlier to London to hear her. She will look fine, a good face and striking eyes.[17]

His enthusiasm was well placed and Edith Evans sang Gutrune in *Götterdämmerung*, just as he had suggested. Pitt also made a shrewd move in casting Walter Hyde (a singer of light opera) as Siegmund and both men stood firm that the American Clarence Whitehill, despite his high fee, must sing Wotan ('Tell Higgins from me: with Wotan stands or falls the English *Ring*'[18]). In December 1907 Richter wrote a letter to Higgins in which he expressed his confidence in the whole project, in the ability of English singers to achieve the high standards set (in his experience) by their orchestral colleagues, and of his hope that the venture would lead to the establishment of a permanent opera company which would perform in English. Much depended on the goodwill of Higgins, Forsyth, and the Syndicate, so Richter made appropriately grateful noises. On 7 January he told Pitt,

I sit here in my lovely apartment and can do nothing better than to give you in writing my impressions of the first ensemble rehearsal. On the whole it was *very* satisfactory; but now the *singers* must be looked after and for that Mr Hedmondt must get on with his work with them as actors in the two days of my absence, without letting them sing. They must speak their phrases. Please tell Mr Higgins that Mr Ar——t is impossible as Donner. God put everything in the man's throat but forgot that a singer must also have something in his head. Impossible! [Charles] Knowles must sing Donner. Would you kindly bring to Mr Forsyth's notice (in my name) that we must have the stage (if it's free) for production rehearsals. . . . We are both responsible for the musical side, and we can accept this responsibility with impunity for we have splendid musical assistance, which is really exemplary. I know of no German theatre who could better this aspect, but on the other hand technical necessities must not be harmed by petty problems with money. In the *Ring* the *aural* experience must not be disturbed by the *visual*.[19]

What Richter would have said to the post-1950 Wagner productions at Bayreuth by the composer's grandson Wieland and his successsors can easily be imagined from this last stricture. Meanwhile the English *Ring* was an unmitigated triumph. As Pitt himself said, 'At a given moment Hans Richter took over altogether and really moulded the whole ensemble into a marvellous state of homogeneity, both musical and otherwise. It was indeed a colossal achievement.'[20] Pitt was brought into the orchestral pit by Richter after *Götterdämmerung* to receive a well-earned ovation from the huge audience. The king and queen attended *Siegfried*, the queen alone

FIG. 15 Richter conducting the first *Ring* in English, *Daily Graphic*, 28 January 1908.

came to *Götterdämmerung*. The press were already wondering after *Walküre* whether the high standards achieved could be maintained for the rest of the cycle. R. A. Streatfeild called it 'an important epoch in the history of English opera' at the conclusion to his glowing notice in the *Musical Times*.[21] After it was all over *The Times* considered that the venture had 'amply demonstrated the possibility of finding good English singers capable of sustaining the different parts just as well as they can be sustained'.[22] Boult noticed that 'Richter was very energetic, even going so far as to wave Alberich across the stage in the middle of the curse [in *Rheingold*]. . . . The whole thing was very creditable and spoke very well of Richter, Pitt, and Mr Hedmondt.'[23] On the day *Rheingold* opened the proceedings Richter ended a note to Mrs Joshua with the words, 'Right then, in the name of Wotan we're off!'[24] When it was all over he wrote to Pitt to 'thank you first of all, for without your sympathy as an artist, without your assistance, perseverance, and energy I should not have been able to carry out the work. *Auf Wiedersehen* until the next English *Ring*.'[25]

Agnes Nicholls, the wife of Richter's future successor at the Hallé, Hamilton Harty, sang Sieglinde in *Walküre* and Brünnhilde in *Siegfried*. Nearly half a century later she recalled the 1908 English *Ring*.

Rehearsals began and what amazed me was the intense desire of those young—many of them untried—singers to give of their very best. They worked long hours on their own so that the ensemble should be perfect. . . . Richter was spending much time in London and was already ready to be of service. He knew everything by memory and would stand up with no score but stick in hand and conduct. He always encouraged these youngsters—indeed all of us—and was delighted with everyone. . . . When [stage] rehearsals began Richter was here, there, and everywhere with helpful advice and colossal patience; but I think the main point was the intense enthusiasm of the company to do the thing really well and to give of their best. . . . The first act of *Valkyrie* went wonderfully and we all enjoyed it. So evidently did the audience for we had ten curtains at the end of that act. I thought they were never going to let us off the stage.

When we were back in our dressing rooms there suddenly came a knock on my door, and when the dresser opened it, lo and behold there was Richter standing there with Percy Pitt. They had come to congratulate us. I could hardly believe it. It seemed incredible that such a great man as Richter should come to us to do that. I was greatly touched, and it always remains in my memory as a very moving incident. . . . We were all very sad when the two cycles were over for we had all enjoyed them and working with such a great man as Hans Richter. How much he taught us and how immense was his knowledge! He always remembered odd things one did. When I used to play the first Rhinemaiden [in concerts with the Hallé], as he passed us to go to his rostrum, he would say to me 'nicht schleppend!' for I had a bad habit of so enjoying the music that I sometimes dragged it out.[26]

31

1908–1909

England

Having concluded his epic English *Ring*, Richter returned north to Manchester, where he found the mood of the orchestra, its administrators, and his public disgruntled at his prolonged absence during the jubilee season and because of the inferior capabilities of his substitute Beidler. He finished the concert season with Bruckner's Te Deum and Beethoven's 'Choral' Symphony on 12 March 1908. After a single concert with the London Symphony Orchestra (23 March) he was supposed to fulfil a long-standing engagement to conduct two concerts (on 26 March and 9 April) for the Philharmonic Society, of which he had been an honorary member since 1905. He had accepted them on 4 July 1907 but wrote to Francesco Berger, the Society's secretary, at the beginning of March 1908 withdrawing from the engagements for family reasons, which would take him to Austria-Hungary at the beginning of April. A flurry of correspondence took place between a tired Richter and the panic-stricken and annoyed committee. They tried to get him to conduct at least the first of the two concerts, but Richter was adamant in his refusal. They were given instead to Henry Wood and another rising young English conductor, Landon Ronald. Richter had been offered £60 per concert (Nikisch was offered £50 for a single one), and Wood and Ronald each received thirty guineas for stepping into the breach. Surprisingly negotiations began immediately to include Richter in the list of conductors engaged for the 1908–9 season, but all the dates were Thursdays, Hallé concert days. Richter did not dare to absent himself more than his plans for another winter series of operas in English at Covent Garden would allow, and so discussions were abandoned. Back in 1887 he had been invited to conduct for the Society but declined out of loyalty to his manager Vert. As a result Hans Richter never conducted for the Philharmonic Society.

The German opera season at Covent Garden in the summer of 1908 consisted of 'half' a *Ring* (*Walküre* and *Götterdämmerung*), *Tannhäuser*, *Tristan*, *Meistersinger*, *Fliegende Holländer*, and two performances of Gluck's *Armide* with Destinn in the title-role. Herbert Thompson described the *Ring* operas as 'the first of a series of Festival performances of Wagner's works, implying not only the abnormally early hours of Bayreuth, but an absence of cuts and the guarantee for the thoroughness of execution which the presence of Hans Richter as the artistic head involves. . . . The orchestra was as usual the feature of the production which would bear comparison with any standard, and the fire and force of its playing under Richter, than whom no conductor realises better the grandeur and breadth of the music, were quite admirable.'[1] Coincidentally the Covent Garden building, like the Hallé orchestra from whom its conductor was absent during February, celebrated its silver jubilee with this summer season. It ended on 10 June, after which Richter travelled to Munich to conduct a single performance of *Meistersinger* on 21 June. It was forty years to the day since its first performance, and thirty-nine years since Richter had conducted in the city. Felix Mottl, Munich's Music Director, had been particularly keen to get him to come and conduct the opera (it was Richter's 125th performance). Not only was it a great occasion, it was also, largely thanks to Mottl, a symbolic political-theatrical bridging of the gulf which had existed for so long between Bayreuth and Munich, for here, with baton in hand at the podium, was Bayreuth's most loyal servant. Richter himself seemed very satisfied with the outcome when he told Percy Pitt, 'It was splendid in Munich: the *enthusiasm*! The performance was also really good, indeed in some places excellent. The orchestral playing was fine, everyone was surprised how beautifully they played *piano* and *pianissimo*. The chorus was very good, beautiful, fresh, and powerful voices. . . . Your Hans Richter, with sciatica in the right leg, is being ill-treated (i.e. massaged) daily.'[2] The journalist Alexander Dillmann began his lengthy review of *Meistersinger* by setting the scene.

6 p.m. A festive and colourful house was packed expectantly shoulder to shoulder in the stalls and balcony. The little door to the orchestra pit was opened and a stocky, elderly man made his way to the podium. It was Hans Richter. A pleasant, intelligent face with its broad, blond-streaked, full beard. Were it not for the sharp lenses in front of the friendly yet serious eyes, one could speak of a Hans Sachs-head. Scarcely had he been recognized by the crowd than he was cheered with such heartfelt feeling as though Hans Richter had stood for the whole of his life at Munich's conductor's desk. From the front of the stalls Felix Mottl and Franz Fischer, those faithful paladin brothers from those past great days, clapped

and called. The recipient of these plaudits was duly grateful in his thanks. The lamps were dimmed and darkness fell over the house. Hans Richter lifted his baton and the first C major chords of the *Meistersinger* Prelude filled the place with its radiant power. Broad and measured, yet fluently, the themes of the *Meistersinger* flow on their way; with tender warmth the muted violins flatter Walter Stoltzing's courting and wooing sounds of love; with short, piercing *staccati* the winds caricature Beckmesser's pompous marking within the wondrous musical fabric, and with its mighty form the Prelude builds up to a fiery yet structured climax and thence to a broad and weighty conclusion. The F trumpets resound, the cymbals strike. The Prelude ends in bright jubilation. The curtain rises and Hans Richter, this time from the conductor's desk, conducts the church ceremony chorus with which, forty years ago, he worked hard as our opera repetiteur when Hans von Bülow held sway over the orchestra with the baton.

. . . Hans Richter conducted with such mighty strokes and yet with that loving emphasis for detail which only someone who is deeply familiar with this work can express. The ovation with which he was assailed for many minutes by this Munich audience may have conveyed to him the depth of heartfelt and sincere warmth at the sight of this man at the podium who is so deeply bound up with the history of this work.[3]

It can only be a matter of deep regret that Richter never recorded, despite living well into the pioneering days of the gramophone. We are reliant on newspaper reviews and eye-witness accounts such as the following by the young Eugene Goossens, who attended a performance as part of the 1909 season of opera in English. He was spirited into the orchestra pit for a *Meistersinger* performance by his double bass teacher.

Fortunately there is a dark alcove behind the bass players in the Covent Garden pit, and here, inconspicuously, I crouched on the floor, the phalanx of basses screening me completely from sight, so that when the 'Old Man' entered and took his place on the rostrum I was able to watch him. . . . As usual he conducted without a score, an easy matter for the man who . . . had copied the full score from Wagner's original manuscript. It was this link with a legendary figure and a legendary past, far more than what seemed then the miracle of Richter's conducting, which invested the occasion for me with considerable significance. He had known Wagner and had lived under his roof, and he stood that night as the incarnation of Wagnerian tradition. The logic of his tempi—the unhurried prelude (a lesson to some of our conductors today [1951]), the ease of the Pogner music (measured and dignified), the clarity of the second act finale (usually a rout), the well-controlled flow of the quintet music (invariably scampered), and the balanced climaxes of the Finale (usually one huge shout)—left me with an indelible and impressive memory.

As a young boy Goossens had attended Richter's concerts when he brought the Hallé orchestra across to his home city Liverpool.

I always found a seat by the organ, on the topmost tier of the orchestra and immediately behind the percussion section, which I appropriated for all the concerts. Up there it was easy to follow the conductor's indications and get a close-up of the players in action, invaluable experience for a boy of my age. Once the cymbal player in my vicinity miscounted his bars in the finale of the *New World* symphony, and the Doctor transfixed him with a stare which endured to the end of the movement, and which, I heard later, resulted in the player's dismissal from the ranks. It is related that the next time the orchestra rehearsed this movement the 'Old Man' stopped the musicians two bars before the unforgettable spot and, in a reminiscent mutter, inquired ominously, 'Iss he still alive?'

A martinet Richter most certainly was, but the men loved him, and the Hallé orchestra under him was without doubt one of the finest in Europe. There is a tendency among my older colleagues to disparage his conducting powers in the light of flashier and more recent stick technique. The Doctor, for all his years, was a musician of pretty catholic tastes; nowadays we are overprone to think of him only as a Wagnerian conductor. Opinions may differ regarding his merits as an interpreter. I was too young at the time to form an estimate, but I *do* remember his stirring versions of the then new Strauss tone poems and much other fairly provocative music. It is quite possible that his rigid beat missed some of the subtler nuances in these works realized by the more fluid indications of modern conductors, but it certainly suited *Die Meistersinger* and works of that category well enough. And the tempi behind the beat were good too!

Richter's stick technique was simplicity itself. He used a short, thick piece of cane with a padded grip, and indulged in few superfluous gestures. The elaborate arabesques of contemporary conducting were totally unknown, and superfluous, to him. The beat was a square one, vehement, simple, and best suited to classic and romantic styles. Especially in long sustained rhythmic patterns did he preserve a marvellous continuity of style.[4]

Second violinist Harold Jones said that when Richter was annoyed his English became virtually unintelligible. When he once had occasion to reprove a flautist, he said, 'Your damn nonsense can I stand vonce or tvice, but sometimes always, by God, NEVER.' Another young musician who came into contact with Richter at this period was Edward (later Sir Edward) Bairstow, composer, conductor, and organist for many years at York Minster. In an unpublished memoir he recalled his first encounter with the great man on 6 December 1907.

In 1907 I was appointed conductor of the Preston Choral Society. . . . It was the custom at Preston to make the first concert of the season largely orchestral. The Hallé orchestra and their conductor Dr Hans Richter were engaged. The chorus sang one short work under my conductorship; Richter had never met me before and knew nothing of me. When he had finished his orchestral rehearsal he said, 'Now you can rehearse your work. It is a ver' good thing for you, young man, to have the chance of conducting a first-rate orchestra, but you must be ver' strict with zem or zay vill make a damn fool of you'. The work was Brahms' *Song of Destiny*. The chorus had had some difficulty with the quick movement where, at the words 'Like water from cliff unto cliff ever dropping', the orchestra are playing three beats in a bar whilst they are singing two. I had accordingly practised beating two with the left hand and three with the right. Richter was rather fascinated by this unusual bit of technique. At the concert in the evening, instead of resting in the artists' room, he came into the hall. After it was over he walked onto the platform and shook hands with me in front of the audience.

He had a very gruff German manner, but a kindly heart. We adjourned to the Park Hotel between rehearsal and concert. He always said on these occasions that he never could eat before a concert, 'I am too excite', but he invariably plodded through the whole menu, drinking a large bottle of Burgundy with it. We then set off for the concert in a four-wheeled cab. The traffic coming up our side of the street had to go past the hall, turn round, and then discharge its passengers at the main entrance. We were rather late so that when we were opposite the entrance I was for getting out and crossing the road, but Richter said 'Do not get out yet, we have to turn over'. After the concert we went back to the Park Hotel. He again told me that the music made him too excited to eat, but he put away a beefsteak and some stewed fruit with another large bottle of Burgundy. It was his custom at Christmastide to invite the orchestra in batches to his home at Bowdon in Cheshire and give them a good meal, but even the burly brass players said that he could out-eat any of them.

He was very kind in giving me tips about the art of conducting. On one occasion he was talking of the necessity of lightning quickness between the thought and the gesture. He said he had been to some scientific society in London where this speed was measured electrically. A number of people sat at a table with their right hands against a contact. Then all sorts of tests were applied. A light was flashed, a weight was dropped, they were touched on the shoulder. They had to bring the right hand down onto the table as quick as thought. He had won all the tests, although he was then nearly seventy, and was very much cock-a-hoop about it.[5]

The amateur musician William Strutt first heard Richter in 1906 and considered that, 'with the possible exception of Manns, he did more for music in England than any other individual'. He heard *Siegfried* and *Meistersinger* in May 1907 and some concerts two months earlier.

Siegfried under Richter, although equalled [was] never bettered. From the sombre prelude to the exultant love duet, the conviction grew on me that Richter and the Covent Garden orchestra formed a combination which could hardly be equalled anywhere else in the world.... Later on that month I heard a wonderful performance of *Die Meistersinger* with Richter in command. If there is any class of persons for whom I am wholeheartedly sorry, it is those lovers of Wagner who have never heard Richter conduct *Die Meistersinger*. All musical London had constant opportunities of revelling in his perfect playing of the overture, but performances of the whole opera have averaged only one or two a year, and even seizing every opportunity I have only heard it some dozen times with the Doctor in the chair. It was impossible for the man to conduct a vulgar or meretricious performance of any work, but there was none in which his absolute authority and supremacy were so indisputable as in Wagner's comic opera. . . . Richter [was] again in command for the festival performances at Covent Garden. They were beyond any question the finest Wagnerian performances I have ever heard.

Ein Heldenleben [was] conducted a trifle sedately by Richter [on 11 March]. . . . Brahms' symphonies have not yet attained the popularity which will one day be theirs, and the fact that they are listened to and loved by the few and tolerated at all by the many is due, like so much else of what is good in English musical taste, very largely to the efforts of Hans Richter.[6]

Richter conducted both cycles of the *Ring* at Bayreuth in the summer of 1908. This was unusual, for he had been limiting himself to conducting just the first cycle and then moving on to Lower Austria for his much-needed annual holiday with his family. In 1906, however, Cosima had relinquished her administrative duties to Siegfried, and Richter offered to lessen the young man's work-load (he was totally responsible for all aspects of an excellent *Lohengrin* that year) by returning to conduct the second cycle. According to Percy Pitt, Richter's opinion of the Festival orchestra in 1908 was that it was 'the best for years', and that the chorus was also 'marvellous'.[7] Herbert Thompson ended his review for the *Yorkshire Post* as follows: 'I have said nothing as yet of the orchestra and perhaps it might suffice to add that it was under the direction of Hans Richter, who conducted the first performance of the Ring in 1876, and now thirty-two years later remains the conductor who best realises the bigness and energy of the music. Others make more of details, but under Richter the music surges along like a resistless torrent, carrying all before it.'[8]

Richter began his 1908–9 season with the Hallé with a mixed bag of popular favourites in Nottingham, and his first Manchester concert was no more adventurous, with the *Meistersinger* Overture, Strauss's *Tod und Verklärung*, Grieg's Piano Concerto (with d'Albert's ex-wife Teresa Carreño as soloist), and Beethoven's Fifth Symphony on the programme. Among the usual favourites a week later (22 October), however, was Debussy's *Prélude à l'après-midi d'un faune*. Regrettably Richter makes no mention anywhere of his opinion of this work, although it can be deduced from the fact that he only conducted it three more times, in Bradford on 30 October and twice at the beginning of the 1910/11 season. Samuel Langford, writing in the *Manchester Guardian*, thought the performance required 'more supple playing. . . . But we would not have Dr Richter do this work. Each to his own; smaller men will come and interpret Debussy for us quite adequately.'[9] Neville Cardus, Langford's successor from 1927, was part of a deputation of young men who met Richter in an attempt to get him to programme contemporary French music. His response was that 'zer ist *no* modern French musik'.[10]

Factions were now forming in Manchester. There were those who openly criticized Richter's programming, mainly led by Jack Kahane of the *Daily Dispatch*. What those critics forgot was that Richter, having given Manchester its first hearing of works, then gave them second and third performances, providing them with firm places in the repertoire. This was particularly true of the Strauss tone-poems, all of which he introduced to the city. It was Cardus who wrote many years later,

Richter's real gift to Manchester and to English music at large came from his presence amongst us—simply that and nothing more. . . . With all [the] clouds of fame about him, with incomparable experiences and gifts, Richter came to Manchester. He had no need to play the politician or propagandist to influence the city's musical culture. In himself he was a sufficient culture and tradition. Merely to see him, to hear his interpretations, was a classical education. He crowned the labours of the Hallé by giving to the orchestra a truly symphonic style and by deepening and dignifying the general taste. . . . As years passed by one or two restive spirits in Manchester chafed under Richter's tendency to ignore all music not made in good German fashion. . . . His genius as a conductor began with a marvellous instinct for the right tempo. . . . Richter had no unnecessary gestures. . . . [He] did his work so thoroughly at rehearsal that on the night he could afford to hold himself in impressive reserve. When he *did* raise both hands above his shoulders, you then heard a crescendo that you would very likely not forget for a long time to come. His interpretations were beautifully graded; he had no use for flashy contrasts of tone, or for a virtuoso display of instrumentation. He was there to unfold the music's beauties. It was his desire to achieve this without seeming to place himself between the audience and the composer. . . . Richter did not consciously set himself to perform any pioneer work during his association with Manchester and the north of England. His influence was cultural not didactic. By coming to Manchester he linked the city up with all the great places of music in which he had made history. He put Manchester on the map which the musician in Vienna, Munich, and Budapest knows. He expelled the merely local spirit which can so easily play havoc with standards and a wide view of things as they are. . . . He belonged to that quaint old epoch in which dignity was considered an indispensable part of art and culture.[11]

Charles Graves, writing in the *Spectator* in 1902, also acknowledged Richter's limitations.

In orgiastic or hysterical music he does not seem at his ease; but then he is an Olympian and we would not have him otherwise. It is said . . . that he is out of touch with certain modern composers who are masters of 'psychology, trigonometry, chemistry' and have every gift but that of melody; but here there are many who will gladly associate themselves in his heresy. . . . If his programmes of late years have suffered from the somewhat monotonous predominance given to certain familiar Wagnerian excerpts, it is an open secret that this does not reflect his own views as to the responsibilities of a conductor in the way of befriending all schools and encouraging young composers of all nationalities. British musicians— instrumentalists and composers alike—have always found Dr Richter a firm friend and a most generously appreciative patron. It is enough to recall his invitation of English players to take part in the orchestra at Bayreuth, his introduction of compositions by British writers into his Viennese programmes, and his efforts to secure a hearing for them when possible at his own concerts.[12]

Elgar now paid Richter the ultimate accolade a composer can afford his champion, the dedication of a major work. On 3 December 1908 Richter revealed Elgar's First Symphony to the world at Manchester's Free Trade Hall and four days later repeated it with the London Symphony Orchestra at the Queen's Hall. At the rehearsal on 6 December for the London première W. H. Reed, playing violin amongst the orchestra, recalled Richter's opening words. '"Gentlemen, let us now rehearse the greatest symphony of modern times, written by the greatest modern composer", and he added, *"and not only in this country"*.'[13] Elgar responded with an inscription written into Richter's copy of the full score, 'To Hans Richter, Mus. Doc. True artist and true friend.' This time Richter was given adequate time with the score, even visiting Elgar in Hereford to discuss it when he should have been conducting the Hallé concert on 5 November (Mendelssohn's *St Paul* which Beidler took over). The two concerts were unmitigated triumphs; critics and musicians were united in recognizing a masterpiece which went straight into the repertoire (Richter conducted it again in London on 19 December and Elgar gave three performances, all at the Queen's Hall, over the next month), and which the public took immediately to their hearts. According to the *Musical Times*, Richter 'realised both its beauty and its importance in the development of modern instrumental music, studied it devotedly, and brought it to a hearing which showed that, whatever may be said of its details, it is a work of beautiful design, high purpose and, above all, strong vitality. It was this last essential quality which each performance under Dr Richter made abundantly evident.'[14] The 19-year-old Neville Cardus stood at the back of the Free Trade Hall for a shilling on the foggy night of 3 December 1908. 'I was one of many who listened with excitement as the broad and long opening melody marched before us, treading its way over a slow steady bass, broad as the broad back of Hans Richter.'[15] Elgar himself was (particularly after the 1904 London Festival in his honour) a great admirer of Richter as a conductor. 'Take the greatest conductor in England, Dr Richter,' he wrote. 'Dr Richter conducted an orchestra of artists, and consequently he had only to give them a lead, explain a piece to them, and they followed him, and you saw in his case absolute dignity in gesticulation, no exuberance of gesture, or anything of that sort. That is what conductors should aim at—the absolute purity of a rendering without any (I would use the word) humbug.'[16]

By the time the symphony was well established in concert programmes around the country comparisons were being made between interpretations by Richter and Elgar himself. The *Monthly Musical Record* commented, 'Dr Richter's readings have not been surpassed for breadth and dignity of conception, but the composer himself, by sweeping the performance along, and imparting to it a warmth of expression that had been wanting previously, made the points where the music seemed to make a little advance far less evident than when a more deliberate and straightforward manner was adopted.' Similarly the *Musical Times* observed that 'each reading has its qualities: one was characterized by greater breadth and force, the other rather by perfection of detail. . . . Richter always emphasized its architectonic character—the big climaxes had wonderful majesty—but his temperament lacked that quick, nervous, rhythmical spring which is always in evidence when the composer is at the desk, and which makes a theme such as that in D minor (immediately following the opening *nobilmente*) like a flaming sword.'[17]

Throughout 1908 Richter was in close touch with Percy Pitt regarding their plans for a repeat of the English *Ring* in Covent Garden's winter season 1908–9. From Bayreuth he begged him not to relax the pressure on Higgins and Syndicate to protect the musical side of the project. 'Give them no rest; don't let up! The success we have achieved must be exploited and built upon.'[18] From his daughter's home at Baracs in Hungary he wrote that he was pleased that Willi Wirk would be returning to stage manage (i.e. direct) the cycle, though he urged Pitt to oversee rehearsals and ensure that Wagner's stage directions were followed to the letter. When it came to replacing orchestral players who could not or did not wish to return for the 1909 season, Richter asked, in the case of string-players only, for young musicians to be recruited from 'both' music colleges (presumably the Royal College and Royal Academy, ignoring both the Guildhall School and Trinity College). Any new wind-players, on the other hand, had to be experienced musicians. He was then quite specific on how to run their audition. Any prepared passages had to be difficult for he knew that English standards of orchestral sight-reading were (and have remained) high. He had particular extracts in mind for the lower strings to read at sight.

Scene four, act two of *Die Walküre* for the violas *and* cellos after 'Schweig und schrecke die Schlummernde nicht' [marked *etwas bewegt, doch nicht zu schnell*]. Then the arpeggios for the violas at the end of act one of *Die Walküre* [presumably the forty bars marked *mässig schnell* at the moment Siegmund pulls the sword from the ash tree]. Also the scene in act two of *Siegfried* between Siegfried and Mime. Solo passages for cello to be taken from act one of *Die Walküre* and *Götterdämmerung* (there are many *expressive* cello solos in the first scene of the first act of *Die Walküre* which can show both sound and performance to advantage). The introduction to *Rheingold*.

Basses: *Lohengrin*, Prelude to act three, and the interlude between the second and third scenes in act three. Prelude to the second act of *Götterdämmerung* (Norns scene). Also see how experienced they are beyond the usual A flat major.[19]

From Bradford he warned Pitt that players and singers coming to London to participate would ask for travelling expenses. 'Quite naturally, for the fee is very moderate and there is a great deal of work. Mr Forsyth [Neil Forsyth, General Manager of the Syndicate] should be magnanimous and just; the English (operatic) heaven will make it worth his while.'[20] Jokes soon became thin on the ground and tempers frayed when Higgins insisted on bolstering up the season with more operas. Besides three *Ring* cycles, four *Meistersingers* were also planned and Pitt was scheduled to conduct Edward Naylor's *The Angelus*, an opera which had won a competition organized by the music publisher Ricordi for a new English opera. Now Higgins wished to add *Madam Butterfly* and *Faust* to the programme. Richter did his best to veto the plan and implored Pitt to talk Higgins out of it, suggesting that more performances of *Meistersinger* and *The Angelus* could be given, or even some single performances in a cut version of the *Ring* operas. The chorus (trained by Emil Kreuz and consisting mainly of students from the music colleges in London) would be overworked as it was, and did not have the routine or experience to cope with more operas. Eventually *Butterfly* (which has very little chorus work) was retained for five performances under Pitt and an extra cut *Walküre* concluded the season on 16 February after exactly a month. Richter understood only too well the enormous risks of financial failure. 'If we fail in our opera', he told Pitt in an undated letter in the New Year of 1909, '(which may be good, even *very* good) it will be a long, *very long*, time before we make good the damage, and our enemies will rejoice at our discomfiture, which will not be our fault but that of people who have no understanding of our art, and who are not patriotic enough to feel that *English opera* has become a matter of honour.' There were a few cast changes in the *Ring*, the most significant being a new Sieglinde (Rachel Fease-Green) and a new Brünnhilde (Minnie Saltzmann-Stevens, who sang the role in all three operas), both American pupils of the tenor Jean de Reszke. Their careers took off as a result of their triumphs in the second English *Ring*.

Once the cycle was under way Richter told Mrs Joshua that he shared her opinion that 'the performance is significantly better than the first [1908 cycle]. Our English singers are feeling more secure and are really going flat out. Certain stage matters must be changed, one can get nowhere with life-expired machinery and an even older machinist. That must and will be changed. Please may I ask you not to send me fruit at present; at this cold time of year it does not agree with me.'[21] W. M. Strutt witnessed Richter's handling of one of those crises for which he was now famous. In Act I of *Götterdämmerung*, Waltraute (sung by Edna Thornton) was not heard off-stage.

It was an awkward moment for both conductor and Brünnhilde, but they rose to the occasion. Richter's arm stopped with a suddenness which reminded one of a pointer which has winded a grouse, two violins played a tremolo and Brünnhilde gazed unmoved at her ring. This lasted for what seemed an eternity, but was in fact about a minute. Then suddenly Waltraute's cry was heard, Richter's arm moved on, the orchestra played, Brünnhilde stood up and the trouble was past. All the same I was glad it was not I who had to explain the matter to the Doctor.[22]

Even as Richter, Pitt, and the singers and players were celebrating at Pagani's restaurant after the last night of the season (16 February), rumours began circulating that the Syndicate had had enough and was about to abandon any further opera in English. Richter and Pitt had met Higgins before the season ended and made a set of proposals regarding the future. Higgins promised to study them, discuss them with his fellow Directors, and report back. The result was the following letter to Pitt.

I have been thinking very seriously over my interview with the Doctor and you on the subject of English opera, and before you invite any of your friends to give yearly guarantees, I urge you both to look at the facts fairly and squarely in the face. On the basis of this year's results, you may take it from me that you will not succeed in raising the curtain under £550 per night, and I think it is more than probable that figure will be increased to £600, because it is very improbable that you will again succeed in securing the services of the artists on such good terms as this year. Furthermore you must recollect that in this calculation there is practically no provision at all for new scenery, costumes etc. etc., and if you are carrying on a seasonal enterprise quite separate from the Syndicate, who provide all these and charge them to their summer expenses, they would amount to a considerable sum. The mounting of *Rienzi* alone would be a serious item. Now although you can get subscriptions for cycles of the *Ring* and special Wagner performances in limited numbers, it is quite certain to my mind, for the other performances such as *Orfeo*, *Samson*, *Fidelio* etc., if your receipts average anything over £300 you will be lucky. This has been the experience not only of this season, but of the Italian autumn seasons, in which the only performances that were ever given at a profit were those in which artists like Caruso, Melba, or Tetrazzini appeared. There is always a market for stars here, and indeed in most countries.

English people only come to the opera to hear something sensational or unusual. *Don Giovanni* with a good ensemble of English artists, but without stars, would not draw £300, and it would cost £500 to give it. After all, what opera houses are there in Europe that do more than just pay their way in spite of subventions? Free opera houses without rent, exemption from taxation, insurance premiums, very often free heating and light—nowhere in Europe, as far as I know, do the receipts average £600 per night excepting of Covent Garden in the summer, and unless they do, you must lose money. It is not a question of language at all. The fact is that unless there is some very special attraction the London public will not come to the opera in sufficient numbers to make it pay. In the summer it is a different story, but if you deducted the private subscriptions from boxes and stalls which depend to a great extent on fashion, then the seats paid for by foreigners, who certainly constitute nearly 40% of our audience, I do not believe that the patronage of the British public *proprement dit* exceeds £300 per night on an average even then.

Under these circumstances I am not prepared to recommend to my colleagues, who rely to a great extent on my judgement, to run their heads against a brick wall. I advise you strongly not to make an appeal for financial assistance without making it quite clear that in all human probability a loss will result. If the Doctor likes, I am quite willing in 1911 to introduce English opera as a feature of the summer season in place of German. It will be an experiment, but one that we can afford to make. If anything will give an impetus to opera in English, that will, but to continue to lose money on English autumn and spring seasons, when there is not even the remote prospect of the enterprise paying its way, is worse than useless and will injure the cause you have at heart. My conviction is that there is very little demand in England for opera at all outside the season, and that outside the small circle of those who have an axe of their own to grind, the idea that a craving exists for opera to be given in English is an absolute delusion. If we can do Wagner better in English than in German, by all means let us do it in English; I don't believe the summer public will care one way or the other. We shall see how they receive the *Valkyrie* in May, but beyond that I am not prepared to go unless someone can give me very good reason for thinking that my conclusions are wrong. Please show this letter to the Doctor.[23]

Richter was devastated by Higgins's letter. Although he conceded that some of its argument was correct, he considered the man to be short-sighted and guilty of underestimating the public's response to opera. 'Our work got full houses and aroused the greatest enthusiasm which promised so much for the future,' he told Pitt from Eastbourne where he was conducting Elgar's new symphony. 'So much work, talent, enthusiasm, and keenness is now to be snuffed out! Higgins' advice not to appeal to rich art-lovers is a sign that there is no basic will to establish an Opera in English. Even the Croats have an opera in their own tongue, and the English?'[24] In fact the season's receipts had averaged out at £558 6s. 1d. per performance, which slightly exceeded Higgins's prerequisite for nightly takings at the box office, but even that cut no ice. There was an underlying reluctance by the Syndicate to break with tradition and cast off from the shores of Italian opera. Clarence Whitehill was unavailable for the two performances of *Walküre* on 29 April and 5 May, and the only substitute was Alfons Schutzendorf, who could only sing it in German. Thus even these two isolated performances reverted to the original language (to the probable benefit of the likes of Hyde, Lunn, and Saltzmann-Stevens, who could now go elsewhere with their roles learnt in two languages). The press were very supportive of Richter and his scheme, the *Observer* describing the Syndicate's decision as 'the severest blow which has ever been struck at our pretensions to rank amongst the musical nations'.[25] It was a depressing end to Richter's dream, and rankled with him for years, as he expressed in a letter to Pitt from Blackpool where he spent the New Year of 1911:

If the people who ruined *our* English opera knew what they had destroyed, they would be ashamed of themselves. This is now the consequence of the stupidity and misdeeds of people who occupy positions of influence. Such office-holders, and I mean the Opera Syndicate, are people of high-minded views with an understanding of Art; if the latter is lacking, the former is at least adequate to support the general good, but only competent artists (specialists) should have the decisive word. People who are society names or who only have money are not suited to be judges and patrons of the Arts. So *that* is the future—probably also the end—of *our* English opera, so gloriously begun and so brilliantly continued the following year! A pity about those talented people, who were so successful, showed so much promise, and worked with us in such *unexpected numbers*! Where would we have been today! I know of no country, no city, where, in such a short time and without many years of preparation, such splendid results would have been achieved as we did two and three years ago. It was a crime of the worst kind when the breath of life was forcibly choked out of our English opera. Show *this* letter to Higgins; my language with which I describe that criminal is too tame.[26]

In his reminiscences of those two seasons of Wagner in English, the orchestral player Hermann Grunebaum wrote of Richter,

All that mattered to him was to give a perfect performance of the music in hand. He strove for no special effects; they had to evolve naturally and at the right moment. The first time I heard him conduct *Rheingold* it sounded like a classical work. The flow of musical line went on steadily, and climaxes came only when the situation required them. Yet it all sounded far from academic or dull. It goes without saying that he knew the *Ring* better, musically, than anyone else. How far he entered into the inner meaning of the work it is difficult to know. There were, however, certain passages that affected him deeply; for example the moment when Siegfried brings the Ring and Tarnhelm out of the cave after slaying the dragon. To Siegfried they are no more than pretty objects. But as the horns steal in with the Rhinegold theme the orchestra beautifully tells us what they are and what they mean. To quote Richter's own words, 'Tears run down my cheeks each time at this point'. The last time he conducted *Götterdämmerung*, the scene between Waltraute and Brünnhilde affected him greatly. The former sings of the gods nearing their end. Did Richter feel that his own career would soon come to an end?

Richter had his genial moods, but it was not thus that he rehearsed opera in the theatre. There he was a tremendous personality, and truly awe-inspiring to a newcomer with his black skull-cap which he wore to protect his head from the theatre draughts. When he sang a cue or corrected a musical phrase he sounded angry, for he had to strain to produce the upper notes. If a member of the orchestra had the misfortune to play a wrong note at a performance it was of no avail for him to get away unobserved at the end. If Richter met him in the street or at a railway station the poor man would be told what note he ought to have played, or have it sung at him there and then. However Richter was never sarcastic or malicious, for all his shouting. He was a generalissimo whom his forces respected and admired, while they lived in awe of him. His tempi varied very little from one performance to another; somehow they had crystallized but were never rigid. He valued good orchestral players highly and did not stint his praise.[27]

Richter's concert season, resumed on 18 February 1909 in Manchester, contained little new, though Leeds and Manchester were treated to the first act of *Walküre* in Jameson's translation with Agnes Nicholls, Walter Hyde, and Robert Radford. Richter paid his last visit to Newcastle for a 'Farewell Concert' with the Choral Union on 24 February, though Thompson's observation that the conductor had now decided not to travel was inaccurate, for he visited many familiar venues during the next two seasons.

There was a note of sadness at last night's concert of the Newcastle Choral Union in the fact that it was described on the programme as Dr Richter's farewell. But though the great conductor has decided to limit his appearances outside Manchester and London, and one cannot but applaud the wisdom of a man who does not postpone his retirement until his friends begin to ask why he does not retire, we may hope that his withdrawal from public life may be a protracted process.... Last night's performance [of Beethoven's Ninth Symphony] was characterised by admirable sanity; the force was there but it was all the more impressive because of the restraint which accompanied it. There was a characteristic deliberation, especially in the scherzo, which might perhaps have had greater swing but would certainly have lost in clearness of point thereby. The adagio was serenely beautiful and highly finished in every detail, the trying horn passages coming out particularly well. The chorus also sang with unflagging energy and unfailing precision; altogether it was a memorable performance of this great work.[28]

Elgar's new symphony was played in Eastbourne, Liverpool, Manchester, Bradford, and London before the season's end in April. After the Bradford concert, which included Casals as soloist, Thompson noted:

Dr Richter has matured his reading of this Symphony and does a justice to its warmth and dignity that was felt in last evening's performance, which was a fine one in every particular.... The second part of the concert, which was nonetheless delightful because no vocalist intruded upon it, began with Dvořák's Cello Concerto, a work which, if not characterised by any great depths, has real charm and is far removed from mere display. The soloist was Mr Pablo Casals, a refined sensitive artist, who made so good an impression at these concerts last season. Wagner's lovely *Siegfried Idyll* was played with the utmost tenderness and sympathy.... A vigorous performance of the *Fidelio* Overture opened the concert which ended with Dr Richter's broad and forceful reading of the ever-popular *Tannhäuser* overture.[29]

Richter's London Symphony Orchestra concerts were more varied. W. H. Bell's *Ballad of the Bird-Bride*, a *scena* for baritone and orchestra with text based on an Eskimo legend, was premièred on 1 March, and Parry's *Orchestral Variations on an Original Theme* were played three weeks later. The composer attended and thought that 'he took the first part too fast, but the impression seemed good and they applauded so long that all the band had to get up and bow'.[30] At a special concert on 7 April to end the season Bach's Triple Concerto was played by Leonard Borwick, Donald Francis Tovey, and York Bowen. The *Musical Times*, meanwhile,

FIG. 16 Frank L. Emanuel's drawing in the *Manchester Guardian* of Richter at rehearsal with the Hallé Orchestra in 1904.

announced that negotiations were under way with several conductors to cover Richter's absences in the forthcoming 1909–10 season, including Weingartner, Mottl, Nikisch, Wood, and Fiedler, but because the English opera project fell through Richter renewed his contract on 22 March for three years and conducted all twenty-one concerts of the next season.

32

1909–1911

England

RICHTER'S despondent mood resulted in his complete withdrawal from music-making in the summer of 1909. He did not go to Bayreuth for the *Ring*, which was conducted by Michael Balling in his place; indeed Richter never conducted the *Ring* at Bayreuth again. He spent the summer months holidaying in Hungary and Lower Austria, returning to England on 10 September. For the rest of the month he was in London or Birmingham for rehearsals for the 1909 Birmingham Musical Festival, which proved to be his last. During the four-day event (5 to 8 October) he conducted *Elijah*, *Gerontius*, Dvořák's *Stabat mater*, Handel's *Judas Maccabeus*, Cherubini's Mass in C, and Berlioz's *Damnation of Faust*, as well as some smaller choral pieces. The orchestral works included Elgar's First Symphony, Parry's *Symphonic Variations*, Strauss's *Till Eulenspiegel*, and Beethoven's 'Eroica' Symphony. The performances all drew critical praise, particularly the Elgar works in which Richter now excelled. It was fitting that his last Festival included a broad and noble performance of the First Symphony followed by a superb rendition of *Gerontius*, the first Festival performance since its disappointing première nine years earlier. Elgar attended both the London rehearsals and the concerts which included his works. 'In *our* Symphony', he asked Richter, 'would you let the violas take the passage in the last movement *f* (bold) at [Fig.] 114 *and* the corresponding passage at [Fig.] 137—(it is my mistake to have marked it *p*)—a lift of the hand at performance. Perhaps the two big chorus[es] in the *first* part of *Gerontius* might go a little more "flowingly"—a *shade* faster. All else is splendid.'[1] His orchestra was essentially a mixture of the Hallé and London Symphony orchestras; the players' names alone confirm how good the musical results must have been. There were the string-players Schiever (the leader), Speelman (one of three of the same name), and Fuchs, the wind-players Needham, Malsch, Schieder, and a galaxy of horn-players, Paersch, Borsdorf, Aubrey Brain (father of Dennis), Thomas Busby, and Van der Meerschen. Most of the rank-and-file names were now British, a tribute to his goal of establishing indigenous talent amidst the profession, and further illustrated by the fourteen vocalists, most of them British, who took solo parts in the Festival.

Even Herbert Thompson took Richter to task for the distinct lack of contemporary and new works in the programme. 'Strauss is the only living continental composer who appears in the programme; Debussy, Sibelius, Reger, Glazunov for instance, who are among musicians who have made a world reputation and are therefore entitled at least to a hearing are not in evidence at all. Of British composers there are, in addition to Bantock and Boughton, only two, Parry and Elgar.'[2] Bantock (professor of music in Birmingham) conducted his own *Omar Khayyám* (Part II and the new Part III) whilst Boughton (a teacher at Birmingham's Institute of Music) directed his choral work *Midnight*. There was to be only one more Birmingham Triennial Festival, that of 1912 under Henry Wood. Most of the choral works were familiar (Verdi's Requiem the exception) but amongst the orchestral works were the names of Sibelius (conducting his own new Fourth Symphony), Delius, Walford Davies, and Scriabin. The First World War put paid to a 1915 event and financial restraints meant its effective demise. Thompson, despite his reservations about Richter's choice of programme, was full of praise for his interpretation of symphonies by Schubert and Beethoven.

When one found that both the *Unfinished* and the *Eroica* were included in the programme, one inclined to wish that one of them at least were a little less familiar, but Dr Richter has, ever since I have known the Birmingham Festivals, had a happy knack of making even hackneyed symphonies memorable by performances of surpassing merit. This was certainly the case this morning. The delicate beauty of the pianissimo he secured from his eighty-four strings in the Schubert was exquisite, and showed that a big orchestra need not mean a loud one, and the refinement of the woodwind was equally remarkable. The lovely melodies were all the more appealing because they were smoothly treated and not tricked out with nuances which only distract the hearer. Of Beethoven we had one of those strong magisterial readings to which Richter has for long accustomed us. The Funeral March . . . moved along as inexorably as fate and even in the bright major of the trio with its gracious melody for the woodwind, a steady emphatic tramp of the staccato basses kept up the rhythmical force of the movement. In the coda the wailing phrases for the oboe together with the clarinet were made wonderfully pathetic by that fine artist Mr Malsch. In some passages in the first movement the woodwind had a better chance of being heard than the composer himself allowed them, the strings being judiciously restrained, and in many details one noticed the thoughtfulness of the conductor, though it was less in the details than in the general effect that the grandeur of the performance lay.[3]

Richter was photographed by his future son-in-law Sydney Loeb as he stood before Birmingham's Town Hall in front of six policemen forming a guard of honour, but Thompson had other advice for the city's police.

At the Birmingham Festival last week a series of performances which I have never known surpassed for all-round excellence were frequently marred by the growing nuisance of the motor horn. Many a time did one get the impression that some member of the orchestra had been afflicted by a temporary dementia when the sound of a distant horn heard during soft music would be sufficiently like the tone of an orchestral instrument to deceive the ear, and also, as was sometimes the case, it happened to be fairly in tune. The funniest unauthorised interpolation was in *Elijah*, where, when the Baal worshippers have been excitedly calling upon their god to give ear to their request, the long pause following their reiterated 'Hear and answer' was filled up by a distinct answer in the shape of a Baalistic hoot. One would think that on such an occasion, when the police are keeping a particular jealous eye on the traffic near the hall, it would be possible to persuade motorists to go at so reasonable a speed that it should not be necessary to clear their path by this barbaric hooting.[4]

When Richter announced his intention of stepping down, Elgar wrote, 'I hear with great grief that you will not conduct a Birmingham festival again: this saddens me more than I can say—you have been *everything* to me and I cannot contemplate the festival without you.'[5] One of the tenors of the chorus, W. T. Edgley, had this last word to say about Richter as Festival conductor for twenty-four years.

Richter was burly in form, heavily bearded, and slow in movement, his manner decisive, brusque, and at times harsh. He was always handicapped through his lack of knowledge of the English language, and his struggle to express his wishes was often apparent. As to his ability as a conductor, there could be no doubt whatever. Was there a fault, he saw it; and he was not satisfied until it had been corrected. At times he would give a preliminary direction, such as occurred at a performance of Bruckner's *Te Deum*. Then to our organist Mr Perkins, he called, 'Have we heard the loudest your organ can do?' 'No sir', came the reply. 'Then let us have it, with all its thunder'. We all remember the mighty mass of tone that accompanied the performance of that great work. Many of the singers will remember too how one of the players failed to get a certain nicety of rhythm. This was mentioned by the conductor. Still the error persisted. Then was seen to descend from his rostrum the bulky figure of Richter, who, on coming to the performer's desk, pointed out and hummed the particular passage. But the uncommon strength of the memory of our conductor caused us most to marvel. He seemed to know all the music by heart, be it an overture, an oratorio, or a symphony; and a very usual practice of his was to conduct without a copy of the score in front of him. It was all in his head, and even in correcting a passage he would unerringly order us back to a certain 'letter-direction' for its inclusion. He seemed to have a particular affection for his fiddlers and blowers, and we singers sometimes felt that his preference was given to them. Well, in spite of the *Gerontius* disappointment in 1900, and that occasion never came near to being a *fiasco* as some of the critics thought it was, we never lost faith in our leader, who, through some of our darker and more difficult moments, had cheered and inspired us, and had often urged us on with his favourite slogans, 'Bolder attack!', 'From the heart!', 'More enthusiasm!', 'Courage!'[6]

Another of Richter's occasional musical accidents took place during the early part of the new season. It occurred during a performance of Brahms's *Triumphlied* at Leeds on 2 November 1909. Thompson wrote next day that 'the unrelieved jubilation certainly calls for a restful episode by way of relief, but in its way it is gorgeous music, and, save for a misunderstanding as to a bass lead which was one of those things that may so easily happen in the absence of a full rehearsal, it was finely sung by the Philharmonic Chorus'.[7] Richter wrote to Thompson immediately to explain that 'the misunderstanding of the bass lead was entirely my fault. I overlooked that the basses of the *first* chorus, *who had to sing*, were on my left, whilst I gave the sign of attack to the basses of the *second* chorus who were on my right. You would extremely oblige me if you would publish these few lines, and so avoid any blame falling on the chorus. The sole culprit was, yours very sincerely, Hans Richter.'[8] His letter was published on the following day.

There were a few new works in the Hallé programmes for the 1909–10 season, such as Bantock's *Pierrot of the Minute* (28 October with a repeat 'by general desire' on 18 November), and Sibelius's *Vårsång* (Spring Song), a tone-poem dating from 1894. Tchaikovsky's *Hamlet* Fantasy, Bach's Sixth 'Brandenburg' Concerto, and Ernest Schelling's *Fantastic Suite* for piano and orchestra were all heard for the first time at the opening concert (6 January) of 1910. The American Schelling, a student of Paderewski and Pfitzner, was soloist in his own work. A week later (13 January) Richter played *Sursum corda*, a fantasia (in fact a rearrangement of part of an early violin concerto) by Alexander Ritter, husband of Wagner's niece Franziska and author of the poem used by Strauss (whom he befriended) as a preface to his *Tod und Verklärung*. At the same concert the tenor John Coates sang two groups of modern songs (accompanied by R. J. Forbes, Brodsky's eventual successor as Principal of the Royal Manchester College of Music), the first by the foreign composers Wolf, Reger, Weingartner, Sibelius, and Debussy, and the second by the British composers Delius, Edward Agate, Julius Harrison, Havergal Brian, and Bantock. Coates joined Richter and the orchestra in the Flower Song from Bizet's *Carmen*. Liszt's *St Elizabeth* was revived after sixteen years on 17 February, and on 3 March Richter performed Haydn's 'Clock' Symphony and Strauss's tone-poem *Don Quixote* for the first time in Manchester (with Simon Speelman and Carl Fuchs promoted from the orchestra for the solo roles).

Coates's inclusion of the song 'In the Seraglio Garden' begs the question of Richter's interest in Delius's work. One letter survives from conductor to composer (10 March 1908) in which Richter regrets not being able to perform *Brigg Fair* in Delius's native Bradford five days earlier. He had been asked so late that the work would have only been rehearsed for the first time on the afternoon of the concert. In the composer's own interest Richter declined the request. Delius appeared to accept Richter's concern, but two years later matters were quite different. His works were among those being championed by Jack Kahane and his newly formed Manchester Musical Society, which blatantly sought to oust Richter from the Hallé ('I'm going to kick Richter out of Manchester. We've had enough of him'[9]), and by Thomas Beecham (who performed *Sea Drift* at the Free Trade Hall on the day after Richter premièred Elgar's first Symphony). Beecham's attempt hardly proved Kahane's case, for there were more performers on stage than listeners in the auditorium. Henry Wood was promoting Holst, Delius, Holbrooke, Bantock, and Vaughan Williams at his London Promenade concerts, and Bantock himself (not hostile to Richter) had a short-lived series of pioneering concerts at New Brighton, much nearer to Richter in Manchester. Delius told Ethel Smyth (who also never had any luck with Richter), 'Wagner they like because Richter plays him, and they like Richter much more than Wagner or any other composer—and the older he gets the more they will like him. Just like Joachim!—and that is the bane of music in England and that is what we must all try to get rid of.'[10]

Another of the new British composers in John Coates's group of songs was Havergal Brian, who earned a living as a music critic. He expressed great admiration for Richter in his writings but nevertheless was part of Kahane's group in Manchester who sought change. In 1936 he compared unfairly Richter's handling of Delius's *Brigg Fair* to his treatment of Wolf's *Penthesilea* years earlier in Vienna. 'Richter, neither knowing nor caring for the Delius score, read through part of it with the orchestra, and then, stopping the rehearsal, had the parts collected and the score retrieved from where he had thrown it. Right or not in detail, the story illustrates the mentality of Richter. However the fight for Delius had begun, and no petulance on the part of Richter could stop it.'[11] In 1909 Brian showed his true colours. 'The Hallé orchestra is well established in Manchester, but the atmosphere there is intensely German, and carries an imaginary notice board, "English composers need not apply". Dr Hans Richter conducts and draws a large salary. With the exception of Elgar's works he doesn't help modern English music at all.'[12] Brian's anti-Richter stance was broadened. 'It is of course very doubtful if many of the musicians and students of Manchester have any desire to hear modern music; indeed there is ground for believing that as a body they have not.'[13] In the final analysis even Brian conceded that it was all a matter for the financial risk-takers and not for a man in Richter's position.

Hans Richter had every reason to be jealous of Wood and the young and fast-rising Thomas Beecham, both of whom were seen as champions of contemporary composers missing from Richter's programmes. He was, however, unstinting in his praise for Wood. 'I would like to say once again', he told Marie Joshua, 'that I would like to meet Henry Wood and his wife [the Russian Princess Olga Ouroussoff]; she is a fine singer, and I consider him to be by far and away the best English conductor; what I have seen and heard of him hitherto I have liked very much indeed. Hopefully he steers clear of the overacting of today's "timebeaters" and has not lost himself in so-called "interpretations" foisted upon the great classical masters. One can be capricious with Beethoven at the Haymarket, but not with the uniquely genuine article. Privatissimum!'[14] Meanwhile Beecham was having a deserved triumph at Covent Garden with the British première of Strauss's *Elektra*. In the wake of ecstatic reviews, Richter, who felt the bitterness of the previous year rekindled as he read the reports in Manchester, wrote to Pitt. Though he attacked Strauss (as an opera composer, not for his orchestral works which he programmed constantly), he was magnanimous towards Beecham, even though the younger man was rarely polite about him.

The reports in the papers are brilliant, but they partly contradict what I have heard privately. I am not surprised that a large section of the public goes along with the celebrating (some of them going against their innermost convictions), if I think how many of them are pleased to praise the man who topples the hated Wagner from his throne. I would not be lying if I admitted to feelings of bitterness creeping over me at the thought of the whole business, when I think that *we* could have taken the whole thing in hand—and that certainly to the benefit of our

Art—if they had not been cowardly. From my heart I do not begrudge Beecham his success, for he is courageous, self-sacrificing, and has proposed good artistic goals. Help him to the best of your powers; I have written in the same vein to [Emil] Kreuz.[15]

Richter's London Symphony Orchestra concerts included Paderewski's B minor Symphony (8 November and 18 December), Bantock's *Pierrot of the Minute* (6 December), Schelling's *Fantastic Suite* together with songs with orchestral accompaniment by Herbert Brewer and Saint-Saëns (12 December), and Stanford's Choral Overture *Ave atque vale* (21 March). On 24 January 1910 Richter shared the conducting of a memorial concert to A. J. Jaeger (*Nimrod* of Elgar's Enigma Variations who died the previous summer) with Parry, Walford Davies, and Coleridge-Taylor, all of whom conducted their own works. Richter conducted Brahms's Alto Rhapsody (with Muriel Foster and men from the Alexandra Palace Choral Society), the Enigma Variations, Hans Sachs's Monologue (Plunket Greene), and the Prelude to *Meistersinger*. Alice Elgar recorded that it was a 'wonderful concert, very large audience. Richter conducted *splendid* performance of variations. He turned to the orchestra, spreading out his arms as if to draw every sound, and made the *Nimrod* gorgeous.'[16] In spite of what was obviously a winding down of his activities, Richter found time to travel to Scotland for concerts in Edinburgh and Glasgow with the Scottish Orchestra on 20 and 21 December 1909, with works from Cherubini to Wagner. His inclusion of Strauss's *Don Quixote* at the Hallé concert on 3 March (and at the pension fund concert a fortnight later) was greeted with enthusiasm by critics such as Thompson and Langford. Carl Fuchs remembered more 'Richterisms' when it came to that work. '"This should be birds not sheeps. Don't talk and disturb my flock."'[17]

At the end of the 1910 season he returned to Covent Garden. He was to have conducted two cycles of the *Ring* and *Tristan*, but at the end of the first cycle it was announced that his health had broken down and conductors were fetched from Germany to replace him. His one cycle of the *Ring* (in German) took place between 25 and 30 April, after which he never conducted at Covent Garden again. Thompson wrote:

Last evening the first of two complete cycles of [the *Ring*] began under conditions which were in more than one respect unusually favourable. Perhaps the first condition of success in London is the appearance as conductor of Dr Richter. Though we are as a people slow to make friends, we are fairly faithful to those whom we have proved, and after thirty years acquaintance with Hans Richter the London public look upon him as a tower of strength in connection with this particular work with which he has been so closely associated ever since its first production at Bayreuth under his direction in 1876. Certainly, though other conductors may make more of individual points, none makes the majesty and breadth of the music so evident, and under no other does it leave so entirely satisfying an impression.[18]

London was in mourning after King Edward VII died on 6 May but nevertheless Richter was not forgotten by the public. His daughter Mathilde wrote in her diary of numerous solicitous enquiries after her father's health. He insisted that she attend the second cycle of the *Ring* under its various conductors, and many people, including total strangers, approached her at Covent Garden anxious to hear news of him. Percy Pitt and Mrs Joshua were constantly in attendance, the latter with her inexhaustible supplies of flowers and cakes, and by 12 May the patient was well enough for a ride in Regent's Park. A week later (19 May) he returned to Bowdon to await better weather and calmer seas before setting off (on 28 May, and 'thanks to the good and strength-giving asparagus'[19] from Mrs Joshua) for the continent and his summer house at Klein Zell near Hainfeld in Lower Austria, where the family arrived on 1 June. There was no Bayreuth Festival in 1910 and by the autumn Richter had made a complete recovery. Thanks to a strict diet of fish, milk, and vegetables prescribed by his doctor, he was fit and fresh for the start of the new season.

It is as well to describe the 67-year-old Richter's final season on the concert platform in detail in order to grasp the enormity of his schedule and its travelling requirements. He began with three concerts on consecutive days in Sheffield, Nottingham, and Manchester. The first two had the same programme, Mozart's *Zauberflöte* Overture, Beethoven's Eighth Symphony, Arensky's F minor Piano Concerto (with Edward Goll), Dvořák's Carnival Overture, Debussy's *L'Après-midi d'un faune* ('perhaps a trifle more robust than the fragile music demands, but very effective and finished in its details'[20]), and three Wagner extracts. The first Manchester concert (20 October) included works by Weber, Elgar, Schumann (his Cello Concerto with Casals), and Beethoven's 'Eroica'. Casals joined him four days later in London for Dvořák's concerto with the London Symphony Orchestra. His second Hallé concert in Manchester included the city's first hearing of the Piano Concerto by Edward Isaacs played by the composer, who was on the staff of the Royal Manchester College of Music. Goldmark's Overture *Sakuntala* was also on the programme, which otherwise was normal Richter fare of Tchaikovsky and Strauss. Kreisler played the Brahms concerto on 3 November and again in Bradford on the following day. Elgar arrived in

Manchester to see Kreisler, who was about to première the new Violin Concerto (10 November) at a Philharmonic Society concert in London with the composer conducting. Elgar called upon Richter at Bowdon but found him out; nevertheless Marie and Mathilde (who described him as 'a great personality'[21]) entertained him.

Earlier in the summer Richter had heard about the new concerto. 'What splendid news!', he wrote. 'The Violin Concerto will be ready in the autumn. Who will have the honour to play it? Kreisler will play in Manchester on November the 3rd. . . . Please let me know as soon as possible who is nominated to introduce your work, because we should like to engage the same artist who is honoured by your confidence. As *conductor* of the Hallé concerts I feel obliged to ask you whether we could have the first performance of the concerto; as *friend* I should not venture to be importunate. . . . What a weather! It seems to be composed and orchestrated by—some of the new German composers.'[22] Elgar, aware of rumours that Richter was on the verge of retirement, felt slightly embarrassed that a major new work of his was about to be presented to the world without the benefit of his baton, and was moved to write in almost valedictory fashion to confirm the friendship and love that existed between the two men. 'I feel a very small person when I am in your company. You who are so great and who has known and been intimate with the greatest. I meet now many men, but I want you to know that I look to you as my greatest and most genuine friend in the world. I revere and love you. . . . Now forgive me for sending this letter which is about nothing—and yet is about everything because it is about you—the man you befriended long ago (Variations) and be assured this man will never forget your kindness, your nobility, and the grandeur of your life and personality.'[23]

Richter then travelled to London for another LSO concert (7 November), which included Brahms's First Piano Concerto (Katherine Goodson) and Mackenzie's *Pastorale* and *Flight of Spirits* (extracted from the composer's *Manfred*). On 10 November he was back in Manchester for a Schumann centenary concert (Harold Bauer soloist in the Piano Concerto), in Liverpool for Beethoven and Wagner on Saturday afternoon (12 November), and in Manchester (Mozart, Beethoven, Dvořák, and Brahms), London (Bach, Mozart, Schubert, and Brahms), Leeds (Beethoven and Wagner), and back to Manchester (Strauss's *Also sprach Zarathustra* and Bruch's Third Violin Concerto with the former leader of the Hallé, Willy Hess) before the month's end. The Leeds concert included Beethoven's Ninth Symphony.

None but men who are in every sense full grown should conduct Beethoven, neither mere pedants nor emotional virtuosi have any business with him. Richter has the controlled emotion, the sense of scale, the strong nervous grip which the music demands, and he gave a noble reading. The scherzo was not made to trip along like a Mendelssohnian fairy piece, its earnestness and force were apprehended, and the nobility of the slow movement and the mystery of the opening were realised, while the finale had at least that energy which it demands. The very human part given to the drums in the scherzo was admirably played by Mr Gezink, the bass trombone 'woke up' with fine effect in the trio and in the adagio Mr Paersch's beautiful style gave charm to the difficult horn passages. The recitatives for the basses were rather hurried, but in this passage they should be more concentrated on the platform. The chorus sang in brilliant style with an absolute minimum of effort.[24]

Richter had eight concerts during the month of December 1910, five in Manchester, two in London, and one in Bradford. They included two Wagner operatic nights, more Schumann celebrations (Frederick Dawson as solo pianist), and a mixed programme of Richter favourites in Bradford with Jean Gérardy as solo cellist in Haydn's D major Concerto and Bruch's *Kol Nidrei*. He conducted the Hallé Choir in Bach's *Christmas Oratorio* (22 December) and, on the following night, in *Messiah*. The year 1911 began, as the season had done, with concerts on three consecutive nights in Manchester, Bradford, and Liverpool. Efrem Zimbalist played Bruch's *Scottish Fantasy* at the first of these concerts, but after his customary violin solos in the second half of the programme (two Hungarian Dances by Brahms) the young Turks of the Manchester Musical Society demonstrated their disapproval of Richter by applauding the violinist to an embarrassing excess. Richter stood his ground and simply waited for the demonstration to run its course before ending the concert with Brahms's Academic Festival Overture. The programme also included a 'memorable and superb reading' of Elgar's *In the South*. 'All those monumental qualities in Richter's art find ample scope in this work,' wrote the *Musical Times*, ignoring mention of Kahane's organized discourtesy.[25] Bradford heard *Don Quixote* for the first time on the following night.

Dr Richter and the Hallé orchestra gave a more than adequate performance of the variations, while as regards a reading, the conductor displayed his idiosyncrasy by realising best the nobler and more seriously intended portions, such as Don Quixote's eloquent advocacy of chivalry and the touching scene of his retirement and death. In the grotesque portions he seemed too much inclined to whitewash Strauss and make him as musically respectable as possible. Thus in the scene of the sheep, the frank realism seemed to be somewhat glossed over and some of the more freakish passages treated in rather too serious a fashion. On the other hand the ride through the air was handled with splendid vigour and here, by the way, the much discussed wind machine proved in practice to be by no means so fearful a wild fowl as in theory it ought to be.[26]

Totally ignoring Kahane and his Society, who had now published a manifesto in the *Guardian* which attacked the Hallé for being artistically pusillanimous, his next Manchester concert was defiantly 'Richterian' by being totally devoted to Wagner operatic extracts. The month ended with the *Christmas Oratorio* in London. Godowsky played Brahms's First Piano Concerto in Manchester on 2 February and a week later Brodsky and Arthur Catterall played Bach's Double Violin Concerto and Catterall Bruch's *Romanze*, Op. 42 (Bruch was currently being performed by many of Richter's soloists in his programmes). At the same concert (9 February) Dukas's *L'Apprenti sorcier* (Germanicized in the programmes as *Der Zauberlehrling*) and Goldmark's Second Symphony were given for the first time. On 4 February Mathilde Richter wrote in her diary, 'Father said at tea he must give up his work.' Two days later she wrote that 'Father was in Manchester and saw Gustav Behrens to tell him about feeling tired and having to give up his work'. 'Father is so quiet,' she wrote on the 7th. On 13 February the story broke in the press, the reason given (and it was correct) that his health was failing and that he wished to retire into private life on the Continent. On that day Richter was in London with the LSO, and he conducted 'with all his usual fire and the wonderful magnetism in which he stands alone'.[27] Despite a great deal of sadness felt by other members of his family, Mathilde observed that her father appeared cheerful and relieved of a great burden when he returned north for *Elijah* on 16 February. A week later Ernst Lengyel played Tchaikovsky's Piano Concerto in a programme which also included Parry's *Overture to an Unwritten Tragedy*, Strauss's *Don Quixote*, and Mozart's 'Linz' Symphony. His own selection of items from the *Ring* was the centrepiece on 2 March, and four days later he went to London (with the Hanley Glee and Madrigal Society as his chorus) to conduct Berlioz's *Damnation of Faust*. His next engagement was also a choral concert, this time his last appearance with the Hallé Choir on 9 March with Bach's B minor Mass. Two days earlier he was presented with a silver cigar box by the choir at a rehearsal. According to the *Manchester Evening News*,

Mr R. H. Wilson, the choir master, spoke feelingly of the regard in which Dr Richter was held by them all. During eleven years they had experienced his terrible frown, and they had also experienced the antidote, his most benevolent smile when they had done satisfactorily. The members of the choir always gave of their best, and whatever the performances had been, they had always been loyal to the Doctor. Mr G. H. Kenyon . . . recalled a remark by Sir Charles Hallé to the effect that if there was a man in Europe whom he would like to stand in his shoes after he had gone that man was Hans Richter.[28]

The following day he took his farewells in Bradford with a Wagner evening, preceded by a luncheon at the Town Hall at the invitation of the Lord Mayor. After the concert he was presented with a monogrammed casket containing an address which expressed regret at his departure, gratitude for his work, and the hope that he might 'long be preserved to transmit by his influence the great traditions he had so nobly upheld'.[29] On 13 March he conducted Schubert, Berlioz, Wagner, and Beethoven in Liverpool, and the next day was his last Hallé subscription concert consisting of Beethoven's Overture Leonore No. 3, the Prelude to Wagner's *Lohengrin*, the Cello Concerto by John Foulds (a rank-and-file player in the Hallé band), Adagio for cello by Woldemar Bargiel (with his principal cellist Carl Fuchs as soloist in both works), and Brahms's First Symphony. Of this last work Langford wrote in the *Guardian*, 'Hearing this sublime work played with spontaneous feeling and intimate understanding of its every detail to a huge audience spellbound from the first to the last, one could not but cast back the mind to cold interpretations and still colder receptions which the work has had in years gone by, and feel that the change was a wonderful testimony to the magnitude of the work Dr Richter has done among us.'[30] Immediately after the concert the conductor sailed for Belfast with the orchestra, where they played traditional Richter fare on the following day. In London on 20 March Bronislav Hubermann was his violinist in the Brahms concerto with the LSO, after which he returned for the Hallé pension fund concert on 23 March; this was Hans Richter's last appearance in Manchester. The programme began with the Overture to *Fliegende Holländer*, Egon Petri played Franck's *Symphonic Variations* and Liszt's *Totentanz*, and the other works were Berlioz's Rákóczy March, Wotan's Farewell and the Magic Fire Music (with the young Bolton-born bass Richard Evans), and Beethoven's Fifth Symphony. 'In such music it is', wrote Langford, 'that the strength of Hans Richter has most clearly spoken to us. Though he goes, its voice will not go with him. Where music speaks the deepest truths, it will remind us most inevitably of him.'[31]

He gave one further concert with the Hallé, crossing the Pennines to Huddersfield with them on 28 March for a concert of nine works ranging from Bach to Wagner. His last three concerts were with the London Symphony Orchestra and all three included works by Elgar; the first was on 30 March for the British Musicians' Pension Fund's fiftieth anniversary and included his last Enigma Variations and the Angel's Farewell from *Gerontius* sung by Muriel Foster with Brahms's Second Symphony to conclude. His official farewell concert with the orchestra took place on 10 April, consisting of the *Meistersinger* Prelude, Bach's Fourth 'Brandenburg' Concerto, Brahms's Haydn Variations, Elgar's *Cockaigne*, and Beethoven's Seventh Symphony. After the rehearsal that day, and before the photographers, Richter was presented with a loving cup and illuminated address by the orchestra. The last concert of his career took place at Eastbourne on 22 April 1911 and repeated the works by Wagner, Brahms, and Beethoven but replaced Bach's concerto with Wagner's *Siegfried Idyll*, and *Cockaigne* with two of Elgar's *Sea Pictures* sung by Grainger Kerr.

The city of Manchester gave Richter a public presentation of parting gifts on 3 April, the day before the conductor's sixty-eighth birthday (he gracefully declined a civic reception with the words, 'the strain and great excitement of such a function would be too much for me in my present state of health, and I dare not risk disobeying my doctor's orders').[32] The gathering took place at the Town Hall where he received a travelling suitcase, six silver candlesticks, and two silver entrée dishes. Marie was presented with a pair of diamond ear-rings, and his daughters Mathilde and Ludovika each received a gold watch bracelet. Richter responded to the various speeches with the following words.

It was a hard struggle before I could make up my mind to retire, to give up a work which I have loved, which was my happiness, but as a true servant of my art I was compelled to do so, having discovered that my services in consequence of the weakness of my health could not be any more useful to my art as I intended, and as they could be in times of good health. The years I lived here were really years of happiness, and I must thank you for the undisturbed sympathy I have enjoyed here. I must especially thank the excellent Committee with which I had to cooperate. My continental experiences of Committees were not always the happiest ones, but here, I must say, the Committee helped me very much in carrying out my artistic intentions. Now may I make two requests? Please support my successor by the same sympathy I undisturbedly enjoyed during twelve years, and please continue to patronise the excellent Hallé orchestra, which is second to none. My deepest gratitude is due to the excellent, most enthusiastic chorus of the Hallé Concerts. How many times, when I came to Manchester, it was rainy, stormy, foggy, most disagreeable weather, and I have said: 'The ladies must remain at home', and yet no place was empty. This was wonderful. With these happy remembrances I part from you, I hope not entirely. My intention is at least to come, as long as I can do it, every year to conduct the Pension Fund concert. These presents will remind me that I *must* come again, and therefore I offer only my deepest and heartiest thanks for all your kindness and say *Auf wiedersehen*.[33]

33

1911–1914

Retirement

RICHTER'S retirement drew many eulogies in the press. Samuel Langford in the *Manchester Guardian* ended his with a rebuke for the conductor's critics in the city, 'When a great artist has proved himself supreme in the works he set himself to do, it is idle to complain because he has not done something else. It is a poor thanksgiving to say our prayers backwards. If Dr Richter had left us because of anything that had been written, there is not the reddest-handed revolutionary who would not have done public penance to bring him back. But no shadow of this kind need mar our farewells; Dr Richter is much too great a man to misunderstand the ardency of youth; was he not himself in the van of a movement that captured whole new provinces for music?'[1] Herbert Thompson took his reader back to Richter's arrival in England in 1877 and traced his influence upon music-making in the land.

The retirement of Dr Richter from the Manchester conductorship has been expected for some little time now, so the announcement is not a startling one, though it is sufficiently serious. There is unfortunately no doubt that of late the programmes of the Hallé concerts have met with a good deal of unfavourable criticism and that a demand has been made for a greater proportion of new or recent music.... The tone of the opposition party indicates rather less than a proper sense of gratitude for what Richter has done for orchestral music in this country since he first came here as Wagner's lieutenant thirty-four years ago. In the Wagner Festival performances at the Albert Hall during that summer he proved the artistic mainstay of the event.... New life appeared to animate the orchestra, every man of whom seemed to be in a measure inspired, but it was not until two years later that his powers became fully manifested in this country. Then were established the orchestral Festival concerts which owed so much to their conductor that they soon assumed the title of Richter Concerts. This was in itself significant for I believe this was the first time in which orchestral concerts took their name from the conductor as such; the Hallé concerts have been instanced as a case in point but it must be remembered that Hallé was founder, organiser and proprietor as well as conductor, and that it was not until the advent of Richter that the personality of the conductor and its possibilities in the direction of individual interpretation were understood in England. There were of course some competent conductors at work, Manns at the Crystal Palace was giving excellent performances and introducing to the British public Schubert and Schumann as well as a host of young native composers, deserving and otherwise. Hallé was doing a similar work for the old friend of his Paris days Berlioz, but the rest were of little account. Cusins plodded dully through the classics at the old Philharmonic concerts and occasionally scandalised our sober patrons by 'bits' of Wagner, concerning which Professor Macfarren in his programme notes was eloquently reticent. Costa, when he ground the German composers under his eye and heel, showed exceeding competence but a minimum of sympathy or comprehension, while Dr Wylde at the New Philharmonic, though he may have been sympathetic, was so notoriously incompetent that every possible anecdote of a conductor's mishaps used to be fathered upon him.

But with the appearance of Richter we turned over a new leaf in orchestral performances. I well remember the impression made by the three orchestral concerts he gave in May 1879, how the audiences grew in size and the enthusiasm increased by leaps and bounds, and how Richter soon had at his feet many of the most prominent conductors of the day, Manns, Hallé, Arditi, Ganz, Carl Rosa, Stanford were among those who frequented the concerts, and the excellence of the performances was the more remarkable since it owed so little to the personnel of the band. For it was the height of the London season when all the best orchestral players were busy and Richter had to recruit his band from every available source. It included some able musicians, . . . But the rank and file were, in the words of a prominent orchestral player of the day, 'third-rate' men, and nothing could have proved Richter's power more strikingly than the influence which he, then almost unknown, so soon exercised on his heterogeneous forces. . . . Beethoven and Wagner generally supplied the nucleus of the programme, other composers the trimmings, and public opinion, which is generally rather narrow in such matters, ratified this arrangement and soon would have no other. Richter was of course in those days regarded as one of the very advanced guard, but 'the whirligig of time brings in his revenge', and while he is still loyal to the great composers to whose music he has so largely devoted his life, a new generation has arisen who know nothing of what he has done and care less, who have but a scant appreciation of what he is doing and incline to regard him as an 'old fogey', while his sympathies, like those of every man who can form an opinion of his own, have their limitations. But for all that it would be difficult to point to a single living conductor who is anything like as uniformly successful in all the different schools. With his bringing up and his artistic convictions it was impossible that he should be a Mendelssohnian, yet he has given us the finest of all readings of the *Midsummer Night's Dream* overture inspired by a warmth and vitality of which it hardly seemed capable. . . . In spite of the deadly force of British tradition regarding a work which it deems its monopoly, he has given some superb performances of *Elijah*, strong and dramatic in character.

At the other end of the scale we have come to Richard Strauss for whose brilliant technique he has respect but for whose music he has notoriously little sympathy. Yet he is too conscientious an artist to allow his opinions to prejudice his interpretations. He has studied these complex scores till he has them at his fingers' ends, and his reading of the abstruse *Zarathustra* is generally admitted to be among the noblest and most impressive of any. If then he has produced such results in music that does not appeal strongly to his sympathies, what should be said of his readings of Mozart, Beethoven, Schubert, Brahms, Tchaikovsky and Wagner with which he is in the closest touch, and how many conductors have shown the power to do equal justice to such a wide range as these names suggest?

One has even heard it whispered that Richter is a self-seeking musician, who has come to this country to make all he can out of it. The accusation is so absurd that I am almost ashamed to reproduce it even for the sake of refuting it. I suppose that even a conductor on whom the chief artistic responsibility hangs deserves a corresponding recompense, and if it is Richter who has been the first to demonstrate the personal influence of the conductor and to make his name an asset in the success of musical enterprise, he may be considered not overpaid if, as one of the first of conductors, he demands something like, let us say, a quarter of the fee which is paid to a vocalist of similar standing. . . . In one respect we have not made the best use of Richter since he came to reside among us. He came with the highest possible reputation as a conductor of opera, and in 1882, when he introduced to this country *Tristan* and the *Meistersinger* and gave fine performances of the *Flying Dutchman*, *Tannhäuser* and *Lohengrin* together with *Fidelio* and *Euryanthe*, he showed us that this reputation was well-deserved. Since then he has conducted notable German performances of the *Ring* at Covent Garden, but the most important because the most promising of all his activities, has been the production of the *Ring* and the *Meistersinger* in English, and with, for the most part, English singers. On this he expended a vast amount of pains and enthusiasm and the vista opened out of a really national opera was one that promised many things. One may recognise some fruits of his labours in the coming performance of the *Ring* at Leeds, Manchester and Glasgow, but so far as London is concerned this splendidly hopeful undertaking has been allowed to fall to the ground. One cannot doubt that it has left its effects and that they will be manifested in the future, when others will reap the reward of the pioneer's work. But the disappointment which Richter undoubtedly experienced at the indifference with which his labours for a worthy 'national' opera were treated, must have been bitter. As regards the provinces, they have apparently still no idea that the greatest of all conductors of

Wagner's dramatic works, which after all are intended for the theatre not for the concert hall, has dwelt in their midst for the past eleven years.[2]

Elgar telegraphed his 'great pain. More than half my musical life goes when you cease to conduct.'[3] At this time the composer was well advanced with his Second Symphony, another work which Richter was never to conduct nor even to hear, for by the time of its première (24 May 1911) he had left England for Bayreuth. Not all the valedictory messages were serious. A long poem by C. L. Graves appeared in the *Spectator*.

> Beloved at once by amateurs and pro's—
> Like W.G., whom ev'rybody knows,
> The other Doctor famed for scores,
> Who, like you, used to count in threes and fours—
> You always kept your band
> In the capacious hollow of your hand.
> For who could challenge orders giv'n by one
> Who knew exactly all that could be done
> By reed or strings or brass,
> And never let a blunder uncorrected pass? . . .
>
> No more, alas! at least in this our isle
> Will rash trombonists, if they miss a cue,
> Be grievously cast down
> By the great Doctor's frown;
> Or if they give it, prompt and clear and true,
> Be raised to rapture by the Doctor's smile.
> No more will our orchestral players find
> A long rehearsal's tedious grind
> Enlivened by the sudden lightning flash
> Of humorous rebuke, or feel the lash
> Of satire stinging in some mordant phrase,
> Or smite the stars uplifted by your frugal praise.
>
> . . . You let no affectation mar your mien
> Grand, leonine, serene;
> But swayed your hearers by the triple dower
> Of sympathy, simplicity, and power.[4]

There were rumours of his return in the autumn of 1911 to conduct German opera at Covent Garden. He met Higgins on 22 April and it was mutually agreed, as he told Marie Joshua, that he would conduct there twice a year. He seized upon this arrangement as an opportunity to revive the idea of opera in English and immediately summoned Percy Pitt.

What you told me yesterday has provided my imagination with a powerful stimulus; at a stroke I am much healthier and stronger because the hope of useful activity and useful work has been revived. All I asked of Fate was a project with sufficient opportunity to take rests and gather my strength in between. Only I must not be inactive! And so my, *our* heart's desire, the lasting establishment of an Opera in English, can now be brought about. It must succeed at the third attempt, and it would be in keeping with all my life's experience to date. Twice negotiations with Wagner came to nought, but at the third attempt it happened. After two postponements I came to Manchester. I could give many more examples, for I've *never* succeeded at the first attempt, it always comes later. My dearest wish would be fulfilled if it were now as it has always been, *success at last!*[5]

Richter left Bowdon for the last time on 1 May 1911 and set off for Hungary via Ballenstedt, where he left his wife and daughters. He then travelled to Vienna to seek medical advice, and whilst there was visited by his son Edgar, who sought his father's advice about his own career. Whilst in the Austrian capital, Richter also put his country house at Weibegg up for sale. He intended to return to England to conduct, first at Covent Garden in the autumn of 1911 and then in Manchester for the Hallé's pension fund concert in March 1912. The Richter family had been very sad to leave, but none more so than Richter himself. England had been his domicile for eleven years, before which he had made annual visits lasting several months each over a period of twenty-one years. As well as the academic honours bestowed upon him by the universities of Oxford and Manchester, he had also been made an honorary Member (fourth class) of the Royal Victorian Order on 8 July 1904 and an honorary Commander of the same order on 21 June 1907. Although he had tended to cling to the German communities in London and Manchester for his social life, he and his family now had many close friends and acquaintances from all walks of life in both cities. His wife Marie was particularly reluctant to leave Bowdon. She told Sydney Loeb, 'I cannot tell you how deeply I feel this bad change! I am quite broken-hearted at the idea of leaving this country and all my beloved friends. You don't know how much I love England. I feel quite homely and must say *Extra Hungaria non est vita*, but if there is, then it is in England. I was broken-hearted when I had to leave my beloved Hungary, but now I feel worse because I have lost the vital power to settle down again.'[6]

At the end of May Richter arrived at Baracs to stay with Richardis and her husband Béla. 'Only when I got to Ballenstedt was the excitement of the last few months released,' he told Mrs Joshua. 'It was not only parting from dear friends, it was also bidding goodbye to *Beethoven*. Every performance of each work was a farewell, a last time. Each time it was as if the idol of my youth was turning away from me. Now I am with my daughter in Hungary. The peace and quiet, the uniformity of *Puszta*-life, the certainty of a regular idleness, all do my nerves good.'[7] After a fortnight of this rural tranquillity he was granted permission by his doctor to go to Bayreuth to begin rehearsals for *Meistersinger*. Marie met him there on 16 June, and together they moved into the Villa Gerber, out of town and away from the Festival crowds. Earlier in the year he had warned Siegfried Wagner, 'Whether or not I am fully able to carry out my task will be made clear at the first rehearsal. I can only attend piano rehearsals if I do not have to play. I must direct the orchestral, stage, and dress rehearsals myself. My prayer is: Wotan, give me health and strength so that I can carry out my plan, which is "to serve, to serve!"'[8]

By 19 June he had conducted four orchestral rehearsals, 'which came off brilliantly', as he told Percy Pitt.

Everyone is delighted by my tried and tested strength, freshness, and staying power, I more than anyone, naturally. The fact that I have complete mastery of the demands of my task has given me a new lease of life, my courage is uplifted, and the gloomy shadows which darkened my spirit have been banished; I have come to life again and am in my element. I am so happy that I plucked up the courage to undertake my task. My work is so arranged that I always have enough time in which to recover. Everyone is doing his best to make my job as easy as possible. The people—on and under the stage—are superb; talented, keen, and inspired. I think something great will happen.... I feel born again and am happy. No risks will be taken, but I feel I am not here in vain.[9]

The five performances were spaced almost a week apart between 22 July and 19 August to preserve the conductor's energy. After the fifth he ended a card to Pitt with the words, 'I wonder if I shall ever step into that "mystical pit" again?'[10] Ernest Newman, no friend to Richter, whom he accused in February 1906, during his brief tenure as the *Manchester Guardian*'s music critic, of being unprepared for a performance of Berlioz's *Roméo et Juliette*, wrote a long and highly critical article about the 1911 Bayreuth Festival in September's *Musical Times*, but even for him *Meistersinger* was a highlight and the orchestra under Richter 'admirable'.

In the August issue it was noted that he was to settle in Bayreuth and establish an operatic school. Nothing came of this scheme, but the town of Bayreuth undertook to house him for the rest of his days. He called his new house, which he occupied from 30 November, Zur Tabulatur, in honour of *Meistersinger*. It had been built in 1740 by the *Markgraf* for one of his court ladies, and, as Richter told Marie Joshua, it reminded him of 'a French chalet [sic] from the time of Louis XV'.[11] The town also undertook alterations and repairs to the house according to the new tenant's wishes. During the *Meistersinger* performances he found time to write a note to Pitt. 'A gentleman from the Wagner Society in London hopes to have finished building a theatre in one of the large parks there in time for 1913; in it he wants to stage *Parsifal* with Bayreuth forces. I think this bold optimist is called Parker. If the man can bring *that* off, I shall publicly convert to Islam in Trafalgar Square and marry the Pope! There are some wonderful dreamers about!'[12] Despite this apparent cheerfulness, his mood was low by the beginning of October. He became worried about the forthcoming move from his lodgings in Münzgasse to the new house and told Pitt that he had refused offers of appearances in St Petersburg, Moscow, Brussels, Antwerp, Vienna, Budapest, and Madrid. There was now no question of travelling to London to conduct at Covent Garden (Franz Schalk took over), and he showed his relief when comparing life in Bayreuth with city life in England. 'No toot-toot, no stink, and one can cross the street safely.'[13]

On 15 October Sydney Loeb proposed to Mathilde Richter and was accepted. He then wrote to her father to ask his permission, which was duly granted. She could expect £60 annually perhaps even rising to £100. Each of his children would inherit one sixth of his assets when he and Marie were dead. He had hoped that the marriage would take place in Bayreuth during the 1912 Festival as a civil ceremony, rather than in the spring in England. He emphasized that, because he had neither racial nor religious prejudice (he told the couple that his Holy Trinity were Bach, Beethoven, and Wagner), he expected that no pressure would be exerted upon his daughter by Sydney or his family to convert her to his Jewish faith. When writing to Pitt in November 1911, Richter defended his future son-in-law against his friend's antipathy towards him. Loeb was, he said, a responsible and honourable man whom he liked very much. His letters to Mathilde became full of fatherly orders and advice; she was not, he insisted, to go to restaurants after concerts or the opera, even in the company of her fiancé, and Sydney was asked to send regular reports on her health. Both were told to write clearly to save Richter's failing eyesight.

He outlined to Pitt his hopes for the deferred engagement at Covent Garden, which meant, in view of his intention to take a five-week cure in the New Year and with his commitments in Bayreuth later in the summer, that it would have to take place in 1913. The year 1913 would, he hoped, be an auspicious one because he would be celebrating his seventieth birthday. More importantly it was the centenary of Wagner's birth. Curiously no Festival was planned for Bayreuth in that year, so he hoped to take his farewell from the conductor's rostrum with the best singers in a Wagnerian blaze at Covent Garden, *Fliegende Holländer*, *Tannhäuser*, *Lohengrin*, *Meistersinger*, *Tristan*, and the *Ring*.

His last letter to Pitt in 1911 begged for help. He enclosed an advertisement in which he endorsed the Orchestrelle Company's Metrostyle pianola. In it he was quoted as saying, 'It was difficult to believe that it was not an artiste performing, for the difference between its playing and that of other such playing devices is so great as to be startling.' 'This English appears so unfamiliar to me,' he told Pitt. '*Perhaps* I said as much, it's even more likely that I wrote it. I can *absolutely* no longer remember making such a remark. Can you find out anything definite? Could you take a look at the letter? I'd like to know the truth, for I don't wish to be used as an advertisement especially as I don't like making music with machines.'[14] In Manchester in 1903 Richter endorsed Schiedmayer pianos with the words, 'I played on it all the evening and am quite delighted with its perfection and exquisite quality of tone,' but his antipathy to making music by mechanical means, as well as the state of his health, may have contributed to his refusal in the spring of 1914 to record himself on film for the Messter Company in Berlin. Their extraordinary request, endorsed by Nikisch, was part of a project in which conductors were filmed conducting a particular work. The idea was then to show this silent film on a split screen simultaneously to an orchestra, who would see the conductor from the front, and to an audience, who would see him from the back as if at a concert. Thus orchestras of the future would be able to give the conductor's interpretation of a particular work long after his death. If only, Messter suggested, this process had been discovered as long ago as 1872, we would then have had Wagner's performance of Beethoven's Ninth Symphony as he conducted it at the foundation-stone laying ceremony at Bayreuth. Now they hoped Richter would bequeath examples of his work to the future and join the ranks of those who had already committed themselves to film, such as Weingartner (Leonore and *Egmont* Overtures) and Oskar Fried (*Symphonie fantastique*). Those who intended to participate included Nikisch, Schuch, Lohse, and Reznicek.

With the arrival of winter in 1911 Richter settled down to the way of life which he would follow for his remaining five years. He was a frequent visitor to Wahnfried and began a regular routine of visiting his daughter Mitzi and her family in Dessau and Richardis in Baracs in Hungary. His last trip to England took place at the end of February 1912, when he and Marie arrived for Mathilde's marriage to Sydney Loeb (now a stockbroker) on 1 March at Paddington Registry Office. Richter gave a reception for seventy-eight friends and relations at the Great Central Hotel. He was reported in the press as being in good health and spirits, but would not stay long enough in the country to conduct the Hallé pension fund concert on 21 March. He told Eva Chamberlain (née Wagner) how moved he had been at his first sight of the white cliffs of Dover. His return journey, on the other hand, was delayed by inclement weather, which reminded him of so many rough crossings of the English Channel. A letter to Mrs Joshua in April, written after his return to Bayreuth, still spoke of plans to conduct at Covent Garden in 1913, though the list no longer contained *Lohengrin* and *Tannhäuser*. He maintained another link with England by subscribing to the *Illustrated London News*.

He undertook the reponsibility of a revision of the text to Wagner's *Ring* at the request of Wahnfried, and found the work both stimulating and enjoyable because he put musical considerations aside to concentrate solely upon the poem. The spring of 1912 was spent with his daughters, after which he took a cure at Wörishofen in Bavaria. His son Edgar, who had already sung the roles of Loge, Siegmund, Lohengrin, and Tannhäuser in various minor theatres in Germany and Austria, was contracted for five years as house tenor in Kassel from the middle of June. Richter returned home on 22 April for his final *Meistersinger* at the Bayreuth Festival. He told many friends that he could not wait to immerse himself in that 'bath of steel in C major', rehearsals for which began towards the end of June. The Loebs were unable to attend the opening on 22 July, because Mathilde was pregnant, but after the fifth and final performance on 19 August 1912, the last of Richter's career, he sent a brief postcard to them which read, 'Yesterday a wonderful conclusion. They wanted to drag me forcibly on to the stage, but I remained resolute and did not go up. I received a beautiful leather armchair to rest in from the orchestra.'[15] His eyes were brimming with tears as he began the third and final act of the opera for the last time. 'Gentlemen,' he said to the orchestra, 'make my farewell really difficult for me.'[16] Mathilde received letters from several friends (including her godmother Marie Joshua) who were present at the last night.

I thought he looked much better than when he was in London last March, but I hear he feels the strain of conducting and that last night's *Meistersinger* was probably his last performance as conductor. It was such a poetic performance that it was a fitting climax to a life of splendid achievement. Nobody who was there can ever forget the enthusiasm after the performance. The whole audience cheered and shouted for quite twenty minutes. When the curtain went up, Siegfried and the whole company and chorus appeared on the stage and implored Dr Richter to come on to the stage, but he was obdurate and would not. I think he did not feel equal to the emotion it would occasion in him. All the performances have been as near perfection as possible, the orchestra quite marvellous in its beauty.[17]

Bayreuth was delightful and the performances better than ever, but *the climax* was the last *Meistersinger*. I can't begin to describe the scene of excitement and enthusiasm at the end and the shouts for Richter, which seemed to go on for quite half an hour—then the curtains parted and we saw the whole company plus Siegfried and the chorus master [Hugo Rüdel] all waving and even calling to your father, who was evidently still at the conductor's stand. They remained like that, making frantic signs for about ten minutes, then your father seems to have bolted out of the theatre, and the curtains closed and we all went reluctantly away. We all realise that it was a very wonderful evening and one will *never* see such a performance or demonstration again. . . . I went to the Orchester-Fest [18 August] but did not get there in time to hear your father's speech—they say he cried when he announced that the next day was *the* last time he would conduct the *Meistersinger* and I can well believe it. Your mother has also cried a lot, I know. She has been through so many emotions—it was a very sad moment for them both.[18]

The etcher William Strang travelled to Bayreuth to draw Richter. On the back of Strang's letter confirming arrangements for the sitting, Richter noted, 'after a not quite successful attempt on 21 July, Mr Strang drew a good picture in a day, on 23 July 1912; in all it took him perhaps somewhat more than three hours'. The conductor was very satisfied with the result, which he hung on the wall of his study 'to remind me of my very happy stay in your hospitable country'. An etching of the portrait was issued in December 1912 in a limited edition of seventy-five signed proofs at six guineas each (half-price to members of the Wagner Society). Strang was paid £31 10*s.* for drawing the portrait plus his travelling costs and hotel expenses totalling £12 7*s.* 8*d.*

After Bayreuth Richter resumed his cure in Bavaria and visited Richardis in Hungary. By the autumn of 1912 he was missing the sound of an orchestra and a chorus; he knew too that the Hallé season had begun under his successor Michael Balling. Alone with his thoughts, he became agitated that *Parsifal* would soon be performed legitimately (though he used the word 'prostituted' in letters to both Mrs Joshua and Percy Pitt) outside Bayreuth after Wagner's thirty-year ban expired in 1913; he believed that the Master's fundamental wish that the opera belonged only to his theatre should be eternally respected particularly by Germans. He ended a letter to Pitt with the words, 'I have burned my batons; there is no turning back. The farewell to the conductor's rostrum was so superbly wonderful that I cannot tempt Fate, but must remain grateful for such benevolence

1911–1914: Retirement 443

and withdraw to my cosy Philistine home.'[19] On the matter of batons, Richter told Sydney Loeb two months later, 'There is no "Tactstock" left, Mädi [his granddaughter by Mitzi] has the stick I used at the last *Meistersinger* performance.' Mädi bequeathed the baton to the Wagner Museum in Bayreuth, where it is exhibited on Richter's handwritten full score of *Meistersinger*. This letter also described the Dutch conductor Willem Mengelberg as 'a fine conductor. He *must* have a glorious success everywhere.'[20]

Once back at Zur Tabulatur he began a regular daily routine. He rose at eight, took an hour to bathe, and undertook his prescribed exercises. He then breakfasted on yoghurt and a large slice of black bread. From ten until midday he took a walk in all weathers, a round trip of over two miles up the hill to the Festival theatre and back. At midday he would occasionally coach singers who came to Bayreuth, many hoping to audition for Siegfried Wagner. The Norwegian soprano Borghild Langaard worked with him on the roles of Senta and Venus in November 1912. 'I shall never forget the delightful time I studied with you,' she wrote. 'Although I have not yet been successful here in Bayreuth, I hope I shall, in the future, not give you reason to be ashamed of your pupil.'[21] Lunch was at one o'clock, after which he read the paper and smoked in his comfortable 'grandfather's chair' (the gift from the Bayreuth orchestra). From four until six in the afternoon he would walk around the Röhren See and back through the Court gardens (Hofgarten). On Fridays, and *only* on Fridays, he went to Grampp's bar for a beer and political discussion. Supper was at 7.30, after which Marie and Ludovika often went to Wahnfried for lectures and soirées whilst Richter stayed at home, reading and smoking his pipe until ten o'clock, when he prepared for bed. Tobacco and snuff were both regularly dispatched to him from London. Ten years earlier the firm of Kapp and Peterson had sent him a meerschaum pipe and 'a quarter pound of mild Coronation to start the colouring of the pipe' as 'a memento of dear old Dublin' after a visit with the Hallé Orchestra.

Although he would often visit Vienna to see former colleagues and old friends on his way to Richardis and her family in Hungary, Richter gave his country house at Weibegg in Lower Austria to his son Hans in January 1913 (having tried unsuccessfully to sell it). It was also at the beginning of 1913 that he received a telegram from an unexpected quarter. 'Dear Master, do come once more to England next month to direct the *Meistersinger*, which you alone know so well how to give in all its beauty. Thomas Beecham.'[22] Richter refused and history repeated itself, for Beecham took over as he had done in 1899 for his father's mayoral concert at St Helens.

Richter's family was now more important to him than ever. Mathilde gave birth to a daughter, Sylvia, on 19 December 1912, but Richter could only dote on photographs of her. All his letters, many addressed to his infant granddaughter, expressed his longing to see her (a three-week trip to England was planned for May 1913), but he never did. Mitzi's children, on the other hand, were often brought from Dessau to Bayreuth, where they brought noise and bustle to the house. In March 1913 he visited Hungary, returning for his seventieth birthday celebrations on 4 April. He received cards, letters, and telegrams from all quarters. Many articles appeared in newspapers and journals, and a medal was struck in his honour in London with a subscription opened for its sale. On the morning of his birthday he was entertained with a programme of music by the band of the seventh infantry brigade of the Bavarian 'Prince Leopold' regiment under their conductor Oscar Jünger. They began with the chorale 'Lobe den Herrn', followed by Wagner's *Huldigungsmarsch*, the Prize Song and chorus from Act III of *Meistersinger*, and ended with the triumphal march on motifs from *Herzog Wildfang* by Siegfried Wagner. Coincidentally Richter had just sent a card to Pitt (on 30 March 1913) on which he listed the instruments present at the first performance of that other celebratory birthday music in which he was involved, the *Siegfried Idyll*. Interestingly he wrote *two* first and *two* second violins, two violas ('I played a part and also the trumpet'[23]), one cello, bass, flute, oboe, and bassoon, and two clarinets, and horns.

For Richter the centenary of Wagner's birth on 22 May 1913 meant more than his own seventieth the previous month. There was nothing lavish, just a small celebration in the old theatre in the centre of Bayreuth, but the town council marked the occasion by granting him and Siegfried Wagner honorary citizenship of Bayreuth and presenting both men with a gold medal. Two weeks later, at a party given for Siegfried's birthday (6 June), he was taken ill in the stifling heat and crush of the 200 guests present, and had to be taken home. This setback caused another postponement of a trip of London to see his new granddaughter when he was already agitated by the news that Mitzi and her children had contracted diphtheria in Dessau. He often wrote letters to his children pointing out his seventy years, forty-one of which had been as a Kapellmeister and sixty-one as a musician. During the rest of the year, however, he was able to make a short trip to Switzerland and visit his family in Hungary and Germany. In November he wrote to Mrs Joshua:

Your report of Elgar's *Falstaff* brought home heavily to me how excluded I feel, despite my joy at the beloved man's success. During the summer Elgar sent me a copy of the *Musical Times* which contained a report of his work together with musical examples. What themes! Unmistakeably and genuinely Elgar! I involuntarily raised my conducting arm and beat time as I read. I can sense how he has developed these themes and how they sound in his characteristic scoring. Admittedly I can only imagine it, for this original master always surprises with his work and even one's boldest imagination proves to be far from reality. He would not be Elgar if one could anticipate his ways. Please give my greetings to the honourable and dear man and his family. I should write to him, but writing a letter in English, with forever having to consult a dictionary, is very troublesome. If only letters and words were notes, matters would be easier. It's hard to understand how one can forget *so little*, for there was never much to my English and my English vocabulary never troubled my brain much. I still read some English books—mostly I leaf through my beloved Dickens—but I soon shut them and am ashamed of my lack of vocabulary.

To conclude, I have another request: Christmas is before us and I know that you are always so generous to me with your presents. I am no longer allowed to smoke heavy Havana cigars. If I must smoke at all, although I am no chimney, I would ask you for some (a few) German cigars of the mildest kind, *those* I am allowed and I would gratefully smoke the health of their honoured benefactress.[24]

When Elgar was told by Mrs Joshua of his friend's enthusiasm for *Falstaff*, he immediately sent him a study score of the work (his last known communication with the conductor). Unfortunately Richter could only read the work with great difficulty due to the small size of the print (the full score was not yet available). Mrs Joshua replied with a long letter, ruing Richter's absence from London's concert life. Elgar's Second Symphony (which 'takes your breath away'[25]) would, she ventured to suggest, benefit more under him than under Wood, Landon Ronald, or even Elgar himself. Pitt, she reported, vouchsafed that the forthcoming première of *Parsifal* at Covent Garden would be better than any production at a German theatre outside Bayreuth (Richter exhorted all his friends and family not to attend any production away from Bayreuth). She also told Richter that Isolde Beidler had written asking for money to help pay legal costs in the court case pending against Cosima concerning her own paternal origins, for Isolde contended that she was Wagner's and not von Bülow's daughter. It was an unpleasant affair and Richter would not be drawn to comment. Though his loyalty was to Cosima and Siegfried, he had also been very close to all the Wagner children from their childhood days, and had installed Isolde's husband Franz Beidler as his deputy with the Hallé orchestra in 1908 and 1909. Mrs Joshua could refuse Isolde's request because of the parlous state of her own finances (she had been a widow since 1906). On a happier note, she told Richter that she had seen his new granddaughter Sylvia, who had beautiful red hair just like the beard of her famous grandfather.

Richter now wrote to Pitt with a surprising admission.

Praise Wotan that I can still sleep well; my dreams are often really vivid, but mostly happy. I often see myself with my English orchestra. This morning I was pleased to discover that I still have the third act of *Tristan* memorised. I conducted it last night and saw the good, trusty faces of my splendid musicians; it was a pleasure to see how the winds paid attention to the many metrical changes in the second scene, and were together in the chording; also how Reynolds played his cor anglais in *one* breath, and how Brough made the Tárogató sound merry. There was Mills [clarinettist] who played his 'Marke' musically and with such apt seriousness, and the place in E major for the horns and the solo violin! The ever-present and beautiful Miss Timothy, as excellent on the harp as on the lute; at this latter point I experienced a tender shiver of excitement. How can I name names when they were all so excellent? . . . But this much is true, as wonderfully beautiful as they played *Meistersinger*, the *Ring*, in short everything, the crowning achievement was for me *Tristan*; I never heard it played anywhere in such an exemplary and perfect manner, not even in Vienna. And the *joy* of the rehearsals, how each work grew on each occasion together with the understanding and the enthusiasm of the orchestra. Bur now I must stop and withdraw soberly to my pensioner's corner. From your old comrade-in-arms, Hans.[26]

34

1914–1916

The Last Years

RICHTER'S letters written after his retirement were primarily domestic in content, though occasionally he would comment on or reminisce about music and musicians. In February 1914 he replied to several questions from Sydney Loeb.

Miss Fanny Davies is an excellent artist whose attainments I remember with pleasure. After many years lying in the archives, where it was forgotten by artists and public alike, I revived *Joseph in Egypt* and performed it in Vienna to great effect. This opera by Méhul is a masterpiece; the Master also loved it very much, which fact can be found in his letters and writings. For today's public, whose ears have been ruined by the motor car and other noises, this music is too refined. Even in Vienna the opera house remained empty despite an excellent performance, and it soon vanished from the repertoire once again. In this case—as so often in others—it is the public which fails, not the opera. Brodsky plays everything superbly, but if you want a special treat, hear him and his quartet playing Schubert, in that he is unique; his warmth, his feeling, his fire resulted in a great success in Vienna. Send him and his splendid colleagues my warmest greetings. [Leonard] Borwick is a refined and stylish artist. Listen to Mengelberg as much as you can, I think very highly of him. Sir [Charles] Villiers [Stanford's] *Irish Rhapsody* is certainly a solid piece of work, just like the earlier ones. Watch out for the boring parts. I have already heard of Albert Coates, but I never saw him. Those reviews of him which I saw were not unfavourable.[1]

Richter's friendship with Stanford had been soured by an unfortunate episode in April 1908. The quick-tempered Irishman had, as on so many occasions, sent a pupil to Richter for help. The young man in question was the pianist and composer James Friskin and he was to attend Richter at the Queen's Hall after a morning orchestral rehearsal. The session finished early to enable the players to fulfil another musical engagement elsewhere later that day, and Richter had to wait half an hour before Friskin arrived punctually at the arranged time. He was shy and nervous in the presence of the great conductor, and, when asked if he was presenting himself as a pianist or composer, replied with the former. Stanford, in his letter of introduction, had particularly asked Richter to hear his compositions. The conductor, on the other hand, felt he could do more for the young man as a performer, though even here there was nothing he could offer him at present. Stanford was furious and reminded Richter that from the start he had warned him about Friskin's shyness and modesty. Richter was particularly upset by his words, 'You began therefore by blaming him for an unpunctuality which did not exist, and to start with an undeserved reproach was not exactly calculated to encourage him or steady his nerve. I am sorry about this, but I cannot say that I am surprised. Your position in this country, gained primarily by your own gifts, was assured by the unflinching support of men like myself, whose goodwill has now become unnecessary to you, and which you therefore have dispensed with.'[2] Richter was stung by this 'unexpected, nay unpleasant surprise'. He corrected Friskin's inaccuracies and rounded on Stanford for his 'undeserved accusation of ingratitude. . . . You insinuate that you have always regarded me as an ungrateful egotist, one who was open to accept favours but never to return them. Now I know it is not necessary for me to tell you that I have never used my friends in this way: on the contrary I have always tried to help them to the best of my ability and as far as it lay in my power. Your letter has robbed me of an illusion, but it has enriched my experience, which is a bad exchange. In spite of the unpleasant effect of your letter, I shall always be at your service musically.'[3]

Stanford's moods were unpredictable (there had been a similar disruption in his friendship with Robin Legge, music critic of the *Daily Telegraph*, with Friskin once again the cause of the argument), but he usually quickly forgave and forgot. Richter, on the other hand, only performed Stanford's music once more after this incident, his Choral Overture *Ave atque vale* with the London Symphony Orchestra on 21 March 1910. Their correspondence stopped but Stanford wrote for the last time at the outbreak of war in August 1914, and sent his letter via Arrigo Boïto in Milan. 'I must send you a handshake in these tragic days. I do not forget the old times, nor everything you have done for music in England. Thank God there are neither political nor military questions in music.'[4] He signed himself as Richter's old admirer and friend, and the conductor told Sydney Loeb that he was glad to hear from the Irishman.

Richter's life in the period shortly before the outbreak of war in August 1914 centred around family and friends. He worried about the Wagner daughters, in particular Daniela Thode's divorce from her husband Henry and Isolde Beidler's rift with her mother. His own son Edgar married in April but his father did not attend the wedding; instead he complained of Edgar's mismanagement of his financial affairs. Mitzi and her husband set off for a tour of Italy by car, an adventure which also filled him with dread. His legs gave him much trouble ('my pedals' as he called them and, after standing before an orchestra for forty-seven years, was not surprised that he was now suffering), and high blood pressure with hardening of the arteries was also diagnosed in May. Another trip to England had to be put off but instead the Loebs (without their daughter) travelled to Bayreuth for the Festival in July. Mathilde thought, 'father looks well but is quiet'.[5] When they left at the end of the month it proved to be a final farewell. He attended a rehearsal for the 1914 Bayreuth Festival but, as he told his former colleague Wilhelm Gericke, 'I'm not going to rehearsals any more; my first attempt turned out to be a miserable affair. I had to leave the hall after twenty minutes, I was so affected by the sound of the music. . . . I must force myself to direct my thoughts elsewhere and to convince myself that I have become estranged from all music and now belong to quite another branch of the profession. . . . I never thought that bidding farewell to the rostrum would affect me *so* badly. I can honestly assure you that this is not vanity, only an urge to work and a pure desire to achieve.'[6]

He had further cause to worry when his two sons Hans and Edgar were conscripted as the international situation deteriorated. After war between Britain and Germany was declared on 4 August 1914 Richter and his London-based family were compelled to exchange letters via Sydney's cousin Caroline Auerbach in Amsterdam. She copied his letters, retained the originals until after the war, and then gave them to Mathilde. In October he sent a message to Elgar. 'Sydney can visit him and tell him that neither bad politics nor the smoke of gunpowder will come between us.'[7] He also told Sydney, 'if you meet Miss Timothy or the gentlemen of my orchestra, please give them my best wishes. I think of them with the greatest of friendship. The time I spent with them will always remain as beautiful, honourable, and happy in my memory.'[8] With the outbreak of hostilities Richter was unable to correspond with Marie Joshua. In February 1914 he thanked her for the piano score of Elgar's Second Symphony ('if I cannot play it because my finger dexterity is less than zero, at least the pleasure of reading it is that much greater if I cannot play any wrong notes').[9] His last letter to her, written in June 1914, described his arteriosclerosis but also included his regret that in the hurly-burly of preparations for the forthcoming Bayreuth Festival he was no longer the recipient of rehearsal schedules.

There then followed an episode which soured Richter's reputation in England, when, in response to the use by English soldiers of dumdum bullets, he renounced his honorary doctorates. *The Times* reported that 'Dr Richter has addressed a letter to the Universities of Oxford and Manchester announcing that he will no longer use the title of honorary Doctor of Music conferred upon him by them, of which he had hitherto been proud. He adds that he has placed his English orders at the disposal of the Red Cross Society, "the beneficial activity of which is also of assistance to wounded English soldiers".'[10] This caused much consternation amongst his English friends (Sydney told Mrs Joshua that 'our friend must have lost his senses'). Richter came to regret his precipitate action and was soon making strenuous efforts to reassure his English friends and former colleagues that it had been taken in the heat of the moment and in the spirit of nationalism which prevailed at the beginning of the war (Max Bruch also renounced his Cambridge doctorate at the time). As *The Times* recalled, 'the most his English friends could do was to adopt the magnanimous attitude of the Committee of the Liverpool Philharmonic Society, which, in February 1915, "decided to take no action for the removal of the portrait of Dr Richter, which hangs in the hall, and which was found with the face turned to the wall"'.[11] Richter was now so concerned for the Loebs that he begged them to leave England for Bayreuth if life in London became unbearable.

In the New Year of 1915 Siegfried Wagner conducted several concerts in aid of the German Red Cross. These took him to Coburg, Darmstadt, and Vienna, and there he encountered Richter, *en route* to Baracs to visit Richardis. Richter slipped into a back row of the concert hall during a rehearsal of one of Siegfried's compositions. Somehow word reached the players that their former chief was present and the rehearsal disintegrated into a chorus of cheers and applause. Richter took matters in hand and restored order with mock outrage. 'Silence!' he boomed. Then after a pregnant pause he continued, 'I see there are still disturbances in the rehearsals of the Philharmonic.' Then he greeted each and every member of the orchestra he still knew. Siegfried, in an attempt to get his rehearsal underway, gently suggested that he return to his hotel in case he became tired. Richter refused. 'It wouldn't occur to me to go. I want to hear my Vienna violins once more. I've been deprived of them for long enough.' A little later Siegfried, now concerned that Richter would become excited and upset, suggested again that he should leave. 'No, dear Fidi,' the old man replied. 'Let me tell you something. With all respect to our Bayreuth orchestra, there's no orchestra in the world like this one. I want to revel in the sound of my Vienna violins. So let me be, I want to enjoy it. Who knows if I will ever do so again in my life.'[12] He stayed until the end and left with tears in his eyes.

At the end of 1915 he was saddened by the death in Hanover of Ernst Schiever, leader of his London orchestra for thirty-three years, but his mood improved when an honorary doctorate of music was conferred upon him by the German University in Prague; once again he could be addressed as Dr Richter. He went to the city to receive it in person in the autumn of 1915. In November he wrote again to Sydney praising England's music and musicians. Loeb passed some of its contents on to Elgar.

I notice that in the repertoire of the English opera season you sent me there is not a single English opera. This made me think of my work. My endeavour was, through the performance of well-translated masterworks, to induce English talent to create national operas. A promising beginning had already been made: Mackenzie, Stanford caused one to hope for the best. I only know *Colomba* from the piano score. I conducted *The veiled Prophet*. I think *Shamus O'Brien* most successful, the score of which the composer himself played to me. The singers are there. What they lack in routine they more than make up with talent and enthusiasm; there are excellent choirs and *the best* [orchestral] *musicians* who more than attain the highest expectations and standards. Even the shabbiest and wickedest politics cannot lessen my gratitude, which fills me at the recollection of artistic work in England. If you meet one of the good people who helped me to further my endeavour to attain my artistic aim, greet him or her from me. If by chance you meet Sir Stanford, give him all the best from me and tell him that his letter from Italy pleased me very much. Is 'Sir Elgar' already writing an opera? From him something great will be expected![13]

In October 1915 Richter suffered the first of a series of strokes which he described to Michael Balling.

On 22 October we came home from holiday. I was well throughout the time away, in spite of the change of food and drink, and indulged in no excess. With no warning I fell unconscious during the night of the 23rd—it must have been after ten o'clock for I heard it strike the hour—and remained thus for thirty-six hours. Kubitz found me senseless in bed early on the morning of the 24th and they sent for Dr Landgraf, who bled me. A nursing sister was also summoned. I knew *nothing* of any of this. I awoke on the morning of the 25th about eleven o'clock and was astonished to find myself still in bed so late, and even more so to find a nun there (the nurse). I had no idea that I had been ill or needed a nurse. I wanted to get up and go out, and not to give in to all this, but the loss of blood had weakened me very much, and my unsteadiness brought home to me at last that something was up. It was a good thing I had asked Kubitz to come, for my people would have let me sleep on and the consequences might have been worse. During the time I was unconscious I must have already been 'on the other side', but Saint Peter gave me an extension of leave—but for how long? Physically I soon recovered, only my eyes are still weak and very unreliable, I am still in the care of my eye specialist Dr Reuter. I still walk three to four hours daily and in all weather. So, now you know it all. The doctors consider that I shall recuperate, we shall see! At first my memory was weak and this worried me, but Landgraf reassured me and it is noticeably better now. I read a lot; music—only Bach! I do not miss new music, I would no longer be capable. Am I missing much?[14]

He found another colleague to whom he could write. Carl Fuchs, his former principal cellist in the Hallé orchestra, was on holiday in Germany at the outbreak of war and, as a naturalized British subject, found himself first interned and then permitted to live with his sister at Jungenheim near Darmstadt until 1919, when he returned to Manchester. Richter told him about his own health problems, playing down the most recent incident, and summed up his view of the war with a quotation from *Meistersinger*. 'I was already on my way to visit my dear old leader and friend Schiever, but "those above" have extended my leave, but for how long? . . . My thoughts and dreams are continually preoccupied with my superb English musicians, choirs, and soloists; they were marvellous, unforgettable times! They will never return, at least not for me. Oh, these hateful diplomats! Will this awful wound ever be healed? "Wahn! Wahn! überall Wahn!"'[15]

Two weeks later, having asked Richter for a reference so that he could play in a Frankfurt orchestra, Fuchs received this reply from his former chief and wrote on the envelope, 'A letter which made me proud.'

It seems strange to me that I am to provide a testimonial for an artist who is the perfect master of his instrument, but if you think it would help you, please use this letter as my recommendation. I have had plenty of opportunities to judge you as soloist, principal orchestral cellist, quartet player, and teacher, and thus I have learned to value you highly. Your playing of the most intricate works from J. S. Bach to the moderns was exemplary and stylistically right. Happy the organization (or organizations) that wins your services. . . . I could not write a long letter as I have to use my weak eyes carefully; but why do you need a testimonial? Just play to the people, that is the very best recommendation.[16]

In the following reply to a letter from Balling, Richter probably referred to the two movements so far written from Elgar's *The Spirit of England* ('To Women' and 'For the Fallen', the 'Fourth of August' following in 1917).

Balling, deprived forever of his post with the Hallé orchestra since the outbreak of war, was now far more anti-English than Richter, who by now had tempered his own feelings.

I cannot believe that Elgar is writing anti-German music. A German (me) freed him from his English problems, German Kapellmeisters (such as Buths in Düsseldorf) performed his works and thereby made his name. He himself often complained bitterly and angrily about the English and considered himself a descendant of the Spanish nation. The name is Teutonic. *Gar* is originally west-Gothic, retained in *Ger*hilde, *Ger*tleite etc., whilst *el* is the Spanish definite article. He was particularly proud of his German successes. He wrote some funeral music (probably a commission) for the Belgian fallen, but I do not believe he is anti-German; you're probably the victim of gossip. . . . Stanford behaved remarkably with a statement in the *Times* against the invasion of *hordes of German musicians and music teachers* since the time of Handel. We can vouchsafe that English orchestral players are excellent, but only honourable and truth-loving English people will concede that they learned that from the Germans. The 81-year-old [Archibald] Ramsden, a pure Yorkshireman, told some fellow who thought it fashionable to complain about me, 'he gave us Elgar and brought/made [*sic*] good music to/in [*sic*] this country'.[17]

The year 1916 began happily for Richter with his wedding anniversary, forty-one years of marriage to Marie, and the news from England in April that the Loebs were expecting a second child. In March and May, however, he suffered several strokes during which he was again unconscious for several hours. His name-day (Johannistag) on 23 June was celebrated with a serenade by a choir of eighty children singing folksongs in the garden of Zur Tabulatur. On 31 July he went to Wahnfried for the last time for a gathering in memory of Liszt, who had died thirty years earlier. At the end of August he reminded his family that it had been forty-seven years since he conducted his first opera, Rossini's *William Tell*. His mind seemed focused both on the distant past and upon his imminent death. His last letter to Ludwig Karpath began with the news that Max Schlosser (the first David in *Meistersinger* in 1868 and the first Mime in the *Ring* in 1876) had died. It was Richter who had been tipped off about Schlosser by an actor and he travelled to Augsburg to audition him. 'His musical security pleased Kapellmeisters, his brilliant acting pleased directors.'[18]

After a further decline in his health in the autumn he bade farewell to Franz Fischer (in handwriting which had sharply deteriorated).

I was unconscious for the whole day on 18 October, you can imagine how worried my wife and daughter are. Yes, this pensioned-off, old Kapellmeister is getting shabby. Nothing lasts forever. The devil which is confining me to bed is called arteriosclerosis. In my fevered imagination I relived my early days as a Kapellmeister, in which you played a part. Our friendship began with *Rheingold* in Munich [1869] and continued with greater warmth in Budapest. Please accept my sincerest thanks for your *loyal* friendship and for the *perfect* artistry with which you supported my work. Your dear, good, honourable friend and colleague.[19]

On 18 October 1916 Mathilde gave birth to her second child, a boy named David Jack. Richter, who, as he described to Balling, was in the grip of another attack, received the news with joy. In his last letter to his daughter and son-in-law, he nevertheless expressed a preference for the names Hans David 'out of consideration for *Meistersinger*, the foundation of my life. The names go so well together, Hans (the old man) and David (the youth).' Back in Bayreuth he was also overjoyed to learn from Siegfried Wagner that his wife Winifred was expecting the Master's first grandson (Wieland, born in 1917 after Richter's death). Meanwhile he gave further signs that he recognized the gravity of his own condition when, in his last letter, he told Mathilde the whereabouts in Vienna of a savings book in her name together with the name and address in the same city of the lawyer with whom his will was lodged. He ended with words that Sydney translated for publication in *The Times* on 8 December 1916. 'Give my good wishes to any you may meet of the friends and artists who worked with me. They are often in my thoughts and in my dreams, bringing back pleasant memories. I think gratefully of the hours I spent with them—the happiest of my professional life.' The paper gave this paragraph the headline 'Dr Richter's Second Thoughts'. The letter ended with greetings for Brodsky and Behrens in Manchester, thanks to Caroline Auerbach (whom he hoped to meet in the New Year) for providing the means to communicate with London, and grateful acceptance of the 'honorary godfather- and godmothership' of the children. 'My eyes are tired, I *must* close with greetings and kisses.'[20]

Richter wrote three more letters before he died. Two of them were addressed to the conductor Otto Lohse (widower of the soprano Katharina Klafsky) in Leipzig. Richter was spending his days studying Bach's cantatas and, in No. 46, 'Schauet doch und sehet', he came across an instrument called the *corno da tirarsi*. He asked Lohse for a description of this instrument (it was a slide-horn) which, in his years of conducting Bach's music, he had never encountered. There was a museum containing instruments from Bach's day in Leipzig, perhaps Lohse could seek the information there. 'Should I be considered sufficiently worthy to enter that part of Heaven where musicians go, and to meet the godfather of music face to face, I shall ask him myself about the *corno da tirarsi*.'[21] A postscript to the letter suggested that Spitta's biography of Bach might be consulted; he would do so himself but the print was too small for his poor eyesight and it would take too long to go through the two volumes. Then, later in November, Richter sent Lohse a postcard. He had, after all, discovered what he wanted to know in Spitta's book (*Johann Sebastian Bach* (Leipzig, 1873–80), ii. 226 n. 175).

My second trumpeter in London, [Walter] Morrow, had a *tromba da tirarsi* which he played in the classical works. The instrument had a slide which he operated with the thumb of his left hand; it had a beautiful tone and an amazing purity. A pity it is no longer used in classical works. It was less practical for fast passage-work, but good for natural tones for which Bach used it. In spite of careful and painstaking searching, I found only one example of its use in one cantata. Best thanks for your kind willingness to help me. The English will not be able to take Bach away from us.[22]

The third letter he wrote was to Joseph Stiegler in Vienna, in which he sent greetings to the Philharmonic orchestra. If, as he had been told, they considered that they had learned and developed under his guidance, he was happy to acknowledge that he had also learned much from them.

There was so much I could dare to do with this exceptionally gifted body of men, for example conducting *alla breve* [as Wagner had urged him to do after the 1876 *Ring*]; yes, a good orchestra can also have a good influence upon a Kapellmeister. If only they would all concede this! Drill sergeants belong on the parade ground; artists, especially young ones, need someone to train and mould them. The gentlemen will receive my picture; I hope I can bring it myself in the New Year. My doctor, who does not pamper his patients at all, gives me hope. It should remind the gentlemen of a man who never portrayed himself as a conductor 'who fell from heaven', but who was always proud that he emerged from the Vienna orchestra.[23]

Hans Richter died at a quarter to midnight on the night of Tuesday 5 December 1916; another of his musical godfathers, Mozart, had died on the same day 125 years earlier. Caroline Auerbach received a telegram on the afternoon of the following day from Marie, but could not get news to the Loebs in London before they read it in the newspapers. In January they received a letter from Marie describing her husband's last hours:

'Those whom the gods love, they take early from us.' They were not early in taking away our beloved one; he was able to achieve old age, but the gods did love him for they took him before he had to endure a long way of suffering. It would have been suffering, if he had been condemned to live on, in a condition of helplessness.

Hofrat [Dr] Landgraf had, already some time ago, diagnosed arteriosclerosis, but in his last days blood clots had formed in his brain. Unfortunately Dr Landgraf only told me on Monday that there was great danger and that I should inform the children, which I did of course, but too late. Not even poor Mitzi, who is nearest and came immediately, had the good fortune to see him alive; at least he spoke and still breathed, but no longer recognized her. . . . It is a blessing my beloved did not suffer—in fact he went to sleep peacefully—and that is a comforting thought for us.

On the 1st [December] he lay quietly in bed, and remained uncomplaining on the 2nd and 3rd. On the afternoon of the 5th my beloved spoke the finest, best, beautiful, grateful words to me. It was almost as if he had relived his whole life; he was serene and lucid. He said, 'You have had much trouble, but it was also wonderful that we could go from triumph to triumph together.' I was terribly upset by these words. I consoled the poor man, and told him that the times had been wonderful, but that he should not get excited. Then he spoke of the children and went peacefully to sleep, without either struggle or pain. Now we must find the courage to have the belief, no the certainty, that our departed one awaits us in the other world, and that death is a release and an awakening for the 'new life'.[24]

After the funeral ceremony, Richter's body was taken to the town of Coburg for cremation and his ashes interred in the cemetery in Bayreuth near to the small chapel containing Liszt's grave. The small gravestone bears a reproduction of his signature, nothing more. The press, both in England and in Germany, were full of tributes at the death of a great musician, whose passing signalled the end of a chapter in the hitherto brief history and art of conducting. Elgar responded to Sydney Loeb's letter enclosing Richter's last message to his English friends with the words. 'I sincerely wish you would allow these extracts to appear in the newspapers. It would do some justice to the revered memory of my dear old friend, and be a sort of antidote to the disgraceful remarks in some of the papers.'[25] The composer was referring to several obituaries which tempered their eulogies with a reminder of Richter's renunciation of his English honours. Blandine Gravina (née von Bülow) wrote to Mathilde from Italy: 'I myself mourn for the Master of Musicians and the dearest, oldest, and truest friend of my family.'[26] From Manchester Gustav Behrens consoled Richter's daughter.

Since 1895 until your dear father finally left England, I had the honour and pleasure of close personal relations with him and am, more than most people, conscious of the great debt of gratitude we owe him for what he has done for us in Manchester and in this country generally. His work with us was, I venture to think, a work of love, for he frequently told me how thoroughly happy he felt with our Hallé orchestra and with our appreciative public.[27]

After Siegfried Wagner had heard Richter's last performance on 19 August 1912, he hoped that he himself at 69 would be as fresh and youthful as Hans Richter (in fact Siegfried only lived to be 61). He joked that the secret of Richter's eternal youth lay in a tree in the garden of Zur Tabulatur. The old man visited it secretly at night for it was Freia's tree in *Rheingold*, and from it he plucked the golden apples. 'We want to let Richter and his tree live,' said Siegfried,[28] but Freia only allowed him four more years after his last *Meistersinger*.

35

Finale

Hans Richter described himself as a Hungarian by birth, an Austrian citizen, and a musician of the German race.[1] As far as England was concerned, he was always the visitor even in his Manchester days, albeit a highly popular one in most quarters. He never accepted English ideals nor sought to identify himself with the English nation. Some thought his inability to master the English language a deliberate attempt to extend his reputation beyond the concert platform, and there were plenty of anecdotes among the obituaries. 'Once he told his cellists to play on the C side,' wrote the *Daily News*. Another was his remark to a brass player named Booth, who lingered too long over a certain passage. 'Mr Booze', he cried, 'do not lean over that bar so long.'[2] On a more serious note, two contrasting obituaries appeared, one by Samuel Langford in the *Manchester Guardian*, the other by Ernest Newman in the *Birmingham Daily Post*. Having charted the course of Richter's life, each wrote his own appraisal of the conductor's contribution towards England's musical life. Langford called Richter 'one of the few men who possess a real genius for penetrating the spirit of great music and for communicating his will to those over whom he stood'.

Much of Wagner's spirit must have passed into Richter before he could carry out, with the zeal of the neophyte who knows no obstacle, those reforms which had caused wise men to shake their heads and proclaim them impossibilities. Today we can hardly imagine how ignorant were some of the singers whom Richter had to train for those early Wagnerian performances. . . . Richter's sympathies were not universal, but wonderfully broad. Entirely unmoved by the popular cry, he never hesitated to show his esteem for unrecognised genius nor his scorn for bad art.

. . . His grip of the orchestra was due partly to the fact that he had been himself an orchestral player for a short time and could approach [it] from the inside. . . . He knew exactly how much [it] can bear at rehearsal without showing the marks of the strain at the evening concert; hence his rehearsals were also lessons in economy of time and energy. By the admirable dignity he preserved when conducting, the clearness, the eloquence of his beat, all who saw him at his work could not fail to be impressed. He could be fiery and impetuous if such was the character of the music, but when a critical moment approached he could always communicate to the players calm and confidence. . . . When he appeared in the orchestra after the customary summer vacation he never failed to express the joy he felt—the joy of the man who prepares to do the work for which nature has best fitted him.[3]

Newman described Richter as an orchestral rather than operatic conductor, who preferred what emanated from the pit rather than what occurred on stage, a surprising statement considering the vast number of times Richter had appeared in the opera-house. After his criticisms of the Hallé conductor in 1906 it was natural that he now attacked his musical tastes. 'What he liked less he was disposed to take less pains with.' He recalled the reasons for Richter's departures from Vienna. 'No one who knows anything of the disgusting back-stairs politics of national and municipal opera-houses will have any difficulty in believing that some intrigue there was; intrigue flourishes in these institutions like fungi in a sodden field.' Nevertheless he considered that Richter had grown diffident in his last years at Vienna and was no match for Mahler's energy and reforms. Newman then acknowledged the 'universal admiration' which the north of England had for Richter when he came to live there. Hallé had become humdrum and Kapellmeister-like in his performances and in some ways Richter became Manchester's Mahler, though in turn he himself was attacked by the young generation of concert-goers whose parents he had nurtured.

In London his influence was not so great at the end of his career as at the beginning. A new race of conductors had sprung up during the last twenty years of his career, each with some peculiar excellence of his own, and the inevitable comparisons made it clear that Richter had decided limitations. . . . As knowledge of the style of other conductors spread, it was seen that there were whole territories of music in which he was far from first-rate. Weingartner, for example, easily surpassed him as a conductor of Berlioz, Nikisch and others as conductors of Tchaikovsky and of certain Wagnerian works such as *Tristan* and so on. One's grievance against some of his performances was not that they were different readings of the work from one's own, but they were no readings at all. Advancing age and a rather full habit of body may have had something to do with these lapses, but the fact remained that many of his later performances could only be regarded as an injustice to the composer. . . . On the whole one cannot help feeling that Richter was never quite equal to the reputation he acquired in England. But with all his limitations—and they were many—he was of the royal line. If the exquisite jewellery-work of the modern conductors was beyond him, he had a wide-reaching and architectural sense that with certain works he gave performances that for strength and grandeur still remain without rivals in our memory. He lacked some gifts, but he had at any rate that rarest quality of all in art— sublimity, and when faced with music that could really be called sublime—there are perhaps hardly more than a score of works in existence that deserve the title—he could always climb to the supreme heights with it.[4]

German obituaries contained few anecdotes, though one which appeared in a Berlin paper concerned Richter at Bayreuth. It was time for the fanfare to call the audience back to the auditorium and Richter, who was passing below the balcony from which these are blown, called up to a man standing there, 'You there, it's time. Blow the signal!' 'I can't', came the reply, 'I'm the Grand Duke of Weimar, but I'm happy to have met you!'[5] Most of the obituaries concentrated on Richter's relationship with Wagner. Karl Klindworth's death had preceded Richter's by a few months and an era was passing. The last of the trio of great Wagner conductors, Levi, Mottl, and Richter, was now dead. Even though he never conducted *Parsifal*, he always defended Wagner's original wish to keep it for Bayreuth. '*Parsifal*', he said, 'is no work for today's rabble in their boxes. In today's theatres it will sound like an *Ave Maria* from the slandering mouth of a rouged street-walker.'[6] Richter the man was Wagner's Kurwenal but Richter the artist was Hans Sachs and *Meistersinger* his Holy Grail. Bayreuth was the shrine at which he could serve his Master (Toscanini and Richard Strauss emulated him by refusing a fee for conducting there). As he demonstrated at his last performance in 1912, under no circumstances would he ever appear on stage there to take applause. Although he joked to the orchestra on that occasion that he could not go before the public in his shirt-sleeves, he added more seriously, 'It is not for me to stand where the Master has stood.'[7]

Herbert Thompson wrote a long obituary in the *Yorkshire Post*, but in 1934, at the request of Mathilde Loeb, he wrote a memoir of Richter which was intended for inclusion in a biography of the conductor.

My introduction to Richter was at the first concerts he gave in London, in May 1879 . . . and when he came, he opened our eyes. We realised very soon that here was a conductor who was not only a master of the technique of his art, but put warmth and dignity into his interpretations. Beethoven and Wagner were the masters whom he specially revealed to us, and they were seldom, if ever, absent from his programmes in the early days. . . . It must be remembered that, while we had enjoyed many opportunities of hearing Beethoven's symphonies (all nine were about that time given on successive Saturdays at the Crystal Palace), Wagner's works after *Lohengrin*, were practically unknown. So, too, we felt, were the Beethoven symphonies when we listened to Richter's magisterial interpretation of them. His habit of conducting everything from memory of course struck me by its novelty. . . . I remember, however, that one writer [Charles Willeby, *Masters of English Music* (London, 1893)] who was eulogising Sullivan could not do so without a mention of people who 'theorise that the chief essentials of a good conductor are that he play a variety of instruments, and have a memory like the well-nigh proverbial cab horse. But they forget that many military bandmasters are possessed of both these acquirements, and are not necessarily good conductors', which was of course interpreted as an oblique thrust at Richter. When Mr Willeby was eulogising Sullivan in 1893, Richter was conductor of the Birmingham Festivals as Sullivan was of the Leeds Festivals, so, absurd as it may appear to us, he no doubt regarded them as rivals, though they were so entirely different in their sympathies and individualities that any comparison was impossible.

. . . In his interpretations one always felt that he gave breadth and dignity to what he conducted. When Tchaikovsky's *Pathétique* symphony was the rage there was no one who made it so serious and splendid as he did and a performance by the Hallé orchestra at Bradford in 1895 will always remain in my memory as an outstanding event . . . He had the limitations of his individuality and nationality; he was never completely at home in French music, but was at his best in the works of the great German school, and in these days of the renewed interest felt in Mozart one cannot forget the prophecy he uttered many years ago, I think after a Birmingham Festival rehearsal of one of the symphonies, 'Gentlemen, there is a future for Mozart'.

I have said that in this country we owe to Richter our introduction to Wagner's later works, and in this connection a word should be said about the season of German opera in 1882. . . . I witnessed nearly all those productions, and, [like] a multitude of music lovers of my generation, [my] allegiance to Wagner, and especially to the *Mastersingers* with which Richter's name is so indissolubly associated, dates from this experience. Some first-rate principals had been secured . . . but it was Richter who wrought them all into a fine ensemble and made this season a new departure in opera. It was announced as a 'First Season', but it had no successor and though it was very well attended, and created great enthusiasm, it would seem that, as is the case with so many operatic ventures, the cost was too great for it to be a financial success.

It was not till Richter came to Manchester as conductor of the Hallé orchestra that I came into personal contact with him. I had converse with him on many occasions and always found him accessible. He was considered brusque in manner, and he certainly did not cultivate any 'airs and graces': he wished to preserve his independence, and I think he found a strange language something of an embarrassment. In early days he no doubt made some linguistic slips, but I have a strong suspicion that not a few of those reported were fathered upon him by the humorous invention of members of his orchestras. He liked to converse with those who could understand his native tongue, and this, I think, made our conversation easier. He loved to be helpful. He took trouble to write to me at some length [3 Feb. 1902] on the question of the derivation of Mozart's name, which some very superficial philologists had referred to as 'Moses', but which he said was a variant of 'Muothart' (or in the form which we associate with lead pencils, 'Hartmuth'). And another query which he answered [29 Oct. 1903] in detail was the responsibility for the orchestration of Liszt's *Hungarian Rhapsodies*, of which, from his Viennese experience, he had inside knowledge. . . . I will attempt a translation.

'How Doppler's name appeared on the title-pages of Liszt's *Hungarian Rhapsodies*, that excellent musician and unsurpassed flute player himself told me. Liszt himself orchestrated the six scores throughout, only the cadenzas for the flute (also those for the clarinet) were, at the Master's own wish, written by Doppler, who at that time was a member of the orchestra of the National Theatre in Pest. Without Doppler's knowledge, the Master, in his gracious way, put D's name on the title, and Doppler, much moved, gave his thanks for this recognition, which appeared to him much too generous and out of proportion to what he deemed his slight services. I may also mention that both the Dopplers (ours, whose name was Franz, was also a gifted composer, the other, Carl, died recently as pensioned Hofkapellmeister in Stuttgart) were among the most faithful followers—I may well say friends—of Liszt. Doppler had been first flautist in the orchestra for some years before I came to Vienna as conductor, and his unsurpassed mastery in interpretation and technique I shall never forget. Yet another charming trait, but one alas which becomes rarer, characterised this excellent man. Although he was a highly gifted and often successful composer, this too modest artist never pushed his own works, but devoted all his energy to the works of great masters such as Berlioz and Wagner. He was an ornament of the Vienna Court orchestra, and a model for the succeeding generation'.

Such a glowing eulogy shows how Richter could overcome his natural reserve when he wanted to do justice to others, and many other instances of his generous enthusiasm might be quoted. . . . I have very pleasant recollections of meetings with Richter at Bayreuth at several of the Festivals, and of having afternoon tea with him and Frau Richter in the interval at the 'Forester's House' in the field at the side of the theatre, where he used to take up his abode. And I shall never forget a glimpse of him I had in the public swimming bath at Bayreuth, where, in a very hot summer, he was endeavouring to find coolness by standing under a jet of water and looked exactly like the figure of an old river god in a fountain.[8]

Because of Richter's distaste for either recording or Messter's films (which came too late for him), we have no visual or aural record of his music-making. The trouble with all the reports of it is that it is impossible to convey the effect of music in anything but a subjective manner, though Sir Adrian Boult's memories must be the closest we can get to the truth. As a conductor he understood Richter's technique, and saw how he achieved his results. Boult studied with and revered Nikisch, but his earliest musical impressions were provided by Richter at the Hallé and in opera in London. The following is drawn from a radio interview he gave in 1980 (three years before his death at the age of 94), and a private conversation in 1972 with Christopher Dyment.

Hans Richter's stick was half an inch in diameter at the butt end, the handle was an inch in diameter. He held it quite firmly but his whole arm was very loose, there was no stiffness about it, but it was a clear 1, 2, 3, 4 and the point of the stick, actuated by his wrist, would clearly mark the number of beats in the bar. Its length was not more than fourteen to sixteen inches, it was wood colour, not white and had a great cork handle, quite firm, which a good fat hand could hold comfortably. He gripped it with his whole fist, *grasped* it with his whole hand, though it was all very loose and easy. The flow of the stick was an irresistible flow as you could hear. The movement of the stick had a very direct effect on the quality of the performance he produced.

There was a sweep and flow in everything Hans Richter did, whether it was in Tchaikovsky or Wagner. He had a tremendous grip of architecture; each piece stood up in front of you most firmly, and you felt that it was a rock. Everything had a steady rhythm and pulse. He would not make any rubatos to speak of except the obvious one in Beethoven's C minor. Compared with Nikisch it was quite ridiculous; it was just absolutely steady. He could do Tchaikovsky No. 6 straight through, but somehow or other it was most telling and dramatic. It might have been due to accentuation, but I'm inclined to think most of it came out of Richter's eye and went straight to the player concerned. Richter's stick always held everything together; one felt that the point of that stick was an emotive force through which everything was going to the orchestra.

Nikisch just used his stick from the two first fingers and thumb; the wrist and forearm were only used for a very wide sweep. The point of the stick was like a painter's brush, moving in that sensitive sort of way. He used a white stick, two feet long with a tapered wooden handle. His hand and stick were half Richter's size. He had an electric mind compared with Richter's solid German one. He used his stick in quite a different way.

You might say that Richter was the last conductor who quite clearly felt himself that the right hand was for rhythm and time, and the left hand was for expression. He only used his left hand for expression when there was anything wrong really. He would sometimes put his hand up to get or anticipate a diminuendo, but generally speaking his left hand only came in almost for an emergency.

I should think that, were he to come alive today, Richter's interpretations would not appear eccentric to us; they might be voted occasionally dull but not often. I mean the mood of *Tristan*, for instance, was nowhere near Richter. Nikisch was *the* man for *Tristan*, Richter was *the* man for *Meistersinger*. In that work he was supreme; I don't think any conductor, before or since, matched Richter in *Meistersinger*. It became congealed somehow. The homage to Sachs was a homage to Richter, the two people seemed to merge as the performance went on. His Beethoven was solid German, absolutely consistent, and only *just* dramatic enough to be exciting. His performances had a natural balance and clarity; there was always movement and impetus behind them.

Richter was a general piece of English furniture. He taught English orchestras a tremendous lot; he had a wonderful ear. He believed so much in balance in the concert hall; of course he always had his first violins on the left and his seconds on the right, but not only that, he had four double basses in the back right corner and four in the left. He separated them; he realised that the visual balance is also the aural balance—it comes to the ear like that.[9]

Boult drew a sketch map of Richter's orchestral layout in his diary. At Covent Garden his violins and cellos were set out on his left in parallel rows all facing across the pit, the violas were directly in front of his podium, and the basses were on the left at the back under the stage facing into the auditorium. Also to be found under the stage were (going from left to right facing the stage) the timpani followed by the harps. To Richter's right and facing the strings were five rows of woodwinds and brass. He had seven desks each of first and second violins, five of violas, and four each of cellos and basses. His layout for the Hallé placed his first and second violins to his left and right respectively (eight desks of each), six desks of violas again directly in front of him, with six desks of cellos behind them. Two rows of winds were set in a crescent shape behind the cellos, flanked on either side by double basses, three desks on one side, two on the other. Timpani and percussion were placed at the back of the platform with the horns to the left and trumpets and trombones to the right. On the matter of Richter's baton, Boult underestimated its length. The only known extant baton owned by Richter is 52 centimetres, or 20½ inches long. In his interview Boult also drew attention to Richter's influence on English orchestras and their playing. On this the 86-year-old violinist, conductor, and composer Benno Hollander recalled (in 1939) playing the violin in the 1877 Wagner Festival orchestra.

Richter saved the concerts. The first rehearsals had been taken by Dannreuther who was hopeless—the sort of conductor who cannot take his head out of the score. Wagner forgot to conduct when things were going well. I remember the *Lohengrin* prelude—the sound was heavenly. Wagner looked charmed and left off giving the beat to take a big red handkerchief out of his pocket and mop his forehead. It was fortunate that Richter was behind him to give the players their cues. Richter was unknown in London, but he had the orchestra at his feet in no time. At his first rehearsal he stopped us almost at once and asked a German violinist how to say B flat clarinet in English. Then he said emphatically, 'Not A! B flat clarinet!' One of the clarinettists had been using the wrong instrument. That sort of thing impressed an orchestra. It was Richter who reformed horn-playing in England. Before that the horn-playing here had been a joke.[10]

In assessing Richter's influence it is worth noting that in later years his name was often mentioned in reviews of conductors such as Toscanini, Furtwängler, or Bruno Walter. In 1933 the latter appeared in London at the Courtauld Sargent concerts and conducted Beethoven's 'Eroica' Symphony. Neville Cardus was very impressed but concluded, 'When shall we ever again hear the mighty wheel of the *Eroica*'s first movement set into motion with the simple inevitability of Hans Richter?'[11] Hollander concluded his *Daily Telegraph* memoir with the view that 'there is no one today who can give a performance of Beethoven's seventh Symphony to compare with Richter's. Weingartner is more like Richter than the others, but Richter had more warmth.'[12] When Koussevitzky performed Tchaikovsky's 'Pathétique' in London in 1935 *The Times* recalled 'that Hans Richter, one of [its] earliest exponents, was so little anxious to worry anything, that, having set the [5/4] time of this movement going, he used to lay down his stick and watch the players do it. This, the moderns would declare, may have been magnificent, but was not conducting. Richter's display of non-conducting, however, was testimony to a principle. Beneath the lilting phraseology of the melody in this movement, the tramp of the five crotchets in a bar is inexorable.'[13]

Albert Coates laboured long and hard over Beethoven's *Missa solemnis* at the Queen's Hall in May 1922 but missed its essential character according to one critic. 'In place of the spirit of passionate adoration which I recall Hans Richter to have evoked when directing a memorable performance years ago, we were offered a fine frenzy.'[14] Furtwängler was likened to 'an artist of more normal stature' compared to the 'giants of the earth' of bygone days (Richter, Mottl, and Steinbach) when he performed Brahms's Third Symphony in 1934, beautiful though his account apparently was; but a more crucial point was made when the critic wrote, 'Undoubtedly the orchestra of today is a far more supple instrument for a conductor to play on than was that of Richter's day.'[15]

Toscanini came to the Queen's Hall in May 1937 to conduct Brahms, and once again Richter was recalled. 'The players said of Hans Richter that there was always room in his beat for all the notes which had to be put into it. The same might be said of Toscanini's beat, but, though memory is insecure at this distance of time, one would add that there is a considerable difference between their tempi for Brahms, and that both were proved right in their results.'[16] When Toscanini conducted Beethoven's Ninth Symphony, Feruccio Bonavia stated that the tempo of the trio in the second movement was exactly the tempo adopted long before by Richter, and Bonavia played in the Hallé's first violins under Richter. 'Toscanini', he went on, 'seems to get out of the orchestra just a little more than others. Richter, too, had such a gift for persuading players to do more than their best, and it is Richter that Toscanini recalls in this as in the considered climaxes which retain the right balance between different families of instruments at their loudest.'[17] Georg Szell was among the guest conductors invited to conduct the Hallé orchestra after Sir Hamilton Harty resigned at the end of the 1934 season. He was interviewed by a reporter between rehearsals and said of the orchestra, 'You can still feel Hans Richter in it. The orchestra still has the good Viennese tradition of Richter.'[18]

There were many players who testified to Richter's greatness, such as this anonymous violinist in an article on conductors.

The great difference between Richter and most (not all) present-day [1927] conductors is that Richter was a plain, blunt musician, without any frills or pretentiousness, but a man who knew his job from A–Z. Richter apparently never indulged in sentiment or in poeticising. He told the players exactly how he wanted everything done, whether a passage should be done with the point of the bow and so on, marking everything precisely and accurately, so that there could be no doubt of his intentions. And on one occasion when my friend, being a young and ardent fiddler, with his fellow desk-man was putting some extra feeling into a passage in Berlioz's *Harold in Italy*, Richter turned round and said to them brusquely, 'None of that stuff here!'[19]

Filson Young, writing in the *Saturday Review* in 1909, dated the emergence of modern conducting ('that is to say the orchestral interpretation of music') from about 1860. He placed Richter with Wagner and von Bülow among the pioneers of the art but it would be more accurate to put Richter at the head of the *second* generation of German conductors. Young acknowledged that Richter's 'method in any other hands than his own would already be regarded as old-fashioned, and those who should imitate his technique would find themselves taking up a wand that in their hands was lifeless'. It was Richter's personality that lay at the heart of his genius. He was a master of his profession and could extricate any player or singer from a crisis in a performance, but how, in Young's view, did Richter's

FIG. 17 *Punch*, 25 March 1903.

"How many instruments do you really play?"
"Only fifteen with impunity," said the Doctor.

personality inspire both them and his audiences? 'There is a massive plebeian impassiveness in the very round of his back that suggests the peace and security, not only of the individual, but of a whole race of men. One might find grander terms for it, and yet do him less justice than by describing his principal attribute as an immense stolidity; stolidity allied to a prodigious slow momentum or power of going unsensitively on to the goal he has in view. Thus he not only arrives himself, but he sees that those under his command arrive with him.' Certainly Richter never seems to have given two beats where one would do, and his baton was a 'wadded pole' compared to Nikisch's 'slender wand'. Both men, in differing ways, trusted their players to phrase, Richter with a broad and spacious beat, Nikisch's small and mercurial. Richter's performances were simple and coherent in providing the shape of a work rather than its finer points of detail. As Newman put it, 'for him the great thing was not the inner life of the line but the general breadth of the page'.[20] What was not missed on the part of his listeners, however, was a definite impression of the composition being played. In the opera pit his impassiveness bordered on the phlegmatic, his Neptune-like figure, 'heavy and motionless, his countenance grave and incurious', in marked contrast to Mahler with his nervous movements and riveting eyes.

In the most tempestuous moments of musical storm and dramatic commotion [Richter] sits at his desk controlling it all, like an old scholar reading in a lamp-lit room. Nothing moves but the arm, except that occasionally the grave countenance and beard slowly revolve in a half circle to left or right; but behind the spectacles are eyes, weary-looking at all other times, that can be trained upon defaulting players with gimlet sharpness. He is utterly indifferent to applause; at the end of a great performance of the *Ring* he will step down from his desk and look up at a house shouting with enthusiasm for him alone, with a countenance no more expressive of emotion than that of a cow looking over a fence. It is at once comic and grotesque and sublime, but it is much more sublime than anything else.

[The opera pit] is the place to study his method and his power of achieving an outline; there is the place to observe his peculiar methods of obtaining a climax, his manner of treating a long crescendo or diminuendo so that there is always a sense of something kept in reserve at the end of it; there, in a word, is the place to make a study of human achievement and mastery, of what can be done by enthusiasm and simplicity of purpose in this world of dissipated forces.[21]

Stanford made some interesting observations of Richter, particularly when comparing him with von Bülow, as the following extracts show. They also include an incident in Vienna concerning Brahms, but it was a hearsay report and the composer never upbraided Richter to his face when it came to performances of his music.

Richter was often stiff in his reading of an unfamiliar score; von Bülow, never. A curious instance of this was a rendering of the slow movement of the First Symphony of Brahms at the Vienna Gesellschaft. So metronomic was it that Brahms, who was listening in a box with a friend, suddenly seized him by the shoulder and said 'Heraus!', hurrying him away. The friend himself told me of his impetuous wrath. When Richter knew a score, he usually got the best out of it. Von Bülow got the best out of it whether it was familiar or not. Both had amazing memories, [and whereas] von Bülow was often extravagant, Richter never [was].

Von Bülow and Richter may be said to be the archetypes from whom modern conducting has descended. Unfortunately more have followed the first than the second.... Richter was all for straightforwardness. He hated extravagance, and even took the *diablerie* out of Berlioz; but his mastery of the orchestra was as great as von Bülow's, and he had authority and instrumental knowledge to back it. He took everything from the standpoint of common sense: for this reason, he was strongest in what he best knew, Beethoven, Weber, and the *Meistersinger*. He was not often eclectic, von Bülow was. He had magnetism, but not so much as von Bülow. He had an even temper, which von Bülow had not. His is the safer ground to follow, but also the less alluring. The perfect conductor will possess a combination of the best qualities of both men: if we cannot have this ideal, we must learn from each separately. But there can be no question that modern conducting sprang from the stock of these two men.

England had been for long in a condition of *mezzoforte* in orchestral playing. Everything was in an irritating *laissez-faire* mood. The best material was there but performances were only pretty good. To make them as super-excellent as the players was the work of an authoritative man such as Richter. He swept away the ridiculous hash of everlasting items from operas—good, bad, and indifferent—of which the evening programmes of Festival Concerts consisted. He restored the orchestra to its proper balance. He had already grasped the weak point of our orchestral armoury, the method and manner of horn playing, and improved it beyond all knowledge. He had insisted upon all *pp*'s and *ff*'s in the performances— the effect of the *pp* in the slow movement of Beethoven's Seventh Symphony had thrilled St James's Hall. He had eliminated mediocrity and incompetence, even when backed by that most difficult of all arguments, length of service. He abolished many absurdities, such as the *pp* opening of 'For unto us' in the *Messiah* and restored many *tempi*, such as the *Alla Siciliana* movements in the same work. With chorus and solo singing he had not much sympathy; he had little to do with the choral training peculiar to this country, and his attitude towards solo singers was always that of the opera conductor—'Niederza schiessen' was the expression which he used to me in Vienna. [This was possibly 'Nieder sollen Sie schiessen' in his dialect, meaning 'You should shoot them down!'] But he knew a great artist when he heard him, and appreciated fully the best type of training and accomplishment. . . . He always wound up, both to chorus and orchestra, by exhorting them to perform 'with all entoosiasm'.

Richter and von Bülow often stopped conducting altogether and left the orchestra alone, but all the time they watched and looked. An organist under Richter—I speak from experience for I played the Mass in D [*Missa solemnis*] twice for him—felt his eye through his spine without looking round for the beat.

The present generation [1922] has brought orchestration to its perfection. As the late Hans Richter said, 'It is no longer a virtue to score well, it is only a vice to score badly.' . . . [He] also once said, in contemplating a new score, 'The greater the number of staves, the fewer the number of ideas.'

It was only the first sound of the orchestra under Richter at the *Meistersinger* [at Bayreuth] which showed every one how common sense triumphed in the end. There was a sureness and a general sense of mastery which put all other conductors there to flight. The very sound was different. Sentimentalism hid its head, and wholesome sentiment took its place.[22]

Another observer of Richter was Eva Ducat, who attended his London Symphony Orchestra concerts from 1904.

It was a thrilling night for me when I first saw him on a concert platform. No sooner did his short, stout figure appear than every man in the orchestra rose to his feet and stood respectfully.... When Richter walked on to the platform he seemed the quietest and most unassuming of men, but once he raised his baton no *generalissimo* marshalling his forces ever exacted a more absolute obedience. He indulged in no excessive movement; he kept his body still, and his gestures would have been stolid had they not been so telling and so expressive. Nothing escaped his vigilance, and his eye, as it darted upon a member of the orchestra, was compelling. Never have I seen such command over an orchestra before or since ... Richter's beat was marvellous in its exactitude and precision; and by this rigid adherence to the tempo, the form in which he built his symphonies was as straight and true as some great work of architecture. As his aim was to play the music as it was written, he did not find it necessary to vary his beat where no alteration of tempo was marked.... Richter's rehearsals were incredibly painstaking; from the first bar to the last every soul playing in the orchestra knew his exact wishes and intentions.... He was not satisfied with contrast, which is enough for most modern conductors, but insisted on the most beautiful grading, and this grading, sweeping unbrokenly from pianissimo to fortissimo, made the magnificence of his climaxes.[23]

Eva Ducat's friend and piano teacher Mimie Shakespeare (daughter of the famous tenor who sang for Richter in the 1880s in London) told of a visit to the Richter home in Vienna.

She told me of a great big room, quite bare except for a grand piano littered with music, and a long table down the middle of the room with benches on each side of it. Here the Richters took their meals and received their friends—not acquaintances—in the most homely fashion, Dr Richter sitting opposite his wife, and the long row of little Richters sitting between them. During the evening the greatest jollity and friendliness prevailed, until suddenly Richter or his wife would say, 'Go to bed children,' and all the children would rise at once and, beginning with the eldest, kiss their father's and mother's hands, Richter saying to each one in turn, with the regularity and intonation of a machine, 'God be with you'.[24]

In 1930 Ernest Newman, in his column 'The World of Music', described a letter he had received from a member of the public questioning any involvement by Richter in the orchestration of *Meistersinger*. Newman correctly refuted any suggestion that the young man did any more than make a fair copy of Wagner's full score for the publisher Schott's use. The first of Newman's two essays, called 'Wagner, Richter, and Die Meistersinger', began with an extract from his correspondent's letter.

'It is difficult to forget the thrilling effect of the first chord under Richter, who made it sound different from other conductors. What did he do exactly? For he did not build it up from the bass, as some modern conductors attempt to do. There was an undoubted *grip* and attack which might be compared to an organist striking the notes of a big chord with hands and feet absolutely together with an apparent increase of sonority'. My correspondent is right as to his facts here, and again when he says that Richter used to get the full electrifying value out of the *cresc*, *ffz*, *dim* in the 89th and 90th bars of the overture. . . . The explanation probably is that the solidity and breadth of the music had their counterpart in Richter's own nature. He was always superb in music that called for answering qualities of this kind in the conductor. I doubt whether we shall ever hear again the *Sanctus* in Bach's B minor *Mass*, or some of the *Brandenburg* Concertos as Richter used to give them.[25]

Newman also dispelled another myth. His correspondent asked if the conductor had been responsible for adding the trill in the tuba part fifty-two bars from the end of the *Meistersinger* Overture, and if Wagner had told Richter that it was Hans Sachs laughing, as the conductor related at rehearsal. Did Richter put it there for fun? Newman denied any suggestion that Richter had added anything to Wagner's score. The second article also rejected any idea that Wagner consulted Richter on matters of instrumentation, in particular the muted trumpets after the Cobblers' Chorus (Act III, scene V) to imitate toy trumpets. Wagner consulted Richter on only one passage (the nasal sound on the horns playing Beckmesser's song in the Finale of the second act) and that was because his amanuensis was a professional horn-player.

John Foulds, the composer and former Hallé cellist under Richter, wrote to the paper in response to Newman's articles.

Here is a story told me by Richter himself, which bears on the Wagner–Richter controversy. Of course Mr Newman is perfectly correct that Wagner, not Richter, wrote the shake in the *Meistersinger*. Richter told me that at an early performance of the Overture under Wagner the tubaist used an instrument upon which it was impossible to play the shake properly, whereupon Wagner told him to leave it out. This was done (possibly the part was altered), and thus the omission carrying Wagner's sanction was perpetuated until Richter discovered it, and very properly insisted upon the shake being restored. Richter's admirers have credited him with an act of creation instead of recreation. Richter, had, however, a great number of cherished instructions, both from Wagner and Brahms, which he brought out in performances of their works. . . . No conductor of today, not even Toscanini, seems to make just these points and one sadly misses them.[26]

Robert Lorenz had also written to the paper selecting one or two such moments in Wagner's *Ring* to which Foulds alluded. 'I dare say a good many of your readers will recall passages that they have never heard equalled since the death of Hans Richter. I wonder whether they remember the swing and precision he put into the introduction to the third act of *Siegfried*, more particularly the part where the strings move up to the Erda theme and down to the theme of Götterdämmerung. Every subsequent performance by other conductors has seemed muddy in comparison.'[27]

Charles Graves, Parry's biographer, sang in the Richter Choir, 'one of the worst and most interesting choirs in London' because it contained in its later years so many old people, a sign of loyalty to the great conductor. The great attraction 'was the privilege which it conferred upon the members of attending the orchestral rehearsals and watching Richter at work with his band—a most illuminating experience. That was something like a revelation, and though in the main a serious business, the proceedings were frequently enlivened by ludicrous *obiter dicta* from Richter, of which the famous comment on a very tame reading of the *Venusberg* music is perhaps the best known: "Gentlemen, you play it as if you were teetotallers—which you are not!"'[28]

Ernestine Schumann-Heink, a singer with whom Richter worked until the end of his career, sang Erda under him at Bayreuth and Magdalene in his last *Meistersinger*.

Richter was a wonder. He was the mainstay at Bayreuth. There were other conductors also, all of them fine, but he was, in my opinion, the greatest of them all. Mahler was the very opposite of Hans Richter. Richter, for example, could sit there in his shirt sleeves at rehearsals and bring out of the orchestra a climax (still looking like a good, nice family father), that nothing in the world could beat. But even so he was always quiet, easy, without any fuss or strain. This was Hans Richter. . . . [He] was very stubborn, an absolute authority. He took great trouble with me always and did many kind things, [and] never had any friction with me. When I sang Erda he always stopped conducting; he let me sing it to the orchestra. . . . He gave me a picture of himself, taken with his hat on, and under it he wrote: 'My dear Heink: Pardon me that I keep my hat on, because for an artist like you one must always say "Hats off!"' That meant a lot to me, because Richter had no patience with *unmusical* singers—he couldn't stand them.[29]

Punch and the *Spectator* would caricature Richter from time to time, but Feruccio Bonavia wrote an amusing short book entitled *Musicians in Elysium* in which Richter (Herr Doctor) is featured in an encounter with an unnamed composer (Mr Maecenas) in the after-life.

D.: London . . . we had some happy times there.
M.: Indeed we had. Covent Garden and the *Ring*.
D.: And Gambrinus [restaurant] after the opera.
M.: Yes, and Gambrinus.
D.: Music is divine, but beer is a necessity.
M.: You liked our England?

D.: Of course I did, . . . [but] they wanted me to keep a player in the orchestra because he had been there thirty years.
M.: What did you do?
D.: I said that if he had been there thirty years it was high time he should go somewhere else.
M.: A little drastic, wasn't it?
D.: In art there is only good and bad.
M.: . . . What of English music?
D.: Elgar, great man.
M.: Any other?
D.: Be content; great men are not born every day.
M.: Not even great conductors?
D.: With all modesty, I confess I agree.
M.: Are you going to do any conducting here?
D.: Alas; they do not need any conducting here! They *know*. At least so I am told. But I am going to hear one of their concerts. I wonder.[30]

Debussy spent a few years at the turn of the century as a writer on music, using the pseudonym Monsieur Croche. He attended Richter's first Covent Garden performances of the *Ring* in 1903 and enthused, almost to the point of blasphemy, about the conductor who, ironically, was later to show so little interest in his own music. His brief description of Richter's baton differs markedly from Boult's.

Recently I attended the performances of *Das Rheingold* and *Die Walküre*. It seems to me impossible to achieve greater perfection. . . . Richter conducted the first performance of the *Ring* at Bayreuth in 1876. At that time his hair and beard were red-gold; now his hair has gone, but behind his gold spectacles his eyes still flash magnificently. . . . If Richter looks like a prophet, when he conducts the orchestra he is Almighty God: and you may be sure that God Himself would have asked Richter for some hints before embarking on such an adventure. While his right hand, armed with a small unpretentious baton, secures the precision of the rhythm, his left hand, multiplied a hundredfold, directs the performance of each individual. This left hand is undulating and diverse, its suppleness is unbelievable! Then, when it seems that there is really no possibility of attaining a greater wealth of sound, up go his two arms, and the orchestra leaps through the music with so furious an onset as to sweep the most stubborn indifference before it like a straw. Yet all this pantomime is unobtrusive and never distracts the attention unpleasantly or comes between the music and the audience.

I tried in vain to meet this marvellous man. He is a sage who shrinks in wild alarm from interviews. I caught sight of him for a moment as he rehearsed Fafner. . . . It is easy to understand my emotion as I watched the conscientious old man bent over the piano while he performed the duties of a mere producer. . . .

Following the strict Wagnerian tradition, there was no applause until the end of each act, when Richter went off contentedly, oblivious of it and impatient, maybe, to find recuperation in a glass of beer.[31]

It must have been a relief to escape into the concert hall from a Manchester fog on a dreary November night in 1910 and hear the music of Beethoven conducted by a man who was born a mere sixteen years after the composer's death; but it must have been little short of magical to hear him conduct Wagner, knowing that he had been doing so for over thirty years, often under the supervision of the composer. Cardus wrote marvellously on Richter, evoking a historical perspective that only a man with such sensitivity as his could do, having seen the great conductor make his way through the orchestra to the podium on so many Thursday nights at the Free Trade Hall. 'To listen to Wagner or Beethoven conducted by Richter was to know that somehow one had been called and chosen out of many. To gaze on Richter was to experience wonder; I once followed in his footsteps as he shambled along the pavement and I tried to fit my boots exactly in the places trodden by him. He had spoken to Wagner; and no composer since has meant so much to the imagination as Wagner meant to those of us who in 1910 had just come of age and were listening to the Hallé orchestra under Richter beginning the *Meistersinger* overture.'[32] Sydney Loeb's photograph of Richter the pensioner shambling down the street in Bayreuth in 1914 with his shopping bag in his hand, or of him, cigar in mouth, entering the artists' entrance of the Free Trade Hall for his last rehearsal with the Hallé orchestra in March 1911, are evocative final glimpses of the man who had so much musical history about him, and imparted it to his fellow man with nothing more than a piece of cane mounted on a stout cork handle.

NOTES

CHAPTER 1

1. *Otto Nicolai's Tagebücher*, ed. B. Schröder (Leipzig, 1892).
2. Otto Strobel, 'Hans Richter' (Bayreuth, unpublished).
3. Ibid.
4. *Musical Times*, July 1899.
5. Ibid, 441.
6. Hans von Bülow to Richard Pohl, 21 Dec. 1868.
7. *Unterhaltungs-Blatt der neuesten Nachrichten*, 103, 24 Dec. 1868.
8. Ibid. 11, 7 Feb. 1869.
9. Josefine Richter von Innffeld, *Neues System/Methodische Entwicklung des Sprachorganismus für den Kunstgesang* (Vienna, 1887).
10. Franz Josef Grobauer, *Die Nachtigallen aus der Wiener Burgkapelle* (Horn, 1954).
11. *Bayreuther Blätter* (1917), 76.
12. T. R. Croger, *Notes on Conductors and Conducting* (London, n.d.).
13. Otto Strobel, 'Hans Richter'.
14. Ibid.
15. *Windsor Magazine*, Sept. 1896.
16. Ernest Newman, *Life of Richard Wagner* (London, 1947).
17. Hans Richter to Josefine Richter, Vienna, 14 Mar. 1865.
18. Hans Richter, 'Conducting Books', (unpublished).
19. Ibid.

CHAPTER 2

1. Richard Wagner to Heinrich Esser, 16 Aug. 1866.
2. Heinrich Esser to Anton Schott; Manfred Eger, *Hans Richter: Des Meisters lieber Gesell* (Bayreuth, 1988).
3. Richard Wagner to King Ludwig of Bavaria, 25 Oct. 1866.
4. Hans Richter to Josefine Richter, 31 Oct. 1866.
5. Hans Richter's diary.
6. Cosima Wagner to King Ludwig, 25 Oct. 1866.
7. Cosima Wagner to King Ludwig, 4 Nov. 1866.
8. Hans Richter to Josefine Richter, 18 Nov. 1866.
9. Hans Richter to Josefine Richter, 25 Dec. 1866.
10. Hans Richter to Josefine Richter, 8 Feb. 1867.
11. Hans Richter to Camillo Sitte, Jan. 1867.
12. Ibid.
13. Hans Richter to Josefine Richter, 9 Feb. 1867.
14. Hans Richter to Camillo Sitte, Jan. 1867.

15. Hans Richter to Josefine Richter, 29 Mar. 1867.
16. Hans Richter to Josefine Richter, undated.
17. Richard Wagner to King Ludwig, 12 June 1867.
18. Hans Richter to Josefine Richter, June 1867.
19. *Die Signale*, 6 June 1867.
20. Hans Richter to Alexander Reinhold, 29 May 1867.
21. Hans Richter to Camillo Sitte, 28 June 1867.
22. Hans Richter to Josefine Richter, June 1867.
23. Hans Richter to Josefine Richter, 19 Jan. 1867.
24. Hans Richter to Josefine Richter, Mar. 1867.
25. Hans Richter to Josefine Richter, undated, *c*.July 1867.
26. Hans Richter to Camillo Sitte, Nov. 1867.
27. Hans Richter to Josefine Richter, Oct. 1867.
28. Hans Richter to Camillo Sitte, Nov. 1867.
29. Ernest Newman, *Life of Richard Wagner*, iv (London, 1947).
30. C. F. Glasenapp, *Das Leben Richard Wagners*, iv (Leipzig, 1912).
31. Carl Gianicelli, *Bayreuther Blätter* (1917).
32. Sir Adrian Boult to Sydney Loeb, 20 June 1951.
33. Hans Richter to Josefine Richter, Oct. 1867.
34. Richard Wagner to Mathilde Maier, 17 Nov. 1867.

CHAPTER 3

1. Carl Gianicelli, *Bayreuther Blätter* (1917).
2. Manfred Eger, *Hans Richter: Des Meisters lieber Gesell* (Bayreuth, 1988).
3. Hans von Bülow to Joachim Raff, 5 May 1868.
4. Richard Wagner to Hans Richter, 26 Apr. 1868.
5. Hans Richter to F. G. Edwards, 10 Dec. 1902.
6. Richard Wagner to Hans Richter, 21 July 1868.
7. Hans Richter to Josefine Richter, Aug. 1868.
8. Hans Richter, 'Conducting Books', (unpublished).
9. Hans von Bülow to Peter Cornelius, 24 Aug. 1868.
10. Hans von Bülow to Emil Bock, 8 July 1868.
11. Hans von Bülow to Joachim Raff, 11 Sept. 1868.
12. Hans von Bülow to Richard Wagner, Dec. 1868.
13. Hans von Bülow to Joachim Raff, 18 Jan. 1869.
14. Richard Wagner to Hans Richter, 25 Nov. 1868.
15. Hans von Bülow to Lorenz von Düfflipp, 7 Apr. 1869.
16. Hans von Bülow to Dr K. Gille, 30 May 1869.
17. Hans von Bülow to Hans von Bronsart, 10 June 1869.
18. Hans von Bülow to Richard Pohl, 23 June 1869.
19. Cosima Wagner's diary, 21 June 1869.
20. Ibid.
21. Ibid.
22. Hans Richter to Richard Wagner, 11 Aug. 1869.
23. Judith Gautier, *Wagner at Home*, trans. E. D. Massie (London, 1910).
24. Ibid.
25. Ibid.

26. Ibid.
27. Ibid.
28. Richard Wagner to Otto Wesendonck, 21 Aug. 1869.
29. Gautier, *Wagner at Home*.
30. *Guardian*, 9 Sept. 1869.
31. Hans Richter to Richard Wagner 28 Aug. 1869.
32. Hans Richter to Richard Wagner, 29 Aug. 1869.
33. King Ludwig to Lorenz von Düfflipp, 29 Aug. 1869.
34. King Ludwig to Lorenz von Düfflipp, 30 Aug. 1869.
35. Richard Wagner to Franz Wüllner, [11] Sept. 1869.

CHAPTER 4

1. Judith Gautier, *Wagner at Home*, trans. E. D. Massie (London, 1910).
2. Cosima Wagner's diary, 12 Sept. 1869.
3. Hans Richter's diary, 22 Mar. 1870.
4. Richard Wagner to Hans Richter, 19 Oct. 1869.
5. Richard Wagner to Hans Richter, 19 Dec. 1869.
6. Ibid.
7. Ibid.
8. Richard Wagner to Hans Richter, 4 Feb. 1870.
9. Richard Wagner to Hans Richter, undated [Mar. 1870].
10. Richard Wagner to Hans Richter, 25 Mar. 1870.
11. Hans Richter's diary, 31 Mar. 1870.
12. Hans Richter to Richard Wagner, [c. 22 May 1870].
13. Richard Wagner to Hans Richter 24 May 1870.
14. Richard Wagner to Hans Richter 20 June 1870.
15. Cosima Wagner's diary, 30 June 1870.
16. Hans Richter's diary, 15 July 1870.
17. Cosima Wagner's diary, 15 July 1870.
18. Ibid. 24 July 1870.
19. Hans Richter to Camillo Sitte, undated [28 Aug. 1870].
20. Cosima Wagner's diary, 16 Dec. 1870.
21. Hans Richter to Theodor Müller-Reuter, 18 Sept. 1909.
22. Hans Richter's diary, 24 Dec. 1870.
23. Cosima Wagner's diary, 25 Dec. 1870.
24. Franz Servais to Hans Richter, 18 Nov. 1870.
25. Cosima Wagner's diary, 5 Jan. 1871.
26. Cosima Wagner to Friedrich Nietzsche; Manfred Eger, *Hans Richter: Des Meisters lieber Gesell* (Bayreuth, 1988).
27. Franz Servais to Hans Richter, [Jan. 1871].
28. Richard Wagner to Hans Richter, 18 May 1871.
29. Richard Wagner to Hans Richter, 29 July 1871.

CHAPTER 5

1. Franz Liszt to Viktor Langer, 25 Aug. 1871.
2. Franz Liszt to Princess Carolyne Sayn-Wittgenstein, 19 Nov. 1871.

478   *Notes*

3. Franz Servais to Hans Richter, [Jan. 1871].
4. Cosima Wagner's diary, 12 May 1873.
5. Richard Wagner to Hans Richter, [c. 28 Sept. 1871].
6. Cosima Wagner's diary, 24 Nov. 1871.
7. Richard Wagner to Hans Richter, 25 Nov. 1871.
8. Richard Wagner to Hans Richter, 4 Jan. 1872.
9. Richard Wagner to Hans Richter, 6 Feb. 1872.
10. Richard Wagner to Hans Richter, 27 Apr. 1872.
11. Hans Richter's diary, 21 May 1872.
12. Anton Seidl to Hans Richter, 6 Oct. 1872.
13. Richard Wagner to Hans Richter, 13 June 1872.
14. Anton Seidl to Hans Richter, 3 Nov. 1872.
15. Richard Wagner to Hans Richter, 16 Dec. 1872.
16. Hans Richter's diary, 7 Jan. 1873.
17. Cosima Wagner to Hans Richter, 1 June 1873.

CHAPTER 6

1. Richard Wagner to Hans Richter, 6 Mar. 1874.
2. Amadé Németh in his book *Ferenc Erkel* (Budapest, 1979) quotes Erkel's written request to the Ministry and dates it Feb. 1874, but Richter's letter dated 27 Jan. 1874 to his future mother-in-law (see n. 7 below) already refers to this appointment as a *fait accompli*: 'As you may already know, the Ministry has entrusted me with the artistic direction of the Opera.'
3. Richard Wagner to Hans Richter, 1 Apr. 1874.
4. Richard Wagner to Hans Richter, 13 May 1874.
5. Hans Richter's diary, 25 May 1874.
6. Hans Richter's diary, 3 June 1874.
7. Hans Richter to Frau Mathilde von Szitányi, 27 Jan. 1874.
8. Hans Richter to Marie von Szitányi, 20 Apr. 1874.
9. Hans Richter to Alexander Reinhold, 18 May 1867.
10. Hans Richter to Marie von Szitányi, 3 May 1874.
11. Hans Richter to Marie von Szitányi, 24 June 1874.
12. Hans Richter's diary, 25 June 1874.
13. Ibid. 29 June 1874.
14. Ibid. 30 June 1874.
15. Hans Richter to Marie von Szitányi, 3 July 1874.
16. Hans Richter's diary, 23 July 1874.
17. Hans Richter to Marie von Szitányi, 17 July 1874.
18. Hans Richter's diary, 25 July 1874.
19. Cosima Wagner's diary, 25 July 1874.
20. Hans Richter to Marie von Szitányi, 18 Aug. 1874.
21. Hans Richter to Marie von Szitányi, 25 Aug. 1874.
22. Cosima Wagner's diary, 16 Oct. 1874.
23. Hans Richter to Marie von Szitányi, 19 Oct. 1874.
24. Hans Richter to Marie von Szitányi, 27 Jan. 1875.
25. Hans Richter to Marie von Szitányi, 17 Dec. 1874.

26. Richard Wagner to Hans Richter, 25 Nov. 1874.
27. Cosima Wagner's diary, 23 Jan. 1875.
28. Hans Richter's diary, 24 Jan. 1875.
29. *Neues Wiener Tageblatt* (W. Frey), 26 Jan. 1875.
30. *Neue freie Presse* (E. Hanslick), 26 Jan. 1875.
31. Hans Richter's diary, 24 Jan. 1875.
32. Ibid. 27 Jan. 1875.
33. Ibid. 5 Feb. 1875.
34. Cosima Wagner's diary, 5 Feb. 1875.
35. Hans Richter's diary, 7 Mar. 1875.
36. Ibid. 8 Mar. 1875.
37. Richard Wagner to Hans Richter, 23 Feb. 1875.
38. Cosima Wagner's diary, 10 Mar. 1875.
39. Hans Richter's diary, 10 Mar. 1875.
40. Ibid. 9 Mar. 1875.
41. Cosima Wagner's diary, 9 Mar, 1875.
42. Cosima Wagner to Marie von Szitányi, 11 Mar. 1875.
43. Manfred Eger, *Hans Richter: Des Meisters lieber Gesell* (Bayreuth, 1988).
44. Hans Richter's diary, 23 Apr. 1875.
45. Ibid. 24 Apr. 1875.

CHAPTER 7

1. Hans Richter to Marie Richter, 22 Jan. 1875.
2. Telegram from Hans Richter to Richard Lewy, 13 Apr. 1875.
3. Telegram from Hans Richter to Richard Lewy, 14 Apr. 1875.
4. Franz Jauner to Hans Richter, 14 Apr. 1875.
5. Hans Richter's diary, 1 May 1875.
6. Manfred Eger, *Hans Richter: Bayreuth, Wien, London und zurück* (Bayreuth, 1990).
7. *Illustrirtes Wiener Extrablatt*, 3 May 1875.
8. Sigmund Bachrich, *Aus verklungenen Zeiten* (Vienna, 1914).
9. Hans Richter's diary, 12 Aug. 1875.
10. Gustav Kietz, *Richard Wagner in den Jahren 1842-49, 1873-75* (Dresden, 1905).
11. Waldemar Meyer, *Aus einem Künstlerleben* (Berlin, 1925).
12. Hans Richter's diary, 1 Nov. 1875.
13. Richard Wagner to Hans Richter, 15 Aug. 1875.
14. Ludwig Karpath, *Richard Wagner Briefe an Hans Richter* (Berlin, 1924).
15. Hans Richter to Richard Wagner, [22 Aug. 1875].
16. Hans Richter's diary, 3 Nov. 1915.
17. Richard Wagner to Hans Richter, [24/25 Aug. 1875].
18. Cosima Wagner's diary, 24 Aug. 1875.
19. Ibid. 25 Aug. 1875.
20. Cosima Wagner to Hans Richter, 25 Aug. 1875.
21. Ibid.
22. Hans Richter to Richard Wagner, [27 Aug. 1875].
23. Richard Wagner to Hans Richter, 2 Sept. 1875.
24. Eger, *Hans Richter: Bayreuth, Wien, London und zurück*.

25. Hans Richter to Marie Richter, 2 Sept. 1875.
26. Hans Richter to Marie Richter, 3 Sept. 1875.
27. John Ella, *Musical Sketches at Home and Abroad* (London, 1878).
28. Eger, *Hans Richter: Bayreuth, Wien, London und zurück*.
29. *Oesterreichische Musiker Zeitung*, 16 Nov. 1875.
30. Cosima Wagner's diary, 7 Nov. 1875.
31. Ibid. 2 Nov. 1875.
32. Ibid. 3 Nov. 1875.
33. Mina Curtiss, *Bizet and His World* (London, 1959).
34. Hans Richter's diary, 1 Nov. 1875.
35. Ibid. 10 Dec. 1875.
36. Richard Wagner to Hans Richter, 11 Oct. 1875.
37. Ibid.
38. Cosima Wagner's diary, 2 Oct. 1875.
39. Ibid. 22 Nov. 1875.
40. Max Millenkovich-Morold, *Wagner in Wien* (Leipzig, 1938).
41. Cosima Wagner's diary, 20 Nov. 1875.
42. Ibid. 15 Dec. 1875.
43. Joseph Sulzer, *Erinnerungen eines Wiener Philharmonikers* (Vienna, 1910).
44. Siegfried Wagner, *Erinnerungen* (Stuttgart, 1923).

CHAPTER 8

1. Cosima Wagner to Hans Richter, 26 Dec. 1875.
2. Joseph Sulzer, *Erinnerungen eines Wiener Philharmonikers* (Vienna, 1910).
3. Hans Richter's diary, 2 and 3 Mar. 1876.
4. Mottl's notes are in his edition (pub. Peters, Leipzig, 1887) of *Lohengrin* and in *Neue Wagner Forschungen*, Autumn 1943.
5. Hans Richter, 'Conducting Books', 7 Mar. 1876.
6. Hans Richter's diary, 3 June 1876.
7. Ibid. 16 June 1876.
8. Richard Fricke, *1876 Richard Wagner auf der Probe* (Dresden, 1906), 8 June 1876.
9. Hans Richter's diary, 11 July 1876.
10. Felix Mottl's diaries, 4 June 1876.
11. Richard Wagner to Hans Richter, 23 June 1876.
12. Hans Richter's diary, 6 Feb. 1897.
13. Fricke, *1876 Richard Wagner auf der Probe*, 2 July 1876.
14. Hans Richter's diary, 10 July 1876.
15. Ibid. 9 Aug. 1876.
16. Ibid. 13 Aug. 1876.
17. Ibid. 14–17 Aug. 1876.
18. Ibid. 30 Aug. 1876.
19. Richard Wagner to Hans Richter, 7 May 1876.
20. *Der Merker*, Oct. 1916.
21. Manfred Eger, *Hans Richter: Bayrenth, Wien, London und zurück* (Bayreuth, 1990).
22. Lilli Lehmann, *Mein Weg* (Leipzig, 1913).
23. Edvard Grieg, *Bergensposten*, Aug. and Sept. 1876.
24. C. V. Stanford, *Pages from an Unwritten Diary* (London, 1914).

25. Wilhelm Ganz, *Memories of a Musician* (London, 1913).
26. *Musical Times*, Sept. 1876.
27. Cosima Wagner's diary, 20 Nov. 1878.
28. Richard Wagner to King Ludwig, 20 Aug. 1865.
29. Cosima Wagner's diary, 9 Sept. 1876.
30. Siegfried Wagner, *Erinnerungen* (Stuttgart, 1923).
31. Richard Wagner to King Ludwig, 22 June 1877.
32. Richard Wagner to King Ludwig, 9 Feb. 1879.
33. Richard Wagner, *Ein Rückblick auf die Bühnenfestspiele des Jahren 1876* (Leipzig, 1883).

CHAPTER 9

1. Richard Fricke, *1876 Richard Wagner auf der Probe* (Dresden, 1906), 2 July 1876.
2. C. F. Glasenapp, *Das Leben Richard Wagners* v (Leipzig, 1912).
3. Hans Richter's diary, 2 Sept. 1876.
4. Richard Wagner to August Förster, 6 Sept. 1876.
5. Marcel Prawy, *Die Wiener Oper* (Vienna, 1969).
6. Richard Wagner to Hans Richter, Mar. 1877.
7. Hans Richter's diary, 28 Mar. 1877.
8. Hans Richter to Marie Richter, 22 Apr. 1877.
9. Richard Wagner to Hans Richter, 26 Apr. 1877.
10. Hans Richter's diary, 20 Apr. 1877.
11. *Athenaeum*, 12 May 1877.
12. Hans Richter to Marie Richter, 4 May 1877.
13. Hans Richter to Marie Richter, 26 Apr. 1877.
14. M. B. Foster, *The History of the Philharmonic Society of London* (London, 1912).
15. Hans Richter to Marie Richter, 29 Apr. 1877.
16. Hans Richter to Marie Richter, 1 May 1877.
17. Ibid.
18. Hans Richter to Marie Richter, 4 May 1877.
19. C. H. H. Parry's diary, 4–18 May 1877.
20. Hans Richter's diary, 9 May 1877.
21. *Musical Times*, 1 June 1877.
22. G. B. Shaw, 'Wagner & Richter', in *How to Become a Musical Critic* (London, 1960).
23. Hermann Klein, *Musicians and Mummers* (London, 1925).
24. *Athenaeum*, 12 May 1877.
25. Ibid. 19 May 1877.
26. *Dramatic Review*, 8 Feb. 1885.
27. Hans Richter to Marie Richter, 24 May 1877.
28. Hans Richter's diary, 19 July 1877.
29. Hans Richter to Marie Richter, 20 July 1877.

CHAPTER 10

1. John Ella, *Musical Sketches at Home and Abroad* (London, 1878).
2. H. Hermann-Schneider, *Status und Funktion des Hofkapellmeisters in Wien (1848–1918)* (Innsbruck, 1981).
3. Ibid.

4. Franz Josef Grobauer, *Die Nachtigallen aus der Wiener Burgkapelle* (Horn, 1954).
5. Ibid.
6. Ibid.
7. *Neue freie Presse*, 3 Jan. 1878.
8. Anton Bruckner to Moritz von Mayfeld, 12 Jan. 1875.
9. Sigmund Bachrich, *Aus verklungenen Zeiten* (Vienna, 1914).

CHAPTER 11

1. Hermann Klein, *Musicians and Mummers* (London, 1925).
2. Hermann Klein, *Thirty Years of Musical Life in London* (London, 1903).
3. Hans Richter's diary, 2 May 1879.
4. Ibid. 12 May 1879.
5. Ibid.
6. C. H. H. Parry's diary, 4–12 May 1879.
7. Herbert Thompson's diary.
8. C. V. Stanford, *Pages from an Unwritten Diary* (London, 1914).
9. Hans Richter's diary, 20 Oct. 1879.
10. *Musical Times*, June 1879.
11. *Athenaeum*, 10 May 1879.
12. Ibid, 17 May 1879.
13. *World*, 14 May 1879.
14. H. Plunket Greene, *Charles Villiers Stanford* (London, 1935).
15. W. Kuhe, *My Musical Recollections* (London, 1896).
16. *Salzburger Volksblatt*, 17 July 1879.
17. Hans Richter to Baron Carl Sterneck, 2 June 1879.
18. A. Dvořák to A Göbl, 23 Nov. 1879.
19. A. Dvořák to F. Simrock, 20 Mar. 1879.
20. C. H. H. Parry's diary.
21. Herbert Thompson's diary.
22. *The Mapleson Memoirs* (London, 1888).
23. Hans Richter's diary, 26 May 1880.
24. Ibid. 29 May 1880.
25. *Musical Times*, July 1880.
26. Ibid. June 1880.
27. Ibid. July 1880.
28. *World*, 2 June 1880.
29. *Musical Times*, July 1880.
30. *World*, 2 June 1880.
31. Ibid.
32. *Athenaeum*, 29 May 1880.
33. Ibid. 5 June 1880.
34. Ibid.
35. *The Athenaeum*, 12 June 1880.
36. Ibid. 19 June 1880.
37. Hans Richter's diary, 11 June 1880.
38. *Musical Times*, July 1880.

CHAPTER 12

1. Hans Richter's diary, 4 May 1880.
2. Edward Speyer, *My Life and Friends* (London, 1937).
3. Hans Richter's diary, 17 May 1881.
4. Ibid. 27 May 1881.
5. Ibid. 30 May 1881.
6. Ibid. 2 June 1881.
7. Hans von Bülow to his mother, [June] 1881.
8. C. V. Stanford to Hans Richter, 3 Oct. 1881.
9. Hans Richter's diary, 13 June 1881.
10. Ibid. 16 June 1881.
11. Ibid. 23 June 1881.
12. Ibid. 27 June 1881.
13. *Musical Times*, July 1881.
14. C. H. H. Parry's diary.
15. Cosima Wagner's diary, 30 June 1881.
16. Richard Wagner to Hans Richter, 1 Sept. 1880.
17. *Wiener Abendpost*, 23 Feb. 1881.
18. Neue freie Presse, 27 Feb. 1881.
19. *Das Vaterland*, 3 Mar. 1881.
20. A. Göllerich and L. Auer, *Anton Bruckner* (Regensburg, 1936).
21. Ibid.
22. Ibid.
23. Ibid.
24. Hans Richter to Anton Dvořák, [Dec. 1880].
25. Anton Dvořák to Hans Richter, 13 Mar. 1881.
26. John Clapham, *Anton Dvořák* (London, 1979).
27. Cosima Wagner to Hans Richter, 19 July 1881.
28. C. H. H. Parry's diary, 24 Oct. 1881.

CHAPTER 13

1. Eugen d'Albert to Marie Joshua, 8 Nov. 1881.
2. Eugen d'Albert to Marie Joshua, 17 Nov. 1881.
3. Eugen d'Albert to Marie Joshua, 2 Dec. 1881.
4. Anna Brodsky, *Recollections of a Russian Home* (Manchester, 1904).
5. Ibid.
6. Herbert Weinstock, *Tchaikovsky* (London, 1946).
7. Eugen d'Albert to Marie Joshua, 12 Dec. 1881.
8. Eugen d'Albert to Marie Joshua, 27 Dec. 1881.

CHAPTER 14

1. Eugen d'Albert to Marie Joshua, 8 Jan. 1882.
2. Frederic Cowen, *My Art and My Friends* (London, 1913).
3. Eugen d'Albert to Marie Joshua, 19 Jan. 1882.

4. Eugen d'Albert to Charles d'Albert, 5 Mar. 1882.
5. Eugen d'Albert to Marie Joshua, 7 Feb. 1882.
6. *Neue freie Presse*.
7. Hans Richter to C. V. Stanford, 24 Oct. 1881.
8. *Wiener Abendpost*, 8 Mar. 1882.
9. Eugen d'Albert to Marie Joshua, 9 Mar. 1882.
10. Eugen d'Albert to Marie Joshua, 16 Feb. 1882.
11. Eugen d'Albert to Marie Joshua, 15 Mar. 1882.
12. Eugen d'Albert to Marie Joshua, 12 Apr. 1882.
13. Hans Richter to Anton Dvořák.
14. Eugen d'Albert to Marie Joshua, 25 Apr. 1882.
15. Eugen d'Albert to Marie Joshua, 28 May 1882.
16. Eugen d'Albert to Marie Joshua, 12 June 1882.
17. Eugen d'Albert to Marie Joshua, 27 Apr. 1882.

CHAPTER 15

1. Hermann Klein, *Thirty Years of Musical Life in London* (London, 1903).
2. *World*, 24 May 1882.
3. *Illustrated Sporting and Dramatic Chronicle*, 24 June 1882.
4. *Musical Times*, July 1882.
5. C. H. H. Parry's diary.
6. Rosa Sucher, *Aus meinem Leben* (Leipzig, 1914).
7. Ibid.
8. Eugen Gura, *Erinnerungen aus meinem Leben* (Leipzig, 1905).
9. Cosima Wagner to Hans Richter, 7 Jan. 1882.
10. Cosima Wagner to Hans Richter, 4 June 1882.
11. *Musical Times*, Aug. 1882.
12. Hans Richter's diary, 4 May 1882.
13. Ibid. 6 May 1882.
14. *World*, 21 June 1882.
15. Hans Richter to Vienna Philharmonic Committee, 8 June 1882.
16. Eugen d'Albert to Charles d'Albert.
17. C. H. H. Parry's diary, 14 Nov. 1882.
18. *Musical Times*, Dec. 1882.
19. *World*, 22 Nov. 1882.
20. Hans Richter's diary, 13 Feb. 1883.
21. Hans Richter to Siegfried Wagner, 14 Feb. 1883.
22. Hans Richter to Marie Richter, 19 Feb. 1883.
23. Richard Wagner to Hans Richter, 28 Jan. 1883.
24. A. C. Mackenzie to Hans Richter, 9 Mar. 1883.
25. *Musical Times*, June 1883.
26. A. C. Mackenzie to Hans Richter, 25 May 1883.
27. *World*, 16 May 1883.
28. Hans Richter to Marie Richter, 4 May 1883.
29. Hans Richter to Wilhelm Jahn, 9 June 1883.
30. C. H. H. Parry's diary, 16 June 1883.

31. Hans Richter to Marie Richter, 19 May 1883.
32. Alfred Hollins, *A Blind Musician Looks Back* (Edinburgh, 1936).
33. Hans Richter to Eva Wagner, 7 Feb. 1884.
34. Hans Richter to Marie Richter, 1 Nov. 1883.
35. Hans Richter's diary, 7 Nov. 1883.
36. Hans Richter to Marie Richter, 8 Nov. 1883.

CHAPTER 16

1. Hans Richter's diary, Feb. 1884.
2. Hans Richter to Marie Richter, 18 Apr. 1884.
3. Pedro Tillett, 'The Richter Concerts', unpublished memoir.
4. Hans Richter's diary, 4 June 1884.
5. Ibid, 22 June 1884.
6. Ibid, 25 June 1884.
7. Ibid, 3 July 1884.
8. Ibid, 9 July 1884.
9. C. V. Stanford, *Pages from an Unwritten Diary* (London, 1914).
10. C. H. H. Parry's diary, 9 July 1884.
11. *World*, 16 July 1884.
12. Lilli Lehmann, *Mein Weg* (Leipzig, 1913).
13. Hermann Klein, *Thirty Years of Musical Life in London* (London, 1903).
14. Hermann Klein, *Musicians and Mummers* (London, 1925).
15. Hans Richter's diary, 5 July 1884.
16. *Musical Times*, June 1884.
17. Percy Young, *Sir Arthur Sullivan* (London, 1971).
18. Arthur Jacobs, *Arthur Sullivan, a Victorian Musician* (Oxford, 1984).
19. Stanford, *Pages from an Unwritten Diary*.
20. *World*, 28 May 1884.
21. Ibid. 11 June 1884.
22. Ibid. 25 June 1884.
23. *Musical Times*, May 1884.
24. C. H. H. Parry's diary, 12 Nov. 1884.
25. *Musical Times*, Dec. 1884.
26. *World*, 26 Nov. 1884.
27. Tillett, 'The Richter Concerts'.
28. Hans Richter's diary.

CHAPTER 17

1. Anton Dvořák to Hans Richter, 20 Oct. 1884.
2. Hans Richter's diary, 25 Apr. 1884.
3. '*University Intelligence*', *The Times*, 25 Apr. 1884.
4. *World*, 20 May 1885.
5. *Musical Times*, July 1885.
6. C. H. H. Parry's diary, 11 May 1885.
7. Ibid. 18 May 1885.

8. Hans Richter's diary, 8 May 1885.
9. Ibid. 18 May 1885.
10. Ibid. 9 June 1885.
11. Ibid. 29 May 1885.
12. C. V. Stanford, *Pages from an Unwritten Diary* (London, 1914).
13. Hans Richter's diary, 28 Aug. 1885.
14. *Musical Times*, Sept. 1885.
15. Ibid. Oct. 1885.
16. Hans Richter's diary, 2 Sept. 1885.
17. A. Göllerich and L. Auer, *Anton Bruckner* (Regensburg, 1936).
18. *Deutsche Zeitung*, 24 Mar. 1885.
19. Göllerich and Auer, *Anton Bruckner*.
20. Ibid.
21. Ibid.
22. Ibid.
23. *Wiener Presse*, 3 Apr. 1886.
24. *World*, 2 June 1886.
25. C. H. H. Parry's diary.
26. *Musical Times*, June 1886.
27. Anton Bruckner to Moritz von Mayfeld, 23 July 1886.
28. Hans Richter's diary, 11 June 1886.
29. *World*, 12 May 1886.
30. C. H. H. Parry's diary.
31. Hans Richter's diary, 17 May 1886.
32. Cosima Wagner to Hans Richter, 25 July 1887.
33. Hans Richter's diary, 26 July 1886.
34. Ibid. 3 Aug. 1886.
35. Ludwig Karpath, *Begegnung mit dem Genius* (Vienna, 1934).
36. Henry Pleasants *The Music Criticisms of Hugo Wolf* (New York, 1978).
37. Hugo Wolf to Theodor Köchert, 21 Sept. 1885.
38. Karpath, *Begegnung mit dem Genius*.
39. Pleasants, *The Music Criticisms of Hugo Wolf*.
40. Frank Walker, *Hugo Wolf* (London, 1951).
41. Hans Richter to Ludwig Karpath, 11 Mar. 1904.
42. Hans Richter to Marie Joshua, 3 Nov. 1886.

CHAPTER 18

1. Hans Richter to Vienna Philharmonic Orchestra, 16 Dec. 1886.
2. Hans Richter to Wilhelm Jahn, 14 Feb. 1887.
3. Theodor Reichmann to Wilhelm Jahn, 17 Feb. 1887.
4. Hans Richter to Anton Dvořák, 31 Mar. 1887.
5. Hans Richter to Anton Dvořák, 13 Apr. 1887.
6. Anton Dvořák to Hans Simrock, 29 May 1887.
7. Hans Richter's diary, 23 May 1887.
8. *Musical Times*, June 1887.
9. A. Göllerich and L. Auer, *Anton Bruckner* (Regensburg, 1936).

10. Eva Wagner to Hans Richter, 4 May 1887.
11. *Allgemeine Musik-Zeitung*, 14/24 (1887), 226.
12. *Kölnische Zeitung*, 1 June 1887.
13. C. H. H. Parry's diary 23 May 1887.
14. Ibid. 18 Apr. 1887.
15. C. V. Stanford, *Pages from an Unwritten Diary* (London, 1914).
16. Hans Richter to Cosima Wagner, 21 July 1887.
17. Cosima Wagner to Hans Richter, 25 May 1888.
18. Hans Richter to Cosima Wagner, 28 May 1888.
19. Hans Richter's diary, 16 May 1888.
20. *The Autobiography of Sir Felix Semon* (London, 1928).
21. Arthur Coleridge, *Reminiscences* (London, 1921).
22. *World*, 13 June 1888.
23. Cosima Wagner to Hans Richter, 18 July 1888.
24. *Musical Times*, Sept. 1888.
25. L. C. Elson, *Reminiscences* (Chicago, 1891).
26. C. H. H. Parry's diary, 31 Aug. 1888.
27. *Musical Times*, Oct. 1888.
28. C. H. H. Parry's diary, 30 Aug. 1888.
29. *Musical Times*, Oct. 1888.
30. C. H. H. Parry's diary, 3 July 1888.
31. Ibid. 22–29 Aug. 1888.
32. *Yorkshire Post*, Sept. 1888.

CHAPTER 19

1. I. Stravinsky, *An Autobiography* (New York, 1936).
2. T. Kretschmann, *Tempi passati* (Vienna, 1910).
3. Joseph Sulzer, *Erinnerugen eines Wiener Philharmonikers* (Vienna, 1910).
4. *The Paderewski Memoirs* (London, 1939).
5. *Neue freie Presse*, 29 Jan. 1889.
6. E. Albani, *Forty Years of Song* (London, 1911).
7. L. P. Lochner, *Fritz Kreisler* (London, 1950).
8. Mark Hambourg, *From Piano to Forte* (London, 1931).
9. Michael Hambourg to Hans Richter, 8 Mar. 1895.
10. *Boston Post*, 15 Apr. 1893.
11. *Boston Journal*, 15 Apr. 1893.
12. *Neue Musik-Zeitung*, 18 (1890).
13. *Boston Post*, 17 Apr. 1893.
14. *Musical Courier*, 3, May 1893.
15. *American Art Journal*, 6 May 1893.
16. *Boston Herald*, 10 May 1893.
17. A. Göllerich and L. Auer, *Anton Bruckner* (Regensburg, 1936).
18. Ibid.
19. Anton Bruckner to Siegfried Ochs, 3 Feb. 1892.
20. Göllerich and Auer, *Anton Bruckner*.
21. Hans Richter to Edward Speyer, 29 May 1897.

22. Hans von Bülow to Hermann Wolff, 21 June 1888.
23. Hans Richter to Carl Gianicelli, 17 Oct. 1894.
24. Hans Richter to Carl Gianicelli, 8 Jan. 1896.
25. *The Memoirs of Carl Flesch* (London, 1957).
26. Ibid.
27. Hans Richter to Carl Gianicelli, [Dec. 1895].
28. Hans Richter to Edward Speyer, 29 May 1897.

CHAPTER 20

1. Minutes of the Vienna Philharmonic Orchestra: GZ 60/99.
2. Hans Richter to Carl Gianicelli, 15 Oct. 1898.
3. F. Zagiba, *Tchaikovsky: Leben und Werk* (Zürich, 1953).
4. Hans Richter to Carl Gianicelli, 10 Nov. 1899.
5. Gustav Mahler to Hans Richter, 12 Apr. 1897.
6. Hans Richter to Gustav Mahler, 19 Apr. 1897.
7. *Neues Wiener Tageblatt*, 28 Sept. 1902.
8. Nellie Melba, *Melodies and Memories* (New York, 1925).
9. L. Karpath, *Begegnung mit dem Genius* (Vienna, 1934).
10. Hans Richter to Gustav Mahler, 25 Feb. 1900.
11. Hans Richter, 10 Jan. 1894.
12. Marie Richter to Ludwig Bösendorfer, 16 Jan. 1901.
13. Hans Richter to his family, 12 Dec. 1910.
14. Hans Richter to Joseph Scheu, 14 Mar. 1900.
15. Hans Richter to Ludwig Bösendorfer, 27 Dec. 1900.

CHAPTER 21

1. Hubert Herkomer to Hans Richter, 3 Dec. 1888.
2. *The Herkomers*, ii (London, 1911).
3. Joseph Barnby to Hans Richter, 14 June 1889.
4. Hubert Herkomer to Hans Richter, 25 Mar. 1889.
5. Hubert Herkomer to Hans Richter, 20 Apr. 1889.
6. Hubert Herkomer to Hans Richter, 6 July 1891.
7. Hubert Herkomer to Hans Richter, 15 June 1900.
8. Hubert Herkomer to Hans Richter, 25 July 1889.
9. Hubert Herkomer to Hans Richter, 1 Feb. 1891.
10. Hubert Herkomer to Hans Richter, 20 Nov. 1903.
11. Hubert Herkomer to Hans Richter, 14 Feb. 1911.
12. C. H. H. Parry's diary.
13. *Star*, 9 July 1889.
14. *Bayreuth 1876-96* (Berlin, 1897).
15. *Musical Times*, Sept. 1889.
16. C. H. H. Parry's diary.

CHAPTER 22

1. *World*, 3 June 1891.
2. Ibid. 14 Feb. 1894.

3. *Musical Times*, Nov. 1891.
4. Ibid.
5. *World*, 14 Oct. 1891.
6. *Yorkshire Post*, Oct. 1891.
7. Ellen Terry to Hans Richter, 4 Oct. 1891.
8. *World*, 15 June 1892.
9. Arthur Goring-Thomas to Hans Richter, 8 Oct. 1891.
10. Ethel Smyth, *As Time went on* (London, 1936).
11. David Bispham, *A Quaker Singer's Recollections* (New York, 1929).
12. *Musical Times*, Dec. 1894.
13. *Yorkshire Post*, Oct. 1894.
14. *Musical Times*, Dec. 1894.
15. *World*, 25 Apr. 1894.
16. *Musical Times*, July 1895.
17. Ibid.
18. Henry Wood, *My Life of Music* (London, 1938).

CHAPTER 23

1. Hans Richter's diary, 25 Oct. 1895.
2. Edith Hall's diary, 16 Oct. 1892.
3. Ibid. 25 Oct. 1895.
4. *Yorkshire Post*, 26 Oct. 1895.
5. *Musical Times*, Dec. 1896.
6. C. H. H. Parry's diary, 31 May 1897.
7. Hans Richter's diary, 24 May 1897.
8. Ibid. 31 May 1897.
9. *Musical Times*, July 1897.
10. Edith Hall's diary, 10 Oct. 1897.
11. *Musical Times*, Nov. 1897.
12. Hans Richter's diary, 26 Sept 1897.
13. Edward German to Hans Richter, 5 Oct. 1897.
14. Edward German to Hans Richter, 6 Oct. 1897.
15. Hans Richter's diary, 27 Sept. 1897.
16. *Musical Standard*, 2, Oct. 1897.
17. Hans Richter's diary, 30 Sept. 1897.
18. C. H. H. Parry's diary, 8 Oct. 1897.
19. Hans Richter's diary, 22 Sept. 1897.
20. *Birmingham Post*, 23 Sept. 1897.
21. Hans Richter's diary, 7 Oct. 1897.
22. *Musical Opinion*. Nov. 1897.
23. *Yorkshire Post*, 6 Oct. 1897.
24. Ibid. 23 Oct. 1897.
25. Hans Richter's diary, 3 and 7 June 1898.
26. *Musical Times*, July 1898.
27. Hans Richter's diary, 20 June 1898.
28. Ibid. 16 June. 1898.
29. *Musical Times*, July 1898.

30. Ibid. Nov. 1898.
31. Hans Richter's diary, 12 Oct. 1898.
32. *Yorkshire Post*, 5 Oct. 1898.
33. Edith Hall's diary, 17 Oct. 1898.
34. Ibid. 20 Oct. 1898.
35. Ibid. 21 Oct. 1898.
36. Hans Richter's diary, 21 Oct. 1898.
37. Hans Richter to N. Vert, 9 Mar. 1899.
38. Hans Richter's diary, 17 June 1899.
39. Ibid. 19 June 1899.
40. C. H. H. Parry's diary, 19 June 1899.
41. A. C. Mackenzie, *A Musician's Narrative* (London, 1927).
42. *Musical Times*, July 1899.
43. Edward Elgar to Herbert Thompson, 26 June 1899.
44. Edward Elgar to Herbert Thompson, 23 Aug. 1899.

CHAPTER 24

1. Hans Richter to Cosima Wagner, June 1892.
2. Hans Richter to Cosima Wagner, 19 Aug. 1892.
3. Cosima Wagner to Hans Richter, 25 Aug. 1892.
4. Hans Richter to Cosima Wagner, 7 Feb. 1895.
5. Hans Richter's diary, 5 June 1895.
6. Ibid. 7 June 1895.
7. *Musical Times*, July 1895.
8. Cosima Wagner to Hans Richter, 6 Jan. 1894.
9. Siegfried Wagner, *Erinnerungen* (Stuttgart, 1923).
10. Hans Richter to Adolf von Gross, 8 July 1895.
11. *Star*, 22–5 July 1896.
12. *Musical Times*, Aug. 1896.
13. Lilli Lehmann, *Mein Weg* (Leipzig, 1913).
14. Carl Fuchs, *Musical and Other Recollections* (Manchester, 1937).
15. Sydney Loeb's diary, 19 Aug. 1896.
16. Ibid. 17 Aug. 1896.
17. Ibid. 18 May 1896.
18. Hans Richter to Cosima Wagner, 24 June 1897.
19. Cosima Wagner to Hans Richter, 21 Apr. 1898.
20. *Musical Times*, Sept. 1899.

CHAPTER 25

1. *Referee*, 19 May 1901.
2. C. H. H. Parry's diary, 5 June 1896.
3. Sir George Grove to Hans Richter, 1 June 1889.
4. Sir George Grove to Hans Richter, 9 Apr. 1897.
5. Sir George Grove to Hans Richter, 30 Sept. 1897.
6. *Journal of Arnold Bennett* (London, 1932–3), 24 Oct. 1899.

CHAPTER 26

1. Hans Richter's diary, 3 Nov. 1899.
2. *Musical Times*, Nov. 1899.
3. Ibid. Dec. 1899.
4. Hans Richter's diary, 17 Nov. 1899.
5. *Yorkshire Post*, 25 Nov. 1899.
6. Hans Richter's diary, 26 and 27 Nov. 1899.
7. Ibid. 30 Nov. 1899.
8. Thomas Beecham, *A Mingled Chime* (London, 1944).
9. *Music and Art*, 13 Dec. 1899.
10. *Yorkshire Post*, 14 Jan. 1900.
11. Ibid. 27 Dec. 1899.
12. Hans Richter's diary, 31 Jan. 1900.
13. Ibid. 3 Feb. 1900.
14. Ibid. 5 Feb. 1900.
15. Ibid. 7 Feb. 1900.
16. Ibid. 8 Feb. 1900.
17. Ibid. 13 Feb. 1900.
18. Ibid. 20 Feb. 1900.
19. *Manchester Guardian*, 29 Nov. 1901.
20. Ibid. 20 Oct. 1897.
21. Hans Richter's diary, 6 Mar. 1900.
22. Ibid. 14 and 15 Mar. 1900.
23. Ibid. 18 June 1900.
24. Ibid. 26 May 1900.
25. Ibid. 1 June 1900.
26. Ibid. 2 June 1900.
27. Hans Richter to Edward Elgar, 21 Sept. 1900.
28. *Morning Leader*, 24 Sept. 1900.
29. *Manchester Guardian*, 25 Sept. 1900.
30. *Birmingham Post*, 25 Sept. 1900.
31. *Musical Standard*, 29 Sept. 1900.
32. *Referee*, 30 Sept. 1900.
33. *Birmingham Post*, 1 Oct. 1900.
34. Ibid. 2 Oct. 1900.
35. Elgar's birthplace, Broadheath.
36. *Yorkshire Post*, 1 Oct. 1900.
37. Ibid. 2 Oct. 1900.
38. Ibid. 4 Oct. 1900.
39. Hans Richter's diary, 3 Oct. 1900.
40. Ibid.
41. W. C. Berwick Sayers, *Samuel Coleridge-Taylor, Musician* (London, 1915).
42. Henry Wood, *My Life of Music* (London, 1938).
43. *Manchester Guardian*, 13 Mar. 1903.
44. Birmingham Oratory.

CHAPTER 27

1. Blanche Marchesi, *Singer's Pilgrimage* (London, 1923).
2. Ibid.
3. *Yorkshire Post*, 3 Nov. 1900.
4. Ibid. 10 Nov. 1900.
5. Ibid. 28 Nov. 1900.
6. F. Busoni, *Letters to His Wife* (London, 1938).
7. *Manchester Guardian*, 13 Mar. 1901.
8. *Yorkshire Post*, 26 Oct. 1901.
9. Cyril Scott, *Bone of Contention* (London, 1969).
10. Cyril Scott to Hans Richter, 14 Dec. 1901.
11. Donald Francis Tovey to Hans Richter, [1902].
12. Mary Grierson, *Donald Francis Tovey* (London, 1952).
13. Rutland Boughton to Hans Richter, 30 Oct. 1902.
14. Lottie Richter to Hans Richter, 30 May 1902.
15. *Yorkshire Post*, 26 Feb. 1902.
16. L. P. Lochner, *Fritz Kreisler* (London, 1950).
17. *Yorkshire Post*, 1 Aug. 1902.
18. Hans Richter to Cosima Wagner, 19 Oct. 1902.
19. *Manchester Guardian*, 21 Nov. 1902.
20. Hans Richter to Cosima Wagner, 26 Nov. 1902.
21. Hans Richter to Cosima Wagner, 21 Dec. 1902.
22. Florence Caulfield to Hans Richter, 7 Nov. 1902.
23. Irmgard Scharberth, *Das Gürzenich Orchester* (Cologne, 1988).
24. Hans Richter to Edward Elgar, 7 Jan. 1903.

CHAPTER 28

1. *Musical Times*, Feb. 1903.
2. Edward Elgar to Hans Richter, 13 Mar. 1903.
3. *Yorkshire Post*, 23 Jan. 1903.
4. Ibid. 12 Feb. 1903.
5. *Manchester City News*, 18 June 1927.
6. Hans Richter to Mrs Joshua, 27 Apr. 1903.
7. *Musical Times*, June 1903.
8. C. H. H. Parry's diary, 16 May 1903.
9. *Yorkshire Post*, 30 Apr. 1903.
10. *Sunday Times*, 10 May 1903.
11. Source unknown, May 1903.
12. *Yorkshire Post*, 13 Oct. 1903.
13. Hans Richter to Edward Elgar, 14 Aug. 1903.
14. R. H. Wilson to Edward Elgar, 4 June 1931.
15. David Ffrangcon-Davies to Edward Elgar, 24 Jan. 1904.
16. *Musical Times*, Nov. 1903.
17. *Yorkshire Post*, 16 Oct. 1903.

18. *Musical Times*, Dec. 1903.
19. Ibid. Mar. 1904.
20. César Saerchinger, *Artur Schnabel* (London, 1957).
21. *Musical Times*, Mar. 1904.
22. Bartók to his mother, 27 June 1903.
23. *Referee*, 20 Mar. 1904.
24. *Musical Times*, Apr. 1904.
25. Edward Elgar to William Henry Elgar, 18 Mar. 1904.
26. *Manchester Courier*, Mar. 1904.
27. *Musical Times*, Apr. 1904.
28. Hans Richter to Carl Gianicelli, 27 Dec. 1900.
29. *Daily Express*, 10 May 1904.
30. Hans Richter to Anton Dvořák, 24 Jan. 1902.
31. Anton Dvořák to Hans Richter, 11, Feb. 1902.
32. Maurice Pearton, *The LSO at Seventy* (London, 1974).

CHAPTER 29

1. Hans Richter to Cosima Wagner, 22 May 1904.
2. *The Times*, 21 Jan. 1959.
3. Lionel Tertis, *My Viola and I* (London, 1974).
4. Bartók to his mother, 18 Sept. 1904.
5. Hans Richter to Edward Elgar, 19 Sept. 1904.
6. Mark Hambourg, *From Piano to Forte* (London, 1931).
7. Karl Eckman, *Jean Sibelius* (London, 1936).
8. *Musical Times*, Apr. 1905.
9. *Illustrated London News*, 10 June 1905.
10. Ibid. 27 May 1905.
11. *Yorkshire Post*, 2 May 1905.
12. *Illustrated London News*, 13 May 1905.
13. Daisy Jordan, 26 Oct. 1905.
14. Hans Richter to Carl Gianicelli, 24 Sept. 1907.
15. Percy Grainger to Hermann Sandby, 7 Dec. 1905.
16. K. Dreyfus (ed.), *The Farthest North of Humanness* (London, 1985).
17. Ernest Newman to Herbert Thompson, 13 Feb. 1906.
18. Sir Adrian Boult's diary, 12 Feb. 1906.
19. *Yorkshire Post*, 9 Mar. 1906.
20. *Musical Herald*, 1 Feb. 1906.
21. C. Rawdon Briggs, unpublished memoir, Jan. 1931.
22. *Standard*, 11 May 1906.
23. Hans Richter to Percy Pitt, 29 Oct. 1906.
24. Hans Richter to Percy Pitt, 12 June 1906.
25. Hans Richter to Percy Pitt, 28 July 1906.
26. Hans Richter to Percy Pitt, 9, Mar. 1906.
27. *Daily Chronicle*, 4 May 1906.
28. *The Times*, 9 May 1906.

CHAPTER 30

1. Hans Richter to Marie Joshua, 6 July 1906.
2. *Yorkshire Post*, 2 Oct. 1906.
3. Ibid. 3 Oct. 1906.
4. Ibid. 6 Oct. 1906.
5. Ibid. 4 Oct. 1906.
6. Ibid. 8. Mar. 1907.
7. *Manchester Guardian*, 22 Dec. 1906.
8. Hans Richter to Otto Schieder, 10 Mar. 1903.
9. Archie Camden, *Blow by Blow* (London, 1982).
10. *Manchester City News*, 18 June 1927.
11. Albert Spalding, *Rise to Follow* (London, 1946).
12. Ibid.
13. Sir Adrian Boult's diary, 8 and 22 May 1907.
14. Hans Richter to Marie Joshua, 22 July 1907.
15. Hans Richter to Carl Gianicelli, 19 Jan. 1908.
16. J. Corredor, *Conversations with Casals* (London, 1956).
17. Hans Richter to Percy Pitt, 22 Nov. 1907.
18. Hans Richter to Percy Pitt, 23 Oct. 1907.
19. Hans Richter to Percy Pitt, 7 Jan. 1908.
20. Daniel Chamier, *Percy Pitt* (London, 1938).
21. *Musical Times*, Mar. 1908.
22. Harold Rosenthal, *Two Centuries of Opera at Covent Garden* (London, 1958).
23. Sir Adrian Boult's diary, 3 and 8 Feb. 1908.
24. Hans Richter to Marie Joshua, 27 Jan. 1908.
25. Hans Richter to Percy Pitt, 20 Feb. 1908.
26. *Radio Times*, 3 Oct. 1957.

CHAPTER 31

1. *Yorkshire Post*, 2 May 1908.
2. Hans Richter to Percy Pitt, 23 and 27 June 1908.
3. *Münchener Neueste Nachrichten*, 2 June 1908.
4. Eugene Goossens, *Overture and Beginners* (London, 1951).
5. Edward C. Bairstow, unpublished memoir.
6. W. M. Strutt, *The Reminiscences of a Musical Amateur* (London, 1915).
7. Hans Richter to Percy Pitt, 27 June 1908.
8. *Yorkshire Post*, 31 July 1908.
9. *Manchester Guardian*, 23 Oct. 1908.
10. R. Daniels, *Conversations with Cardus* (London, 1976).
11. *Radio Times*, 31 Oct. 1930.
12. *Spectator*, 29 Nov. 1902.
13. W. H. Reed, *Elgar* (London, 1946).
14. *Musical Times*, Jan. 1909.
15. Neville Cardus, *Autobiography* (London, 1947).
16. *Musical Standard*, 30 July 1904.

17. C. Redwood (ed.), *An Elgar Companion* (Ashbourne, 1982).
18. Hans Richter to Percy Pitt, 27 June 1908.
19. Hans Richter to Percy Pitt, 13 Sept. 1908.
20. Hans Richter to Percy Pitt, 30 Oct. 1908.
21. Hans Richter to Marie Joshua, 27 Jan. 1909.
22. Strutt, *The Reminiscences of a Musical Amateur*.
23. Henry Higgins to Percy Pitt, 16 Feb. 1909.
24. Hans Richter to Percy Pitt, 20 Feb. 1909.
25. *Observer*, 18 Apr. 1909.
26. Hans Richter to Percy Pitt, 3 Jan. 1911.
27. *Musical Times*, June 1951.
28. *Yorkshire Post*, 25 Feb. 1909.
29. Ibid. 6 Mar. 1909.
30. C. H. H. Parry's diary, 22 Mar. 1909.

CHAPTER 32

1. Edward Elgar to Hans Richter, [3] Oct. 1909.
2. *Yorkshire Post*, 14 Aug. 1909.
3. Ibid. 9 Oct. 1909.
4. Ibid. 15 Oct. 1909.
5. Edward Elgar to Hans Richter, 20 Mar. 1910.
6. J. Sutcliffe Smith, *The Story of Music in Birmingham* (Birmingham, 1945).
7. *Yorkshire Post*, 3 Nov. 1909.
8. Hans Richter to Herbert Thompson, 3 Nov. 1909.
9. Gerald Cumberland, *Set down in Malice* (London, 1918).
10. Frederick Delius to Ethel Smyth, 17 Feb. 1909.
11. *Musical Opinion*, Dec. 1936.
12. *Staffordshire Sentinel*, 30 Sept. 1909.
13. *Musical World*, 16 Nov. 1907.
14. Hans Richter to Marie Joshua, 8 Dec. 1909.
15. Hans Richter to Percy Pitt, 22 Feb. 1910.
16. Alice Elgar's diary, 24 Jan. 1910.
17. *Manchester City News*, 18 June 1927.
18. *Yorkshire Post*, 26 Apr. 1910.
19. Hans Richter to Marie Joshua, 26 May 1910.
20. *Yorkshire Post*, 19 Oct. 1910.
21. Mathilde Richter's diary, 3 Nov. 1910.
22. Hans Richter to Edward Elgar, 1 July 1910.
23. Edward Elgar to Hans Richter, 9 Nov. 1910.
24. *Yorkshire Post*, 24 Nov. 1910.
25. *Musical Times*, Feb. 1911.
26. *Yorkshire Post*, 24 Jan. 1911.
27. *The Times*, 14 Feb. 1911.
28. *Manchester Evening News*, 8 Mar. 1911.
29. *Yorkshire Post*, 11 Mar. 1911.
30. *Manchester Guardian*, 17 Mar. 1911.

31. Ibid. 24 Mar. 1911.
32. Hans Richter to Charles Behrens, 15 Mar. 1911.
33. *Manchester Guardian*, 4 Apr. 1911.

CHAPTER 33

1. *Manchester Guardian*, 14 Feb. 1911.
2. *Yorkshire Post*, 17 Feb. 1911.
3. Edward Elgar to Hans Richter, 13 Feb. 1911.
4. *Spectator*, 18 Mar. 1911.
5. Hans Richter to Percy Pitt, 30 Mar. 1911.
6. Marie Richter to Sydney Loeb, 14 Feb. and 14 Mar. 1911.
7. Hans Richter to Marie Joshua, 31 May 1911.
8. Hans Richter to Siegfried Wagner, 20 Feb. 1911.
9. Hans Richter to Percy Pitt, 19 June 1911.
10. Hans Richter to Percy Pitt, 20 Aug. 1911.
11. Hans Richter to Marie Joshua, 13 Dec. 1911.
12. Hans Richter to Percy Pitt, 15 Aug. 1911.
13. Hans Richter to Percy Pitt, 26 Oct. 1911.
14. Hans Richter to Percy Pitt, 21 Nov. 1911.
15. Hans Richter to Sydney Loeb, 20 Aug. 1912.
16. Manfred Eger, *Hans Richter: Bayreuth, Wien, London und zurück* (Bayreuth, 1990).
17. J. R. Brotherton to Sydney Loeb, 20 Aug. 1912.
18. Marie Joshua to Sydney and Mathilde Loeb, 25 Aug. 1912.
19. Hans Richter to Percy Pitt, 7 Oct. 1912.
20. Hans Richter to Sydney Loeb, 16, Dec. 1912.
21. Borghild Langaard to Hans Richter, 26 Nov. 1912.
22. Thomas Beecham to Hans Richter, 8 Jan. 1913.
23. Hans Richter to Percy Pitt, 30 Mar. 1913.
24. Hans Richter to Marie Joshua, 15 Nov. 1913.
25. Marie Joshua to Hans Richter, Dec. 1913.
26. Hans Richter to Percy Pitt, 13 Jan. 1914.

CHAPTER 34

1. Hans Richter to Sydney Loeb, 23 Feb. 1914.
2. C. V. Stanford to Hans Richter, [Apr. 1908].
3. Hans Richter to C. V. Stanford, 22 Apr. 1908.
4. C. V. Stanford to Hans Richter, 15 Aug. 1914.
5. Mathilde Loeb's diary, 12 July 1914.
6. Hans Richter to Wilhelm Gericke, 9 July 1914.
7. Hans Richter to Mathilde Loeb, 1 Oct. 1914.
8. Hans Richter to Sydney Loeb, 29 Oct. 1914.
9. Hans Richter to Marie Joshua, 6 Feb. 1914.
10. *The Times*, 23 Sept. 1914.
11. Ibid. 7 Feb. 1916.
12. Manfred Eger, *Hans Richter: Bayreuth, Wien, London und zurück* (Bayreuth, 1990).

13. Hans Richter to Sydney Loeb, 27 Nov. 1915.
14. Hans Richter to Michael Balling, 13 Jan. 1916.
15. Hans Richter to Carl Fuchs, 12 Dec. 1915.
16. Hans Richter to Carl Fuchs, 29 Dec. 1915.
17. Hans Richter to Michael Balling, 20 Jan. 1916.
18. Hans Richter to Ludwig Karpath, 12 Sept. 1916.
19. Hans Richter to Franz Fischer, 31 Oct. 1916.
20. Hans Richter to Sydney Loeb, 15 Nov. 1916.
21. Hans Richter to Otto Lohse, 17 Nov. 1916.
22. Hans Richter to Otto Lohse, 28 Nov. 1916.
23. Hans Richter to Joseph Stiegler, 24 Nov. 1916.
24. Marie Richter to Mathilde Loeb, 21 Jan. 1917.
25. Edward Elgar to Sydney Loeb, 9 Dec. 1916.
26. Blandine Gravina to Mathilde Loeb 9 Apr. 1917.
27. Gustav Behrens to Mathilde Loeb, 8 Dec. 1916.
28. Peter Pachl, *Siegfried Wagner: Genie im Schatten* (Munich, 1988).

CHAPTER 35

1. Hans Richter to Sydney Loeb, 14 Apr. 1915.
2. *Daily News*, 8 Dec. 1916.
3. *Manchester Guardian*, 7 Dec. 1916.
4. *Birmingham Daily Post*, 7 Dec. 1916.
5. *Neueste Nachrichten*, Hamburg, 6 Dec. 1916.
6. 'Hans Richter's letzte Rede', *Allgemeine Musik-Zeitung*, 29 Dec. 1916.
7. Zdenko von Kraft, *Der Sohn* (Graz, 1969).
8. Herbert Thompson, 'Some Recollections of Richter' (unpublished, 1934).
9. BBC, *Collectors' Corner*, 8 Apr. 1980, and Christopher Dyment's interview with Sir Adrian Boult, 18 Sept. 1972.
10. *Daily Telegraph*, Feb. 1939.
11. *Manchester Guardian*, 9 Apr. 1933.
12. *Daily Telegraph*, Feb. 1939.
13. *The Times*, 15 June 1935.
14. *Sunday Times*, 14 May 1922.
15. *The Times*, 26 Jan. 1934.
16. Ibid. 29 May 1937.
17. *Daily Telegraph*, June 1937.
18. *Manchester Guardian*, Oct. 1934.
19. *Illustrated London News*, 12 Nov. 1927.
20. *Sunday Times*, 26 Oct. 1930.
21. *Saturday Review*, 30 Oct. 1909.
22. C. V. Stanford, *Interludes* (London, 1922).
23. Eva Ducat, *Another Way of Music* (London, 1928).
24. Ibid.
25. *Sunday Times*, 26 Oct. 1930.
26. Ibid. 16 Nov. 1930.
27. Ibid. 9 Nov. 1930.

28. *Royal College of Music Magazine*, 9 3 (1913).
29. Mary Lawton, *Ernestine Schumann-Heink: Last of the Titans* (New York, 1928).
30. Feruccio Bonavia, *Musicians in Elysium* (London, 1949).
31. Claude Debussy, *Monsieur Croche the Dilettante Hater* (London, 1927).
32. Neville Cardus, *Second Innings* (London, 1950).

APPENDIX I
Works Conducted by Hans Richter

AN ANALYSIS OF RICHTER'S OPERATIC REPERTOIRE

1868–1912: 2263 performances of 94 operas

Composer	Opera	Number of performances
Adam	*Le Postillon de Lonjumeau*	3
Auber	*Le Bal masqué*	3
	Le Domino noir	2
	Fra Diavolo	7
	Le Maçon	6
	La Muette de Portici	1
	La Part du Diable	3
	Le Premier Jour de Bonheur	6
Bachrich	*Muzzedin*	3
Beethoven	*Fidelio*	79
Bellini	*Norma*	39
	La sonnambula	1
Bizet	*Carmen*	137
Boïeldieu	*La Dame blanche*	7
	Jean de Paris	2
Brüll	*Das goldene Kreuz*	40
Cherubini	*Les Deux Journées*	30
	Medea	3
Cornelius	*Der Barbier von Bagdad*	13
Delibes	*Le Roi l'a dit*	7
Dittersdorf	*Das Rothkäppchen*	7
Donizetti	*Dom Sébastien*	12
	L'elisir d'amore	1
	La favorita	1
	La Fille du régiment	5
	Lucrezia Borgia	1
Flotow	*Martha*	1
Fuchs	*Die Königsbraut*	2
Gluck	*Le Cadi dupé*	1
	Orfeo ed Euridice	1
Goetz	*Der Widerspenstigen Zähmung*	1

Goldmark	Die Königin von Saba	1
Gounod	Faust	24
	Roméo et Juliette	72
	Le Tribut de Zamora	18
Grisar	Bonsoir Monsieur Pantalon	18
Hager	Marffa	2
Halévy	La Juive	11
Heuberger	Mirjam	5
Kauders	Walther von der Vogelweide	2
Krempelsetzer	Der Rothmantel	4
Leoncavallo	Pagliacci	56
Leschetizky	Die erste Falte	2
Lortzing	Undine	2
	Der Waffenschmied	1
Marschner	Hans Heiling	30
Mascagni	L'amico Fritz	35
	Cavalleria rusticana	1
Massenet	Le Cid	18
Méhul	Joseph	3
Meyerbeer	L'Africaine	58
	Dinorah	19
	L'Étoile du Nord	27
	Les Huguenots	19
	Le Prophète	28
	Robert le Diable	18
Mozart	Bastien und Bastienne	6
	La clemenza di Tito	5
	Don Giovanni	97
	Die Entführung aus dem Serail	13
	Le nozze di Figaro	19
	Der Schauspieldirektor	1
	Die Zauberflöte	11
Nessler	Der Trompeter von Säkkingen	17
Nicolai	Die lustigen Weiber	16
Pfeffer	Harold	1
Ponchielli	La gioconda	5
Rossini	The Barber of Seville	54
	William Tell	52
Smareglia	Cornelius Schut	5
	Der Vasall von Szigeth	12
Smetana	The Kiss	4
Stanford	Savonarola	1
Thomas	Mignon	3

Verdi	*Aida*	22
	Rigoletto	5
	La traviata	27
	Il trovatore	23
Volkmann	*Richard III*	8
Wagner	*Der fliegende Holländer*	55
	Götterdämmerung	78
	Lohengrin	198
	Die Meistersinger	141
	Das Rheingold	63
	Rienzi	15
	Siegfried	84
	Tannhäuser	85
	Tristan	57
	Die Walküre	123
Weber	*Euryanthe*	5
	Der Freischütz	40
	Oberon	6
	Preziosa	6
Zenger	*Ruy Blas*	1

AN ANALYSIS OF RICHTER'S SYMPHONIC REPERTOIRE

Composer	Work	Number of performances
Beethoven	Symphony No. 1	10
	Symphony No. 2	16
	Symphony No. 3	97
	Symphony No. 4	32
	Symphony No. 5	91
	Symphony No. 6	46
	Symphony No. 7	85
	Symphony No. 8	35
	Symphony No. 9	56
Berlioz	*Harold in Italy*	19
	Symphonie fantastique	11
Brahms	Symphony No. 1	28
	Symphony No. 2	18
	Symphony No. 3	19
	Symphony No. 4	20
Bruckner	Symphony No. 1	1
	Symphony No. 2	1
	Symphony No. 3	5
	Symphony No. 4	4
	Symphony No. 7	5
	Symphony No. 8	1
Dvořák	Symphony No. 5	1
	Symphony No. 6	2
	Symphony No. 7	1
	Symphony No. 8	2
	Symphony No. 9	19
Elgar	*Cockaigne*	15
	Enigma Variations	33
	Froissart	1
	In the South	14
	Introduction and Allegro for Strings	6
	Pomp and Circumstance March No. 1	2
	Pomp and Circumstance March No. 2	2
	Symphony No. 1	13
Franck	Symphony in D minor	1
Glazunov	Symphony No. 5	1
	Symphony No. 6	10
	Symphony No. 7	1
Mendelssohn	Symphony No. 3	12

Mozart	Symphony No. 4	12
	Symphony No. 31	9
	Symphony No. 33	6
	Symphony No. 34	2
	Symphony No. 35	1
	Symphony No. 36	5
	Symphony No. 38	9
	Symphony No. 39	14
	Symphony No. 40	16
	Symphony No. 41	14
Rimsky-Korsakov	*Sheherazade*	5
Schubert	Symphony No. 8	34
	Symphony No. 9	21
Schumann	Symphony No. 1	16
	Symphony No. 2	13
	Symphony No. 3	12
	Symphony No. 4	15
Sibelius	Symphony No. 2	1
Strauss	*Also sprach Zarathustra*	14
	Don Juan	25
	Don Quixote	5
	Ein Heldenleben	3
	Sinfonia domestica	3
	Till Eulenspiegel	41
	Tod und Verklärung	19
Tchaikovsky	Symphony No. 3	1
	Symphony No. 4	7
	Symphony No. 5	11
	Symphony No. 6	75

Richter conducted sixty-four performances (including eight public dress rehearsals) of Beethoven's 'Choral' Symphony between 9 April 1873 and 23 November 1910 in the following cities: Aachen (2), Antwerp (2), Berlin (2), Birmingham (3), Blackpool (1), Bradford (1), Budapest (6), Leeds (2), Liverpool (1), London (21), Manchester (6), Newcastle (2), Paris (1), St Petersburg (2), Stuttgart (2), Vienna (10).

AN ANALYSIS OF RICHTER'S CHORAL REPERTOIRE

(Excluding Sunday Court Chapel services in Vienna)

Composer	Work	Number of performances
Bach	*Christmas Oratorio*	3
	Magnificat	4
	Mass in B minor	8
	St John Passion	1
	St Matthew Passion	6
Beethoven	*Missa solemnis*	12
Berlioz	*The Damnation of Faust*	2
	Grande Messe des morts	15
	Symphonie funèbre et triomphale	1
	Te Deum	3
Brahms	German Requiem	7
Bruckner	Te Deum	3
Dvořák	*Spectre's Bride*	2
	Stabat mater	3
Elgar	*The Apostles*	3
	The Dream of Gerontius	10
	The Kingdom	1
Gounod	*Mors et vita*	2
Handel	*Joshua*	1
	Judas Maccabeus	1
	Messiah	11
	Saul	2
	Theodora	1
Haydn	*The Creation*	4
	The Seasons	1
Liszt	*St Elizabeth*	17
Mendelssohn	*Elijah*	20
	St Paul	1
Mozart	Requiem	4
Schubert	Mass in E flat	4
Schumann	*Faust*	1
	Paradise and the Peri	1
Verdi	Requiem	3

APPENDIX II
Cities and Towns where Richter Conducted

EUROPE

Aachen, Amsterdam, Antwerp, Basle, Bayreuth, Bologna, Berlin, Berne, Bratislava, Breslau, Brno, Brussels, Budapest, Cluj, Cologne, Copenhagen, Cracow, Düsseldorf, Freiburg, Geneva, Graz, Hamburg, Hanover, Katowice, Leipzig, Linz, Ljubljana, Lyons, Milan, Moscow, Munich, Paris, Poznań, Prague, Raab, St Petersburg, Salzburg, Stockholm, Strasbourg, Stuttgart, Trieste, Turin, Venice, Vienna, Wiesbaden, Zürich.

GREAT BRITAIN

Belfast, Birmingham, Blackpool, Bradford, Brighton, Bristol, Burnley, Bushey, Cambridge, Cardiff, Cheltenham, Doncaster, Dublin, Dundee, Eastbourne, Edinburgh, Glasgow, Halifax, Hanley, Huddersfield, Hull, Leeds, Liverpool, London, Manchester, Middlesbrough, Newcastle, Nottingham, Oxford, Preston, Rotherham, Sheffield, Southport.

INDEX OF NAMES

Abt, Franz 67
Adam, Adolphe 28, 67, 196
Agate, Edward 424
Ahna, Heinrich de 58
Akeroyd, V. V. 316, 318
Albani, Emma 187, 192, 206, 219, 245, 254, 283, 288, 294, 300, 301, 320, 350
Albert, Charles d' 167, 176, 180, 196
Albert, Eugen d' 166–72, 174–86, 194, 196, 215–16, 218, 225, 227, 234, 238, 241, 253, 254, 255, 338, 358, 369, 409
Alexandra, Empress of Russia 255
Alexandra, Princess of Wales, later Queen 143, 152, 159, 192, 369, 370, 371, 374, 401
Alma-Tadema, Sir Lawrence 142, 156, 158, 159, 193, 324
Ambros, August Wilhelm 88
Anderton, Thomas 220
Apponyi, Count 82
Arditi, Luigi 140, 434
Arensky, Anton 427
Armbruster, Karl 142, 156, 188, 191, 206, 280
Assmeyer, Ignaz 5
Auber, Daniel 9, 28, 52, 139, 213, 241
Auer, Leopold 173, 174
Auerbach, Caroline 449, 454, 455
Augusz, Baron 78–9
Austin, Ambrose 243
Austin, Frederic 354

Bach, Johann Sebastian 21–2, 44, 54, 82, 99, 118, 135, 137, 147, 150, 158, 169, 172, 176, 184, 200, 214, 222, 236, 238, 241, 245, 288, 290, 301, 327, 337, 343, 344, 351, 366, 368, 374, 379, 381, 383, 386, 391, 392, 399, 418, 423, 428, 429, 430, 431, 432, 439, 452, 454, 455, 471
Bache, Walter 156, 193
Bachrich, Sigmund 88–9, 139, 198, 229, 230, 231, 362
Backhaus, Wilhelm 355, 380
Bairstow, Sir Edward 407–8
Baker, F. A. 317
Balling, Michael 330, 377, 420, 442, 451, 452, 453, 454
Bantock, Sir Granville 359, 391, 399, 421, 423, 424, 426
Bargiel, Woldemar 431
Barnby, Sir Joseph 210, 281, 327
Barry, Charles Ainslie 35, 193, 226, 238, 285, 324
Barth, Heinrich 150, 171, 327
Bartók, Béla 369, 370, 378, 381, 382
Bauer, Harold 254, 353, 428
Bauer, L. 317
Baugham, E. A. 341
Bechstein, Hans 115, 388, 398, 400
Becker, Hugo 254, 338
Beecham, Sir Joseph 334
Beecham, Sir Thomas 334–5, 424, 425, 426, 443
Beeth, Lola 241, 248
Beethoven, Ludwig van 15, 19, 22, 28, 31, 38, 44, 46, 47, 48, 49, 50, 52, 54, 55, 58, 59, 63, 71, 73, 74, 75, 80, 82, 98, 99, 133, 135, 140, 141, 142, 143, 144, 145, 147, 149, 150, 153, 154, 155, 158, 159, 160, 161, 162, 163, 164, 166, 172, 177, 178, 182, 183, 186, 188, 197, 198, 201, 202, 205, 206, 210, 212, 213, 215, 216, 217, 218, 219, 220, 221, 225, 227, 235, 238, 239, 241, 242, 243, 245, 247, 248, 253, 254, 264, 266, 267, 270, 274, 278, 282, 286, 287, 288, 289, 290, 292, 293, 296, 299, 300, 301, 303, 304, 305, 308, 313, 314, 325, 328, 329, 332, 333, 334, 336, 339, 347, 348, 349, 350, 352, 355, 360, 362, 366, 367, 372, 374, 379, 383, 386, 387, 391, 392, 394, 395, 398, 399, 404, 409, 418, 420, 421, 425, 427, 428, 431, 432, 434, 435, 437, 439, 440, 460, 461, 463, 464, 465, 466, 468, 469, 474
Behrens, Gustav 227, 297, 298, 306, 308, 336, 340, 430, 454, 456
Beidler, Franz 377, 398, 399, 404, 411, 445
Bell, William Henry 418
Bellini, Vincenzo 52, 55, 216, 250
Benedict, Sir Julius 142, 218
Bennett, Arnold 330

Bennett, Joseph 116, 156, 211, 281, 299, 304–5, 310, 326, 327
Bennett, William Sterndale 288, 356, 366
Berger, Francesco 404
Beringer, Oskar 193
Berlioz, Hector 3, 15, 40, 54, 63, 103, 135, 142, 143, 161, 164, 166, 180, 214, 218, 238, 242, 245, 246, 285, 288, 289, 301, 334, 339, 358, 360, 368, 369, 378, 382, 391, 399, 420, 430, 431, 434, 438, 459, 462, 466, 468
Bernhardt, Sarah 324
Bertram-Meyer, Marie 17, 18, 19
Betz, Franz 17, 19, 26, 30, 31, 36, 37, 41, 47, 59, 66, 72, 90, 91, 93, 108, 109, 112, 118, 193, 250
Bianchi, Bianca (Bertha Schwarz) 170, 241
Bibl, Rudolf 131
Bignio, Louis 101
Bispham, David 292, 301, 320
Bizet, Georges 40, 87, 97, 100, 103, 139, 236, 424
Black, Andrew 286, 294, 301, 303
Blau, Julius 11
Blauvelt, Lillian 334, 352
Blume, Ludwig 14
Boccherini, Luigi 118
Boïeldieu, F. A. 28
Boïto, Arrigo 448
Bonavia, Ferruccio 466, 472–3
Borodin, Alexander 253, 368
Borsdorf, Adolf 317, 375, 421
Borwick, Leonard 254, 286, 294, 418, 447
Bösendorfer, Ludwig 171, 180, 182, 183, 277, 278
Boughton, Rutland 354, 421
Boult, Sir Adrian 24, 383, 398, 402, 463–4, 473
Bowen, York 383, 393, 418
Brahms, Johannes 15, 19, 55, 63, 86, 99, 100, 118, 134, 135, 136, 137, 138, 139, 147, 148, 150, 154, 157, 158, 160, 161, 162, 163, 164, 169, 170, 171, 172, 175, 176, 177, 180, 182, 183, 184, 185, 188, 203, 204, 205, 216, 218, 222, 224, 225, 226, 229, 230, 231, 232, 233, 235, 238, 241, 242, 248, 254, 255, 259, 261, 263–4, 266, 268, 280, 284, 286, 287, 288, 291, 293, 299, 302, 304, 309, 314, 315, 317, 326, 329, 332, 333, 344, 350, 353, 357, 360, 367, 368, 369, 387, 391, 397, 399, 408, 409, 426, 427, 428, 429, 430, 431, 432, 435, 465, 466, 468, 471
Brain, Aubrey 421

Brain, Dennis 421
Brandt, Karl 31, 32, 38, 57, 64, 65
Brandt, Marianne 71, 113, 160, 161, 165, 188
Brassin, Louis 40, 204
Brema, Marie 294, 301, 304, 319, 320, 332, 341
Brenner, Ludwig von 131
Brereton, W. H. 294
Breuer, Hans 320
Brewer, Herbert 426
Brian, Havergal 424–5
Bridge, Sir Frederick 245
Briggs, Christopher Rawdon 384–7
Brodsky, Adolph 172, 173, 174, 180, 185, 193, 306, 308, 340, 355, 357, 370, 379, 380, 385, 423, 430, 447, 454
Brodsky, Anna 172–3
Bronsart, Hans von 29, 30
Bronsil, Hans 317
Brossa, F. 336–7
Browning, Robert 123
Bruch, Max 134, 156, 205, 238, 242, 248, 254, 264, 338, 355, 360, 368, 428, 429, 430, 450
Bruckner, Anton 6, 76, 86, 99, 115, 131, 133, 134, 135, 136, 137, 138, 139, 148, 161, 162, 163, 164, 175, 195, 203, 222, 223, 224, 225, 226, 238, 241, 253, 255, 256, 259–63, 268, 280, 287, 309, 368, 369, 381, 393, 399, 404, 422
Brüll, Ignaz 118, 139, 161, 204, 254, 285
Bryn, Borghyld 400
Buchta, Alois 331
Bülow, Blandine von (later Countess Gravina) 12, 21, 45, 196, 456
Bülow, Daniela von (later Daniela Thode) 12, 21, 45, 92, 238, 449
Bülow, Hans von 4, 8, 12, 13, 15, 16, 17, 18, 20, 21, 22, 24, 26, 27, 28, 29, 30, 31, 35, 36, 37, 45, 62, 63, 98, 116, 117, 120, 147, 158, 163, 178, 182, 183, 191, 192, 194, 212, 221, 225, 237, 264, 314, 315, 324, 392–3, 406, 445, 466, 468–9
Bülow, Marie von 4
Burger, Anton 389
Burgstaller, Alois 319
Burmester, Willy 254
Burne-Jones, Sir Edward 324
Burrian, Carl 377, 388
Busby, Thomas 375, 421
Busoni, Ferruccio 233, 254, 278, 303, 304, 332, 350–1, 368, 370, 379
Buths, Julius 453

Butt, Clara 332, 344, 350, 371
Byrd, William 159, 344

Camden, Archie 378, 394–6
Campbell, Francis 200
Cardus, Neville 337, 410, 412, 465, 474
Carreño, Teresa 253, 409
Caruso, Enrico 415
Casals, Pablo 399, 418, 427
Catterall, Arthur 379, 380, 395, 430
Chandon, Josef 123
Chappell, Arthur 142, 234
Cherubini, Luigi 54, 72, 133, 139, 161, 262, 292, 293, 360, 361, 420, 426
Chopin, Frederic 177, 254, 255
Chorley, Henry 35
Christie, Winifred 157
Clapham, John 165
Coates, Albert 447, 465
Coates, John 360, 423, 424
Coburg, Duchess of 326
Coburg, Duke of 326–7, 328
Coleridge, Arthur 243
Coleridge-Taylor, Samuel 328, 343, 345, 352, 426
Colonne, Edouard 375
Cooper, Joseph 309
Cornelius, Peter (composer) 28, 250, 369, 387
Cornelius, Peter (singer) 388, 398, 400
Costa, Sir Michael 151, 210, 212, 294, 301, 302, 366, 368, 392, 434
Cowen, Sir Frederic 160, 178–81, 210, 220, 226, 239, 280, 288, 298, 299, 301, 306, 307, 308, 335, 337, 339, 349, 355, 375, 379
Cramer, Pauline 286
Cronje, General Piet 338
Crossley, Ada 301
Cusins, William George 243, 392, 434
Czasensky, Albert 2
Czerny, Carl 230

Dale, Benjamin 378
Damrosch, Leopold 116
Damrosch, Walter 218, 324
Dannreuther, Edward 121, 122, 123, 124, 142, 149, 150, 156, 160, 191, 196, 205, 464
Dartmouth, Lord 325
Darwin, Charles 158
Davenport, Dr 340
Davies, Ben 301, 327, 350
Davies, Fanny 245, 254, 350, 447
Davies, Sir Henry Walford 421, 426

Davison, James 122, 156, 211
Dawson, Frederick 297, 429
Dayas, William 385
Debussy, Claude 409, 410, 421, 424, 427, 473–4
Deichmann, Carl 325, 328, 398
Delibes, Leo 140, 196, 334, 352
Delius, Frederick 334, 421, 424, 425
Demuth, Leopold 321, 322
Dessoff, Otto 85, 88, 98, 99, 100, 103, 136, 146, 194, 275
Destinn, Emmy 372, 373, 377, 389, 398, 405
Dickens, Charles 445
Diez, Sophie 26
Dillmann, Alexander 405–6
Dingelstedt, Franz 43, 84, 85, 101
Dittersdorf, Carl Ditters von 28
Dohnányi, Ernst von 254, 266, 267, 270, 305, 332, 355, 369
Döhnhoff, Countess 70
Dolmetsch, Arnold 301
Dolmetsch, Elodie 301
Donizetti, Gaetano 52, 55, 71, 250, 275
Door, Anton 171, 183
Doppler, Carl 462
Doppler, Franz 72, 462
Drexler, Josef 107
Dubez, Peter 72
Ducat, Eva 469–70
Düfflipp, Count Lorenz 24, 27, 29, 36, 37
Dukas, Paul 430
Duparc, Henri 46
Duse, Eleonore 324
Dustmann, Louise 85
Dvořák, Antonin 99, 135, 137, 148, 149, 150, 153, 164, 165, 185, 195, 203, 215, 220, 229, 234, 236, 237, 241, 245, 253, 254, 255, 263, 266, 285, 286, 289, 291, 294, 298, 299, 302, 303, 304, 309, 339, 345, 352, 354, 355, 356, 360, 361, 370, 374, 396, 399, 418, 420, 427, 428
Dyck, Ernest van 248, 249, 304, 365
Dyment, Christopher 463

Eberle, Ludwig 25
Eckart, Richard du Moulin- 312
Eckert, Karl 66, 91, 92, 98
Edgley, W. T. 422–3
Edward, Prince of Wales (later King Edward VII) 142, 143, 242, 327, 371, 374, 379, 401, 427
Edwards, Frederick George 27, 309
Egressi, Béni 6

Ehnn, Berta 85, 97, 101, 119, 171
Eilers, Albert 71
Elgar, Lady Alice 359–60, 426
Elgar, Sir Edward 157, 252, 280, 288, 308–10, 327, 331, 336, 340–6, 347, 348, 350, 356, 358, 359, 360, 366, 367–8, 370, 371, 372, 374, 375, 378, 379, 382, 383, 390, 391, 393, 399, 411, 412, 418, 420, 421, 422, 424, 425, 426, 427, 428, 429, 431, 432, 436, 444–5, 449, 451, 452, 453, 456
Elizabeth, Empress 120, 132, 277
Ella, John 98–9, 133
Elman, Mischa 383, 391
Elson, Louis C. 244
Engel, Louis 144, 152–3, 188, 194, 197, 199, 208, 211–13, 217, 225, 226, 243, 285, 287
Epstein, Julius 183
Erdtmann, Baroness Dorothea von 348
Erkel, Elek 53
Erkel, Ferenc 43, 53, 54, 65, 73
Erkel, Gyula 53
Erkel, László 53
Erkel, Sandor 53, 82, 265
Ernst, Heinrich 203
Esser, Heinrich 7, 8, 9, 10, 60, 84, 85, 274, 275
Essipoff, Annette 177–8, 252, 287
Esterházy, Count Nikolaus 1
Ettling, Henry 340
Evans, Edith 400–1
Evans, Richard 431

Faistenberger, Johann 14
Fassett, Isabel 213
Fauré, Gabriel 40
Fease-Green, Rachel 414
Feustel, Friedrich 60, 61
Ffrangcon-Davies, David 366, 367
Fiedler, Max 375, 419
Fillunger, Marie 285, 286
Fischer, Franz 60, 108, 115, 318, 322, 357, 388, 405, 453–4
Fischer, Wilhelm 27
Fitzwilliam, Lady Alice 333
Fitzwilliam, Lord 332–3
Flesch, Carl 266
Flotow, Friedrich 101, 250
Forbes, R. J. 423
Förster, Dr August 119
Forsyth, James 202, 297, 334, 340
Forsyth, Neil 362, 401, 413
Foster, Muriel 333, 347, 368, 391, 426, 432

Foster, Myles Birkett 122
Foulds, John 431, 471
Franck, César 40, 381, 399, 431
Frank, Ernst 67
Franke, Hermann 126, 141, 142, 143, 149, 150, 155, 156, 158, 159, 165, 187, 190, 192, 193, 206, 211, 218, 226, 234, 240, 241
Frantzen, Theodore 206, 326, 398
Franz Josef I, Emperor 5, 66, 84, 85, 132, 133, 140, 235, 240, 260, 261, 275
Franz, Robert 135, 220, 222, 229, 245, 294, 301, 368
Frederick the Great, Emperor 70
Fremstad, Olive 365
Fricke, Richard 112, 119
Fridberg, Franz 7
Fried, Oskar 440
Friedländer, Thekla 149, 150
Friedrich III, Emperor 242
Friedrich Franz II, Grand Duke of Mecklenburg-Schwerin 114
Friedrichs, Fritz 320, 322
Friskin, James 447, 448
Fröhlich, Louis 332
Fuchs, Carl 317, 318, 319–20, 339, 340, 360, 362, 368, 370, 380, 381, 396, 421, 424, 426, 431, 452
Fuchs, Johann Nepomuk 177, 230, 268, 272, 273, 275, 277
Fuchs, Robert 118, 137, 149, 150, 153, 161, 204, 214, 218, 235, 241, 250, 254, 280, 303, 304
Fuller-Maitland, John 301, 336
Furtwängler, Wilhelm 110, 465

Gabrilowitsch, Ossip 254, 298, 299
Gade, Niels 229
Gadski, Johanna 322
Gallrein, Alfred 317
Ganz, Wilhelm 115, 327, 434
Garcia, Manuel 35
Gaupp, Gustav 210
Gautier, Judith 32–3, 34, 35, 39, 40, 46
Gautier, Théophile 33
George, Prince of Wales (later King George V) 374
Gérardy, Jean 254, 264, 429
Gericke, Wilhelm 135, 165, 170, 213, 226, 229, 252, 449
German, Edward 299–300
Gezink, Willem 429
Gianicelli, Carl 23, 45, 46, 264, 265, 266, 270, 369, 372, 382, 399

Giraud, Ernest 97
Gladstone, William Ewart 331
Glasenapp, Carl Friedrich 45, 46, 357
Glaser, Emanuel 39, 40
Glatz, Dr Franz 76, 78, 81
Glazunov, Alexander 232, 308, 348, 355, 359, 360, 383, 421
Glinka, Mikhail 218, 308, 380
Gluck, Christoph Willibald von 24, 38, 84, 142, 143, 161, 171, 216, 250, 405
Göbl, Alois 148
Godowsky, Leopold 430
Goetz, Hermann 67, 85, 197, 252, 368
Goldmark, Karl 54, 85, 107, 136, 139, 161, 175, 241, 246, 250, 254, 286, 296, 298, 368, 380, 427, 430
Goll, Edward 427
Göllerich, August 163, 222, 226, 260
Gomes, Antonio 200
Goodson, Katherine 428
Goossens, Sir Eugene 406–7
Götze, Emil 225
Gounod, Charles 19, 40, 52, 55, 69, 84, 102, 103, 139, 161, 170, 175, 197, 200, 220, 257, 334, 350
Grace, Dr W. G. 436
Grädener, Hermann 171, 222, 280
Grainger, Percy 382
Grammann, Karl 107
Grange, Anna de la 6
Graves, Charles 411, 436, 472
Gravina, Count Biagio 196
Greene, Harry Plunket 146, 301, 309, 334, 341, 344, 379, 426
Grengg, Karl 250
Grieg, Edvard 115, 245, 253, 325, 352, 399, 409
Grisar, Albert 197
Gross, Adolf von 198, 244, 316, 318
Grove, Sir George 157, 193, 216, 246, 329–30, 369, 392, 393
Grün, Jacob 132, 147, 171
Grunebaum, Hermann 417
Grüning, Wilhelm 319, 320
Grützmacher, Friedrich 58, 102
Gudehus, Heinrich 206, 207, 209, 226, 238, 244
Gulbranson, Ellen 319, 320, 388, 398
Gura, Eugen 62, 66, 67, 188, 189, 190, 191, 192, 283
Gutmann, Albert 163, 185, 233
Gye, Ernest 187

Hackenberger, Franz 317

Halévy, Fromental 52, 55, 161
Haley, Mrs Edward 355
Halir, Carl 222
Hall, Edith 296–7, 299, 307–8
Hall, Marie 399
Hallé, Lady (see Norman-Neruda, Wilma)
Hallé, Sir Charles 143, 146, 149, 150, 153, 156, 157, 202, 205, 210, 226, 227, 249, 264, 278, 296, 297, 306, 307, 323, 325, 337, 340, 347, 349, 383, 399, 430, 434, 459
Hallwachs, Reinhold 31, 32
Halslinger, Johann (Johannes Hager) 228
Hambourg, Mark 254, 255–6, 379
Hambourg, Michael 256
Handel, George Frederick 107, 135, 142, 159, 211, 219, 236, 245, 248, 301, 344, 350, 356, 420, 453
Hanslick, Eduard 27, 35, 44, 77, 100, 120, 122, 137, 138, 139, 148, 162, 173, 178, 180, 181, 182, 209, 223, 224, 225, 226, 230, 233, 241, 253, 260, 261, 271, 278
Harford, Francis 353
Harris, Sir Augustus 185, 291, 387
Harrison, Julius 424
Hartvigson, Fritz 193, 201
Harty, Sir Hamilton 337, 402, 466
Hatton, Walter 317
Hauk, Minnie 55, 85, 157
Haupt, Marie 72
Hausmann, Robert 242, 248
Haydn, Joseph 1, 3, 11, 44, 54, 75, 99, 132, 133, 135, 150, 160, 161, 217, 218, 229, 236, 254, 262, 263, 266, 352, 358, 386, 424, 429
Haydn, Michael 277
Haynald, Archbishop 74, 77, 78
Heap, Dr Charles Swinnerton 301, 342, 344, 366
Hecker, Emil 59
Heckmann, Robert 238
Hedmondt, Charles 400, 401, 402
Hedmondt, Emil 325
Heermann, Hugo 183
Heinrich, Franz 269
Heinrich, Max 284, 286
Heissler, Carl 6
Hellmesberger, Georg 98
Hellmesberger, Josef jun. 132, 147, 175, 225, 233, 270, 272
Hellmesberger, Josef sen. 6, 55, 58, 101, 102, 107, 131, 132, 133, 136, 137, 138, 139, 171, 229, 264
Helm, Theodor 99, 163, 222

Hempel, Frieda 388, 398
Henley, William 336
Henschel, Sir George 54, 142, 143, 150, 156, 159, 165, 286, 287, 294, 324
Henschel, Lillian 286
Henselt, Adolf von 254, 379
Herbeck, Johann 81, 85, 86, 87, 88, 97, 98, 131, 134, 135, 136, 137, 138, 194
Herkomer, Sir Hubert 193, 280–4, 286, 323–4
Hess, Willy 338, 360, 428
Hessen-Darmstadt, Grand Duke of 328
Heuberger, Richard 149, 234, 235, 250
Higgins, Harry 291, 362, 388, 400, 401, 412, 413, 414–16, 417, 436
Higginson, Colonel Henry 256, 258, 326
Hill, Carl 72, 92, 112, 122, 123, 124, 125, 225
Hiller, Ferdinand 58
Hofmann, Josef 254, 324
Hofmann, Baron Leopold von 86, 101, 149
Hoffmann, J. 339
Hoffmann, Josef 64, 120
Hohenlohe-Schillingsfürst, Prince Konstantin zu 82, 101, 132, 227
Holbrooke, Josef 391, 424
Hollander, Benno 464–5
Hollins, Alfred 201
Holmès, Augusta 39, 40
Holst, Gustav 424
Hölzl, Gustav 29, 204
Hoose, Elison van 332
Hubay, Jenö 267
Huber, Károly 53
Hubermann, Bronislav 431
Hueffer, Francis 156
Hülsen, Botho von 107
Hummel, Johann Nepomuk 3, 176
Hummer, Reinhold 248
Humperdinck, Engelbert 165, 195, 250, 313, 357, 393
Humphreys, Granville 384
Hunt, William Holman 158
Hyde, Walter 401, 416, 417

Ihnffeld, Anton von 4, 19, 42
Irving, Sir Henry 227, 283, 330
Isaacs, Edward 379, 380, 427
de l'Isle-Adam, Villiers 33
Isnardon, Jacques 283

Jachmann-Wagner, Johanna 59
Jaeger, Augustus Johannes 309, 310, 426
Jäger, Ferdinand 140

Jahn, Wilhelm 107, 159, 171, 182, 194, 195, 196, 198, 200, 203, 213, 227, 236, 237, 241, 248, 249, 250, 259, 265, 268, 271, 272, 312
Jaide, Luise 71, 113
James, Edwin 317
Jameson, Frederick 149, 400
Jank, Angelo 31
Janson, Agnes 292
Jauner, Franz 81, 82, 87, 88, 89, 90, 94, 96, 101, 102, 103, 105, 106, 119, 120, 140, 149, 150, 159, 175
Joachim, Joseph 35, 58, 136, 141, 157, 171, 226, 242, 248, 288, 289, 290, 292, 329, 354, 355, 356, 379, 399, 424
Johnstone, Arthur 337–8, 342, 345, 350, 357, 368
Johnstone, George Hope 343, 367
Jones, Harold 407
Jordan, Daisy 381
Joshua, Marie 142, 156, 157, 159, 167, 174, 176, 177, 180, 182, 184, 186, 190, 194, 200, 216, 234, 239, 308, 323, 324, 325, 326, 327, 328, 362, 380, 390, 398, 402, 414, 425, 427, 436, 437, 441–2, 444, 445, 449, 450
Joukowsky, Paul 198
Jowett, Benjamin 216
Jünger, Oscar 444

Kahane, Jack 410, 424, 429, 430
Kahl, Oscar 47, 48
Kalbeck, Max 224
Karajan, Herbert von 110
Karpath, Ludwig 94, 233, 453
Kastner, Emmerich 61
Kauders, Albert 250
Kennedy, Michael 298, 309
Kenyon, G. H. 430
Keppler, Dr Friedrich 198
Kerr, Grainger 432
Kiel, Friedrich 135
Kienzl, Wilhelm 198
Kietz, Gustav 90, 91
Kindermann, August 28, 36
King, Frederic 213
Klafsky, Katharina 266, 454
Klein, Hermann 126, 141, 188, 209–10, 211, 364–5
Kleinecke, Wilhelm 6, 11
Klengel, Julius 254
Klindworth, Karl 35, 45, 46, 357, 460
Knapp, August 72
Knappertsbusch, Hans 110

Index

Kniese, Julius 312
Knöbel, Theresia 1
Knowles, Charles 401
Kocian, Jaroslav 356
Koessler, Hans 267
Koussevitzky, Sergei 465
Krancevics, Dragomir 82
Kraus, Emil 188, 190
Kraus, Ernst 322, 388, 398
Krauss, Dr Emil 76
Kreisler, Fritz 254, 255, 355, 356, 379, 427, 428
Krempelsetzer, Georg 28
Kremser, Eduard 162
Kretschmann, Theodor 251–2
Kreuz, Emil 400, 413, 426
Krug, Arnold 135
Kruiss, Theodor 71
Kubelik, Jan 340, 356
Kuhe, Wilhelm 146
Kupfer-Berger, Mila 102
Kurz, Selma 372

Labatt, Leonhard 101, 102, 119, 169
Lablache, Luigi 208
Labor, Josef 253
Lachner, Franz 7, 11, 25
Lalande, Désiré 317, 318
Lalo, Edouard 40, 254
Lammert, Minna 108
Lamond, Frederic 254, 291
Lamoureux, Charles 305, 320, 329, 335
Landgraf, Dr Karl 70, 451, 452, 456
Landskron, Leopold 14
Langaard, Borghild 443
Langer, Viktor 48, 52
Langford, Samuel 337, 393–4, 410, 426, 431, 433, 458–9
Lanner, Joseph 236
Lardner, Edward 316
Lassalle, Jean 283
Lassen, Eduard 36
Laubach, W. 317
Leffler-Burckhard, Martha 388
Legge, Robin 448
Lehmann, Lilli 108, 115, 187, 206, 207, 208–9, 216, 241, 249, 319, 320, 388
Lehmann, Liza 336
Lehmann, Marie 59, 108, 216, 236, 319
Lehmann, Rudolf 123
Lenbach, Franz von 70
Lengyel, Ernst 430
Leo, Hermann 347
Leoncavallo, Ruggero 195, 250

Leschetizky, Theodor 149, 167, 169, 184, 197, 255, 369
Lesimple, August 121, 239
Levi, Hermann 35, 36, 116, 194, 198, 210, 221, 222, 223, 240, 242, 244, 260, 264, 291, 295, 312, 314, 315, 352, 388, 460
Lewis, Sir George 157
Lewis, Henry 317
Lewy, Gustav 86, 87
Lewy, Richard 86, 87
Lichtenberg, Professor 259
Lichtenfeld, Adolf 14
Liechtenstein, Prince 271
Lieven, Josefine (Josefine Richter) 4
Lind, Jenny 98, 146, 157
Lindemayr, Karl 14
Liszt, Franz 2, 15, 21, 22, 23, 24, 33, 34, 35, 36, 42, 43, 48, 52, 53, 54, 55, 56, 62, 63, 64, 71, 72, 73, 74, 76, 77, 79, 80, 81, 82, 95, 98, 99, 120, 135, 136, 142, 143, 146, 150, 154, 155, 160, 167, 172, 175, 179, 185, 186, 196, 204, 206, 213, 214, 216, 218, 221, 228, 230, 234, 238, 241, 242, 243, 250, 254, 264, 266, 267, 295, 303, 313, 314, 330, 332, 339, 350, 353, 357, 360, 370, 374, 378, 381, 382, 385, 398, 424, 431, 453, 456, 462
Litchfield, Mrs 158
Litolff, Henry 356
Little, Lena 216, 257, 286
Lloyd, Edward 210, 213, 219, 243, 245, 284, 286, 294, 301, 327, 341, 350
Loeb, David 454
Loeb, Sydney 24, 320, 341, 422, 437, 439, 440, 441, 443, 447, 448, 449, 450, 451, 453, 454, 455, 456, 474
Loeb, Sylvia 444, 446
Lohse, Otto 440, 454, 455
Lorenz, Robert 471–2
Lortzing, Albert 66, 67, 114, 228, 250
Löwe, Ferdinand 163, 222, 233, 263, 269, 275
Lucca, Pauline 85, 97, 161, 170, 175, 176, 187, 216, 225, 241
Ludwig, Joseph 281
Ludwig I of Bavaria, King 10, 11, 13, 14, 16, 17, 18, 26, 27, 29, 30, 32, 33, 35, 36, 37, 38, 41, 43, 44, 45, 46, 61, 64, 111, 114, 116, 117, 228
Lully, Jean-Baptiste 381
Lunn, Louise Kirkby 365, 366, 373, 389, 398, 416

McCormack, John 399

MacCormack, Sir William 218
Macfarren, Sir George 434
MacIntyre, Margaret 288, 327
Mackenzie, Sir Alexander 198, 199, 205, 220, 234, 242, 253, 283, 288, 289, 310, 326, 327, 360, 375, 428, 451
Mahler, Gustav 83, 97, 136, 138, 158, 195, 196, 249, 259, 268-78, 280, 291, 308, 320, 351, 372, 459, 468, 472
Maier, Mathilde 24
Malibran, Maria 208
Mallinger, Mathilde 11, 17, 18, 19, 26, 30, 44
Malsch, W. M. 326, 421, 422
Malten, Therese 188, 226, 244
Mancinelli, Luigi 286
Maney, C. F. 317
Manns, Sir August 143, 146, 149, 154, 160, 165, 199, 201, 210, 221, 326, 341, 361, 392, 393, 408, 434
Mapleson, Colonel James 150
Marchesi, Blanche 332, 347-8
Maria Theresia, Archduchess 261
Mariani, Angelo 57
Marion, Biro de 206
Mark, Paula 250, 255
Marschner, Heinrich 101, 161, 276, 277, 292
Marteau, Henri 242
Mascagni, Pietro 195, 250
Masini, Angelo 140
Massenet, Jules 40, 195, 222, 241, 249
Maszkowski, Rafael 264
Materna, Amalie 85, 86, 87, 90, 101, 102, 109, 112, 113, 115, 119, 120, 121, 123, 124, 140, 165, 169, 195, 201, 241, 248, 249
Matt, Albert 317
Max, Archduke 6
Mayerhofer, Karl 140, 216
Mayfeld, Moritz von 138, 226
Medez, Josef 120
Meerschen, Henri van der 375, 421
Méhul, Etienne 118, 447
Meissner, Anton 262
Melba, Nellie 275, 415
Mendelssohn, Felix 19, 54, 55, 103, 106, 135, 137, 147, 149, 172, 200, 203, 204, 211, 219, 220, 229, 238, 241, 250, 290, 292, 294, 360, 361, 390, 391, 395, 397, 411, 435
Mendès, Catulle 33, 40, 46
Mengelberg, Willem 443, 447
Menzel, Adolph von 91
Merian, Emil 21

Merrick, Frank 356
Messager, André 388
Meyer, Waldemar 91
Meyerbeer, Giacomo 8, 19, 28, 40, 52, 55, 73, 74, 84, 118, 139, 161, 197
Meysenburg, Malwida von 47
Mihalovich, Ödön 55, 63, 82
Mildenburg, Anna Bahr- 270, 389
Mitterwurzer, Anton 44
Molique, Bernhard 229
Moore, Jerrold Northrop 309
Morold, Max 105
Morrow, Walter 379, 455
Moscheles, Charlotte 157
Moscheles, Felix 255
Mosenthal, Salomon 101, 172
Mosonyi, Mihály 73, 266
Mosshammer, Otto 317, 318
Moszkowski, Moritz 254
Mottl, Felix 60, 106, 108, 115, 137, 146, 201, 214, 221, 228, 240, 241, 242, 244, 264, 270, 271, 274, 294, 295, 304, 312, 313, 314, 315, 318, 319, 329, 352, 357, 405, 419, 460, 465
Mozart, Wolfgang Amadeus 9, 11, 15, 16, 19, 28, 54, 55, 66, 70, 71, 75, 84, 99, 102, 132, 133, 135, 137, 143, 147, 149, 154, 158, 161, 164, 172, 175, 180, 181, 198, 216, 221, 229, 240, 241, 242, 250, 254, 262, 264, 274, 277, 285, 293, 301, 305, 309, 338, 340, 344, 352, 355, 359, 360, 368, 372, 374, 377, 381, 386, 387, 391, 393, 398, 427, 428, 430, 435, 455, 461, 462
Mracek, Franz 79
Muck, Karl 223, 312, 352, 356, 377
Mühlfeld, Richard 317
Müller, Georg 102
Müller-Reuter, Theodor 47
Muncker, Theodor 58
Murska, Ilma di 73, 85

Nachbaur, Franz 26, 188
Nalbandian, Johannes 316
Nápravník, Eduard 250
Naval, Franz 255
Naylor, Edward 413
Needham, V. L. 379, 421
Neruda, Olga 296
Nessler, Viktor 225
Neumann, Angelo 106, 187, 192, 321
Newman, Ernest 34, 35, 37, 38, 337, 381, 382-3, 393, 438, 458, 459-60, 467, 470-1

Index 515

Newman, Robert 295, 375
Nicholas II, Tsar of Russia 255
Nicholls, Agnes (Lady Harty) 333, 356, 379, 391, 398, 402–3, 417
Nicodé, Jean Louis 352
Nicolai, Otto 2, 24, 25, 98, 99, 101, 139, 216, 275
Nicolini, Ernest 159
Niemann, Albert 17, 59, 94, 108, 112, 121, 187
Nietzsche, Friedrich 46, 48, 49
Nikisch, Arthur 52, 83, 103, 121, 242, 256, 258, 265, 266, 295, 326–7, 328, 335, 375, 378, 383, 404, 419, 440, 459, 463, 464, 467
Nikita (Louisa Nicholson) 255
Nilsson, Christine 140, 150, 151, 152, 192
Nollet, Georg 102
Nordica, Lillian 291–2, 304
Norman-Neruda, Wilma (Lady Hallé) 150, 296, 350, 359, 397
Nottebohm, Gustav 229

Offenbach, Jacques 7, 40, 47, 175
Ondricek, Frantisek 203, 356
Onslow, George 241
Orczy, Baron Felix von 50, 52
Orridge, Ellen 159, 201
Oudin, Eugène 292, 324, 326

Pachmann, Vladimir de 243
Paderewski, Ignaz 252, 255, 287, 423, 426
Paersch, Franz 339, 421
Paganini, Nicolò 98, 254, 266, 340
Palestrina, Giovanni 292, 293
Palliser, Esther 336
Papier, Rosa 204, 216, 225, 248, 250, 271
Parry, Sir Charles Hubert Hastings 123–5, 142, 149, 153, 160, 166, 189–91, 193, 196, 200, 205, 208, 213, 218, 225–6, 227, 239, 245–7, 284–5, 286, 288, 289, 299, 301, 309, 328, 334, 341, 344, 363, 418, 420, 421, 426, 430, 472
Pasdeloup, Jules 35, 40, 41
Patey, Janet 219, 243, 245, 246
Patey, John 246
Patti, Adelina 85, 140, 146, 159, 187, 192, 227
Pauer, Ernst 167
Paumgartner, Hans 162
Paur, Emil 326
Pedro II of Brazil, Emperor 51, 114
Penny, Dora 309
Perfall, Baron Carl von 4, 24, 28, 30, 32, 34, 35, 37, 38
Perkins, Charles 329, 330, 422
Perron, Carl 286, 312, 318, 319, 320
Petri, Egon 382, 431
Petri, Henri 254
Pfeffer, Karl 106, 236, 237
Pfitzner, Hans 423
Pitt, Percy 362, 365, 375, 380, 387, 388, 391, 398, 400, 401, 402, 403, 405, 409, 412, 413, 414, 416, 425, 427, 436, 438, 439, 440, 442, 444, 445, 446, 448
Planer, Minna 16
Plank, Fritz 238, 244
Pohl, Richard 4, 30, 35
Poldini, Eduard 387
Pollini, Bernhard 187, 192, 270
Ponchielli, Amilcare 200, 213
Popper, David 148
Praag, Meyer van 317
Price, Daniel 333
Proch, Heinrich 6, 85, 98
Prout, Ebenezer 220, 368
Pugno, Raoul 381
Purcell, Henry 288, 301, 336, 347, 367, 381
Pyk, Louise 159, 160

Rachmaninov, Sergei 360
Radecke, Albert 91, 143
Radford, Robert 372, 417
Raff, Joachim 28, 54, 63, 102, 136, 198, 205, 222, 254
Ramesch, Professor 6
Ramsden, Archibald 326, 327, 453
Raupp, Wilhelm 182
Redfern, E. S. 379
Reed, Joseph 333
Reed, William Henry 411
Reeves, Sims 243
Reeves, Wyn 375–6
Reger, Max 421, 424
Reichenberg, Franz von 133
Reicher-Kindermann, Hedwig 113, 187
Reichmann, Theodor 187, 195, 206, 216, 236, 237, 240, 244, 248, 255
Reinhardt, Max 241
Reinhold, Alexander 18, 68
Reinhold, Hugo 177
Reinl, Josephine 389
Renard, Marie 249
Reszke, Edouard de 304, 321
Reszke, Jean de 283, 304, 320, 321, 414
Reubke, Julius 330
Reuling, Wilhelm 98
Reuss-Koestritz, Prince Henry 232

Reznicek, Emil 254, 440
Ribary, Josef 78
Ricci, Luigi 129
Richter, Anna 1
Richter, Anton (father) 1, 2, 3, 4, 5, 67, 68
Richter, Anton (grandfather) 1
Richter, Antonia (sister) 2
Richter, Edgar (son) 184, 201, 323, 341, 384, 437, 441, 449
Richter, Eleonore (granddaughter) 202
Richter, Georg 1
Richter, Gustav 108
Richter, Hans (son) 176, 201, 341, 443, 449
Richter, Josefine, née Czasensky (mother) 2–6, 8, 11, 14, 16, 17, 19, 20, 22, 28, 33, 38, 39, 42, 43, 44, 49, 67, 74, 76, 78, 114, 120, 164, 170, 171, 172, 180, 201, 311
Richter, Joseph (brother) 2
Richter, Josephus 1
Richter, Lotte 354
Richter, Ludovika (daughter) 120, 311, 323, 331, 341, 432, 443
Richter, Marie (daughter Mitzi) 331, 341, 378, 390, 440, 443, 444, 449, 456
Richter, Marie (granddaughter Mädi) 378, 443
Richter, Marie (sister) 2, 11
Richter, Marie, née Szitányi (wife) 67–75, 77, 78, 79, 81, 82, 86, 89, 90, 92, 93, 94, 95, 97, 100, 107, 112, 120, 128, 161, 164, 168, 174, 176, 177, 179, 180, 183, 198, 199, 201, 202, 204, 205, 225, 242, 262, 266, 277, 281, 282, 311, 315, 325, 341, 369, 380, 428, 432, 437, 438, 439, 440, 442, 443, 453, 455–6, 462, 470
Richter, Mathilde (daughter) 165, 311, 320, 323, 331, 341, 369, 378, 381, 384, 427, 428, 430, 432, 439, 440, 441, 444, 449, 450, 453, 454, 455, 456, 460
Richter, Melchior 1
Richter, Melzer 1
Richter, Pius 132
Richter, Richard (son-in-law) 341
Richter, Richardis (daughter) 100, 105, 106, 120, 161, 174, 180, 273, 311, 315, 340, 341, 390, 437, 440, 442, 443, 450
Richter, Willibald 354
Rietz, Julius 66
Rimsky-Korsakov, Nikolai 254, 303, 309, 331, 350
Risegari, Carlo 9, 325, 334, 380, 384
Risler, Edouard 267
Ritter, Alexander 423
Ritter, Christian 391
Ritter, Josef 188, 250
Rodewald, Alfred 227, 308, 325, 340, 368
Ronald, Sir Landon 375, 404, 445
Rooy, Anton van 304, 320, 321, 322, 365, 373, 377, 389, 398
Rosa, Carl 151, 188, 189, 199, 206, 434
Rosé, Arnold 132, 161, 214, 231, 234, 241, 286
Rosenthal, Moritz 214, 253, 278, 295
Rösler, Johann 253
Rossini, Gioachino 24, 28, 46, 52, 55, 66, 69, 73, 84, 118, 139, 213, 453
Rott, Hans 136, 137
Rotter, Ludwig 132
Roze, Marie 177
Rubini, Giovanni 208
Rubinstein, Anton 86, 118, 140, 146, 158, 172, 185, 254, 324, 355
Rubinstein, Josef 60, 71
Rüdel, Hugo 442
Rudolf, Crown Prince 277
Rumford, Kennerley 332
Russell, Ella 332

Sadler-Grün, Friederike 70, 123, 124
Safonov, Vasily 375
Saint-Saëns, Camille 35, 36, 40, 46, 140, 149, 150, 157, 171, 200, 241, 242, 253, 254, 264, 270, 304, 323, 338, 347, 368, 396, 399, 426
Salieri, Antonio 133, 136
Saltzmann-Stevens, Minnie 414, 416
Salvi, Matteo 7, 84, 85
Sandby, Herman 382
Santley, Sir Charles 219, 243, 245, 246, 247, 336, 350
Sarasate, Pablo de 148, 220, 241, 270, 333, 355, 395, 397
Sauer, Emil 254
Sauret, Emil 254
Sayn-Wittgenstein, Princess Carolyne 23, 53
Scaria, Emil 71, 85, 87, 96, 102, 111, 120, 140, 171, 187, 195, 201, 228, 250
Scarlatti, Alessandro 336
Schäfer, Reinhard 114
Schalk, Franz 138, 277, 439
Schalk, Josef 138, 222
Scharwenka, Xaver 148, 149, 150
Schefsky, Josephine 188
Scheidemantel, Karl 206, 207, 244
Schelle, Eduard 44
Schelling, Ernest 423, 426

Index

Schelper, Otto 31
Schembera, Viktor 128, 129
Scheu, Joseph 100, 278
Schieder, Otto 394, 421
Schiever, Ernst 24, 141, 150, 156, 303, 325, 328, 340, 420, 451, 452
Schleinitz, Countess 33
Schlesinger, Maximilian 123
Schlosser, Carl (Max) 4, 26, 31, 36, 113, 114, 123, 187, 453
Schmedes, Erik 321
Schmid, Adolf 317
Schmitt, Friedrich 4, 44
Schnabel, Artur 369-70
Schnorr von Carolsfeld, Ludwig 17
Schobinger-Amrhyn, Lorenz 45
Schönberger, Benno 171
Schopenhauer, Artur 45, 96, 172, 227, 273
Schöpfleuthner, Johann 2
Schott, Franz Philipp 10, 24, 50, 470
Schrammel, Johann 235
Schrammel, Josef 235
Schrödter, Friedrich 133
Schubert, Andreas 161
Schubert, Franz 6, 19, 54, 86, 118, 135, 147, 150, 153, 154, 161, 162, 172, 214, 242, 254, 263, 266, 289, 296, 301, 303, 323, 325, 337, 339, 344, 370, 386, 421, 428, 431, 434, 435, 447
Schuch, Ernst von 96, 142, 242, 375, 440
Schuch-Proska, Clementine 142, 143, 147, 206
Schultz-Curtius, Alfred 141, 294
Schumann, Clara 286
Schumann, Georg 393
Schumann, Robert 15, 54, 85, 103, 135, 142, 143, 147, 149-50, 153, 168, 175, 201, 205, 221, 238, 241, 254, 292, 293, 327, 344, 355, 360, 381, 399, 427, 428, 429, 434
Schumann-Heink, Ernestine 304, 319, 320, 322, 472
Schütz, Heinrich 135
Schutzendorf, Alfons 416
Schwendner, Abbot 77
Schytte, Ludvig 254
Scott, Cyril 353-4, 382
Scriabin, Alexander 421
Sechter, Simon 6
Seidl, Anton 59-62, 72, 108, 122, 187, 192, 321-2
Seilern von Aspang, Count Karl 101
Sembrich, Marcella 187
Semon, Sir Felix 242, 326

Semper, Gottfried 14, 15, 101
Seppilli, Armando 327
Serov, Alexander 35
Servais, Franz 40, 48, 49, 53
Sgambati, Giovanni 204
Shakespeare, Mimie 470
Shakespeare, William 159, 193, 324
Shaw, George Bernard 125-6, 127, 285, 287, 288, 289, 291, 294, 318-19
Sherwin, Amy 213
Sibelius, Jean 380, 421, 423, 424
Siehr, Gustav 111, 118
Sigl, Eduard 28, 29
Siloti, Alexander 355, 360
Simon, Henry 297
Simrock, Fritz 148, 237, 298
Sinding, Christian 355
Sistermans, Anton 321, 322
Sitte, Camillo 14, 15, 17, 18, 21, 22, 47, 59, 114
Smareglia, Antonio 250
Smetana, Bedřich 178, 195, 250, 253, 294, 325, 339, 368, 380
Smyth, Dame Ethel 291, 424
Soldat, Marie 216
Solomon, John 375
Somervell, Arthur 299, 300
Sommer, Karl 198, 250
Sophie, Archduchess 5
Sophie Charlotte, Archduchess 16
Sourek, Otakar 165
Spalding, Albert 396-8
Speelman, Simon 421, 424
Speidel, Ludwig 194
Speyer, Edward 157, 159, 263, 266
Spies, Hermine 238, 286
Spitta, Philipp 455
Spohr, Louis 24, 31, 54, 150, 204, 254
Squire, William Barclay 246
Stainer, John 193, 246
Standhartner, Dr Joseph 58, 96, 183
Stanford, Sir Charles Villiers 115, 143, 146, 157, 158, 159, 160, 181, 184, 190, 193, 196, 199, 204, 206, 208, 210, 211, 218, 219, 220, 225, 226, 227, 239, 242, 264, 280, 287, 288, 289, 294, 299, 305, 309, 347, 357, 360, 366, 375, 379, 391, 426, 434, 447, 448, 451, 453, 468-9
Staudigl, Josef 240
Stavenhagen, Bernhard 248, 286
Steinbach, Fritz 168, 358, 375, 383, 465
Stenhammar, Wilhelm 355, 360
Stepanoff, Frau von 183
Stephanie, Crown Princess 252, 261

Stewart, Frank 317
Steyer, Ignaz 2
Stiegler, Joseph 455
Stierl, Helene 70
Stimpson, James 219, 220
Stocker, Jakob 45
Stocker, Stefan 241
Stocker, Verena (Vreneli) (née Weidmann) 12
Stockley, William 219–20, 289, 294, 343, 344
Stradella, Alessandro 54
Strang, William 442
Straus, Ludwig 157, 165
Strauss, Franz 20, 26, 27, 117
Strauss, Johann 77, 223, 235
Strauss, Richard 20, 27, 254, 262, 298, 299, 311, 315, 335, 339, 350, 355, 356, 358, 361, 368, 369, 380, 382, 383, 391, 393, 395, 399, 407, 409, 410, 420, 421, 423, 424, 425, 426, 427, 428, 429, 430, 435, 460
Stravinsky, Igor 250–1
Streatfeild, Richard 402
Strutt, William 408–9, 414
Such, Henry 254
Sucher, Josef 5, 77, 103, 114, 178, 190, 202
Sucher, Rosa 5, 188, 189, 190, 191–2, 206, 207, 238, 244, 319, 320
Sullivan, Sir Arthur 157, 167, 210, 211, 245, 283, 288, 294, 335, 359, 368, 372, 461
Sulzer, Joseph 103, 105–6, 253, 362
Suppé, Franz von 7
Sutro, Ottilie 327
Sutro, Rose 327
Svendsen, Johan 241, 304, 309
Svendsen, Oluf 241
Swert, Jules de 118, 205
Szell, Georg 466
Szitányi, Béla von 315, 437
Szitányi, Mathilde von (née von Montbach) 67, 69, 73, 74, 79, 81, 90, 234
Szitányi, Wilhelm von 67, 69
Sztankovits, Bishop Johann 2

Tamagno, Francesco 327
Tamburini, Antonio 208
Tappert, Wilhelm 138
Taubert, Wilhelm 91, 148
Tausch, Julius 238
Tausig, Carl 8, 22, 50, 185
Taussig, Leo 317
Tchaikovsky, Modest 270
Tchaikovsky, Pyotr Il'ych 99, 115, 118, 172, 173, 185, 186, 253, 254, 263, 265, 270, 292, 296, 297, 298, 299, 302, 303, 304, 308, 329, 330, 333, 334, 336, 338, 339, 340, 344, 355, 356, 357, 360, 361, 370, 379, 382, 383, 386, 391, 395, 423, 427, 430, 435, 459, 461, 463, 465
Ternina, Milka 304, 363, 377
Terry, Ellen 290–1, 330
Tertis, Lionel 378
Tetrazzini, Luisa 415
Tetzlaff, Karl 201
Thern, Károly 54
Thode, Henry 449
Thomas, Ambroise 84
Thomas, Arthur Goring 245, 288, 291, 294
Thompson, Herbert 142, 149–50, 153, 247–8, 289–90, 292–3, 297, 301–3, 306–7, 310, 333, 335–6, 343–5, 348, 349, 350, 352–3, 356, 363–4, 366–7, 368, 382, 383, 390–3, 405, 409, 417–18, 421–2, 423, 426–7, 428–9, 433–6, 460–2
Thornton, Edna 398, 414
Tichatschek, Joseph 17, 18, 44
Tillett, Pedro 205, 214
Timanova, Vera 118
Timothy, Miriam 446, 449
Tirindelli, Pier Adolfo 291
Toscanini, Arturo 460, 465, 466, 471
Tovey, Sir Donald Francis 354, 393, 418
Trebelli, Zélia 140, 219, 243
Trust, H. 317
Turgenev, Ivan 35
Türr, General 74
Twain, Mark 266–7

Uhlig, Theodor 274
Unger, Georg 67, 72, 111, 112, 113, 123, 124

Valleria, Alwina 243
Verdi, Giuseppe 19, 52, 55, 66, 85, 87, 99, 100, 104, 140, 161, 195, 213, 249, 275, 284, 334, 350, 356, 421
Vertigliano, Narciso (N. Vert) 198, 238, 239, 240, 256, 309, 323, 324, 325, 326, 327, 331, 348, 404
Viardot-Garcia, Pauline 35
Victoria, Princess (daughter of King Edward VII) 374
Victoria, Queen 3, 192, 239, 350
Vieuxtemps, Henri 358
Vivian, A. P. 317
Vogl, Heinrich 4, 17, 18, 30, 112, 187, 240, 319, 320
Vogl, Therese (née Thoma) 17, 30, 187

Index

Volkmann, Robert 54, 55, 75, 147, 150, 153, 183, 204, 254, 360, 370
Voss, Egon 110
Vragassi, Dr 218

Waddington, Sidney 400
Wagner, Cosima 12–14, 16, 20, 23, 26, 30–2, 40, 44–9, 53, 58–9, 62–3, 69–70, 72–4, 78–81, 88, 90, 92–7, 100–1, 105, 107, 110, 112–13, 116, 128–9, 137, 160, 165, 192–3, 198–9, 202, 214, 221, 227–8, 238, 240, 243, 250, 265, 285, 292, 304, 305, 311, 312–13, 314, 315, 318, 319, 320, 321, 322, 327, 348, 352, 353, 356, 357, 377, 378, 390, 392, 400, 409, 445
Wagner, Eva (later Eva Chamberlain) 12, 20, 179, 214, 238, 352, 441
Wagner, Isolde (later Isolde Beidler) 12, 46, 129, 214, 238, 377, 445, 449
Wagner, Richard 4, 6–38, 40–50, 52–82, 84–102, 105–14, 116–30, 136–43, 145, 147, 150–1, 156, 158–62, 164–6, 168, 172, 176, 177, 179, 183, 184, 186, 187, 189, 192–5, 205, 206, 208, 209, 211, 212, 214, 216, 217, 218, 221, 224, 226, 228, 229, 230, 238, 241, 242, 251, 256, 257, 262, 263, 264, 273, 274, 283, 284, 287, 288, 291, 293, 294, 295, 296, 299, 302, 303, 308, 313, 314, 315, 316, 317, 319, 320, 321, 325, 327, 332, 333, 338, 339, 341, 344, 347, 349, 352, 355, 357, 360, 361, 362, 363, 364, 365, 366, 367, 369, 372, 374, 380, 383, 386, 390, 392, 399, 400, 405, 406, 409, 412, 416, 418, 423, 424, 425, 426, 427, 428, 429, 430, 431, 432, 433, 434, 435, 437, 439, 440, 442, 444, 445, 447, 455, 458, 459, 460, 461, 462, 463, 464–5, 466, 470, 471, 474
Wagner, Siegfried 12, 30, 66, 92, 104, 117, 179, 197, 295, 308, 312, 313–15, 317, 319, 320, 322, 352, 357, 369, 409, 438, 441, 442, 443, 444, 445, 450–1, 454, 457
Wagner, Wieland 315, 401, 454
Wagner, Winifred 454
Walker, Edyth 250, 255
Walker, Frank 232, 233
Walter, Bruno 277, 465
Walter, Gustav 85, 225, 241

Weber, Carl Maria von 8, 19, 52, 54, 55, 66, 85, 102, 106, 135, 139, 161, 164, 177, 183, 206, 234, 238, 242, 325, 339, 352, 360, 361, 394, 427, 468
Weber, Gustav 78, 79, 82, 83, 204
Weiglein, Ludwig 134, 248
Weingartner, Felix 242, 254, 260, 285–6, 358, 375, 419, 424, 440, 459, 465
Weinwurm, Rudolf 131
Wesendonck, Mathilde 47
Wesendonck, Otto 34, 47
Whitehill, Clarence 398, 401, 416
Widor, Charles Marie 329
Wieniawski, Henri 254
Wilhelm I, Kaiser 114
Wilhelm II, Kaiser 357
Wilhelmine, Markgravine of Bayreuth 70
Wilhelmj, August 59, 65, 67, 91, 92, 120, 122, 123, 125, 141, 324, 327, 336, 340
Willeby, Charles 461
Williams, Anna 219, 246, 286, 294, 301
Williams, Ralph Vaughan 424
Wilson, R. H. 337, 360, 366, 367, 380, 430
Wilt, Marie 85, 225, 241, 250
Winkelmann, Hermann 133, 188, 189, 190, 192, 195, 197, 201, 216, 248, 252
Winterbottom, C. 317
Wipperich, Emil 317
Wirk, Willi 412
Wlassack, Eduard 271
Wolf, Hugo 137, 228–33, 380, 424
Wolff, Hermann 264, 339
Wood, Sir Henry 295, 305, 335, 336, 345, 374–5, 404, 419, 421, 424, 425, 445
Wood, Lady Olga (née Ouroussoff) 425
Wüllner, Franz 38, 43, 116, 239, 358
Wurda, Josef 3
Wylde, Dr Henry 434

Young, Filson 466–8
Ysaÿe, Eugène 254, 291, 349, 350, 355, 358

Zenger, Max 28
Zichy, Anton von 43
Zichy, Count Geza 179
Zimay, László 55, 75, 82
Zimbalist, Efrem 429
Zumpe, Hermann 61, 62, 107, 115, 304